Anupam Shukla, Ritu Tiwari, and Rahul Kala

Towards Hybrid and Adaptive Computing

Studies in Computational Intelligence, Volume 307

Editor-in-Chief
Prof. Janusz Kacprzyk
Systems Research Institute
Polish Academy of Sciences
ul. Newelska 6
01-447 Warsaw
Poland
E-mail: kacprzyk@ibspan.waw.pl

Further volumes of this series can be found on our homepage: springer.com

Vol. 285. Roberto Cipolla, Sebastiano Battiato, and Giovanni Maria Farinella (Eds.)
Computer Vision, 2010
ISBN 978-3-642-12847-9

Vol. 286. Zeev Volkovich, Alexander Bolshoy, Valery Kirzhner, and Zeev Barzily
Genome Clustering, 2010
ISBN 978-3-642-12951-3

Vol. 287. Dan Schonfeld, Caifeng Shan, Dacheng Tao, and Liang Wang (Eds.)
Video Search and Mining, 2010
ISBN 978-3-642-12899-8

Vol. 288. I-Hsien Ting, Hui-Ju Wu, Tien-Hwa Ho (Eds.)
Mining and Analyzing Social Networks, 2010
ISBN 978-3-642-13421-0

Vol. 289. Anne Håkansson, Ronald Hartung, and Ngoc Thanh Nguyen (Eds.)
Agent and Multi-agent Technology for Internet and Enterprise Systems, 2010
ISBN 978-3-642-13525-5

Vol. 290. Weiliang Xu and John Bronlund
Mastication Robots, 2010
ISBN 978-3-540-93902-3

Vol. 291. Shimon Whiteson
Adaptive Representations for Reinforcement Learning, 2010
ISBN 978-3-642-13931-4

Vol. 292. Fabrice Guillet, Gilbert Ritschard, Henri Briand, Djamel A. Zighed (Eds.)
Advances in Knowledge Discovery and Management, 2010
ISBN 978-3-642-00579-4

Vol. 293. Anthony Brabazon, Michael O'Neill, and Dietmar Maringer (Eds.)
Natural Computing in Computational Finance, 2010
ISBN 978-3-642-13949-9

Vol. 294. Manuel F.M. Barros, Jorge M.C. Guilherme, and Nuno C.G. Horta
Analog Circuits and Systems Optimization based on Evolutionary Computation Techniques, 2010
ISBN 978-3-642-12345-0

Vol. 295. Roger Lee (Ed.)
Software Engineering, Artificial Intelligence, Networking and Parallel/Distributed Computing, 2010
ISBN 978-3-642-13264-3

Vol. 296. Roger Lee (Ed.)
Software Engineering Research, Management and Applications, 2010
ISBN 978-3-642-13272-8

Vol. 297. Tania Tronco (Ed.)
New Network Architectures, 2010
ISBN 978-3-642-13246-9

Vol. 298. Adam Wierzbicki
Trust and Fairness in Open, Distributed Systems, 2010
ISBN 978-3-642-13450-0

Vol. 299. Vassil Sgurev, Mincho Hadjiski, and Janusz Kacprzyk (Eds.)
Intelligent Systems: From Theory to Practice, 2010
ISBN 978-3-642-13427-2

Vol. 300. Baoding Liu (Ed.)
Uncertainty Theory, 2010
ISBN 978-3-642-13958-1

Vol. 301. Giuliano Armano, Marco de Gemmis, Giovanni Semeraro, and Eloisa Vargiu (Eds.)
Intelligent Information Access, 2010
ISBN 978-3-642-13999-4

Vol. 302. Bijaya Ketan Panigrahi, Ajith Abraham, and Swagatam Das (Eds.)
Computational Intelligence in Power Engineering, 2010
ISBN 978-3-642-14012-9

Vol. 303. Joachim Diederich, Cengiz Gunay, and James M. Hogan
Recruitment Learning, 2010
ISBN 978-3-642-14027-3

Vol. 304. Anthony Finn and Lakhmi C. Jain (Eds.)
Innovations in Defence Support Systems, 2010
ISBN 978-3-642-14083-9

Vol. 305. Stefania Montani and Lakhmi C. Jain (Eds.)
Successful Case-based Reasoning Applications, 2010
ISBN 978-3-642-14077-8

Vol. 306. Tru Hoang Cao
Conceptual Graphs and Fuzzy Logic, 2010
ISBN 978-3-642-14086-0

Vol. 307. Anupam Shukla, Ritu Tiwari, and Rahul Kala
Towards Hybrid and Adaptive Computing, 2010
ISBN 978-3-642-14343-4

Anupam Shukla, Ritu Tiwari, and Rahul Kala

Towards Hybrid and Adaptive Computing

A Perspective

Springer

Anupam Shukla
ABV-IIITM Gwalior
Morena Link Road
Gwalior
474010 Madhya Pradesh
India
E-mail: dranupamshukla@gmail.com

Rahul Kala
ABV-IIITM Gwalior
Morena Link Road
Gwalior
474010 Madhya Pradesh
India
E-mail: rahulkalaiiitm@yahoo.co.in

Ritu Tiwari
ABV-IIITM Gwalior
Morena Link Road
Gwalior
474010 Madhya Pradesh
India
E-mail : tiwariritu2@gmail.com

ISBN 978-3-642-14343-4 e-ISBN 978-3-642-14344-1

DOI 10.1007/978-3-642-14344-1

Studies in Computational Intelligence ISSN 1860-949X

Library of Congress Control Number: 2010930136

Typeset & Cover Design: Scientific Publishing Services Pvt. Ltd., Chennai, India.

Printed on acid-free paper

9 8 7 6 5 4 3 2 1

springer.com

Preface

Soft Computing today is a very vast field whose extent is beyond measure. The boundaries of this magnificent field are spreading at an enormous rate making it possible to build computationally intelligent systems that can do virtually anything, even after considering the hostile practical limitations. Research over the years by numerous scholars has made this field today stand like a giant in solving all kinds of problems. At one end, the positive developments are a source of great encouragement and motivation for all whose fantasies of future are slowly turning out to be realities of today. At the other end the enormous research along with the gigantic literature is a scary sight for many young researchers who will to master this game. The dynamic nature of the field further ensures a person once on the cliff of awareness soon happens to come at the level of the shore unless updated. The other characteristic of the domain is its multi-disciplinary nature that involves knowledge from different fields to effectively make and deploy a system based on soft computing principles. This has again been the prime reason for the presence of people from different domains who make their contributions into the field in their own way. Further this result in a unique multi-directional development in a collaborative manner. In this context, this book addresses the various issues, problems, proposed solutions and driving factors behind the changes. We not only discuss the evolution of the various modern systems, but also give an in-depth analysis into their practical usage as effective problem solving tools.

Speech forms a very characteristic biometric identity that is widely used for various applications of identification and authentication. Working with speech involves the use of all signal processing and analysis techniques that aid in their applications. Signal Processing is a valuable tool that is used for the purpose of pre-processing, segmentation and extraction of the features from the speech input. Good speech processing techniques aid in making effective speech identification and authentication systems that are extensively used in the modern world. Speech is particularly a signal that is badly affected by the presence of noise. This leads to a difficulty in their operation and hence necessitates the need of good noise removal techniques. Similar is the case with feature extraction where keeping the time and frequency level details are both necessary for good application. The use of Fast Fourier, Short Time Fast Fourier, Wavelet and other transforms provide promising techniques.

The growing automation of the society and rapid rise in automated systems urges the need of intelligent systems to be built with speech. These intelligent systems make extensive use of various statistical and soft computing techniques for the purpose of identification and authentication. The developments in the various models and most recently the customized hybrid models ensure a very promising future in the good identification and authentication with speech based systems and high accuracy levels. Good machine learning techniques with high generalizing ability or high classificatory powers are the keys to good systems.

Soft computing primarily comprises of three different approaches namely Artificial Neural Networks (ANN), Fuzzy Inference Systems (FIS) and Evolutionary Algorithms (EA). Of these the ANNs are excellent means of machine learning. They can summarize a large amount of data and store it in forms of interconnected neuron weights or connection strengths. The beauty of the ANNs does not only lie in the fact that it can refer to these summarizations or weights to reproduce the outputs whenever needed, rather it is able to give correct outputs to unknown inputs called as the generalizing ability of the ANN. The ANN learning may be supervised, unsupervised or by reinforcement. The problems may be generally classified into the heads of functional prediction and classification. A related term is pattern matching. Depending upon the problem and learning a variety of ANNs exist like Multi-Layer Perceptron with Back Propagation Algorithm, Learning Vector Quantization, Self Organizing Map, Radial Basis Function Network, Hopfield Neural Networks, Recurrent Neural Networks, etc. Each of these networks works on a set of principles that govern its general working as well as learning. It is very important to know this concept well before anything is physically implemented using ANNs. Another major problem with the use of ANNs is the network parameters. These parameters need to be judiciously set so as to optimize the overall performance. The network performance largely depends upon a good choice of parameters. Many newbie's when asked to train an ANN get completely lost with the parameters and their values. Many experienced are certainly much more sound in the art of parameter setting, but the optimality of choice is always a question. Towards the end what ultimately matters in the ANN is to get the correct set of values corresponding to the weights and biases. The hunt for this set is the big problem. In this book we discuss both the meaning and importance of every parameter along with associated practical issues, as well as the problem of parameter setting.

The Fuzzy Inference Systems (FIS) are commonly used systems because of their immense applicability to model the problem in the form of logic or a set of linguistic rules. FIS hence find a variety of application ranging from robotic control to vulnerability assessment. An increasing demanding is their use in many functional prediction and classification problem. Being rule based, they are simpler to frame, interpret and understand. However unlike the ANN that trains using its own set algorithms, the FIS are very flexible where each membership function and rule needs to be defined by the user. This may require a lot of practice and understanding of the system. Many times the system may still not be able to give a decent performance. The inability of the human to set rules or to judge the correct number of rules, membership functions, etc. can lead to a poor performance. Here as well it is important to know the importance of every parameter and the practical issues and tradeoffs associated between them. Even then it is virtually impossible for a human to figure out a system that gives the best performance. This book addressed and explains all these issues. As we more further through the text more and more issues would be raised, discussed and solved.

The Evolutionary Algorithms (EA) are tools of optimizations. These algorithms can vary the parameters to optimize the value of the set objective function. The intelligent modeling and representation of problem further makes these algorithms even more effective in solving a variety of problems. The EAs try to generate a generation of solutions from the previous generations. The newer generations keep getting better and hence the objective function value keeps improving. These systems further present a pool of parameters which are difficult to understand and play a vital role in the optimization process. The knowledge of the manner in which the parameters work, along with the practical considerations, may only provide a part help as the real parameters may be very difficult to fix that give the best performance. Another interesting fact is that it may never be assumed that the best solution comes from parameters that are constant over time. They may be needed to be made variable during the whole process.

Ever since the preliminary work was proceeding into each of these spheres of soft computing, a limitation of the individual systems resulted in mixing more than one systems to solve the problems. As the problems start displaying complex characteristics, the basic methods may fail to effectively solve them. As a result we require complex problem solving methodologies. This is done by the use of hybrid soft computing methods. Here one or more methods are fused for problem solving. The advantage of one eliminates the disadvantages of the other methods and vice versa. The resultant system is hence much better. The increasing growth of the hybrid systems is an indication of their paramount importance and applicability to most real life scenarios and problems. The high accuracies of these systems make them instrumental tools for industrial usage. While these systems sound huge and complex, all of their concepts come from the basic systems and hence knowing the basic tools makes it easy to master these systems.

Evolution is a fundamental that has changed the outlook of people towards systems. We now talk about evolving systems that iteratively adapt themselves towards the changing or constant scenarios. The evolution lets a system attain the optimal architecture and values of the various parameters. The system may be grown to attain better forms as the generations increase. The wide variety of systems may require customization of the existing evolutionary algorithms for better adaptation to the problem. It is a common feature to design evolutionary operators and have a customized individual representation as per the problem requirements. These play a vital role in effective convergence and exploration by the Evolutionary Algorithms. The wide application of evolutionary techniques has further seen development of numerous algorithms derived from evolutionary algorithms.

Swarm Intelligence is now a commonly used technique that derives its inspiration from the flocking birds, ant colonies and other swarms. These systems present simple, but powerful optimization techniques. A number of techniques have been devised using different analogies from the biological world. Two of the commonly used techniques include Particle Swarm Optimization where each particle has a velocity and position and uses its experience along with the global information to make movements towards the global minima. The other commonly used technique is Ant Colony Optimization that uses the principle of pheromone deposition by multiple ants for the purpose of solution building. Genetic Programming is another positive development to evolution that uses a tee-based representation for the evolution of individuals in the population pool. The use of specialized operators of crossover and mutation make generation of effective architectures of the solution possible. As a result Genetic Programming finds extensive use in numerous real life applications as optimizing or evolving agents. Grammatical Evolution presents another promising framework that uses Grammar and the notion of language for the optimization. The individual here represent a series of steps which may be converted to the actual solution by using the grammar. As a result the individuals are much smaller and well suited especially for variable length solutions with variable architectures.

Evolutionary Strategies are good systems that carry forward the genetic reproductions with much limited individuals that reflect the population. The use of simple operators aids in making effective computations for generation of population of a higher generation. The re-combination of parent chromosomes, mutation for all child chromosomes along with the selection of parent or parent and child population is carried out that gives these algorithms the optimization powers. Other evolutionary topics of interests include Co-evolution that encourages each and every individual to aid in the evolution of other individuals leading to effective individual of the population; Differential Evolution that carries forward evolution based on individual differences to drive the evolution in the correct direction; Immunization Systems that derive their analogy from the natural immunization processes; Cultural Algorithms that use belief space to affect the evolution process; etc.

One of the most commonly used hybrid systems includes the Adaptive Neuro Fuzzy Inference Systems (ANFIS). These are Fuzzy Systems modeled in Neural Network architecture. This enables the use of ANN training algorithms to tune the FIS. These systems overcome one of the major problem of fixing the various parameters of the FIS especially those of the membership functions. These systems enable us to give a training data set and the FIS is initialized and trained automatically based on back propagation or hybrid training algorithm. As a result the generated system is much better able to cater the needs of the problem and gives far better results as compared to the traditional system that used to be manually trained to get a decent performance. ANFIS has thus resulted in engineering Fuzzy Systems without having to crunch ones head figuring out and guessing the system parameters and rules. This goes a far way in avoiding human errors and enhances performance, especially in unknown data.

Evolutionary Artificial Neural Networks are good means of problem solving with the use of Artificial Neural Networks where the neural parameters are set with the help of Evolutionary Algorithms. These networks evolve along with time to attain the optimal architecture and parameters. These networks hence remove the limitation of the human errors in parameter setting that largely lead to sub-optimal performance. There is a large amount of use of these systems in learning of fixed architecture neural networks, evolving the architecture along with network parameters and evolving the learning rule of the ANNs. Many newer models and recent literature makes use of innovative representation and specialized use of the evolutionary operators for good convergence. The use of swarm intelligence, grammatical evolution and other invariants of genetic algorithm are also promising steps towards the use of evolutionary ANNs. The evolutionary ANNs are far better in attaining global minima as compared to the traditional ANNs. Usually coupled with a local search strategy, the evolution of ANNs represents a very complex optimization problem. The present approaches hence take a long time for the training and evolution.

Similar to the evolutionary Artificial Neural Networks, the Evolutionary Fuzzy Inference Systems are good ways to evolve the Fuzzy Inference Systems with the use of evolutionary operators. These systems enable effective and optimal generation of fuzzy systems that solve the problem in a much optimized manner. As a result the performance of the entire systems to known and unknown inputs is much better than the conventional systems. The evolution of the fuzzy inference systems may involve parameter optimization or the entire evolution of the architecture of the FIS along with the parameters. This evolution as well is a high dimensional complex problem that is slow to evolve. The bulk of evolutionary algorithms ranging from genetic to co-evolution provide promising solutions to the problem.

Modular Neural Networks (MNN) are other promising agents that make use of modularity to fragment the problem in modules to be solved by the ANNs. The decomposition results in a lot of improvement of performance that arise out of high data size with a very little loss of generality. As a result the Modular Neural Networks are able to give a very high performance to a large amount of data that usually hinders performance in the conventional ANNs. The high modularity of the problem plays a major role behind these ANNs. MNNs first decompose the problem into modules. Each module may be handled separately and in-parallel by the various ANNs. These ANNs perform independent of each other and convey their output to the integrator. The integrator collects all the outputs to give the final output of the network. In this manner the final answer to the problem is computed as a collective performance of independently functioning modules. MNNs attract the interests of numerous researchers over various problems because of their high performance in data sets where the conventional ANNs fail or give poor performances. The task of breaking a problem or data into modules is a common feature in most of the systems of use today.

As a result of the hybrid approach, we have numerous hierarchical models that may be regarded as systems working over systems. The hierarchical models are used to solve numerous problems especially the complex ones where one hierarchy caters to the needs of the other hierarchy. Hierarchal Genetic Algorithms and Nested Evolutionary Strategies are such algorithms that use hierarchies of similar algorithms for optimal performance. Here the parameters of the GA or Evolutionary Strategy are set using the higher hierarchy that results in better performance by parameter setting in real time.

Evolution does not restrict itself to the simple ANN and FIS. Evolution of ANFIS is a unique solution to evolve the ANFIS along with time and generations. This system further makes use of all key algorithms for problem solving with one algorithm overcoming the problems of the other algorithm. Modular ANNs present another innovative approach that use evolutionary algorithms as a means for evolving modular ANNs. These systems make effective use of modularity as well as optimize this modularity by system evolution. These hence mark complex but highly effective systems that build great applications of high performance. Fuzzy Neural Networks and Rule Extraction of ANNs are other concepts of varied uses and applications. A lot of work in these domains as well has led to their increasing importance in real life systems.

Parameters play a key role in deciding the fate of any Soft Computing system. The strategy to keep the parameters constant largely keeps systems sub-optimal. As a result there is a need identified to make the systems as dynamic as possible. We strive to make systems where the system itself is able to judge the optimal value of the parameters at each and every instant of time. This gives rise to the adaptive systems that are able to remove a big limitation of static nature of the parameters and add a lot to the overall system flexibility. These systems once engineered are a big boon to the performance. The big question however lies in defining the correct strategy to vary these parameters as a function of time, situation, or likewise. This further brings us closer to the need of parameter-less systems where the parameters automatically attain their optimal values. Every attempt to kill parameters ultimately results in higher complexities and addition of more number of parameters as illustrated by the No Free Lunch theorems. But the ultimate goal may be re-defined to make the parameters as passive as possible, so that the ultimate system always has best possible conditions for optimal performance.

One may sometimes wonder the myriad of options available today in solving any problem. These are numerous models each of ANNs, FISs, EAs. Also they may be fused in multiple ways to generate even a larger variety of solutions to the problem. In this manner the number of available models is immense and is increasing rapidly in the soft computing literature. To choose the correct model may hence be a difficult question. Solutions may exist to mix models and later even models of models, but increased hierarchies are known to increase complexity and computation that has adverse effects on the systems. Evert system has some advantages and disadvantages. The models are highly data and problem specific. Numerous comparisons of various methods over various problems have been made, but no unanimous decision to the choice of method exists. As a matter of fact no promising solution is likely to emerge in the near future as well that exceed all methods so far under all problems and data. The book presents the concepts to enable readers understand the situations of the use of each and every method discussed. The various advantages and disadvantages present a tradeoff between the use of various systems in various conditions. We have not yet been able to produce systems that can take problem and themselves select the most optimal model and solution. This would be an open question that is yet awaiting an answer.

The book presents the basic concepts of the parameters of the basic systems namely ANN, FIS and EA that enable better understanding of the systems at the later stage and appreciate the hybridization. Later the book explores the various hybrid approaches one by one. This includes ANFIS, Evolutionary ANN and FIS, MNNs, Hybrid GAs including hierarchical GAs, Hybrid Genetic Algorithm

Particle Swarm Optimization, etc. In all the above system we try to explain the basic need of the system, as well as the design issues that would ultimately lead to optimal performance. Since evolution is a major base of the entire book, we discuss many traditional and modern evolutionary approaches including GA, GP, Swarm Intelligence, Grammatical Evolution, Co-Evolution, Differential Evolution, Evolutionary Strategies, etc. We try to focus on the basic working philosophies of these systems to enable effective system design and decision regarding choice of systems.

The key features of the book include:

- Understanding regarding basic working philosophies of the basic systems that lead to foundations of hybrid systems
- Understanding system parameters role, dos and don'ts
- Using Speech as a real life example for system design and problem solving
- Understanding reason behind fusion and hybridization of systems, their applicability, advantages, disadvantages and tradeoffs
- Presenting the power of evolutionary approaches in effective system evolution and optimization
- Presenting and discussion over system design issues

Contents

Part I: Simple Computing Techniques

1 Introduction ... 3
1.1 Recognition Systems ... 6
1.1.1 Data Acquisition ... 7
1.1.2 Pre-processing ... 8
1.1.3 Segmentation ... 9
1.1.4 Feature Extraction ... 9
1.1.5 Pattern Matching ... 10
1.2 Biometric Systems ... 10
1.2.1 Physiological Biometrics ... 11
1.2.2 Behavioral Biometrics ... 11
1.2.3 Fusion Methods ... 12
1.3 Types of Problems in Soft Computing ... 12
1.3.1 Classification ... 13
1.3.2 Functional Prediction ... 14
1.3.3 Optimization ... 16
1.4 Machine Learning ... 16
1.4.1 Supervised Learning ... 17
1.4.2 Unsupervised Learning ... 18
1.4.3 Reinforcement Learning ... 19
1.5 Expert Systems ... 20
1.6 Problems in Soft Computing Systems ... 22
1.7 Generalization ... 24
1.8 A Design Perspective ... 27
1.9 Chapter Summary ... 27
References ... 28

2 Artificial Neural Networks ... 31
2.1 Multi Layer Perceptron ... 33
2.1.1 Back Propagation Algorithm ... 36
2.1.2 Number of Layers and Neurons ... 38
2.1.3 Learning Rate ... 40
2.1.4 Momentum ... 40
2.1.5 Epochs ... 40

2.2 Radial Basis Function Networks....43
2.2.1 Working....43
2.2.2 Conceptualization....44
2.2.3 Number of Neurons and Spread....45
2.3 Learning Vector Quantization....47
2.3.1 Working....48
2.3.2 Conceptualization....48
2.4 Self Organizing Maps....49
2.4.1 Self Organizing Feature Map....49
2.4.2 Clustering....50
2.5 Adaptive Resonance Theory....51
2.5.1 Working....51
2.5.2 Conceptualization....52
2.6 Recurrent Neural Networks....53
2.6.1 Working....54
2.6.2 Conceptualization....55
2.7 Problems with Neural Networks....55
2.8 Chapter Summary....56
References....57

3 Genetic Algorithm....59
3.1 Simple Genetic Algorithm....62
3.1.1 Conceptualization....63
3.2 Individual Representation....64
3.2.1 Role of Good Encoding Technique....66
3.3 Scaling....66
3.3.1 Proportional Scaling....67
3.3.2 Rank Scaling....67
3.3.3 Top Scaling....68
3.4 Selection....68
3.4.1 Roulette Wheel Selection....69
3.4.2 Stochastic Uniform Sampling....69
3.4.3 Tournament Selection....70
3.5 Crossover....70
3.5.1 Bit String Representation Individuals....71
3.5.2 Numeral Vector Representation....73
3.6 Mutation....73
3.6.1 Uniform Mutation....74
3.6.2 Gaussian Mutation....75
3.7 Other Genetic Operators....76
3.7.1 Elite....76
3.7.2 Insert and Delete....76
3.7.3 Repair....77
3.7.4 Hard and Soft Mutation....77
3.8 Number of Individuals....77

3.9 Stopping Criterion 78
3.10 Diversity 79
3.11 Chapter Summary 80
References 81

4 Fuzzy Logic 83
4.1 Fuzzy Sets 84
4.2 Fuzzy Logical Operators 87
4.2.1 AND Operator 88
4.2.2 OR Operator 89
4.2.3 NOT Operator 90
4.2.4 Implication Operator 91
4.3 Aggregation 93
4.4 Defuzzification 94
4.5 Fuzzy Inference Systems 95
4.5.1 Number of Membership Functions 96
4.5.2 Type of Membership Functions 98
4.5.3 Rules 99
4.6 Comparisons between Fuzzy Systems and Neural Nets 99
4.6.1 Differences 100
4.6.2 Similarities 100
4.6.3 Comparisons with RBFNs 101
4.6.4 Remarks 102
4.7 Limitations of the Fuzzy Systems 102
4.8 Type-2 Fuzzy Systems 102
4.9 Other Sets 104
4.9.1 Rough Sets 105
4.9.2 Vague Sets 105
4.9.3 Intuitionistic Fuzzy Sets 105
4.10 Chapter Summary 106
References 107

Part II: Simple Intelliegent Systems

5 Speech Signal Analysis 111
5.1 Feature Extraction 112
5.1.1 Good Features in Classification 112
5.1.2 Need of Features 113
5.1.3 Who Decides the Features? 114
5.1.4 Analogies with the Human Brain 114
5.1.5 Number of Attributes and Classification 115
5.2 Spectral Analysis 116
5.2.1 Number of Zero Crossings 117
5.2.2 Duration of Utterance 117
5.2.3 Pitch 117

5.2.4 Amplitude 117
5.2.5 Spectrogram Analysis 118
5.2.6 Power Spectral Density 118
5.3 Bank of Filters 119
5.3.1 Fourier Transform 121
5.3.2 Short-Time Fourier Transform 122
5.3.3 Wavelet Transform 123
5.4 Linear Predictive Coding 124
5.5 Chapter Summary 127
References 127

6 Intelligent System Design in Speech 129
6.1 Recognition Systems Using Speech 130
6.1.1 Input 130
6.1.2 Preprocessing 131
6.1.3 Segmentation 132
6.1.4 Dynamic Time Warping 133
6.1.5 Feature Extraction 133
6.1.6 Pattern Recognition 134
6.2 Types 134
6.2.1 Speech Recognition 135
6.2.2 Speaker Recognition 136
6.2.3 Speaker Verification 136
6.2.4 Language Identification 137
6.3 Further System Classifications 137
6.3.1 Text Dependence 137
6.3.2 Speaker Cooperation 138
6.3.3 System Robustness 139
6.4 Template Based Matching 139
6.5 Speaker Verification System Design 141
6.5.1 Enrolment 141
6.5.2 Verification 142
6.5.3 System Update 142
6.5.4 Verification Ratios 142
6.5.5 Matching Threshold 143
6.5.6 Mimic Resistance 144
6.6 Recognition with Closed and Open Data Sets 145
6.7 Making of a Good Database 146
6.8 Example of Speaker Recognition 146
6.8.1 Making up of a Database 146
6.8.2 Speech Processing 147
6.8.3 Inputs and Outputs 147
6.8.4 Training 148
6.8.5 Testing 148
6.9 Chapter Summary 149
References 150

7 Intelligent Systems Design in Music 153
7.1 Genre Classification....................154
7.1.1 Feature Extraction....................155
7.1.2 Pattern Recognition....................157
7.1.3 n-Fold Cross Validation....................158
7.1.4 Experimentation....................158
7.2 Musical Classifications....................158
7.2.1 Feature Extraction....................159
7.2.2 Dimensionality Reduction....................160
7.2.3 Pattern Recognition....................161
7.2.4 Experimentation....................163
7.3 Music Composition by Interactive Evolutionary Computation....................165
7.3.1 Interactive Evolutionary Computation....................166
7.3.2 Genetic Architecture....................167
7.3.3 Individual Representation....................167
7.3.4 Genetic Operations....................168
7.3.5 Evaluations....................169
7.4 Computer Assisted Music Composition....................169
7.4.1 Thematic Bridging....................169
7.4.2 Individual Representation....................170
7.4.3 Fitness Evaluations....................171
7.5 Chapter Summary....................171
References....................171

8 Intelligent System Design in Face 175
8.1 Face as a Biometric....................176
8.2 Visual Dimensionality Reduction....................177
8.2.1 Principle Component Analysis....................178
8.2.2 R-LDA....................179
8.3 Morphological Analysis....................180
8.4 Pattern Recognition....................181
8.4.1 Multi-layer Perceptron....................182
8.4.2 Radial Basis Function Networks....................182
8.5 Experimentation....................182
8.6 Chapter Summary....................183
References....................183

Part III: Evolutionary Computing

9 Swarm Intelligence 187
9.1 Particle Swam Optimization....................188
9.1.1 Terms and Concepts....................188
9.1.2 Algorithm....................189
9.1.3 Maximum Velocity....................190

9.2 Ant Colony Optimization ... 191
9.2.1 Algorithm ... 192
9.3 Artificial Bee Colonies ... 194
9.3.1 Employed ... 194
9.3.2 Onlooker ... 196
9.3.3 Scouts ... 196
9.3.4 Algorithm ... 197
9.4 Stochastic Diffusion Search ... 198
9.5 Probability Based Incremental Learning ... 200
9.5.1 Representation ... 200
9.5.2 Generation of Samples ... 201
9.5.3 Training ... 201
9.6 Biogeography-Based Optimization ... 202
9.6.1 Representation ... 203
9.6.2 Habitat Modification ... 203
9.6.3 Mutation ... 204
9.7 Chapter Summary ... 205
References ... 205

10 Genetic Programming ... 187
10.1 Analogy with Genetic Algorithms ... 210
10.2 Problem Representation ... 212
10.3 Fitness Function ... 214
10.4 Selection ... 216
10.5 Crossover ... 217
10.6 Mutation ... 218
10.7 Other Commonly Used Operators ... 220
10.7.1 Elite ... 220
10.7.2 Replication ... 220
10.7.3 Architecture Altering ... 220
10.8 Convergence ... 221
10.9 Code Growth ... 223
10.9.1 Parsimony ... 223
10.9.2 Code Modification ... 224
10.9.3 Parameter Modification ... 224
10.10 Generating Complex Programs ... 224
10.10.1 Automatically Defined Functions ... 225
10.10.2 Emergent Library Builder ... 225
10.10.3 Comparisons ... 226
10.11 Grammatical Evolution ... 228
10.11.1 Grammar ... 229
10.11.2 Individual Representation ... 230
10.12 Chapter Summary ... 232
References ... 232

11 Evolutionary Strategies 235
11.1 Adaptation 236
11.2 Individual Representation 237
11.3 Evolution 238
11.3.1 (1+1) ES 238
11.3.2 ($\mu/\rho+\lambda$) and ($\mu/\rho+\lambda$, k) Representations 239
11.3.3 (μ/ρ, λ) Representation 239
11.4 Recombination 240
11.5 Mutation 241
11.6 Chapter Summary 245
References 246

12 Other Evolutionary Concepts 249
12.1 Differential Evolution 250
12.1.1 Attribute Difference D_{abcd}^{j} 251
12.1.2 Mutation 251
12.1.3 Crossover 253
12.1.4 Selection 253
12.1.5 General Algorithm 254
12.2 Artificial Immune System 254
12.2.1 Representation 255
12.2.2 Detectors 256
12.2.3 Comparing Particles 257
12.2.4 Training 257
12.2.5 Memory Storage 259
12.3 Co-evolution 259
12.3.1 Decomposition 260
12.3.2 Fitness Evaluation 261
12.3.3 Replenishment 261
12.3.4 Other Evolutionary Operators 262
12.4 Cultural Algorithms 263
12.4.1 Belief Space 263
12.4.2 Belief Space Representation 264
12.4.3 Acceptance 265
12.4.4 Influence 265
12.4.5 Adjust 265
12.5 Cellular Automata 266
12.5.1 States 267
12.5.2 Neighborhood 268
12.5.3 Transition Rules 268
12.5.4 Space Time Diagrams 270
12.6 Chapter Summary 271
References 271

Part IV: Hybrid Computing

13 Evolutionary Neural and Fuzzy Systems 277

13.1 Issues in Use of Neural and Fuzzy Systems 278

13.1.1 Fixed Architecture 278

13.1.2 Variable Architecture....................................279

13.2 Evolutionary Systems 279

13.3 Evolving Fixed Architecture Neural Network 280

13.3.1 Individual Representation 281

13.3.2 Genetic Operators 281

13.3.3 Local Search 282

13.3.4 Fitness Evaluations 283

13.4 Connectionist Approach in Neural Network Evolution 283

13.4.1 Problems with Large Number of Weights 284

13.4.2 Limited Connection Neural Network....................................285

13.4.3 Individual Representation 286

13.4.4 Genetic Operators 287

13.4.5 Fitness Evaluation....................................287

13.5 Incremental Evolution of Neural Network....................................288

13.5.1 Individual Representation 288

13.5.2 Increase in Complexity 289

13.5.3 Crossover 290

13.5.4 Mutation....................................290

13.5.5 Grow 290

13.5.6 Elite....................................291

13.5.7 Add 291

13.5.8 Fitness Evaluation....................................292

13.6 Grammatical Evolution of Neural Network....................................292

13.6.1 Grammar 293

13.6.2 Genetic Operators 294

13.6.3 Fitness Evaluation....................................295

13.7 Evolution of Fixed Architecture Fuzzy System 295

13.7.1 Individual Representation 296

13.7.2 Genetic Operators 296

13.7.3 Fitness Evaluation....................................297

13.8 Evolving Variable Architecture Fuzzy System....................................298

13.8.1 Individual Representation 298

13.8.2 Genetic Operators 301

13.8.3 Fitness Function....................................301

13.9 Evolving a Learning Rule....................................301

13.9.1 Learning Rule Modeling....................................302

13.9.2 Evolutionary Optimizations....................................303

13.10 Chapter Summary 304

References 305

14 Modular Neural Networks 307
14.1 Modular Neural Networks 308
14.1.1 Problem Division 309
14.1.2 Modules 309
14.1.3 Integration 310
14.2 Ensemble 310
14.2.1 Ensembles for Classification 310
14.2.2 Ensembles for Functional Approximation 313
14.2.3 Integrator 314
14.3 Modularity in Computation 320
14.3.1 Divide and Conquer 320
14.3.2 Hierarchy of Experts 321
14.4 Modularity in Input Space 322
14.4.1 Division of Input Space 323
14.4.2 Training and Testing 324
14.4.3 Classification 324
14.4.4 Functional Prediction 326
14.4.5 Problem Generality 327
14.5 Evolutionary Modular Neural Network 327
14.5.1 Individual Representation 328
14.5.2 Genetic Operators 329
14.5.3 Fitness Evaluations 330
14.6 Co-evolution of Modular Neural Network 330
14.6.1 Module Level 331
14.6.2 Network Level 332
14.7 Chapter Summary 332
References 333

15 Hybridizing Neural and Fuzzy Systems 337
15.1 Adaptive Neuro Fuzzy Inference Systems 338
15.1.1 Layered Architecture 340
15.1.2 Training 344
15.2 Evolutionary Adaptive Neuro Fuzzy Inference Systems 346
15.2.1 Interaction between the Evolutionary Approaches 347
15.2.2 Genetic Programming 347
15.2.3 Genetic Algorithm 350
15.3 Fuzzy Neural Networks 350
15.3.1 Fuzzy Arithmetic 352
15.3.2 Training 355
15.4 Rule Extraction from Neural Networks 356
15.5 Chapter Summary 357
References 358

16 Parallel and Hierarchical Evolutionary Algorithms 361
16.1 Fitness Landscape Complexity 362
16.1.1 High Dimensionality 362

16.1.2 Sensitive Fitness Landscape 363
16.1.3 Parameter Setting 363
16.1.4 Diversity 364
16.2 Island Model Algorithm 365
16.2.1 Islands 365
16.2.2 Information Exchange 366
16.2.3 Algorithm Outline 367
16.3 Hierarchical Fair Competition 368
16.3.1 Sub-populations 369
16.3.2 Transfer of Characteristics 370
16.3.3 Algorithm Outline 371
16.4 Nested Evolutionary Strategies 372
16.4.1 Inner Evolutionary Strategies 372
16.4.2 Outer Evolutionary Strategies 372
16.4.3 Algorithm Outline 373
16.5 Hybrid Genetic Algorithm Particle Swarm Optimization 376
16.5.1 Genetic Algorithm 376
16.5.2 Particle Swarm Optimization 377
16.5.3 Algorithm Outline 377
16.6 Chapter Summary 378
References 379

Part V: Hybrid Intelligent Systems

17 Fusion Methods in Biometrics 385
17.1 Fusion of Face and Speech 386
17.2 Pre-classification Fusion 387
17.2.1 Fusion Process 387
17.2.2 Inputs 388
17.2.3 Classification 389
17.2.4 Experimentation 390
17.3 Post-classification Fusion 390
17.3.1 Problem Division 390
17.3.2 Module Inputs 391
17.3.3 Modules 393
17.3.4 Integrator 393
17.3.5 Experimentation 394
17.4 Division of Input Space 394
17.4.1 Problem Division 395
17.4.2 Modules 396
17.4.3 Experimentation 396
17.5 Chapter Summary 397
References 397

18 Multimodal Biometric Systems 401
18.1 Biometric Modalities 402
18.1.1 Bi-Modal Systems..402
18.1.2 Tri-Modal Systems 403
18.1.3 Higher Modality Systems 403
18.2 Fusion of Speech, Lip and Face..404
18.2.1 Face..405
18.2.2 Speech..405
18.2.3 Lip..406
18.2.4 Fusion 406
18.2.5 Experimentation..407
18.3 Fusion of Face, Fingerprint and Speech..408
18.3.1 Face..408
18.3.2 Fingerprint 409
18.3.3 Speech..411
18.3.4 Fusion 411
18.4 Fusion of Face, Fingerprint and Hand Geometry..412
18.4.1 Face..412
18.4.2 Fingerprint 413
18.4.3 Hand Geometry..413
18.4.4 Fusion 413
18.4.5 Experimentation..416
18.5 Chapter Summary 416
References 416

Part VI: Other Supplementary Topics

19 Adaptive Systems 421
19.1 Parameters 422
19.1.1 Role..422
19.1.2 Parameter Setting by Evolutionary Algorithms 423
19.1.3 Towards Parameter-Less Systems 424
19.2 Classification of Adaptation 424
19.3 Types of Adaptation 425
19.3.1 Static Systems..426
19.3.2 Deterministic Systems 426
19.3.3 Adaptive Systems 427
19.3.4 Self-adaptive Systems..428
19.4 Level of Adaptation..428
19.4.1 Environment 428
19.4.2 Population..429
19.4.3 Individual..429
19.4.4 Component..430
19.5 Chapter Summary 430
References 430

20 A Taxonomy of Models **433**

21 A Programmer's Approach **441**
21.1 Neural Systems 443
21.1.1 Training and Testing Data Sets 443
21.1.2 k-Fold Testing 445
21.2 Modular Neural Networks 447
21.2.1 Ensembles 447
21.2.2 Modularity in Computation 450
21.3 Evolutionary Neural Networks with Fixed Architecture 453
21.4 Evolutionary Neural Networks with Connectionist Architecture 458

Part I
Simple Computing Techniques

Chapter 1
Introduction

Abstract. Soft Computing Systems have undertaken a radical change largely attributed to their widespread use along with a vast research community that has developed over the years. Before beginning the ballad of development of this domain, it is important to get the stage set. This chapter explores the various concepts and terms used in computationally intelligent systems of today. We give a brief introduction to recognition systems, machine learning, expert systems and biometric identification. The major focus of the chapter is upon presenting the application of Soft Computing systems, the manner in which Soft Computing approaches contribute towards the application, and the various problems and issues that the application presents. These issues open gateways for a lot of research for the research community. While the sophisticated Soft Computing systems of today may be able to effectively solve a wide variety of problems, the data availability and computational constraints would always be a limitation for the flawless growth of the soft computing systems.

Soft Computing is an exciting field of today that has marked its presence in almost all application domains that one may visualize. The applications include biometric identification systems (Kala *et al* 2010b; Shukla *et al* 2009c), robotics (Kala *et al* 2009a; Shukla *et al* 2009a), biomedical engineering (Kala *et al* 2009b; Shukla *et al* 2009b), optical character recognition (Shukla *et al* 2009d), wireless communication (Shukla *et al* 2009e), series forecasting (Giles, Lawrence and Tsoi 2001), etc. Each of these fields stands tall as a well-studied disciple of today with innovative use of Soft Computing as problem solving technique. Soft Computing systems further present promising applications for the future that would be much more robust and scalable to the needs. The complete paradigm shift from theory to application in a real life scenario is itself a challenging task that puts forth us numerous issues and problems. Numerous soft computing systems have already made it into the industrial use that we come across in form of automatic document processing, fuzzy controllers, interactive voice based response systems, etc. As the applications and related issues of these systems are increasing, the Soft Computing systems are becoming more and more sophisticated in nature.

Early motivation of the researchers was to solve simple problems using the soft computing techniques. The success in effectively solving these problems was phenomenal and the high accuracies were great source of inspiration for the subsequent development. As the research continued, the problems started becoming more and more complex (Intrator 1992). The complexity was in terms of lesser

A. Shukla et al.: Towards Hybrid and Adaptive Computing, SCI 307, pp. 3–30.
springerlink.com

assumptions, more requirements from the system and factors related to scalability and robustness. The success and widespread applicability further attracted numerous people from diverse streams who further contributed towards enabling the domain solve complex problems. Today the Soft Computing community incorporates numerous people associated with plentiful of application areas across different laboratories of the world. While some concentrate over the theoretical aspects of Soft Computing, the others are busy in customizing the solutions towards their application areas.

Soft Computing may be conceived as the *computing* that gives *soft* answers to the presented problems. By computing we are basically referring to problem solving where the input is presented and the output needs to be computed based on the system state (Kasabov 1998). The notion of giving soft answers is that we are never cent percent sure of the validity of answers even under the applicable assumptions. The answers may not necessarily be precise. In this regard soft computing differs from hard computing where the outputs are always well-determined given that the underlying assumptions hold. Soft Computing systems are capable of solving problems that cannot be solved by the hard computing methodology. Using the hard computing methodology either the systems would get so complicated that we cannot model and build them, or the assumptions get so large that they never hold. The beauty of soft computing lies in the ease with which it is able to solve really complex problems. We are hence able to get the desired system up and working with a little loss to the precision. This however does not guarantee that there would be no errors by the soft computing systems. Even after extensive research, we have not yet reached a landmark where soft computing systems give 100% correct answers. The reasons for the underperformance are many, most of which would be discussed in the subsequent chapters of the book.

Noise and uncertainties are an indispensable part of almost any system of today (Acero and Stern 1990). Many times the data is recorded and is found to be corrupted by noise. In many other scenarios the data may not be available at all, or may be available with very low certainty of its validity. Another major inspiration in the use of Soft Computing is the capability of these techniques to handle noise and uncertainty. Soft computing systems can perform reasonably well even in the presence of a large amount of noise or missing data (Konar 1999). This enables their use in multiple domains where noise removal may not be possible. There are various reasons of noise in data like noisy environment and recording conditions, recording instrument limitations, faulty processing techniques, etc. While some of these may be rectified by a noise removal technique, others are completely impossible to remove. The noisy nature of data is still a major challenge and hurdle in the soft computing systems that are noise resistant by virtue.

The entire field of Soft Computing is highly interdisciplinary in nature that actively takes inputs from the other fields. This further calls upon a lot of collaborative work between people of different disciples. A common example of this is the biometric recognition system with the use of face (Shukla *et al* 2009f; Shukla *et al* 2009g). The entire task of capturing of image, detection of features and then recognition of face and speech involves a lot of use of image and signal processing. The final output of this processing is fed into the Soft Computing systems for the task of recognition. The inter-disciplinary nature of the domain has been largely responsible for the relatively restricted development of this field where someone is expected to master an additional domain in order to develop an application. On the whole the inter-disciplinary nature plays a key role in the multi-dimensional development of the domain as a result of the collaborative development by people throughout the globe.

Traditional soft computing systems incorporate Artificial Neural Networks (ANN), Fuzzy Logic (FL) and Evolutionary Algorithms (EA). ANNs are inspiration from the human brain and are promising agents of learning from the historical data and generalizing their learning to new data (Lippman 1987). Fuzzy Inference Systems are logic based systems that use fuzzy rules similar to production rules for the task of problem solving with known relations between inputs and outputs (Kasabov 1998). EAs are an inspiration from natural evolution process (Mitchell 1998). These systems are excellent optimizing agents where the higher generation individuals are always more fit and perform better than their parents. These systems have been used since a very long time for solving the various kinds of problems. Each of these would be discussed separately as a chapter of its own.

Besides these systems, there is a lot of use of heuristic techniques for problem solving. Heuristics has always played a key role in driving the soft computing systems (Pearl 1984). In many systems heuristics are the sole agents of problem solving. Heuristic is an indication of the closeness of a state of the problem to the goal state. It is used in the search problems where they denote the indicative distance between any of the nodes to the goal. Heuristic Search techniques try to use this indicative distance or the heuristic value to aid in the search operation. A commonly used search algorithm is the A* algorithm (Shukla *et al* 2008). This algorithm maintains an open list of nodes opened and to be visited, and a closed list of completely visited nodes. The nodes in the open list are expanded in priority of their total cost and then added to the closed list from the open list. The total cost here is the sum of historic cost that is well determined and the heuristic cost that is only an indicative measure. The contribution of the historic and the heuristic cost may be varied by assigning them weights in the addition process. The heuristic search techniques perform much better than the non-heuristic search algorithms that include Breadth First Search, Depth First Search, Dijkstra's algorithm etc (Cormen *et al* 2001).

Heuristics are set by logic of the problem being undertaken. The certainty of the heuristic behavior largely determines the performance of the search algorithm. The best scenario happens in case where the heuristic is precisely known. Then the indicative heuristic cost corresponds to the actual cost and the search becomes highly predictive in nature. However in case the heuristic changes sharply or shows abnormal behavior, the search operation is more or less random in nature. For such scenarios a modified version of A* algorithm was presented by Shukla and Kala (2008) that expanded a collection of nodes from the open list with a range of heuristic values from good to bad. This was names as Multi-Neuron Heuristic Search (MNHS). This algorithm performs better than the conventional A* algorithm for solving maze-like problems. The heuristics only provide an indicative measure of the closeness of the goal. This measure needs to be used intelligently and cautiously to make a system driven by heuristics. The use of heuristics in problem solving is a field in itself (Chavatal 1979; Bonet and Geffner 2001).

The traditional soft computing systems comprising of ANNs, Fuzzy Logic and EAs are good problem solving agents. But the complexity of problem many times exposes their weaknesses in problem solving. The demand for higher performance urges the minimization of the limitations of each of these systems. It is fascinating observation that one of these systems can effectively couple with another one to remove its limitation. The advantages of both the systems hence get added up and the disadvantages are minimized to a great extent. The heuristic techniques further come into play to contribute towards this fusion process. This results in creation of sophisticated systems well-suited for the complex problems. These are the hybrid systems that mark the future of soft computing (Bunke and Kandel 2002; Jain and Martin 1998). The high performance of these hybrid systems has cast a deep impact in scope of application of soft computing. We hence find a large amount of hybridization between systems of various kinds. The fusion of systems with other systems is a complex phenomenon that needs to be well-studied and understood for the maximization of advantages. The large scale hybridization presents a thousand opportunities to make effective systems that remove the limitations of the traditional systems. However, every act of hybridization makes the system extremely complex that may have its own limitations in terms of computational or data requirements.

Nature may itself be seen as a fascinating system that is still a mystery for the millions of researchers around the globe. The ease with which the parent bird identifies its child out of all the birds, or the ease with which the flocking birds coordinate with each other is both an inspiration as well as mystery. Every natural process in nature seems to be carved with great ability that the human made artificial machines would take a long time to imitate. Brain is the most phenomenal unit that does large volumes of calculations that is responsible for the human intelligence. The evolution and adaptation of life, immunity system, human cognition, etc. have played a key role in the development of artificial systems on the lines of natural systems. Nature still has a million possibilities that attract people from all across the globe. A clearer understanding of these systems, innovative modeling and building up of artificial counterparts holds the key for a lot many efficient problem solving systems (Marrow 2004).

1.1 Recognition Systems

One of the prominent areas of application of Soft Computing is in recognition systems. These systems carry forward the task of recognition. Recognition may be of face (Shukla *et al* 2009f, 2009g), signature (Lam and Kamins 1989) or even character (Shukla *et al* 2009d). Another commonly used term is pattern recognition that deals with the recognition of patterns in the mentioned forms. This differs from the conventional recognition since the pattern to be identified is available in form of high dimensional audio, visual or other form of information. Besides, there is a possibility of a large amount of noise in the system. The computational limitations are another characteristic of these systems. Hence there needs to be specialized technique to solve the problems of recognition for these kinds of data. In this section we give an overall overview of problem solving in pattern recognition.

The problem is usually studied in two separate heads. These are identification and verification (Reynolds 1995). Identification determines which of pattern, out of all the available kind of patterns, the input unknown pattern corresponds to. Say the system was of farce recognition. In this problem the face of a person would be given as the input and the system would be asked to identify the person. The other related problem is of verification. Here an unknown pattern is given, which claims a certain identity, and the system is supposed to verify whether the unknown pattern is the same as per claim. In this problem the face of the person would be given to the system as an input and the system would be asked to verify whether it is face of the same as claimed. Both these systems are extensively used in many practical applications.

The entire task of problem solving with recognition systems may be easily broken down into a variety of steps. These are data acquisition, pre-processing, segmentation, feature extraction and pattern matching (Shukla *et al* 2010). The overall system hence formed is given in figure 1.1. We discuss each of these steps one by one in the next sub-sections.

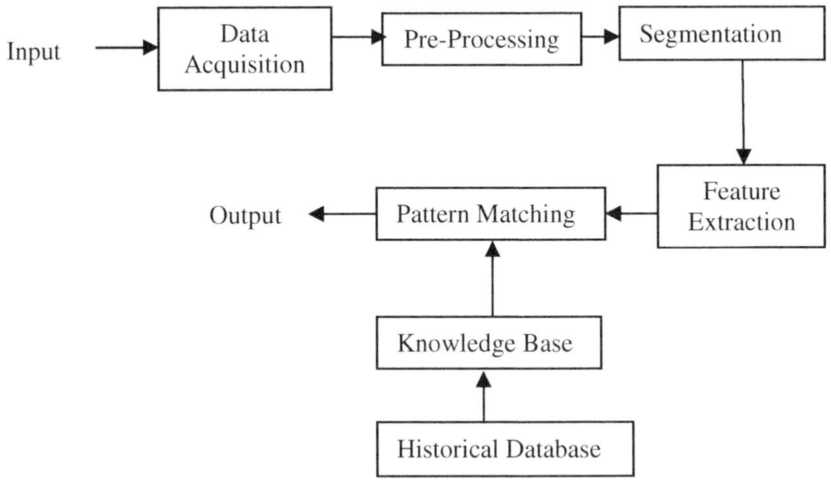

Fig. 1.1 General Architecture of Recognition System

1.1.1 Data Acquisition

Data Acquisition is the first step towards recognition systems. This step deals with capturing of the data and presenting it to the system for further processing. In other words it may be said that this provides the input to the system. The data for any recognition system needs special devices that would be specific for the system. For face recognition system, the input is in form of an image taken by a still or video camera (Shukla *et al* 2009f), for speaker recognition it is in form of a speech signal recorded by a microphone (Shukla *et al* 2009c), for iris recognition it is the image recorded by special cameras (Wildes 1997). Other systems would require scanners or special sensors for inputs. The recognition system may be offline or on-line (Plamondon and Srihari 2000). This depends upon the manner in which it is being used. An online system takes the input and simultaneously processes it. This is unlike the off-line system where the input is taken previously and is later input to the system in different form. The input device plays different role in these contexts as well. Consider the problem of signature recognition. An offline signature recognition system takes as its input an image that is the signature of the person. A common way of giving the input is by the use of scanners. The person must have already signed before on the paper. On the other hand an online signature recognition system makes use of light pens. The person is made to sign using this pen. His motion, speed and movements are recorded by the system while he signs. The system then carries forward the recognition task.

Noise is a major factor that largely comes in this step (Dirks, Morgan and Dubno 1982). Hence the recording needs to be done using the selected device in a manner and in a surrounding such that the noise is minimal. It may however not always be possible to stop noise. Since the system needs to be built for real life scenarios, the presence of noise and poor recording instruments is natural. This is one of the major problems that come in designing effective systems. Better devices and noiseless recording environments are ideal requirements of a recognition system.

1.1.2 Pre-processing

The acquired input is usually prone to noise and is hence not very valuable for most of the recognition purposes. The first major task before the recognition system is to minimize the noise in the recording. The method of removal of noise depends upon the type of noise present which is a characteristic of the system and the surroundings in which the recording was done. We consider the input in form of an image (or video) or speech. Noise is usually in the form of Gaussian noise, salt and pepper noise or impulse noise. The noise is removed by the application of special filters. The image or speech signals are passed through these filters that allow the information to pass through but do not allow the passage of noise. Commonly used filters include low pass filter, high pass filter and median filter. Salt and pepper noise is commonly found in images in the form of black and white pixels that occur randomly in the entire image. This noise is removed with the help of median filter. This filter replaces any pixel in the distorted image by the median pixel value calculated between all neighboring pixels. Gaussian noise is the noise in which the amplitude of distribution of the noise varies in a Gaussian manner. The noise removal technique largely depends on the assumption of the noise. It is not possible to completely remove nose from a randomly given signal. The general tendency is to model the noise based on the surrounding on recording and then to design a filter that carries a converse effect (Brailean *et al* 1995). This converse effect of the filter removes noise only if the noise behavior was as per the model. Noise removal works well to remove many general noises from the data. However the problems with most recognition systems are the irregular noises that are in the surroundings and influence the main signal. For a speaker recognition system this would be the noise of trains or vehicles that are passing by, if the recording was in open space.

1.1.3 Segmentation

Once the data with the absence of noise is available, the next task is segmentation. Segmentation deals with the separation of the needed part of the data from the rest of the recorded data. We try to identify the Region of Interest (ROI) which is worked upon further by the recognition system. Segmentation of the speech signals may trivially be done based on the amplitude values. In the problem of character recognition, segmentation would be supposed to separate lines, words and finally the character from the rest of the areas (Pal and Pal 1993). The segmented part of the entire image is processed further in the subsequent steps. Segmentation makes use of edge detection to carry forward the task. Edge detection is a technique to identify the edges in the entire image that uses the change in intensity across the image pixels as a measure of edge. Edges denote a sharp change in intensity value. Various types of image matrix operators may be used for the task of edge detection (Canny 1985). This includes the first order edge detection operators, second order edge detection operators and the canny edge detection. The higher order operators give an idea of both the edge as well as the direction of the object. Segmentation is used to divide the image into segments based on objects, connected components etc. It is a relatively difficult task to identify the correct region of interest in the entire image that may house multiple objects. It may be simple for most of the recognition systems which have approximate prior known location of objects like face recognition when only the facial image is given as an input.

1.1.4 Feature Extraction

The segmented data is of very high dimensionality. Dimensionality refers to the volume of input that the input represents or the number of input attributes. Such large number of attributes cannot be given to the recognition system, as it would take immense amount of time for processing. We hence need to reduce the number of input attributes by a technique known as feature extraction (Trier, Jain and Taxt 1996). This step extracts the useful part of the data out of the entire data. In other words the large volume of data is summarized into a small set of values that can be easily processed by the recognition system. The large volumes of data have another disadvantage. This is the requirement of large historical database that the system can train itself with. The larger number of attributes requires larger systems with more number of parameters which ultimately requires a larger training database. This may not always be available.

Feature represents an input attribute of the system that in some way represents the original high dimensional input. The set of all features should effectively represent the entire input considering the problem requirements. The number of features hence should be low enough for computational reasons and high enough to uniquely represent the data class. For a signature recognition or object recognition system, the features are normally in form of frequencies that make the signature. For a face recognition system, it may be in form of distance between eyes, mouth and eyes, etc (Shula *et al* 2009f, Shukla and Tiwari 2008).

1.1.5 Pattern Matching

The next major task that computes the system output is pattern matching. The various features extracted by the system are given as input for this algorithm. The system also has a historical database that is a collection of historical information about a lot of collection of inputs and the associated outputs. This database stores a number of instances of every class. Each instance has its feature set representing the input. The task of the pattern matching technique is to compute the correct output to which the given input belongs to by using the information already stored in the historical database. A number of approaches are used for this task of pattern matching that incorporates both statistical and soft computing approaches. Some of the techniques include Bayes' theorem (Kumar and Shukla 2008), Neural Networks (Shukla *et al* 2009c), template based matching, clustering (Kala, Shukla and Tiwari 2010b) etc. Many times the historical database may be too large in size. This is especially the case when the system continuously records data and keeps adding it in the historical database (Kala *et al* 2009c). In such a case it would be computationally very expensive to match the given input with each of the available instance in the historical data set. Also the data would never have a precise match. In such a case we need an intelligent technique to decide the output class by looking at the trends of the historical data set. This makes the entire task of pattern matching un-trivial and difficult in nature. Further it is a prime requirement that the system gives high performance for unknown data. The data is hence stored in summarized form where the form depends upon the chosen method of pattern matching.

1.2 Biometric Systems

One of the most extensively used recognition systems is the biometric systems. These systems try to identify or verify an individual on the basis of their biometric identity. Biometric modalities have been known to be unique to a person for a very long time. The face is the most widely used modality for everyday recognition of people by others. Again we can identify people by listening to them which advocates the uniqueness of voice. The signature is used as proof for all official purposes. Similar is the case of fingerprint, iris, etc. The biometric recognition systems try to use the same concept of automatically identifying people by any of their biometric modalities. The common uses of these systems include the security checks, biometric login facilities, etc. These systems are more secure as compared to the paper and manual protection schemes which can be forged. Besides, these systems are more easy to operate and give a faster service as compared to their counterparts where checks are very slow and time consuming. Imagine a fingerprint controlled entry system where a person is supposed to swipe his finger to get access. It would be very easy and swift to allow the entry of a large queue as a result of automation of the system by use of biometrics. Biometrics is hence becoming very popular and is being extensively used in various scenarios (Jain, Ross and Prabhakar 2004; Phillips *et al* 2000). The system may be classified in three broad heads depending upon the type of modality being used. These are physiological biometrics, behavioral biometrics and fusion methods. Each of these is discussed in brief in the next sub-sections. This would form the major application area of the book that we would study in the subsequent chapters. The various biometrics are summarized in table 1.1

Table 1.1 Types of Biometrics

Physiological Biometrics	Behavioral Biometrics	Fusion Methods
Face	Speech	Speech and Face
Palm	Gesture	Face and Lips
Fingerprint	Keystroke	Fingerprint and Palm
DNA	Gait	Face, Fingerprint and Speech
Iris	Signature	Etc.
Ear	Etc.	
Lips		
Etc.		

1.2.1 Physiological Biometrics

Physiological biometrics deals with the use of modalities that do not change with time. These represent the physical characteristics of the biometric modality which are constant with time. This is hence also called as passive biometric. The common examples of these modalities include face (Shukla *et al* 2009f, 2009g), iris (Wildes 1997), hand geometry etc. The task of recognition with these modalities deals with capturing the data as a single instance, which is mostly in the form of image. Using a single instance at a snapshot of time may many times be argued as a quicker and easier way of using biometrics. Once the capturing has been done, the standard image processing techniques may be applied for the recognition purposes. Since there is no dependency on time, the amount of data may usually be low which may be conveniently processed.

1.2.2 Behavioral Biometrics

The other type of biometrics is behavioral biometrics. In these systems the modalities change with time. The recording is done for some duration of time. These systems do the task of recognition by observing the behavior over a period of time. The identification process involves the study of the manner in which the recorded data changes along with time. This is hence called as active biometric. It may be visualized as an added dimension of time over the physiological biometric system. This further raises numerous issues related to time varying change of acquired data. The commonly used behavioral biometric systems include speech (Shukla *et al* 2009c), gesture, signature, etc.

1.2.3 Fusion Methods

Biometric identification with a single biometric modality gives a limited performance. This is due to the effect of noise, loss of information in feature extraction and numerous other reasons. As a result identification with a single modality gives limited performance. The need of better recognition scores emphasizes on the use of multiple modalities at the same time for the task of recognition. Here the various modalities contribute towards the recognition decision and hence make effective systems of large use. The recognition is greatly improved by the combined use of biometric modalities. The most commonly used fused biometric systems are the bi-modal systems that fuse two modalities for the recognition purposes. The fusion involves the pooling of features of both the modalities before giving it to the system for the task of recognition. The pooling of the features may many times suffer from the disadvantages of large number of features in the system. The system hence selects only some vital features from both modalities to make a small and effective pool of features. This largely improves the recognition score as well as the training time and training ability (Shukla and Tiwari 2008c). Another method is to classify the inputs and make decision individually with every modality and then devise a mechanism for making the overall decision by considering all the modalities based on the individual modality decisions (Prabhakar and Jain 2002). The commonly used fusion methods include fusion of face and speech, face and ear, face and iris, etc. One of the major factors in deciding the modalities being fused is the ease of use of the final system. Fusion of face and speech may be easily visualized as a simple to use system where the user comes and speaks the text. While the microphone records the speech; the video camera would record the face. Then the recognition system would carry forward its task of recognition. This may however not be the case when speech is fused with fingerprint. For the user it would mean two independent identification systems that need to be operated one after the other. Hence such a system is not used.

1.3 Types of Problems in Soft Computing

Soft Computing is being used as an effective system of solving a variety of problems. These problems may broadly be classified into three separate heads. Each of the problems where soft computing finds application belongs to either of these classes. It may however be clarified that the problems may use a combination of these categories. The motivation behind hybrid systems is the use of different techniques for attacking the different issues with the problem. These types are classification, functional prediction and optimization. Each of these is discussed one by one.

1.3.1 Classification

Classification deals with the determination of the correct class to which the input belongs. Here the output is always a discrete value denoting the class name to which the input is said to be classified to. Consider that problem takes inputs with *N* attributes. We plot a *N* dimensional space, called as the input space, with each axis denoting each of these attributes. The input represents a point in this space. Also the entire historical data set represents points in this data set. Let each of the possible output class be designated by a label. Further let the entire historical data set be plotted in this input space with the point being denoted by the label to which the input belongs. The task of classification is to designate the correct label to the input in this input space based on the labels and locations of the other points. This task is carried forward by the construction of decision boundaries. These are the boundaries that separate the points of various classes from each other. As a result the points that lie in the same class form a separate region of their own what is bounded by the decision boundary. The entire task of the system is to construct these decision boundaries in the input space, given the historical database. Afterwards whenever any new input is given to the system, it may easily map it to the correct class by locating the region in which the input lies.

Ideally the input attributes are chosen in such a manner that the inter-class separation is high and the intra-class separation is low. This means that the various instances of the same class must lie as close as possible to each other. The various classes on the other hand must be distant apart in the input space. In such a scenario it becomes very simple to construct the decision boundaries. This is because there is a lot of scope to easily fit these boundaries. They may further take any shape or comfortably move about in the input space. The problem comes when the inputs corresponding to the different classes start inter-mingling with other. Now it becomes extremely difficult and many times impossible to draw clear decision boundaries. This may be due to the presence of noise or even the bad selection of attributes in the system. The problem further comes in the classification of inputs that lie very close or sometimes on the input boundaries. Now classifying the input to any of the classes is not an easy decision due to the high possibility of error. These inputs greatly influence the system performance and make the task of classification difficult.

Another invariant of the problem may be when only the inputs are known that is the case with unsupervised learning. In such a case the system tries to find patterns in the inputs represented in the historical data and tries to associate the unknown input to any of these patterns. We would study this in the subsequent section 1.4.2.

Consider a 2-dimensional view of an arbiter input space given in figure 1.2. Here we may easily observe two things. Firstly the decision boundaries cannot be easily drawn. This is due to the high mixing of the inputs from different classes.

We hence try to figure out a decision boundary that most nearly classifies the data correctly. The second important thing we notice is that the decision boundary constructed is of nearly smooth shape. It would have been possible to better classify the data by making very rough decision boundaries that would have taken large number of steep turns. But we prefer a simple decision boundary as we know there is always a possibility of noise. The points lying near the decision boundaries might have been wrongly recorded. Making the decision boundary complex for these inputs would affect the system, making it lose generalization or capability to correctly classify new data. This would further make the system too sensitive in nature where little movement in the input space causes a complete change in output. Most of the cases the data should be able to separate in simple decision boundaries.

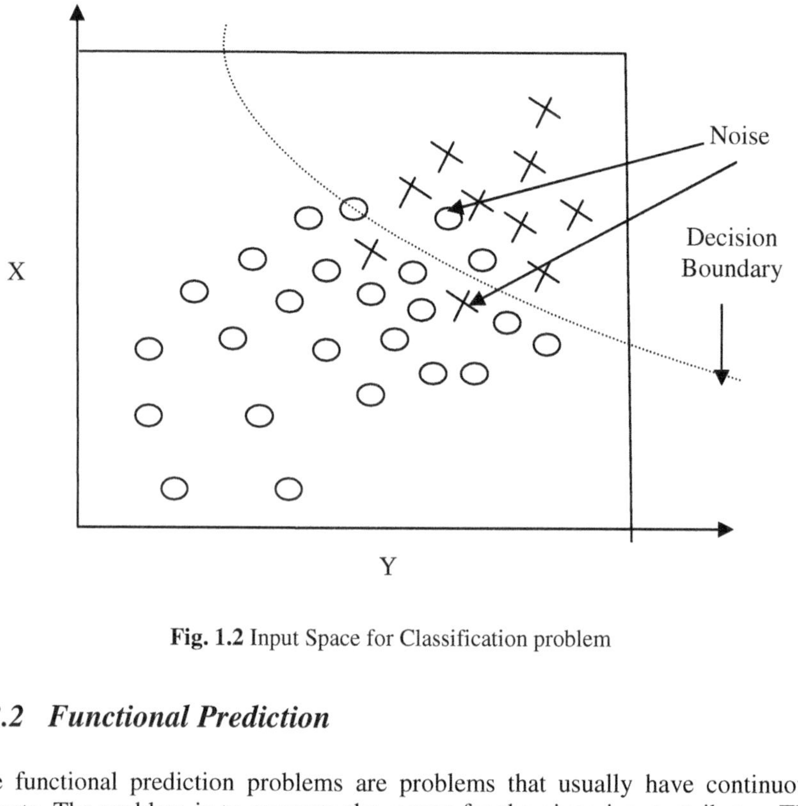

Fig. 1.2 Input Space for Classification problem

1.3.2 Functional Prediction

The functional prediction problems are problems that usually have continuous outputs. The problem is to compute the output for the given input attributes. The output is always some function of the inputs that the system is supposed to determine. The system is given the historical database that is a collection of a large amount of input and output combinations. The system is supposed to imitate a function that maps the inputs to the outputs in the historical data set. Now whenever a new input comes, it uses this imitated function to compute the output. Again consider the input has N input attributes. For simplicity we assume that the system has 1 output. In this case we plot a $N+1^{th}$ dimensional space. Here the N

dimensional space corresponds to the system input attributes as was the case with classification. This forms the base of the plot whose vertical axis is used to plot the output.

This gives us a complex surface in the multi-dimensional space that the system imitates. The historical data set are points in this surface. In other words we are given points in this high dimensional space. We need to construct a functional surface that best covers all these points. It may be seen that there would be multiple ways in which this can be done. We would again prefer to select the way in which these points can be appreciably fitted with the simplest surface. This is accounting for the presence of noise as well as the simple nature of the assumed problem. Further too complex fitting would make the function very sensitive in nature that would have huge changes in the output values for very small changes in the input attributes. This would mean a loss of generalizing capability by the system.

It may be noted that it is not mandatory that historical database would be available and needs to be used for the training of the system. In case of the absence of the same, this section should be restricted to the general concept of the term rather than the curve fitting notion presented.

Consider figure 1.3 where the system has been assumed to take just a single input. The output is plotted in the vertical axis and the input is plotted in the horizontal axis. It may be easily seen that the simpler line is a better approximation of the problem, since it would give predictable outputs to the unknown inputs. The more complex fitting curve may give completely un-imaginable outputs which may be wrong.

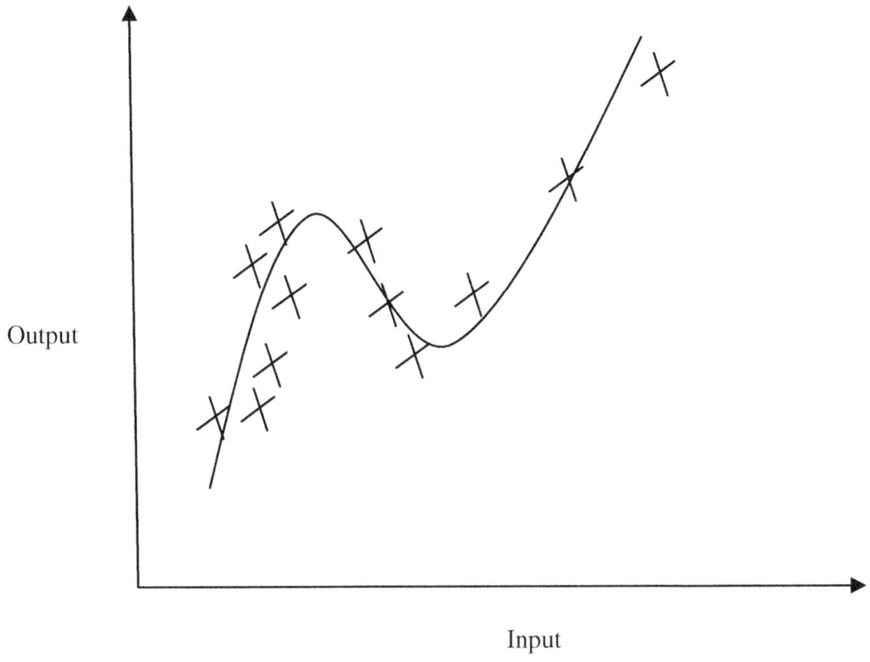

Fig. 1.3 Functional Approximation in Input Space

1.3.3 *Optimization*

Optimization is another fundamental problem of study. This problem deals with finding the best solution to a problem out of a large number of solutions possible. The goodness of the solution is measured by a function called as the objective function. The value of this function may also be regarded as the fitness value for a particular solution. The system has a set of parameters that can be modified. The system is supposed to find the best combination of parameter values that maximized the objective function. Each combination of parameters forms a solution of the problem.

Consider that the objective function is given by $F(x_1, x_2, x_3, \ldots x_n)$. Here $x_1, x_2, x_3, \ldots x_n$ are the parameters which may be modified. Each one of these has got a lower bound and an upper bound value. The system may set any combination of values of the modifiable parameters. The aim is to minimize F and find out the combination of values of all x_i for which the minimal value of F is recorded. In order to better understand the concept, we plot the function F for all values of $x_1, x_2, x_3 \ldots. x_n$. This would be a $N+1$ dimensional space with the base as the N dimensional space with each axis corresponding to a modifiable parameter. This space is commonly known as the fitness landscape. Now every combination of modifiable parameters, along with the objective value is a point in this landscape. The aim is to find the point that is the least in value. The problem is usually caused by the fact that the entire landscape is complex and filled with many hills and valleys. This makes the task of search very complex.

One such fitness landscape for two modifiable parameters is given in figure 1.4. Here the presence of multiple local minima may easily be seen. These are the points which possess a low objective value as compared to all their neighbors. Many search algorithms are prone to get struck at such local minima. Since we can explore only some part of the search space because of computational reasons, the entire task becomes challenging. The end result must ideally be the global minima that lie at some specific location in the fitness landscape. The complexity of the fitness landscape largely determines the ease of search. A highly complex landscape has numerous hills and valleys. This is especially for problems with high dimensionality or high number of modifiable parameters.

1.4 Machine Learning

Machine Learning does the task of learning of the historical data. We discussed earlier that recognition systems cannot perform the task of matching of given unknown input with every possible instance in the historical database. Also the historical database is in form of raw data that might not always be useful. Machine learning does the task of extraction of patterns, trends and knowledge from this historical database. This summarizes the historical data into a set of principles that guide the mapping of the inputs to outputs. Machine Learning mainly deals with the learning of historical database. This means that whenever any instance from this database would be again given to the system, its output would be correctly

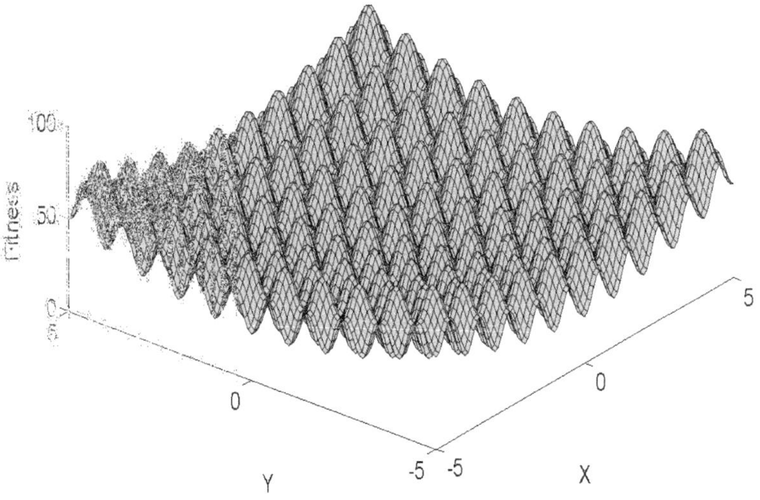

Fig. 1.4 Fitness Landscape of Optimization Search Space

computed and returned by the system. A good modeling of the system involves the selection of correct inputs and outputs to the problem, such that there is a relation between the inputs and the outputs. These relations are determined by the system in the process of machine learning. This enables the system to generalize the findings from the historical database into new data as well. Hence whenever any new input is recorded and given to the system, it would be able to compute the correct output. This however holds only when a similar input is given as recorded in the historical database and the new data obeys similar rules, patterns or trends that were shown by the historical data shown previously to the system. Machine Learning is classified into three major heads depending upon the form of inputs and outputs. These are supervised learning, unsupervised learning and reinforcement learning.

1.4.1 Supervised Learning

Supervised learning is the learning in which both the inputs and the output are given to the system. The system adjusts its state such that the outputs to the inputs keep getting closer and closer to the desired outputs or the targets (Konar 1999; Graupe 2007). This kind of learning is analogous to the learning where there is a teacher who trains the students by constantly giving the questions and correcting the answers. Each of the instances of the input is given to the system for the computation of the output as per the system state. The system computes the output to the applied input. The error is then measured as the difference between the obtained output and the target. The error is a measure of the performance of the system. The system then tries to adjust the various system parameters in such a manner that the error is reduced next time the same input is applied. This is repeated for all the inputs in the training data set. The entire data set may be passed for a number of times called as the number of epochs. Normally the error keeps reducing from one epoch to the other. In this manner after sufficient number of epochs, the system is said to have been trained. In other words the various system parameters are optimized to their optimal values. Now whenever any of the inputs from the training data set is again given to the system, its output is computed with lesser errors. The system may further be able to give the correct output to the new inputs as well. This depends upon the generalizing ability of the system that is a characteristic of the system configuration of use, design as well as the problem being considered. Noise many times plays a major role in reducing the generalizing capability. This is because of the fact that the supervised training tries to adjust the system parameters as per the given target which is wrong. This results in wrong adjustment of the system and hence the overall performance of the system reduces. The concept of supervised learning may be summarized by figure 1.5.

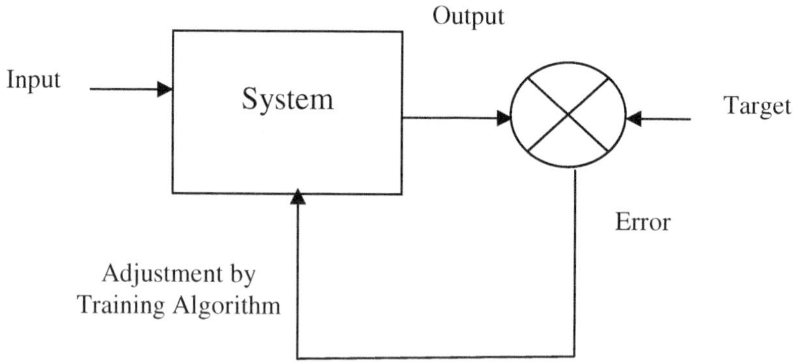

Fig. 1.5 Supervised Learning

1.4.2 Unsupervised Learning

The other type of learning is the unsupervised learning. In this type of learning only the inputs are given to the system for the task of learning. Outputs are not given (Konar 1999; Graupe 2007). This may be due to the unavailability of outputs or choice of learning technique. The system is supposed to adjust the system parameters by the inputs alone. The use of this type of learning involves finding patterns in the input data itself and association of any subsequent data using these patterns. The system tries to learn the input behavior using the data in the historical data set. Using these learning the system tries to figure out the probable behavior of the unknown input whenever it may be applied. The mapping of the input to the output takes place by consulting the outputs of the similar inputs in the historical data set. This forms the basis of working of these systems employing unsupervised learning. Clustering of the input data by various means is a common method of working of these systems. The various data may be grouped into big and small groups.

Now any input may be dealt by computing the group to which it belongs to. The output of this input would be similar to the outputs presented by members of this group. The unsupervised learning finds immense applications in classificatory problems. The various classes lie at different regions of the input space. Inputs belonging to the similar classes are found together. These problems may be easily solved by clustering of the input space into regions which normally separate the classes. A major problem is usually the large size of the historical data and limited size of working memory. Further the system must be trained in a manner to enable it perform well to unknown inputs as well to give it a high generalizing power. The concept of unsupervised learning is given in figure 1.6.

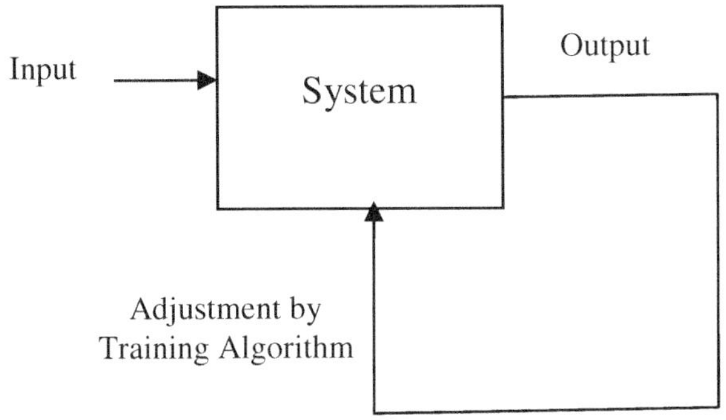

Fig. 1.6 Unsupervised Learning

1.4.3 Reinforcement Learning

The last type of learning is called as the reinforcement learning. Here we do not give the outputs along with the inputs just like unsupervised learning. However the performance of the system can be computed (Sutton and Barto 2004). Hence the system knows how well it has computed the output to the given inputs. This type of learning happens by a series of reinforcements. The system is given the inputs and the corresponding outputs are computed and judged. If the system shows good performance, it is rewarded with a positive reinforcement. On the other hand if the performance was bad or the output was faulty, the system is penalized with a negative reinforcement. The aim is to modify the system parameters in such a manner that the total reinforcement is maximized. By doing so the overall system performance is enhanced and the trained system performs close to the ideal system. This process of giving inputs, calculating the outputs and rewarding the reinforcements is done numerable number of times with every instance of the input in the historical database. As the training goes on, the rewards given to the system is maximized and the penalty is minimized. Reinforcement learning is used in many scenarios. Consider the case where we have a system that models robot behavior of overcoming obstacles and moving towards the goal. In such a case, we may let the system move a number of times on predefined maps. If the robot reaches the

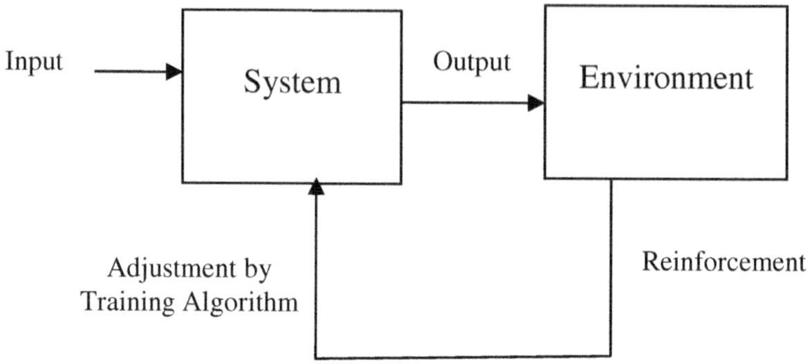

Fig. 1.7 Reinforcement Learning

goal, it is rewarded with the reward being inversely proportional to the duration of journey. A high negative penalty is given in case the robot collides with an obstacle. By using this reinforcement learning phenomenon, we can easily train the robot to reach the goal in any map by avoiding the obstacles. The concept of reinforcement learning is summarized in figure 1.7.

1.5 Expert Systems

Expert systems are intelligent systems that behave just like human experts in problem solving. These systems take the inputs from the user, process it, give the correct output and hence behave just like experts (Kasabov 1992). They are further able to learn from their errors and adapt themselves to the new inputs. This is just like an expert constantly modifying the systems for better performance to the changing scenarios. The expert systems can solve the designated problems with great accuracies and are hence used in numerous domains for automated intelligent processes. Biomedical disease diagnosis systems, financial series forecasting systems, biometric identification systems, etc. are some of the examples of expert systems that we come across regularly. The previous sections have spoken about the various concepts that drive the expert systems. In this section we try to build a bigger picture that presents the entire model of Expert Systems as a whole. The general structure of expert systems may be given by figure 1.8. This comprises of a user interface, knowledge base, database, inference engine, knowledge acquisition system and explanation. Each of these is discussed one by one.

- **User Interface:** The user interface provides a means for the user to interact with the system. All the inputs are taken or recorded by the user interface and similarly all outputs are given to the user interface. The user interface is designed as per the ease of use of the user. This calls for a proper manner of taking inputs, proper presentation of the outputs, ensuring the ease of understanding of the outputs by the user, etc. The user interface interacts with the inference engine at the other end that is responsible for the processing of the information.

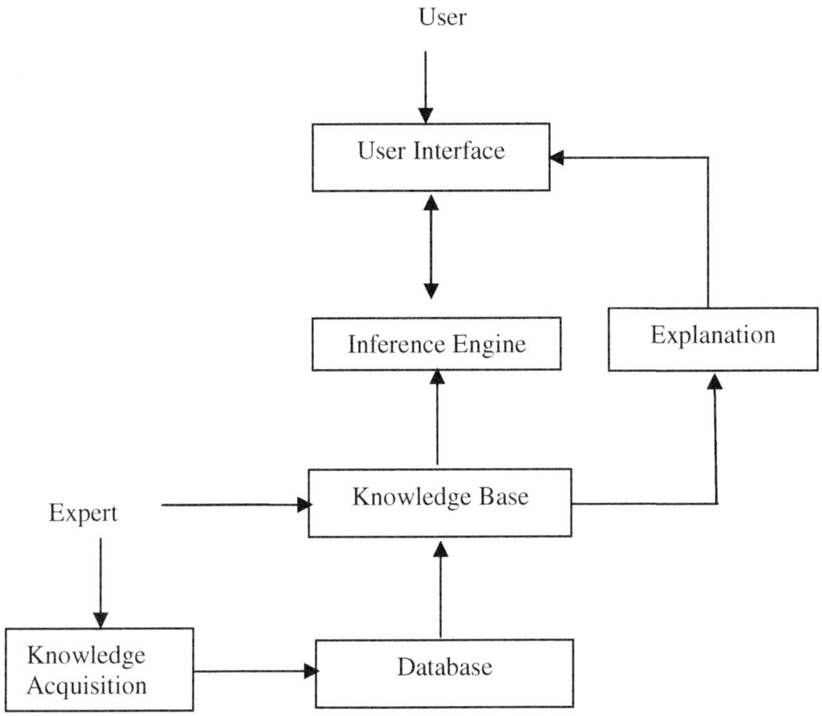

Fig. 1.8 General Architecture of Expert System

- **Data Base:** The database is a collection of all the historical recordings of the system. We discussed earlier that systems many times try to extract hidden information or facts from historical database. This database is developed over the time by the expert system as more and more data is fed into it. The database plays a key role in the overall performance of the system. The system is able to perform well or give the correct output only of inputs that behave in a manner similar to the data in the data base. It is hence important for the database to be large and diverse in nature.
- **Knowledge Base:** The knowledge base is a collection of knowledge available with the system. It is usually made by the learning of the data in the data base and consists of patterns, trends or rules that summarize the data. The knowledge base has a major role to play as the system cannot always consult the historical database for computing outputs due to computational limits. Also the learning by the system needs to be in such a manner that the system performs well to both the historical data as well as new data that might be entered into the system. The knowledge base may hence be regarded as the summarized version of the database, where the summary is stored in various forms that depend upon the system being used.

- **Inference Engine:** The inference engine does the overall processing needed by the system. Its major task is to provide the user interface with the needed outputs and acquire and process the inputs. This module of the expert system interprets the knowledge of the knowledge base in order to compute the correct output to the given input. The knowledge of the knowledge base is thus used to generate vital information needed by the user.
- **Knowledge Acquisition:** The purpose of knowledge acquisition system is to enable the system capture new data that come before it along with time. The data acquired by the user interface is captured by this module and added to the database of the system. The acquisition of new data is very important for the system. We know that the systems run on the patterns, trends and logic acquired by them from the historic data. As the time progresses, there might be a change in these trends and the new data might not follow the trends being followed by the old data. Hence it is very important to keep recording the new data so that the system is aware of the changes that are needed in the system. This module updates the database. The knowledge base may be externally updated at certain internals, or may be adaptively updated regularly. This is done by the re-learning of the historical database.
- **Explanation:** This module of the expert system does the task of reasoning regarding the computation of output from the input. It traces the manner in which the different modules work for the generation of the correct output. It further traces the manner in which the information is processed, what rules were used and how did they affect the generation of outputs from the inputs. The working of the system may hence be explained by this module.

1.6 Problems in Soft Computing Systems

Soft Computing has been under constant research and development for a very long time now. The success of this field has been phenomenal, where it has influenced almost any disciple where computing can find a place. The widespread use of Soft Computing systems is an evidence of its applicability in multiple domains. There are still many issues that challenge the researchers and pose a serious limitation over the use of these systems. Even after a bulk of research, we have not been able to attain 100% accuracy for any of the biometric systems. The same is the case of character recognition, where small children are able to recognize characters much better than the most sophisticated machines. Any series prediction systems works well only to a small degree. In this section we discuss some of these issues that need to be addressed for a flawless growth of Soft Computing systems. As the technology is progressing, there is a promising development in the minimization of each of these issues. This showcases a wonderful future of these systems with more enhanced accuracies and performances to a more robust and scalable data.

- **Noise:** The data recorded may many times not be the actual data. This is due to the presence of noise that is caused by numerous reasons that we discussed in the previous sections. Noise greatly deteriorates the performance of the systems. This is due to the reason that the systems try to adjust themselves considering the given target output which is faulty and hence the system gets driven in the wrong direction. In the context of performance of the Soft Computing systems, the noise may be broadly of two types. The first is the impulse noise where a negligible fraction of the data is noisy. However the amount of noise is very high. The other kind of noise is the uniform noise that affects almost the entire data with small magnitudes. In any real life system, there might be a combination of both types of noise in the data with varying magnitudes. The Soft Computing systems are usually resistant to impulse noises, as long as they are a relatively small fraction of the entire database. This is because the system adjustments are largely driven by the affect of the other data items and the noisy data produces relatively small effects. However the system performance greatly reduces for even a little increase in the number of noisy instances. The uniform noise on the other hand has a general reduction in system performance. The reduction again depends upon the magnitude of noise. By their very nature, the systems can resist small noises.

- **Uncertainty of data:** A similar concept to the presence of noise is the uncertainty of data. Here we may have some data with us, but we may not be sure of its validity. In many other scenarios the data may not be available at all. Consider a system in which a thermometer was used for measuring the temperature. Further consider that the thermometer was not working and the data could not be recorded. Now we may have one attribute unmeasured in the entire list of inputs. Not using the data at all may sound natural, but many times data can be precious and we may not be able to afford the loss of a single data instance. This causes a problem with the Soft Computing systems. Again consider that three different thermometers were used in the same experiment and all recorded different temperatures. Again we may be in a dilemma to decide on the actual temperature. This adds the uncertainty associated with the recorded data. Such instances are a problem in the effective use of Soft Computing systems. The systems are much resistant to uncertainties as well. These systems can take uncertain data and try to the best out of them.
- **Computation:** The training of the systems requires a lot of computation. This is especially in case the historical database is large. The processing machines have a computational limitation that prohibits them from making immense computations in a short span of time. The systems hence need to cut down on the computational requirements. The design of systems requiring too much of computation would be unrealizable, as these systems might require an unbearably large amount of time for the training purposes. In many other scenarios we may have to cut down the total number of training epochs to complete the process in finite time. This restricts the system performance. Very complex problems require larger computations. They have larger system parameters to tune which cannot be done in small number of iterations or epochs. Hence the computational limitation of the present processors has an impact on the training data size, number of inputs and outputs, as well as the complexity of problem.
- **Data Requirements:** The increasing demand from the Soft Computing systems is giving rise to complex relationships between the inputs and the outputs. This requires large and complex systems. These systems further have a large number

of modifiable parameters that are usually modified with the help of training algorithm. The training algorithm for such large systems would require a large amount of data for the optimal tuning of the parameters. Further the data needs to be diverse in nature so as to cover all the possible trends and patterns. It may not always be possible to record such large volumes of data. For many problems data is very hard to get or expensive to record. These systems impose a limitation in the effective use of Soft Computing systems. In many cases a large volume of data may be available but there would not be diversity of data. In such a case the system would be as small as a simple system that imitates the small set of behavior that the given database possesses. Hence the availability of sufficient volume of high quality data is a prime requirement for the systems which is again a limitation for many problems. It may be recalled that large volume of data has computational costs and hence we may not blindly add large chunks of data in the historical database in any problem of study. Only data showing unknown trends by the system may be given for the purpose of learning.

1.7 Generalization

Generalization is the property of a system to compute the correct output of the new inputs that have not been presented to it before. We have already discussed that the system tries to find rules or patterns in the historical data that summarize it. If these rules are generalized enough, then these can be extended towards the unknown inputs as well. Hence the system is capable of giving correct output to the unknown inputs. In this section we discuss some of the vital concepts that would be the base of discussion for the rest of the book. In the rest of the book we would make numerous references to generalizing and localizing of systems that we discuss here.

A generalized system views the entire input space and then tries to form global rules or figures out global trends that are applicable to a large variety of data. The formation of these global rules is a difficult task as one needs to fit in a large amount of data available in the database. As a result the training time of these systems is quite large. Even after training the system may not be able to effectively fit the different data items with a small set of rules. Hence the performance of the system towards the data in the historical database would generally be poor. However since the rules are global, the unknown input given is likely to perform better. This is because of the fact that the global trends that well given the rest of the system should also apply to the new input and we would have a sufficiently large idea of the output of the new input. Most of the systems known have a fair degree of generalization that enable their modeling by using simple set of rules that drive the entire system. Also if some of the data items are noisy in the database, it would not greatly affect the overall performance of the system. This is because the computation of output to any input depends upon a large number of data items in the database rather than just a few of them. The noise of a few data items is minimized by the other data items in the database. Further it may not always be possible to form such global rules. In such a case we try to loosen the criterion and try to walk towards the other end that comprises of the localized systems.

The localized systems are the ones where we try to see a small set of data and then try to form rules that fit these data elements. This may be seen as the problem of working in a small portion of the input space. The output of any data item in such systems is largely dependent on the adjacent data items. It may be easily visualized that if the adjacent data items are noisy, the affect would be large on the given input. Hence these systems are much more prone to noise. The training of these systems is usually easy as we have a small set of data to cater to. In other words there are a large number of parameters, each controlling some aspect of the performance of the algorithm. These may be easily tuned by the system as per the requirements of the particular input. After minor tuning the system may be able to exactly fit the system outputs to the desired targets. As we increase the size of the input space considered to decide the output or reduce the number of rules to model the system, the problem starts getting more and more generalized in nature.

The design of any system demands the correct setting of the parameters of the system. The aim is to have as high generalizing ability as possible. The generalizing ability depends upon the system, problem under consideration and the historical database. If the database has a large number of distinct trends that influence the mapping of the input to the outputs, the systems needed would be more complex and accordingly the generality would be lower. This is when a small set of simple rules would fail to appreciably map the inputs to outputs. Hence the system would fail to perform unless it possesses the needed complexity. Makin the systems too complex with plentiful of parameters and rules might reduce the generalizing capability. In such a context the systems would perform well on the data they are trained with, but would not be able to perform well on new data. The tradeoff between generality and localization, training accuracy and testing accuracy, training time and testing accuracy is always a difficult question from the design perspective.

Consider the problem of functional approximation where the system is supposed to imitate a function. Consider that there is a single input attribute whose input space is given by figure 1.9. In this case if the system knows only the data points shown in region *A*, it might wrongly predict the output of point *a* shown in figure. This is because the system is unaware of the global trends of the curve. It would predict the curve given by solid line which is evident by the training data of that region. This graph may also be interpreted as a global curve with some noise causing deterioration. If the same system is trained with the entire plot depicted in figure, it might correctly compute the value of the point *a*. Here it would predict the curve shown by the dotted line. Consider that the series was a financial time series, the importance of both generality and locality may be seen.

Now consider a classification problem with two attributes whose input space is plotted in figure 1.10. In this problem even if the system is shown a small part of the input space given by region *A*, it may be able to correctly compute the correct output of point *a*. Hence the correct output at the point *a* largely depends upon the outputs at the neighboring points. This argues the fact that the classification problems are much localized in nature as compared to the functional prediction problems. However the importance of generalizing ability may still not be neglected. This would be needed in case the algorithm is asked to classify the entire input space. Many points would require high generalizing ability for better performance.

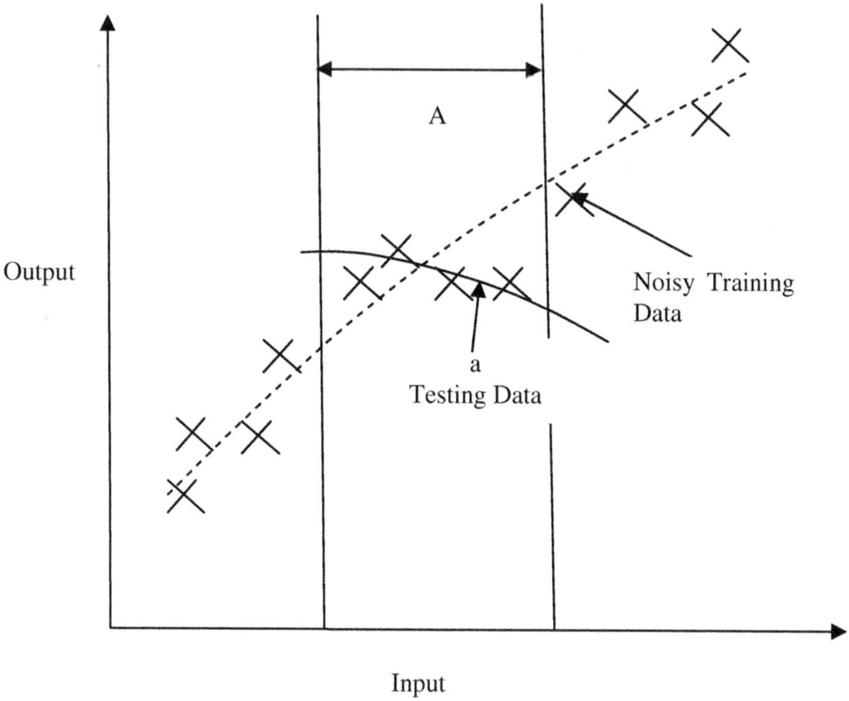

Fig. 1.9 Loss of Generality in Functional Prediction Problems

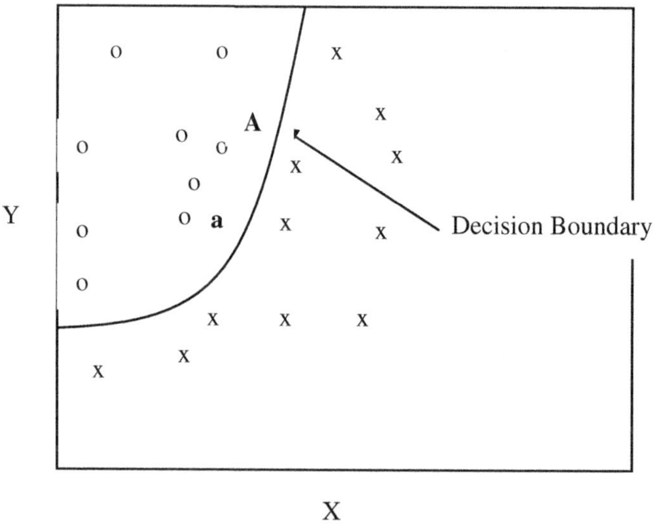

Fig. 1.10 Generality in Classification Problems

1.8 A Design Perspective

Soft Computing systems are effective problem solving agents that can model and solve almost any problem of practical use and importance. The major task is to design a system that solves the problem. The task starts with the modeling of the problem inputs and outputs. This usually happens after the realization of the system requirements and its behavior. Soft Computing systems may not be able to take any kind of input. The task of modeling deals with the determination of inputs, outputs along with their ranges. Then we need to select the correct system that would solve the problem. The wrong choice may cause a poor performance. Many times the problem may be trivially solved by the use of non-Soft Computing techniques. Using Soft Computing techniques may not only be useless in such contexts, but may also give poor performance. Hence the need of the problem and the offering of the system need to be well established before use.

The various systems have a bulk of parameters that need to be manually set. The correct setting of parameter is of ample importance for the system. Hence the understanding of the various parameters and their effect on the overall system is an important task. The individual systems may still not be able to solve the problem. This may be due to the limitations of the single Soft Computing systems. In such a context we may need to hybridize the systems with other systems. In hybridization as well it is important to understand the advantages and disadvantages of the individual systems before fusing them to form a hybrid system. A wrong use of hybrid system may reduce the performance than the individual system due to the added complexity and associated computational cost.

All this makes soft computing more of a design oriented approach where the novelty of lies in making effective design and effective modeling of the problem. The selection of correct models and correct parameters out of the available or newly created models is a challenging task that requires sound understanding of the various aspects of the system behavior. This makes the soft computing engineer a designer by profession and the various models as designing tools which need to be understood and then effectively used to make the design of the overall final model. Designs are always prone to errors due to human limitations. This is the motivation behind adaptive systems that can adjust themselves as per the state and problem needs and behavior. Efforts exist in making the entire task as automated as possible. A system must hence be able to take the problem and adjust itself accordingly to give the output without useless burden of design into the hands of the developer which may be erroneous. We can still not figure out a single system that effectively solves all problems. Literature is full of different models that give good performance in different contexts and different types of data. The search for a model that solves all kinds of problems is still open and calls for the attention of researchers.

1.9 Chapter Summary

In this chapter we discussed a variety of terms that are frequently used in problem solving with hybrid and adaptive computing. The other aim of the chapter was to present the various concepts that must be understood well for effective understanding of the intelligent systems. The chapter explored the various aspects of soft computing. The discussion included neural networks, evolutionary algorithms, fuzzy logic and heuristics. The motivation of fusion of these systems to form hybrid systems was also stated. Further the various problems prevalent in the various individual systems were stated. Recognition is a fundamental problem that is studied in numerous forms. The chapter hence stated the basic scheme used by any recognition systems. Discussion over various stages of data extraction, preprocessing, segmentation, pattern matching etc. was done. Biometric recognition was specifically taken for discussion in the chapter. We classified the biometric recognition systems into physiological, which do not change their characteristics with time; behavioral, that change their characteristics with time; and fusion where two or more modalities are mixed. The chapter further explored classification, functional approximation and optimization as the basic problem solved using Soft Computing techniques. We saw the mechanism using which each of these problems are solved. The problems associated with localized training and loss of generalization was also presented. We saw this in reference to both classification and function prediction problems. The chapter also explored the concepts of machine learning and the various types of machine learning including supervised that is assisted by the output, unsupervised where the output is not known and reinforcement where the environment provides some reinforcement. The other part of discussion was the complete design of the expert system, where we saw how the various modules carry out the different tasks. We further discussed some limitations of the soft computing approaches.

References

[1] Acero, A., Stern, R.M.: Acoustical Pre-Processing for Robust Spoken Language Systems. In: Proceedings of First International Conference on Spoken Language Processing, Kobe, Japan, pp. 1121–1124 (1990)

[2] Bonet, B., Geffner, H.: Planning as heuristic search. Artif. Intell. 129(1-2), 5–33 (2001)

[3] Brailean, J.C., Kleihorst, R.P., Efstratiadis, S., Katsaggelos, A.K., Lagendijk, R.L.: Noise reduction filters for dynamic image sequences: a review. Proc. IEEE 83(9), 1272–1292 (1995)

[4] Bunke, H., Kandel, A.: Hybrid Methods in Pattern Recognition. World Scientific, Singapore (2001)

[5] Canny, J.: A Computational Approach to Edge Detection. In: Fischler, M.A., Firschein, O. (eds.) Readings in Computer Vision: Issues, Problems, Principles, and Paradigms, pp. 184–202. Morgan Kauffman Publishers, San Francisco (1985)

[6] Chavatal, V.: A Greedy heuristic for the set covering problem. Math. Oper. Res. 4(3), 233–235 (1979)

[7] Cormen, T.H., Leiserson, C.E., Rivest, R.L., Stein, C.: Introduction to Algorithms. MIT Press, Cambridge (2001)

[8] Dirks, D.D., Morgan, D.E., Dubno, J.R.: A Procedure for Quantifying the Effects of Noise on Speech Recognition. J. Speech Hear. Disord. 47, 114–123 (1982)

[9] Giles, C.L., Lawrence, S., Tsoi, A.C.: Noisy Time Series Prediction using Recurrent Neural Networks and Grammatical Inference. Mach. Learn. 44(1-2), 161–183 (2001)

[10] Graupe, D.: Principles of Artificial Neural Networks. World Scientific, Singapore (1999)

[11] Intrator, N.: Feature Extraction Using an Unsupervised Neural Network. Neural Comput. 4(1), 98–107 (1992)

[12] Jain, A.K., Ross, A., Prabhakar, S.: An Introduction to Biometric Recognition. IEEE Trans. Circuits Syst. Video Technol. 14(1), 4–20 (2004)

[13] Jain, L.C., Martin, N.M.: Fusion of Neural Networks. In: Fuzzy Systems and Genetic Algorithms: Industrial Applications. CRC Press, Boca Raton (1998)

[14] Kala, R., Shukla, A., Tiwari, R.: Fusion of Evolutionary Algorithms and Multi-Neuron Heuristic Search for Robotic Path Planning. In: Proceedings of the IEEE 2009 World Congress on Nature & Biologically Inspired Computing, NABIC 2009, Coimbatote, India, pp. 684–689 (2009a)

[15] Kala, R., Shukla, A., Tiwari, R.: Comparative analysis of intelligent hybrid systems for detection of PIMA indian diabetes. In: Proceedings of the IEEE 2009 World Congress on Nature & Biologically Inspired Computing, NABIC 2009, Coimbatote, India, pp. 947–952 (2009b)

[16] Kala, R., Shukla, A., Tiwari, R.: Fuzzy Neuro Systems for Machine Learning for Large Data Sets. In: Proceedings of the IEEE International Advance Computing Conference, Patiala, India, pp. 541–545 (2009c)

[17] Kala, R., Shukla, A., Tiwari, R.: A Novel Approach to clustering using Genetic Algorithm. Intl. J. Engg. Res. Indu. Appls. 3(1), 81–88 (2010b)

[18] Kala, R., Vazirani, H., Shukla, A., Tiwari, R.: Fusion of Speech and Face by Enhanced Modular Neural Network. In: Proceedings of the International Conference on Information Systems, Technology and Management, ICISTM 2010, Bankok, Thailand, March 2010. CCIS, vol. 54, pp. 363–372 (2010b)

[19] Kasabov, N.K.: Foundations of Neural networks. In: Fuzzy Systems, and Knowledge Engineering. MIT Press, Cambridge (1998)

[20] Konar, A.: Artificial Intelligence and Soft Computing: Behavioral and Cognitive Modeling of the Human Brain. CRC Press, Boca Raton (1999)

[21] Kumar, A., Shukla, A.: Improving Document Classification Using Anchor Text. In: 2008 International Conference on Information and Knowledge Engineering, Las Vegas, Nevada, USA, pp. 90–95 (2008)

[22] Lam, C.F., Kamins, D.: Signature recognition through spectral analysis. Pattern Recognit. 22(1), 39–44 (1989)

[23] Lippman, R.P.: An introduction to computing with neural nets. IEEE ASSP Magazine 3(4), 4–22 (1987)

[24] Marrow, P.: Nature-Inspired Computing Technology and Applications. BT Technol. J. 18(4), 13–23 (2004)

[25] Mitchell, M.: An Introduction to Genetic Algorithms. MIT Press, Cambridge (1998)

[26] Pal, N.R., Pal, S.K.: A review on image segmentation techniques. Pattern Recognit. 26(9), 1277–1294 (1993)

[27] Pearl, J.: Heuristics: Intelligent search strategies for computer problem solving. Addison-Wesley, Reading (1984)

[28] Phillips, P.J., Martin, A., Wilson, C., Przybocki, M.: An Introduction to Evaluating Biometric Systems. Comput. 33(2), 56–63 (2000)

[29] Plamondon, R., Srihari, S.N.: On-Line and Off-Line Handwriting Recognition: A Comprehensive Survey. IEEE Trans. on Pattern Anal. Mach. Learn. 22(1), 63–84 (2000)

[30] Prabhakar, S., Jain, A.K.: Decision-level fusion in fingerprint verification. Pattern Recognit. 35(4), 861–874 (2002)

[31] Reynolds, D.A.: Speaker identification and verification using Gaussian mixture speaker models. Speech Commun. 17(1-2), 91–108 (1995)

[32] Shukla, A., Kala, R.: Multi Neuron Heuristic Search. Intl. J. Comput. Sci. Netw. Secur. 8(6), 344–350 (2008)

[33] Shukla, A., Tiwari, R.: A Novel Approach of Speaker Authentication by Fusion of Speech and Image Feature using ANN. Intl. J. Inf. Commun. Technol. 1(2), 159–170 (2008)

[34] Shukla, A., Tiwari, R., Kala, R.: Mobile Robot Navigation Control in Moving Obstacle Environment using A* Algorithm. In: Intelligent Systems Engineering Systems through Artificial Neural Networks, vol. 18, pp. 113–120. ASME Publications (2008)
[35] Shukla, A., Tiwari, R., Kala, R.: Mobile Robot Navigation Control in Moving Obstacle Environment using Genetic Algorithms and Artificial Neural Networks. Intl. J. Artif. Intell. Comput. Res. 1(1), 1–12 (2009a)
[36] Shukla, A., Tiwari, R., Kala, R.: Real Life Applications of Soft Computing. CRC Press, Boca Raton (2010)
[37] Shukla, A., Tiwari, R., Kaur, P.: Intelligent System for the Diagnosis of Epilepsy. In: Proceedings of the IEEE World Congress on Computer Science and Information Engineering, CSIE 2009, Los Angeles/Anaheim, USA, pp. 755–758 (2009b)
[38] Shukla, A., Tiwari, R., Meena, H.K., Kala, R.: Speaker Identification using Wavelet Analysis and Modular Neural Networks. J. Acoust. Soc. India 36(1), 14–19 (2009c)
[39] Shukla, A., Tiwari, R., Ranjan, A., Kala, R.: Multi Lingual Character Recognition using Hierarchical Rule Based Classification and Artificial Neural Network. In: Yu, W., He, H., Zhang, N. (eds.) ISNN 2009. Part II, LNCS, vol. 5552, pp. 821–830. Springer, Heidelberg (2009d)
[40] Shukla, A., Tiwari, R., Rungta, S., Kumar, M.S.: A New Heuristic Channel Assignment in Cellular Networks. In: IEEE World Congress on Computer Science and Information Engineering, CSIE 2009, Los Angeles/Anaheim, USA, pp. 473–478 (2009e)
[41] Shukla, A., et al.: Face Recognition using morphological method. In: Proceedings of the IEEE International Advance Computing Conference, IACC 2009, Patiala, India, pp. 529–534 (2009f)
[42] Shukla, A., et al.: Intelligent Biometric System using PCA and R-LDA. In: Proceedings of the IEEE Global Congress on Intelligent Systems, GCIS 2009, Xiamen, China, pp. 267–272 (2009g)
[43] Sutton, R.S., Barto, A.G.: Reinforcement Learning: An Introduction. MIT Press, Cambridge (2004)
[44] Trier, O.D., Jain, A.K., Taxt, T.: Feature extraction methods for character recognition-A survey. Pattern Recognit. 29(4), 641–662 (1996)
[45] Wildes, R.P.: Iris Recognition: An Emerging Biometric Technology. Proc. IEEE 85(9), 1348–1363 (1997)

List of Figures

Fig. 1.1. General Architecture of Recognition System
Fig. 1.2. Input Space for Classification problem
Fig. 1.3. Functional Approximation in Input Space
Fig. 1.4. Fitness Landscape of Optimization Search Space
Fig. 1.5. Supervised Learning
Fig. 1.6. Unsupervised Learning
Fig. 1.7. Reinforcement Learning
Fig. 1.8. General Architecture of Expert System
Fig. 1.9. Loss of Generality in Functional Prediction Problems
Fig. 1.10. Generality in Classification Problems

List of Tables

Table 1.1. Types of Biometrics

Chapter 2
Artificial Neural Networks

Abstract. Artificial Neural Networks (ANN) are an inspiration from the human brain. These systems contain a large number of neurons that work in a parallel architecture. Each neuron takes its input directly from system or from other neurons. The information is processed and given to the other neurons. This is the basic phenomenon that makes possible all simple and complex problem solving ability of these networks. The chapter discusses the various models of neural networks that include multi-layer perceptron with back propagation algorithm, radial basis function networks, learning vector quantization, self organizing maps and recurrent neural networks. We discuss the basic philosophies and problem solving approach of these networks. A lot of emphasis is given on the various system parameters and their role and importance in the overall system design. We further illustrate the various limitations of the different models. This forms the motivation behind the use of hybrid systems that we present in the subsequent chapters.

One of the most phenomenal aspects of Soft Computing is the ease with which we can solve very complex problems, which would have been impossible to solve using other conventional ways. Most of the problems have a huge volume of data available, which is a recording of the past inputs and outputs. The beauty lies in the simplicity in which we can make automated systems that extract all the complex or simple manner in which the input maps to the outputs. This mapping is done by learning the historical behavior of the data available in the historical database. The Artificial Neural Networks (ANNs) are such agents that can employ machine learning for the task of desired mapping of the inputs to the outputs. These networks can hence automatically assume any shape that carries forward the task of determination of the outputs to the presented input. Any problem has predefined inputs and outputs. The relation between the inputs and the outputs is done by a set of rules, formulae or known patterns. These networks do the task of predicting these rules such that the overall system performs better when any of the data from the historical database is again presented. These networks are hence widely used for the problems of functional prediction or classification. Since we are not expected to perform much of the task related to the identification of the system behavior, the overall system emerges naturally without human aid. This makes these systems much robust against the inputs due to absence of human errors, as long as the data obeys the same laws showcased by the historical data. In case this does not hold well, these systems would fail completely due to the absence of any kind of logic in them (Graupe 1999; Kasabov 1998).

A. Shukla et al.: Towards Hybrid and Adaptive Computing, SCI 307, pp. 31–58.
springerlink.com

The ANNs are an inspiration from the human brain. The human brain comprises of as many as 10^{11} biological neurons. Each of the biological neuron is connected to the other neurons making massive 10^{22} connections between the various neurons. Each of the biological neuron is an information processing unit in itself. All the various neurons operate in parallel (Konar 1999). They take their input from the body inputs or from the other neurons. This input is processed by them. The processed information goes to the other neurons or as the final output. In this way the information receipted by the body is continuously processed by the various neurons one after the other. After a large volume of information, the final output is computed. The information is in the form of electrical signals that keeps getting transmitted between biological neurons through the nerve fiber. The dendrites receive the information or signal. These signals are passed through a little thick fibre, called dendron. As the information in form of electrical signals is transferred between the neurons, there is some loss depending upon the fiber. The received signals collected at different dendrites are processed within the cell body and the resulting signal is transferred through a long fibre named axon. At the other end of the axon, there exists an inhibiting unit called synapse. This unit controls the flow of neuronal current from the originating neuron to receiving dendrites of neighborhood neurons.

The ANNs by their basic architecture represent the human brain. They consist of a set of artificial neurons. The various artificial neurons are joined or connected to each other by connections. These connections aid the flow of information or data between the neurons. In this manner the information or data given to the system as inputs is processed again and again by the various neurons and the results are exchanged. The output of one neuron becomes the input for the other neuron. This process goes on and finally the computed answer is returned by the system. In this manner these systems are able to take the inputs, process them with the aid of multiple artificial neurons and generate the outputs. Artificial neurons behave similar in concept to their biological counterparts. The task of any fundamental artificial neuron may be divided into two parts. The first part does the weighted addition of the inputs presented to it. Here each connection has a weight associated with it. As the input arrives through the connection, it is multiplied by the corresponding weight. The addition of all such inputs is performed. The second part of the neuron consists of an activation function. The weighted addition of the first part is passed through the activation function. This is the final output of the system. The activation function is usually non-linear to enable the ANNs solve non-linear problems.

The necessity of non-linear activation function is natural for the functional approximation problems as the curve to be approximated in the input space would always be non-linear. For the classification problems the context may be slightly different. Most of the classical problems had classes arranged in a manner such that a simple line could separate them in the input space. This is natural in an idealistic scenario where the inter-class distances are high and the intra-class distances are low. However consider the *XOR* problem where the system is asked to imitate the *XOR* operator for two inputs (Wasserman 1989). The input space is given in figure 2.1. In this problem it can be easily seen that there is a necessity of

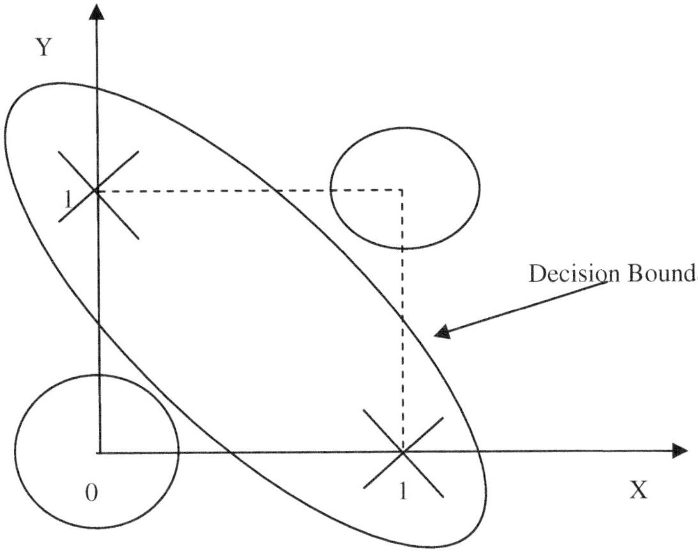

Fig. 2.1 Input Space for XOR problem

a non-linear decision boundary that effectively separates the two classes. This is obtained by keeping a non-linear activation function. Such problems are commonly known as problems of non-linear separability.

Numerous models of ANNs exist in literature (Lippman 1987; Windrow and Lehr 1990). They all use different fundamentals of problem solving. Many of these models use supervised or reinforcement learning whereas the others use unsupervised learning. Some are suited to the functional prediction problems, whereas the other are more suited towards the classification problems. Besides each of these problems have numerous issues that the ANN models crack by their own means. In the upcoming sections we discuss some of the commonly used ANN models. In each of these the methodology would be to understand their working philosophy rather than extensive mathematical formulation. This would enable us understand their presumptions and limitations that encourage their hybridization as hybrid systems. The models that we discuss here are Multi-Layer Perceptron with Back Propagation Algorithm, Radial Basis Function Networks, Learning Vector Quantization, Self Organizing Maps and Recurrent Neural Networks.

2.1 Multi Layer Perceptron

The first and the most basic model we study is the Multi Layer Perceptron model (MLP). From our analogy of the biological neuron we know that an artificial neuron or a perceptron does the task of taking weighted sum of the inputs and passing it through an activation function (Rosenblatt 1968, 1961; Rumelhart and McClelland 1986c). A single neuron is usually not able to solve the problem, especially if the problem is non-linear in nature. This requires the use of multiple neurons in a layered architecture one after the other. This model is called as the MLP and is presented in figure 2.2. Here all the perceptrons are arranged in an orderly manner in the form of layers. The first layer is the input layer where the inputs are applied. The number of neurons of this layer corresponds to the total number of input attributes in the system. Each input attribute is a neuron of this layer. This layer is also called as a passive layer, since it does not do any kind of information processing. The inputs are taken and given to the next layer for processing as it is. The last layer visible in the system is the output layer. The total number of neurons in this layer corresponds to the total number of outputs in the system. Each neuron represents some output attribute. Unlike the input layer, this layer is active and does the task of information processing. All the layers between the input and the output layer are the hidden layers that consist of variable number of neurons. Each of the neurons in these layers perform the task of processing. They take the inputs from the previous layer and give the computed output to the next layer. In this manner layered computation of the information takes place starting from the input layer to the output layer through the hidden layers.

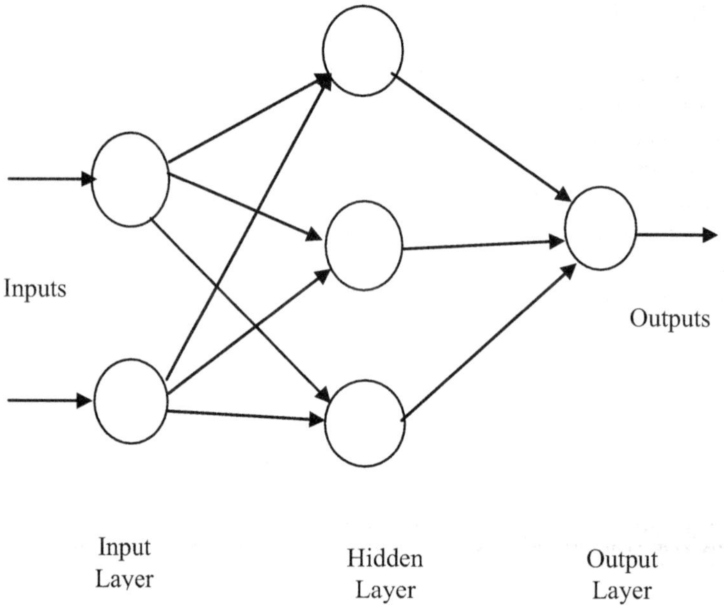

Fig. 2.2 Multi Layer Perceptron

Each of the circles in the output and the hidden layers, represented in figure 2.1, is an artificial neuron that does the task of information processing. The inputs are given to these neurons in the form of neural connections in-between the neurons. Each connection has a weight associated with it. Each neuron first computes the weighted average of the input and then applies the activation function. This

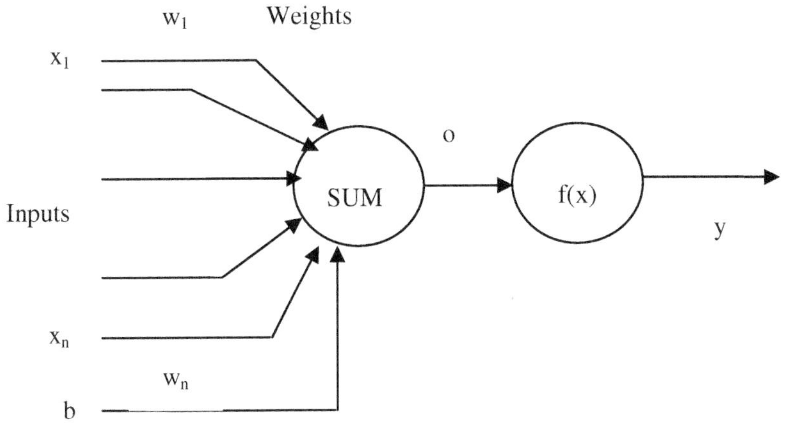

Fig. 2.3 Processing by a single neuron

concept has been presented in figure 2.3. The equations corresponding to the same are given by equation (2.1) and (2.2). Here each x_i is the input, *f(x)* is the activation function and *y* is the output of this neuron. *b* is the bias that is an extra input added to the system with a weight of unity.

$$o = \sum_{i=1}^{n} x_i w_i + b \tag{2.1}$$

$$y = f(o) = f(\sum_{i=1}^{n} x_i w_i + b) \tag{2.2}$$

For performance reasons, it is always preferred to keep all input and outputs of the ANN the range of -1 to 1. This step is called as normalization where the various attributes are normalized by using the normalization equation. Another important concept is the role of adding an extra input bias in the system. Consider the same classificatory problem shown in figure 2.1. We may need to move the decision boundary in the input space to optimally fit it in such a way that the classes are separated. While the activation function enables the decision boundary to assume any non-linear structure and weights decide the shape, the bias controls its mutual movement in the input space. The role of bias in functional prediction problems is similar where they are used for same purposes with the functional surfaces in the input space. This also plays a vital role in enabling ANNs solve problems. Bias is hence an added term to every neuron that enables it to control its output and match it as close as possible to the desired target (Windrow 1962; Windrow and Hoff 1960; Windrow and Winter 1988).

We next present the Back Propagation Algorithm (BPA) used to train the network. We then present the role and importance of the various parameters (Shukla 2010). This would be followed by the overall problems with the BPA and the MLP.

2.1.1 Back Propagation Algorithm

The Back Propagation algorithm does the task of tuning of the ANN to enable it perform as desired. Here the algorithm sets the various weights and biases of the ANN to their most optimal values. The aim of training is to ensure that the network gives the closest possible outputs, as compared to the target, for the data that it was trained with. The basic methodology for this algorithm is the application of inputs, computation of outputs and comparison of the outputs to the target that gives us the errors. The errors are back propagated in a backward direction starting from the output layer to the input layer (Barron and Barron 1988; Grossberg 1982; Rumelhart and McClelland 1986; Rumelhart *et al* 1986). The propagation of errors is used by all the neurons of the various layers to carry forward adjustments in their weights and parameters. The adjustment is done such that the next time this input is presented, the error approaches zero or that the output closely follows the target. This entire process is repeated for all the inputs in the training data set and for multiple epochs. In this manner the algorithm sets the various weights and biases of the system. The basic algorithm is given in figure 2.4.

Consider the error space in case of the training of the ANN with BPA. This would be a highly dimensional space with the error plotted against all the weights and biases as a single surface. Any state of the ANN corresponds to a set of weights and biases having a particular error value. This is a point in the mentioned error space. The BPA starts with a random state of the ANN which may be taken as a random position in this error space. The BPA has an approach of the steepest gradient. In this approach the algorithm tries to jump to the local minima around the present point. For doing this the algorithm calculates the tangent to the error surface at the resent point of location. Since the error surface mathematically complex, an approximation technique is used. The algorithm uses this to calculate the position of the minima pointing in the direction of the slope (Owens and Filkin 1989; Pineda 1987). The plot of the error surface for 3-dimensions is given in figure 2.5.

One of the problems with the strategy would be the very sensitive nature of BPA. Since we have only approximated the location of the minima, the assurance of its position can never be made. If we allow the system to jump to the location, it might overshoot the position. It would then try to again compute the location again. This might result in the training algorithm being unstable with large hops in search of the local minima. It may easily be imagined that a good strategy would be to slowly march towards the minima with small steps, so as to reach it. This is done by the learning rate that shortens the steps of the BPA movements. Another problem with the algorithm would be that it always converges to local minima. This problem is solved with the help of momentum. The momentum ensures that the algorithm always moves ahead, despite the guidance of the slope. This stops the algorithm from converging to any point. As the iterations grow and the algorithm starts converging towards the local minima by virtue of its converging strategy, it is pulled out and guided ahead by the momentum. The net movement of the

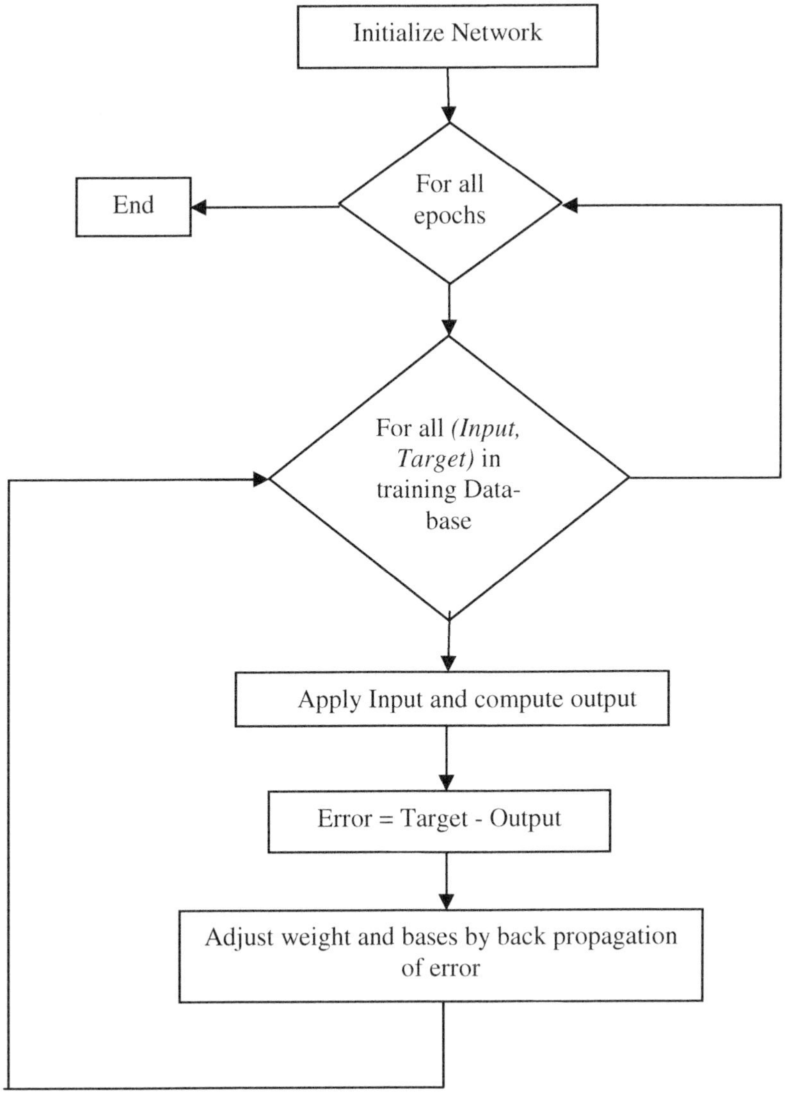

Fig. 2.4 The Back Propagation Algorithm

algorithm depends both upon the learning rate as well as the momentum. The equation governing the change of weight for any interconnection weight of the algorithm is given by (2.3). This statistically clarifies the presented concepts.

$$\Delta w_{jk}(t+1) = \eta \delta_k^p y_j^p + \alpha \Delta w_{jk}(t) \tag{2.3}$$

Here $\Delta w_{jk}(t)$ is the change in weight for weight from j^{th} neuron to k^{th} neuron at iteration of t.

η is the learning rate $(0 \le \eta \le 1)$
α is the momentum. $(0 \le \alpha \le 1)$
δ_k^p is the slope or the error at layer p
y is the output at layer p

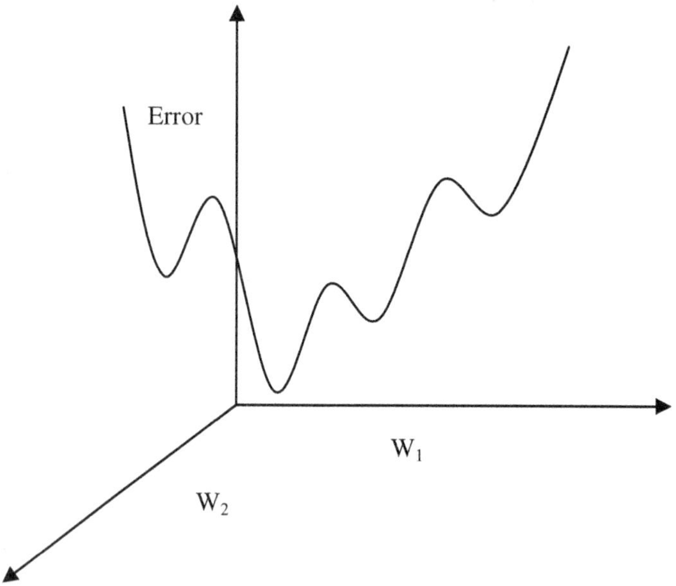

Fig. 2.5 Error Surface of Neural Network

2.1.2 Number of Layers and Neurons

The first and foremost parameter that we need to fix at the time of design of the ANN is the number of hidden layers and the number of neurons in each of these hidden layers. Both the number of hidden layers as well as the number of neurons denote the total number of weights and biases of the final network. As the number of layers or neurons increase, there is an increase in the modeling complexity which denotes a more localized behavior and a lesser generalizing capability. All the disadvantages of drop of generality hold that we discussed in chapter 1. Hence the system must be trained using as less number of hidden layers and neurons as possible. The increase in the hidden layer usually has a much larger increase in complexity than the increase of the number of neurons. This is due to the large number of addition of connections that have to be added. For most of the problems the number of hidden layers is hence kept as unity and the number of neurons is altered for performance appraisal. Every problem represents some basic complexity. The system must hence possess some least number of neurons below which the training and hence testing will not be optimal. As we further add neurons, the fitting of the training data will start improving due to the added modifiable parameters in the system. This would improve the training accuracy, but the testing accuracy would naturally drop. This is due to the loss of generalizing parameters.

Another way to view the same effect of the addition of the number of layers and neurons is by the input space. Every combination of weight and bias corresponds to a characteristic plot of the input space where the outputs are plotted against the inputs. A small network having lesser number of neurons and layers has a smaller number of parameters that the training algorithm is supposed to modify. Modification of these parameters changes the input space. The input space for the desired output is known in the form of a collection of points. We need to fit these points in the input surface as wisely as possible. Having a few parameters to modify enables a training algorithm capable of only a few adjustments. This restricts the training algorithm for fine curve fitting. The functions imitated by the ANN for smaller sizes are simpler than the ones with large sizes. As a result the training algorithm does a simple curve fitting that has high generality. If however the number of neurons or layers is kept very high, the training algorithm would be able to make complex surfaces of the input space that appreciably fit the points in the training data set, but take complex curves in between. This is because the algorithm has a large number of parameters and hence the overall curve can have complex shapes. All this may be summarized by the fact that the number of neurons denotes the number of turns, or local minima that the input surface may have. Higher turns makes the function sensitive, when a slight change in input has a paramount change in output. This may be summarized in figure 2.6 that plots the input space using two neural networks of high and low number of neurons. The training points have been indicated separately.

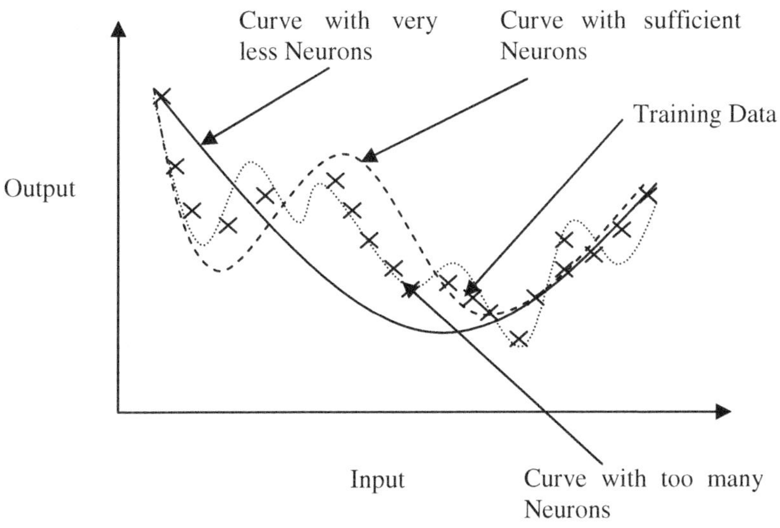

Fig. 2.6 Training with different number of neurons

2.1.3 Learning Rate

The other major parameter that we set at the time of network training is the learning rate. Learning rate determines the speed with which the network gets trained. In other words it determines the speed with which the state of the ANN approaches the local minima in the error space. Learning rate needs to be optimally set as both high and low values may produce poor results. Very large value of learning rate leads to a rapid change in the ANN state towards the local minima. As a result it may sometimes overshoot the minima. This would then result in the training algorithm trying to reverse the direction in order to get to the minima. It may again over-shoot the minima. Hence the algorithm keeps oscillating in the error space between large error values. The magnitude of error would depend upon the value of the learning rate. Very small learning rate would result in the system marching towards the local minima in very small steps. As a result it would take very long time for the system to attain the local minima, despite the gradient descend approach. In such a context the system may not attain optimal performance level even after a large number of training epochs or after sufficiently large time. These two conditions are shown in figure 2.7.

2.1.4 Momentum

Momentum does the task of pulling networks out of local minima, in order to enable it achieve global minima. The magnitude of the momentum decides the total force with which this phenomenon takes place. It is similar to the concept of momentum term used in physics where a vehicle continues to move in the same direction even after the removal of applying force. Similarly the ANN is moved in the direction in the same direction in which it was moving with a certain force irrespective of the direction and magnitude that the gradient computes. The momentum again cannot be too large or too small. Large values of the momentum would result in the ANN constantly moving in a direction without trying to converge to minima. This would make the algorithm give high error values. Very low values of momentum will make the algorithm converge towards the local minima. Momentum would be unable to effectively pull out the state of the ANN and drag it outside the minima. Hence momentum needs to be judicially kept understanding and trying to figure out both these trends in the training curve.

2.1.5 Epochs

Epochs are the total number of iterations in the training algorithm. Even this parameter affects the overall performance of the ANN. Too large and too small number of epochs may both be sub-optimal in the performance of the ANN. As the ANN is trained with the help of BPA, the errors in the training data reduce with time. The network further requires a fine number of epochs to effectively train the various weights and biases. This is especially when the training rate is

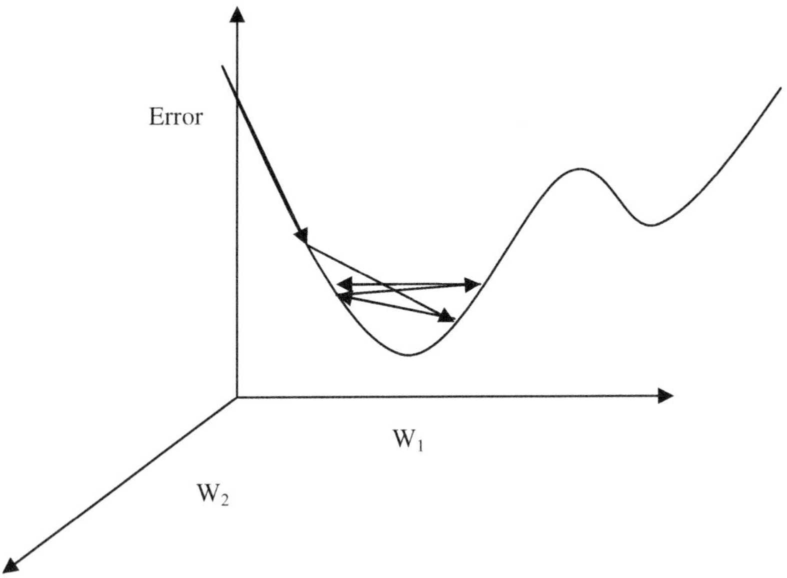

Fig. 2.7(a) Oscillation with large value of learning rate

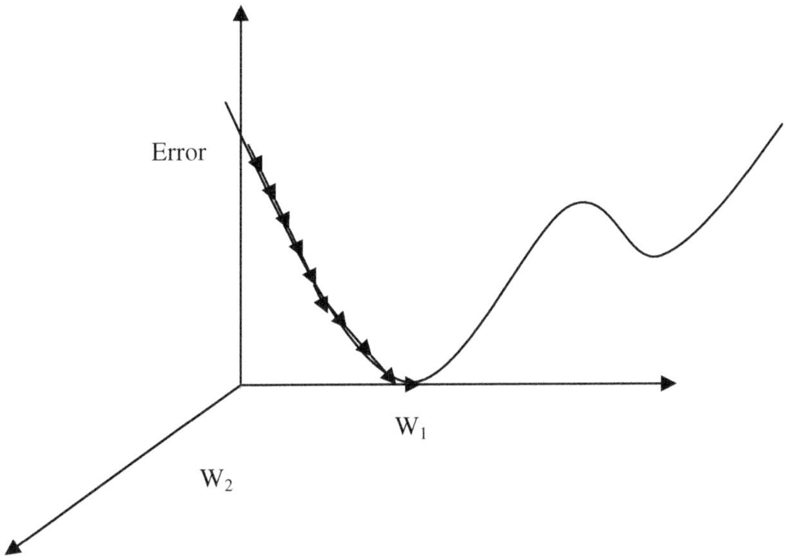

Fig. 2.7(b) Slow learning with small value of learning rate

kept low as the network state slowly marches the minima. Hence small values of epochs are un-desirable. The large values of epochs may further cause problems of performance in the testing data or the new data that is presented to the system. The large training cycles would naturally result in lowering of the errors of the training inputs. This is because of the constant alteration of the weights and biases to behave as ideal as possible to the training data set. However prolonged training causes the network to over-learn. This is when the system forcefully starts fitting the training data onto the curve to get as small errors as possible. This phenomenon is also known by the term over-fitting. As a result the curve starts getting complex shapes and loses generality. This badly affects the new data that cannot be correctly computed by the system for the generation of correct outputs. This reduces performance. This is especially true of networks with larger number of neurons than the ones needed to solve the problem.

In order to overcome this problem of the ANNs, we adopt the principle of early stopping. The entire data set is divided into three parts. The first part is used for the training purposes called as the training data set. This is the data which is given to the training algorithm for adjusting or tuning of the system by varying the weights and biases. The other data is the testing data. This is the data especially reserved for the overall testing of the system. The performance of the algorithm in the testing data set is termed as the overall performance of the system. This data is not shown to the system during any of the stages of learning. The last data set is the validation data set. After every epoch the performance of the system is measured for the training and the testing data set. The performance of the training data set keeps increasing along with epochs. However the performance against the validation data set may not always increase. This is because no system adjustment was being made as per the validation data set. Initially as the system starts to learn, the performance of both training and the validation data set decrease. After some times the performance of the training data set decreases while that of the validation data set increases. This is the stage when over-learning is said to have started. We hence stop training the algorithm at this point of time itself. A typical training curve between training and validation performance is given in figure 2.8.

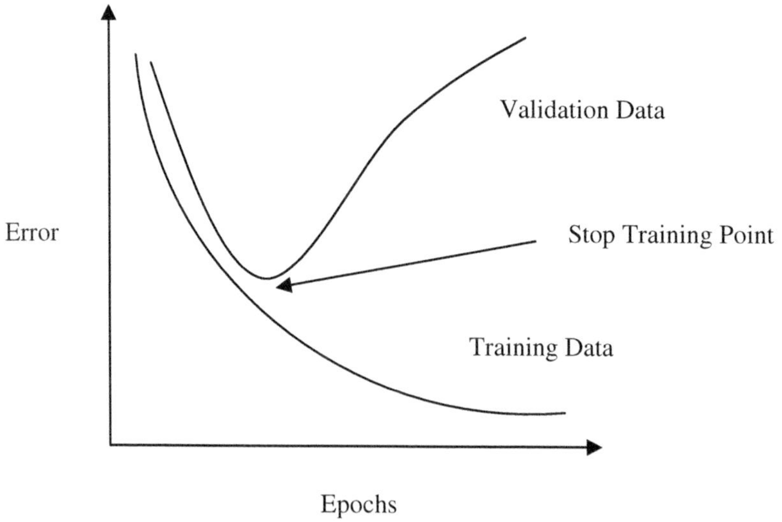

Fig. 2.8 Early Stopping of neural network with validation data

2.2 Radial Basis Function Networks

The other commonly used model of ANNs is Radial Basis Function Networks (RBFN). This model has a simple 3 layer ANN architecture. The first layer is the input layer where the inputs are applied. The second layer is the hidden layer. The last layer is the output layer where the system delivers its output. The architecture of these networks is given in figure 2.9. These networks however differ in the manner in which they process information for the generation of the outputs from the inputs.

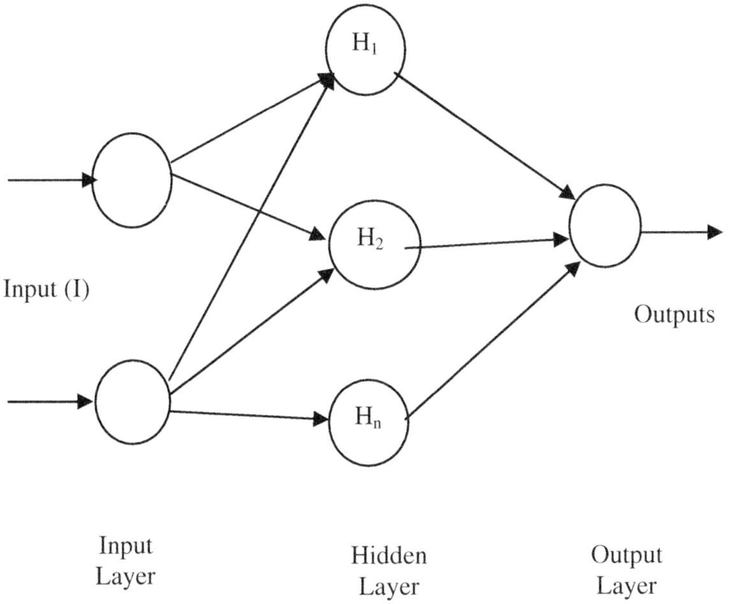

Fig. 2.9 Architecture of Radial Basis Function Network

2.2.1 Working

Consider the input space of the problem being considered. Each neuron of the hidden layer in the RBFN corresponds to a location in the input space. The various neurons are spread all round the input space. The input itself is a point in this input space consisting of some value of different attributes. At the time of processing of the inputs, each of these neurons calculates its distance from the input. Suppose the input applied was $I<i_1, i_2, i_3, \ldots, i_n>$. Here n is the number of input attributes or dimensionality of the problem. This distance passes through a non-linear activation function. The output p_k of the kth hidden neuron, found at a location of H^k $<h^k_1, h^k_2, h^k_3, \ldots h^k_n>$ is given by equation (2.4).

$$p_k = e^{-\beta_k \|I - H^k\|^2} \tag{2.4}$$

Here ||x|| can be any norm which may usually be taken as the Euclidean norm.

The outputs of the various neurons of the hidden layer serve as the inputs of the output layer. This layer is a linear layer and simply computes the weighted sum of the various hidden layer neurons. Each connection of this layer corresponds to the weight that is multiplied by the associated hidden neuron. In this way the system generates the final output. The output for any output node o_i may be given by equation (2.5). It may be seen that since this layer is linear, it is computationally much less expensive and contributes towards the algorithm speed and simplicity.

$$o_i = \sum_j \lambda^i_j p_j \tag{2.5}$$

Here λ^i_j are the weights from the ith output node to the jth hidden node.

2.2.2 Conceptualization

Using the analogy of the input space and the mathematical formulation we discuss the basic working methodology of the RBFNs in this section. Unlike the MLP with BPA discussed in section 2.1 that tried to form a unanimous function of specific complexity to map the inputs to outputs, these networks partition the input space. Each of the neuron of the hidden layer can be called as the center of one such partition. The partitioning is however not discrete in nature, it is rather fuzzy. This means that any point in the input space actually belongs to all the partitions with varying degrees. We rather say that every input is associated with every hidden neuron with varying degree. The degree varies with the distance of the input from the hidden neuron in a Gaussian or exponential manner. As we move away from any neuron, the association decreases following equation (2.4). This may be graphically be visualized in figure 2.10 that denotes the drop of neural activation or drop of association (in our analogy) with the increase of distance. The distance to which the association is effective is decided by the radius or spread of the neuron (β_k) which is specific to the neurons. There are multiple hidden neurons spread at different locations of the input space. At the application of any input they get activated to varying degrees with the activation usually being largest for the neuron that lies closest to the input. For the ease of understanding we neglect the very low activations of the hidden layer neurons, as their contribution would almost be zero. In such a context an input effectively activates only neurons that lie in its close vicinity. Other neurons are inactive. Hence we say that the input space has been partitioned. This may again be regarded as localizing of the problem where only the neighboring nodes are consulted for computing the output to any applied input.

Now the various hidden nodes are activated to various degrees, depending upon their closeness to the input applied. Each hidden neuron has some idea of the output that is prevalent in the region it represents. If the problem is simple, we may assume that the output does not change appreciably with small changes in inputs.

Hence the entire region around the hidden neuron would have similar outputs. These outputs are stored as weights of the connections between the hidden and the output layer. These get multiplied by the activation of the corresponding hidden neuron. This may be taken as a competition between the various hidden nodes to influence the final output of the system. A neuron with higher activation has more say to decide the final output than the neuron with lower activation. This competitive model is averaged by the output nodes and in this manner the final output is evaluated. This is a linear addition of the weighted sums. The various weights along with the centers of various nodes may be tuned by the application of training algorithm. Various training algorithm may be used including statistical methods, BPA, gradient descent approach etc.

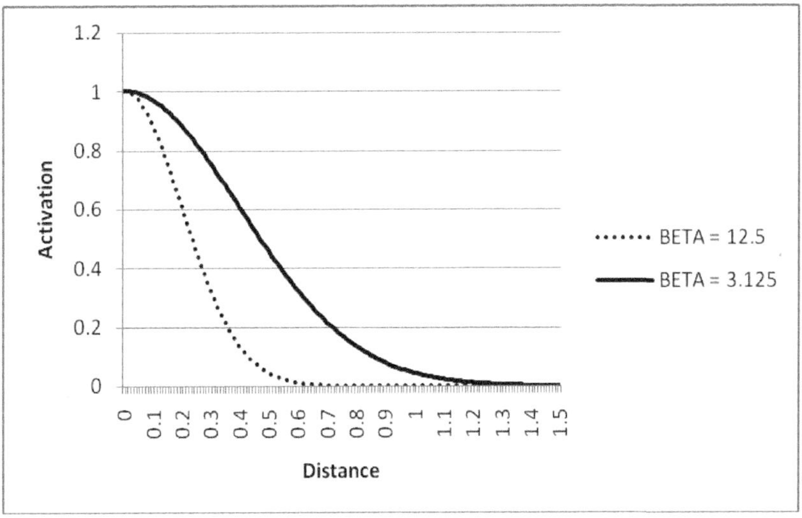

Fig. 2.10 Loss of activation with increasing distance in Radial Basis Function Networks

2.2.3 Number of Neurons and Spread

A major factor in the algorithm is the number of neurons in the hidden nodes that may be varied at the time of network construction. The larger number of neurons denotes a partitioning of the input space. The more the input space gets partitioned; there are associated problems of loss of generality, drop of resistance to noise, etc. However the training time reduces. Hence the number of neurons should be as less as possible, just was the case with MLP with BPA.

Another important parameter of the algorithm is the spread or the radius of the hidden neurons. The larger the spread more is the reach or influence of the hidden neuron. Larger spread of the various hidden neurons would result in a large intermixing in the input space. Hence any point in the input space would be largely affected by many neurons rather than just a very few neighboring neurons. This

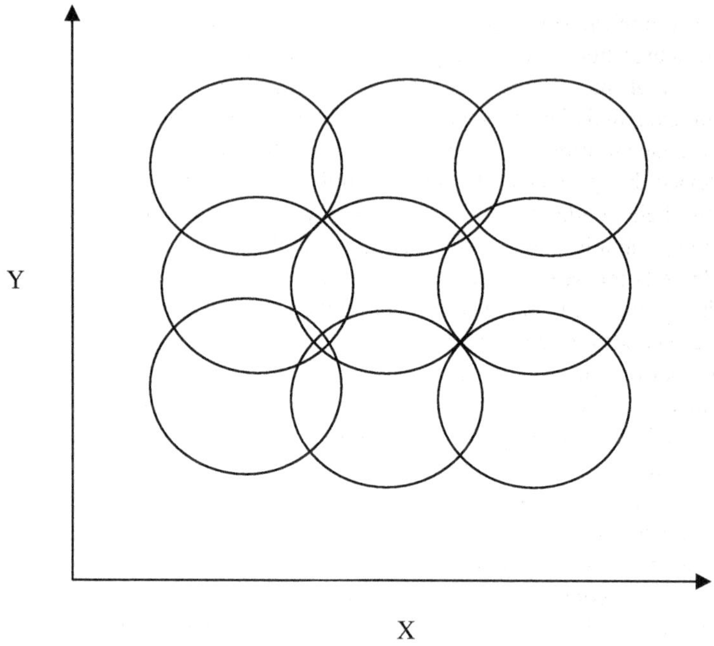

Fig. 2.11(a) Radial Basis Function Network with small spread

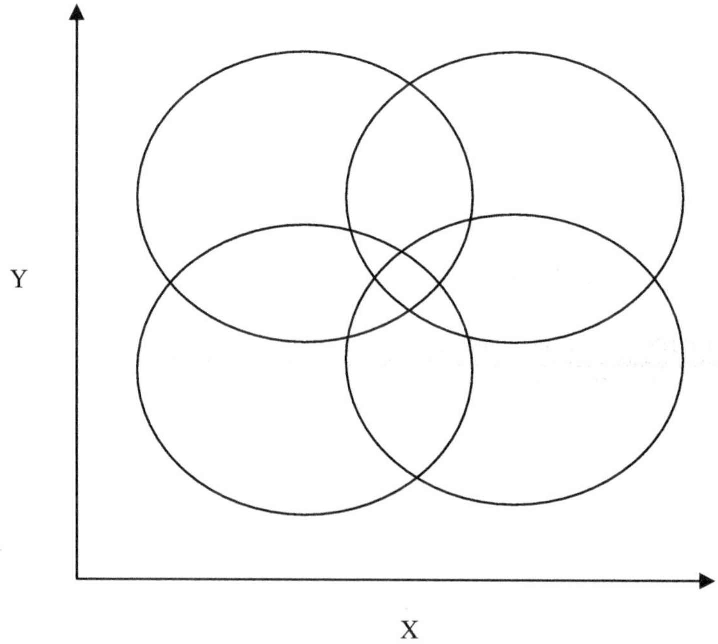

Fig. 2.11(b) Radial Basis Function Network with small spread

would add a lot to the increase of generalization. In other words a lot of neurons are being consulted for the computation of the output belonging to the different regions in the input space. Reducing the spread or the number of neurons would make the problem or the network very localized in nature. Here a neuron affects only its immediate neighboring points. This concept can be easily understood from figure 2.11 that shows a 2 dimensional input space with the hidden neurons along with their influential regions. It may be noted that we used a Gaussian function that marks a gradual change in activation value from 1 to 0. Clear boundaries can hence never be plotted as shown in figure. This is however only for general understanding the concept of spread and its influence of generality.

2.3 Learning Vector Quantization

Learning Vector Quantization (LVQ) is another model of ANN that is extensively used for the purpose of classification. These are also three layered networks consisting of the input layer, hidden layer and the output layer (Kohonen 1988; Kohonen 1990; Hammer and Villman 2002). The input layer gives the input to the system. The hidden layer performs the processing of this input. The output layer consists of as many neurons as there are classes in the problem of study. Each neuron of this layer corresponds to a class. Once any input is applied to the system,

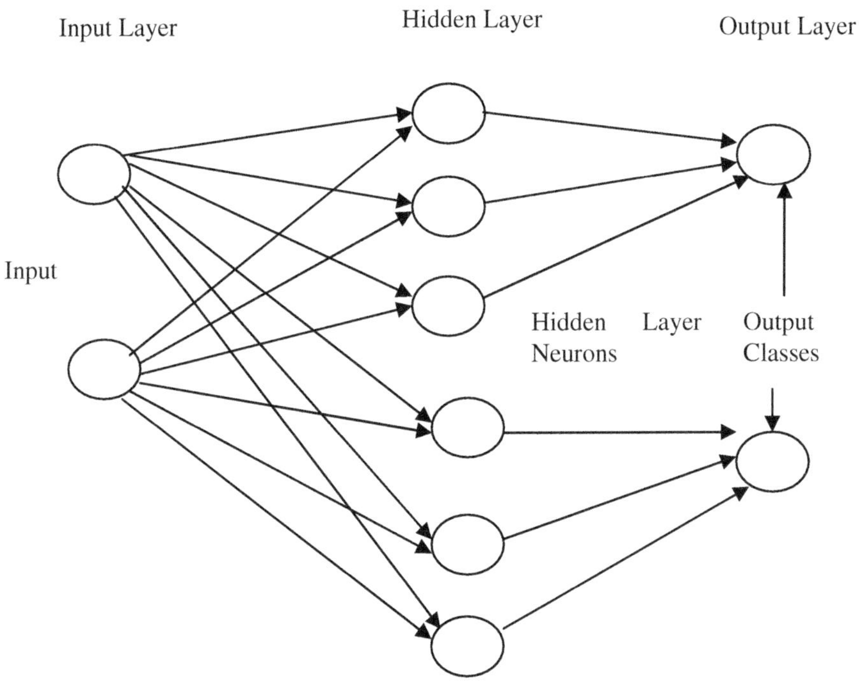

Fig. 2.12 Learning Vector Quantization

only one of the output neurons is active that gives an output. The class corresponding to this neuron is stated as the final output class of the network. The other difference between these networks and the other discussed networks is that there are limited connections between the hidden layer and the output layer. All the connections are of same unity weight. The general architecture of these systems is given by figure 2.12.

2.3.1 Working

Just like RBFN, each of the neuron of LVQ corresponds to some point in the input space. Each neuron does the task of computing the distance between itself and the input neuron. Using the same notations with input as $I<i_1, i_2, i_3, \ldots, i_n>$ and any hidden neuron as $H^k <h^k_1, h^k_2, h^k_3, \ldots h^k_n>$, this distance p_k may be given by equation (2.6).

$$p_k = \| I - H_k \| \tag{2.6}$$

LVQ follows a winner-takes-all criterion in its working. Here the different neurons in the hidden layers compete with each other for influencing the output in their favor. The final winner is the one that has the best score. This is the neuron with the least distance computed from equation (2.6). Using the winner-takes-all criterion, only the winning neuron is made active. All the other neuron of this layer is rendered as inactive. The winning neuron produces an output that goes to the next layer. At the next layer, there is a connection between this neuron to the neuron corresponding to the classes to which this neuron belongs to. The output generated by this neuron goes to the stated class of the next layer or the output layer. This is returned as the output of the system. In this manner the computation of output from input is done by the winner-takes-all criterion in LVQ.

2.3.2 Conceptualization

Here we have discussed that each neuron in the output layer is a class and each neuron in the hidden layer is a point in the input space that corresponds to some output class. Hence we can easily group the various neurons in the hidden layers by their classes. We keep the neurons that correspond to the same class together. Each neuron here corresponds to some point in the input space. Each of this group is a class that competes to be declared as the final output class. As per the working fundamental of LVQ we take the norm and the closest neuron is declared as the winner. This means that the entire input space may be easily broken down into completely segmented regions, each being under the control of some neuron in the hidden layer. In order to do this we simply check which of the neuron is being activated for all the points in the input space. Each neuron would bind a large part of the input space that lies in its vicinity. Each of these neurons may hence be stated as the sub-classes that belong to the designated class. When a particular neuron activates by the winner-takes-all criterion, we state that the input is classified to the sub-class represented by that input. Hence the sub-classes compete with each other for activation or classification. Now we know that classification is much more a localized problem where the output at a point largely depends upon the outputs at the neighboring points. This is because of the fact that little change in inputs normally have no change in outputs, especially when the inter class distances are high and the inter class distances are low. In such a context, ideally there must be a single neuron in the hidden layer per class, or a single sub-class per class. However for classification it is even possible for a class to lie in multiple distinct regions. This necessitates the presence of multiple sub-classes. Another major reason for the presence of sub-classes is the non-idealistic nature where the different classes mingle with each other. This necessitates the need of multiple sub-classes arranged in a judicious manner that assures a high generalization. Hence the training algorithm of LVQ tries to place the various neurons of a class as equitably as possible in the region enclosed by the input space.

2.4 Self Organizing Maps

Self Organizing Maps (SOM) are classifiers that undergo unsupervised learning. Here only the inputs are given to the system at the time of training, the outputs are not given. SOMs are able to cluster the given input data and represent them in a simple and less memory intensive manner (Kohonen 1989; McCulluch and Pitts 1943; Hebb 1949; Fukushima 1975). These clusters are then used to find out the correct output class to any of the applied input. These networks are hence also called as natural clustering agents. The working of these networks is exactly the same as the LVQ. The inputs are applied at the input layer. The hidden layer consists of a set of neurons. Each neuron computes its distance from the applied input. The philosophy of winner-takes-all is followed here. The class corresponding to the winning neuron is the final output of the system. The basic difference comes in the unsupervised mode of training in which every neuron tries to associate itself with the applied input. In this manner continuous training makes certain neurons get inclined towards certain like inputs. In this manner the phenomenon of clustering happens.

2.4.1 Self Organizing Feature Map

The hidden neurons in SOM are arranged in an ordered manner that is called as the Self Organizing Feature Map (SOFM). The manner or the topology in which this arrangement is done is specific to the implementation. One of the common topologies is the rectangular topology in which the neurons are found at the corners of a rectangle. This is shown in figure 2.13. This means that the hidden layer would contain m x n neurons where m and n are the length and breadth of the rectangle. These neurons represent the various points in the input space. In this manner we may easily visualize that the high dimensional input space has been concentrated to lie in a 2-dimensional network of neurons given in figure. This is the manner in which these networks perform by mapping up of the high dimensional input into a characteristic feature map. Now every time an input is given to the network, this feature map is used for the determination of the output by computing the closest lying neuron and its associated class.

Another major point here is that the mapping may not be equitable. In other words the various neurons may not lie equitably in the input space as one might actually expect. In an ideal scenario the inter-class distances are high and neurons are placed at the regions where most of the data for a class is found. Hence a large number of neurons are found at close locations in the input space and some other locations of the cell might be completely left. In a non-idealistic scenario as well, there is a large placement of neurons where bulk of the data is found. The sparsely populated regions of the input space are quite left out with small number of neurons in the vicinity. A focus here would be the data items that lie near the decision boundaries. The major challenge here, just like in LVQ, is the effective placement strategy of the neurons that is done by the training algorithm. The unequal placement strategy makes the algorithm better in efficiency, even though the basic methodology is the mapping up of the input space to the feature space.

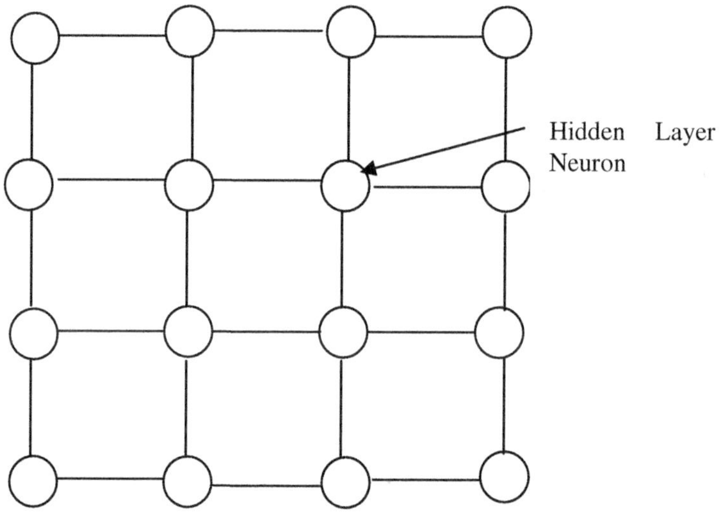

Fig. 2.13 Self Organizing Feature Map

2.4.2 Clustering

SOMs are natural clustering agents. The basic methodology of SOM to get trained in an unsupervised manner is their ability to form clusters of the inputs and map them up in the feature space. In this section we briefly study this clustering mechanism and methodology as it forms the basis of classification. Effective classification can be done by forming clusters of input space as per the training data and storing the clustering centers and then using these centers as templates for the task of pattern matching. Each neuron of SOM represents a cluster center in the input space. We discussed the same concepts as sub-classes while discussion LVQ. The uniqueness of SOMs is that the clusters can be easily seen in the feature map which maps the entire input space. The difference with LVQ is that the SOM clustering is independent of outputs and hence the clusters are made only based on their attribute values. The training algorithm does the role to distribute the input space into the set number of clusters decided by the number of neurons in the hidden layer or the number of neurons in the feature map. The training algorithm iterates over all the inputs and tries to attract winning neuron and neurons in the neighborhood towards itself. It further tries to separate the winning and neighboring neurons from the rest of the neurons in the input space. Continuing in this manner gets the entire data set gets partitioned or clusters. This may be referred by the twin terms of clustering or learning. The update of the position of any hidden neuron $H^k <h^k_1, h^k_2, h^k_3, \ldots h^k_n>$ for the application of any input $I<i_1, i_2, i_3, \ldots, i_n>$ at any iteration t is given by equation (2.7)

$$H^k(t+1) = H^k(t) + \eta(t)(I - H^k(t)) \tag{2.7}$$

Here $\eta(t)$ is the learning rate that decreases with time.

2.5 Adaptive Resonance Theory

Adaptive Resonance Theory (ART) is another model of the ANN that carries the task of classification to map the inputs into the output classes. The major problem with most of the ANN models is plasticity (Carpenter and Grossberg 1987a, 1987b; Cohen and Grossberg 1983; Grossberg 1974, 1987). Here the models forget the data that was given long back and chiefly remember the data that was given recently. This happens when there is a lot of data to learn and there is less number of neurons. The network, just like human brain, remembers the recent data well. The recent data causes the network learning to completely forget the old data. Hence the old features of the network are lost. This problem has a large impact on the system in terms of generalization. ARTs are an attempt to solve this problem of plasticity. These networks make the ANN learn new data without forgetting the old one. The new data may not be learnt perfectly making the errors very small, but the retention of old data is an added advantage.

2.5.1 Working

The ART working is similar to the earlier discussed models with respect to the performance. However here we also need to take care that the network does not forget the old trends that were extracted from the historical data not presented to the system for a reasonably long time. This is done by the addition of an orientation system in the architecture of the ANN. The modified ANN model hence consists of the input layer, the hidden layer or the comparison layer and the output layer or the competitive layer. The inputs are applied at the input layer. The hidden layer consists of hidden neurons. Each of the hidden neurons is a point in the input space. The output or the competitive layer does the task of selecting one of

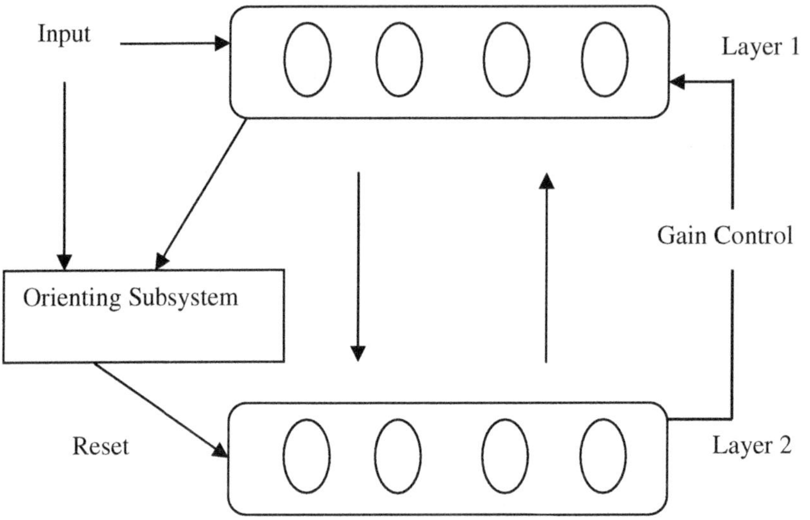

Fig. 2.14 Adaptive Resonance Theory Networks

the output classes. The orientation system prohibits the forgetting of the ART that preserves the past trends. This is also known as plasticity control. The basic architecture of this system is given by figure 2.14.

The orientation system monitors the training process that changes the location of the hidden neurons in the input space. It traces a learning action that can cause forgetting or disruption of a vital neuron. All such actions are prohibited to take place by the orientation system. When the inputs are applied to the system, the hidden neurons compute their output which goes to the output class for competitive output evaluation. As learning fundamental the input applied tries to attract the winning neuron towards itself by pulling the neuron in its direction in the input space. This modifies the location of the winning neuron. In case of ART the backward phase is executed where the input vector and the location of the neuron are matched. If the difference between their locations is small, the learning continues. However if there is a large gap between them, then the learning is inhibited. This is done by the orientation system that resets the network to its initial state. The winning neuron is then deactivated and the entire process is repeated for the rest of the neurons. Again the winning class and neuron is identified and the change is made which is later verified for its closeness. If again the neuron lies far enough, the training is cancelled. We keep continuing the process till a close neuron is matched. In this manner we preserve a neuron from destroying its learning. Learning can only happen when the change is small enough.

2.5.2 Conceptualization

ART mark an interesting fundamental of plasticity control as compared to the other models which is maintained by the phenomenon of measuring the deviation in the location of the neuron in terms of its location in the input space. This naturally prohibits the large movements which can be easily visualized as forgetting of old trends. The small movements on the other hand denote tuning in the location of the neurons which is the basic motive of learning. Besides the concepts that have been discussed in the previous sections, one of the major points to understand is the manner in which the restriction helps in forgetting of old data while training. So far while training we assumed to some extent that the training data was a well shuffled collection of data items. Hence the different data items coming as inputs in different times targeted and moved different neurons. This did not affect the overall neuron placement to go contrary to some class, unless the system had less neurons than the number actually required for learning. The overall effect was the clustering of data. But suppose that the data is not well shuffled and different segments of the training data are rich in different classes that lie at similar locations in the input space. For simplicity we consider that the entire database is filled with a number of classes to some extent which is followed by an entirely new class later in the database. Now the entire network might have been tuned by the different classes and the various neurons must be held with different classes as sub-classes. Further these neurons must already be arranged in a non-equitable distribution to aid in classification of classes already presented to the ART. In such a scenario a new class arrives with a new location in the input space. It is highly likely that the inter-class distances be sufficiently high as the problem is a classification problem. Now the new input has no neurons that lie close to it. But the classification needs to be carried out. Hence the general notion would be to attract the closest neuron. This may snatch away key neuron with a sub-class by the new class. This snatching would further be facilitated by the other occurrences of the class in the input training data set that follow in the training procedure. In this way the network ultimately loses the neuron. This loss would be more prominent for classes whose data items have not been shown to the training for a very long time. This is due to the multiple movements of the possessed neurons by multiple data items in the training procedure over the course of time.

The ART stops such actions by the restrictive training discussed earlier. It may be noted that the inputs go through a state change by the multiplication with weights and then the closest match is selected for the hidden neurons for the rest of the computation. This is unlike the earlier presented models where the hidden layer took a norm of distances with the neurons representing points in the search space. The same is used for matching and other purposes.

2.6 Recurrent Neural Networks

The Recurrent Neural Networks (RNN) are another major type of neural networks that are used for problem solving. These networks are cyclic in nature unlike the other presented models that were arranged in a manner that cycles can never be formed (Hopfield 1982, 1984; Werbos 1990; Williams and Zipser 1989). The conventional ANNs have a big limitation of their static nature. The information flow is only forward where the predeceasing layers process data and forward it to the next layers. These networks allow backward connections where every neuron gets the feedback from the forward layers as well as itself. This allows the ANN to again process data and again transmit the output for further processing by the other layers in backward and forward direction to which it is connected. In this manner there is a lot of dynamism which drives these networks. Further the algorithm operates in timestamps or iterations where a unit processing is performed by each of the neurons in a single iteration. The output of the system continuously changes with time as the layers undergo changes driven by the feedback connections.

2.6.1 Working

By architecture these networks are conventional MLPs with feedback connections that again feedback the information to the other neurons for further processing. A general structure of these networks is given in figure 2.15. In these networks the input is first applied at the input layer. The input layer is a passive layer that transmits the applied inputs for the further processing. The various neurons then process the inputs and transfer the computed outputs to the other layers for the further processing. This way the outputs of all the neurons, including the ones at the output layer are determined. This forms the first iteration of time. The various neural outputs of various neurons are now prepared to provide feedback to the system. All neurons now again process the information in a layered manner. The feedback of the various neurons plays a major role in driving the system. This makes a new set of outputs of all the neurons in the system. We continue the process for a few iterations. As we proceed, the output starts converging to some value. This is the final system output.

Training of the recurrent neural network involves the setting of the various weights. Weights exist between the forward as well as the feedback neurons. The standard BPA algorithm may be used for the setting or training of the weights. The major difference is that the time iterations factor is to be incorporated into the working of the training algorithm. It may be visualized that these networks may be expanded in a manner that there are only forward network connections by replication of the layers. This is referred as the unfolding with time. This corresponds to

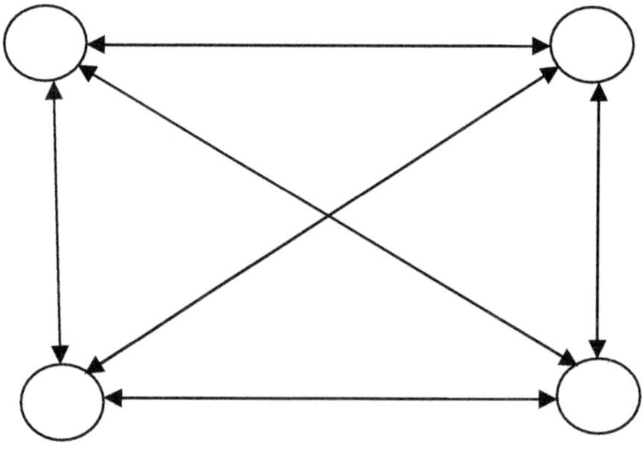

Fig. 2.15 General Architecture of Recurrent Neural Network

a standard MLP which can be trained with BPA. The restrictions of similarity of the unfolded weights would however apply. This again represents the additional dynamism found in these networks.

2.6.2 *Conceptualization*

The additional dynamism introduced in these networks by the addition of the recurrent neural network connections mark a great boon for the extensive use of these networks in a variety of applications related to series forecasting, speech recognition, etc. The fundamental aspect of these networks is that the output of these networks not only depends upon the input, but also the previous outputs or the previous state that is fed as feedback in the form of recurrent connections in between neurons. An important aspect of systems is that the input may not always be as desired. In such a context it is important for the system to have some mechanism to rectify the inputs. This is only possible by a feedback from the other neurons. The generated output acts as a means to drive the input and rectify it towards the generation of the ideal output. Hence there exist feedback connections. In order to make the systems completely dynamic with a lot of processing and rectification there exist feedback connections in all neurons at all the layers. This adds a lot of processing that is performed in these networks. The bulk of processing further makes these systems computationally more expensive as compared to their counterparts without recurrent connections. But the added feedback drives the systems in correct direction which is not prevalent in the counterparts.

2.7 Problems with Neural Networks

We have discussed numerous models and manner of working of ANNs. These networks are being used since very long and are effective problem solving agents for machine learning, functional prediction and classification. These are easy to use and implement in any problem. The various models further give a rich variety to the designer for use in any real life application. Different models give better performance which largely depends upon the problem, requirements and data. There are still numerous problems in the conventional use of these networks that motivate their hybridization and the making up of hybrid systems that have a potential for better performance in most of the complex scenarios. We illustrate a few of these problems in this section

One of the major problems with ANNs is training. The training algorithms are iterative in nature and try to minimize the error to as small value as possible. This may lead to the system to converge to local minima. The ultimate objective is always to reach the global minima. Consider the BPA for example that is trained using the gradient descend approach. Even though momentum avoids the convergence to local minima, its capability is limited in nature. It is quite possible that the network ejects from local minima and again gets trapped at some other local minima. The entire search space for the training algorithm would be heavily large and cannot ideally be explored for the location of global minima in such a manner. This requires a much more sophisticated search operation.

Another major shortcoming of ANNs is that the architecture needs to be fixed by the designer. This is highly likely to be sub-optimal. The various parameters of the ANN as well as the training algorithm need to be optimally set for good performance. Wrong setting at any of the stages may lead to bad results and an unoptimal solution. Now the task of figuring out the most optimal design is highly cumbersome where there are bulks of parameters to handle and any combination of them gives a different overall performance. Again and again training with different parameters is done that is both sub-optimal and time consuming.

The problem being considered may have a large complexity due to the complex mapping of the inputs to the outputs. In such a scenario there might usually be generation of very large networks. The large networks further have problems of time, training and data requirement. In many scenarios the training data might completely refuse to train itself at all. In such a case we are left with no options as adding neurons would result in lower number of epochs. The data itself might allow few epochs. The network may not be able to train itself due to computational constraints. This is another problem with the use of ANNs.

The ANNs are a network that take input and process it to generate the correct output. No inference about the manner in which the inputs generate the output can be made from these networks. It is completely unknown what logic or rules enable generate the output or what is the system behavior. The entire ANN in other words acts as a big black box in problem solving. In case the system showcases some new behavior not found in the historical database, there is no means to specify or model the behavior. In such a case there is a problem in their use. Similarly if the problem logic is partially known, it cannot be specified in these networks. As a result the overall data requirement by these networks is very large which is needed to tune the weights. As the problem logic is completely unknown, the number of weights is large with not even a slightest of idea regarding their possible values.

Online learning, plasticity control, large time of training and training v/s memory tradeoff are other common problems of the ANNs.

2.8 Chapter Summary

The chapter explored the variety of ways in which problems could be solved with the help of neural networks. The chapter explored the basic use of these systems for machine learning and generalizing the learning to the unknown data as well. Both the functional prediction and classification problems were dealt with. The chapter introduced the use of non-linearity with the XOR problem. We further moved to the various neural models. A lot of emphasis was given on model conceptualization and understanding the role of the various parameters. The first model to be discussed was the Multi-Layer Perceptron. We saw how the artificial neurons could be arranged in a layered manner to regress over the training inputs. The Back Propagation Algorithm was discussed as a training algorithm. The other model included Radial Basis Function Networks which makes use of different neurons as templates at different regions of the input space. The discussion over classificatory models of neural networks included Learning Vector Quantization and Self Organizing Maps that used supervised and unsupervised learning respectively. Here the neurons stored prototypes in their own way. We further discussed about the Adaptive

Resonance Theory neural networks and plasticity. The other neural model was recurrent neural network where the neurons had additional feedback connections. These provided greater dynamism for better performance of the neural network.

References

[1] Barron, A.R., Barron, R.L.: Statistical learning networks: A unifying view. In: 1988 Symp. on the Interface: Statistics and Computing Science, Reston, VA, pp. 192–203 (1988)

[2] Carpenter, G.A., Grossberg, S.: A massively parallel architecture for a self-organizing neural pattern recognition machine. Comput. Vis. Graph. Image Process. 37, 54–115 (1987a)

[3] Carpenter, G.A., Grossberg, S.: ART-2: Self-organizing of stable category recognition codes for analog input patterns. Appl. Opt. 26, 4919–4930 (1987b)

[4] Cohen, M., Grossberg, S.: Absolute stability of global pattern formation and parallel memory storage by competitive neural networks. IEEE Trans. Sys. Man and Cybernet. SMC 13, 815–826 (1983)

[5] Fukushima, K.: Cognitron: A self-orgainizing multilayered neural network. Biolog. Cybern. 20, 121–136 (1975)

[6] Graupe, D.: Principles of Artificial Neural Networks. World Scientific, Singapore (1999)

[7] Grossberg, S.: Classical and instrumental learning by neural networks. Progress in Theoret. Biol. 3, 51–141 (1974)

[8] Grossberg, S.: Competitive learning: From interactive activation to adaptive resonance. Cogn. Sci. 11, 23–63 (1987)

[9] Grossberg, S.: Learning by neural networks. In: Grossberg, S. (ed.) Studies in Mind and Brain, pp. 65–156. D. Reidel Publishing Co., Boston (1982)

[10] Hammer, B., Villmann, T.: Generalized relevance learning vector quantization. Neural Networks 15(8-9), 1059–1068 (2002)

[11] Hebb, D.: The Organization of Behavior. John Wiley, New York (1949)

[12] Hopfield, J.: Neural nets and physical systems with emergent collective computational abilities. In: Proc. of the National Academy of Sciences, vol. 79, pp. 2554–2558 (1982)

[13] Hopfield, J.J.: Neural networks with graded response have collective computational properties like those of two state neurons. In: Proc. of the National Academy of Sciences, vol. 81, pp. 3088–3092 (1984)

[14] Kasabov, N.K.: Foundations of Neural networks. In: Fuzzy Systems, and Knowledge Engineering. MIT Press, Cambridge (1998)

[15] Kohonen, T.: Learning Vector Quantization. Neural Networks 1, 303 (1988)

[16] Kohonen, T.: Self-organization and Associative Memory. Springer, Heidelberg (1989)

[17] Kohonen, T.: Improved versions of learning vector quantization. Intl. Jt. Conf. Neural Nets 1, 545–550 (1990)

[18] Konar, A.: Artificial Intelligence and Soft Computing: Behavioral and Cognitive Modeling of the Human Brain. CRC Press, Boca Raton (1999)

[19] Lippman, R.P.: An introduction to computing with neural nets. IEEE ASSP Magazine 3(4), 4–22 (1987)

[20] McCulluch, W.S., Pitts, W.: A logical calculus of the ideas imminent in nervous activity. Bull. Math. Biophys. 5, 115–133 (1943)

[21] Owens, A.J., Filkin, D.L.: Efficient training of the back propagation network by solving a system of stiff ordinary differential equations. In: Proc. Intl. Joint Conf. on Neural Networks, Wash., DC, vol. 2, pp. 381–386 (1989)

[22] Pineda, F.J.: Generalization of backpropagation to recurrent neural networks. Phys. Rev. Lett. 18, 2229–2232 (1987)
[23] Rosenblatt, F.: The perceptron, a probabilistic model for information storage and organization in the brain. Psychol. Rev. 65, 386–408 (1958)
[24] Rosenblatt, F.: Principles of Neurodynamics. In: Perceptrons and the Theory of Brain Mechanisms. Spartan Press, Washington (1961)
[25] Rumelhart, D.E., McClelland, J.L.: An interactive activation model of the effect of context in language learnin. Psychol. Rev. 89, 60–94 (1986)
[26] Rumelhart, D.E., Hinton, G.E., Williams, R.J.: Learning representations by backpropagation errors. Nature 323, 533–536 (1986)
[27] Rumelhart, D.E., McClelland, J.L.: Parallel Distributed Processing: Exploring in the Microstructure of Cognition. MIT Press, Cambridge (1986c)
[28] Wasserman, P.P.: Neural Computing Theory and Practice. International Thomson Computer Press (1989)
[29] Shukla, A., Tiwari, R., Kala, R.: Real Life Applications of Soft Computing. CRC Press, Boca Raton
[30] Werbos, P.J.: Backpropagation through time: What it does and how to do it. Proc. of the IEEE 78(10), 1550–1560 (1990)
[31] Widrow, B.: Generalization and information storage in networks of ADALINE neurons. In: Yovits, M.C., Jacobi, G.T., Goldstein, G.D. (eds.) Self-Organizing Systems, pp. 435–461 (1962)
[32] Widrow, B., Hoff, M.E.: Adaptive Switching Circuits. In: 1960 IREWESCON Convention Record, New York, Part 4, pp. 96–104 (1960)
[33] Williams, R., Zipser, D.: A learning algorithm for continually running fully recurrent neural networks. Neural Comput. 1, 270–280 (1989)
[34] Windrow, B., Lehr, M.A.: 30 Years of Adaptive Neural Networks: Perceptron, Madaline, and Backpropagation. Proc. IEEE 78(9), 1415–1442 (1990)
[35] Widrow, B., Winter, R.: Neural nets for adaptive filtering and adaptive pattern recognition. Comput. 21, 25–39 (1988)

List of Figures

Fig. 2.1. Input Space for XOR problem
Fig. 2.2. Multi Layer Perceptron
Fig. 2.3. Processing by a single neuron
Fig. 2.4. The Back Propagation Algorithm
Fig. 2.5. Error Surface of Neural Network
Fig. 2.6. Training with different number of neurons
Fig. 2.7(a). Oscillation with large value of learning rate
Fig. 2.7(b). Slow learning with small value of learning rate
Fig. 2.8. Early Stopping of neural network with validation data
Fig. 2.9. Architecture of Radial Basis Function Network
Fig. 2.10. Loss of activation with increasing distance in Radial Basis Function Networks
Fig. 2.11(a). Radial Basis Function Network with small spread
Fig. 2.11(b). Radial Basis Function Network with small spread
Fig. 2.12. Learning Vector Quantization
Fig. 2.13. Self Organizing Feature Map
Fig. 2.14. Adaptive Resonance Theory Networks
Fig. 2.15. General Architecture of Recurrent Neural Network

Chapter 3
Genetic Algorithm

Abstract. One of the most exciting aspects of life is its evolutionary nature where the individuals keep improving along with generations. Genetic Algorithms are an inspiration from this natural evolution and find themselves as powerful optimizing agents for solving numerous real life applications. These algorithms can model complex problems and return the optimal solution in an iterative manner. This chapter presents the manner in which we model and solve the problem using this evolutionary technique. The role of the various parameters and optimal parameter setting as per the problem requirements would be discussed. The chapter would present mutation, selection, crossover and other genetic operators. Evolution forms the base for most of the complex systems that are designed to evolve with time. In this chapter we hence first study the basic concepts and then take an inspiration towards evolving systems. At the same time we present the limitations of evolution that marks a threshold to massive potential of problem solving that these algorithms have.

Human beings have assumed a very complex shape today. We are highly intelligent beings that can solve very complex problems in fractions of seconds. We further possess great powers in regard to intellect, behavior and adaptation to changing world which is both inspirational as well as a mystery. The majesty of the human being was not a sudden phenomenon that led to our creation. We evolved from our ancestors and the history takes us long back. The earlier life forms were much simple as compared to our present state. As a matter of fact life started as simple uni-cellular organisms that performed basic functions for living. Since then the growth and development is a fascinating story of evolution.

Considering the human race is another fascinating story. Each generation of the mankind results in the generation of a new generation. The newer generation is found to be better and fit as compared to the previous generation. The newer individuals are able to perform many more complex tasks and possess a higher level of intelligence as compared to the ancestors. In this manner the race keeps improving along with time and generations. Another major aspect here is the change. The environment is constantly changing that calls for a change in the behavior and characteristics of the individuals as well. The individuals that were fit earlier may no longer be fit as per the change in the environment. In such a context the individuals keep adjusting to the changes in the environment. Every new generation adapts itself as per the changing needs. Hence the individuals keep developing

A. Shukla et al.: Towards Hybrid and Adaptive Computing, SCI 307, pp. 59–82.
springerlink.com

good characteristics as per the demand of the environment. This makes them fit to the changing scenarios as well.

Evolutionary Algorithms (EA) are an analogy to this evolution found in nature. These algorithms try to generate numerous solutions to a problem and iterate through them in time and generations for the continuous improvement in the optimality of the solutions (Mitchell 1999; Davis 1987). The optimality is called as the fitness of the solution. A solution may be called as an individual. The complete collection of individuals at any generation or point of time constitutes a population. This completes the analogy of these systems with the natural world. Now we have a population that consists of a set of individuals (or solutions). These solutions vary in their fitness (or optimality, usually measured by a function called as the fitness function). The entire algorithm runs in generations. As the generation proceed, the individuals (or solutions) got fitter (or more optimal) as per the requirements. In this manner the algorithm is run for a few generations. The best individual out of the entire population is picked up upon the algorithm termination as the best individual that solves the presented problem. The environment is usually kept constant in the entire experimentation. This means that the behavior of the fitness function is the same in the entire algorithm run. However this algorithm is able to give a good performance against a changing environment as well where the fitness function behavior changes along with time and generations.

EA is used as a search algorithm to solve problems with a massive search space. These are the problems which cannot be effectively solved by the conventional exhaustive search mechanism or any other technique due to the computational limits. These algorithms are able to give a great performance in such problems by making the best possible search as per the time and computational limitations. The search results of EA are usually not the most optimal statistically. Many times the solutions may be almost the most optimal with the solution lying very close to the optimal solution. Many other times the solutions are far from the most optimal solution but the value is still optimal in nature. There is a tradeoff between the time and optimality that is best attacked by these algorithms. Another common statement is that EA is used for solving the NP-hard problems. These are the problems whose solution is not known in any polynomial time. The time requirements to these problems increase exponentially with the increase in the input size. Most of the real world problems are of large size. Hence the use of conventional techniques fails in such problems. It may take years to do an exhaustive search or search based on any other heuristics. In such scenarios the EA is able to give decent solutions in finite amount of time. This forms the basis of their widespread use in different problems and disciples.

The entire space in which the EA works is referred as the fitness landscape or more commonly as the search space. Image the problem is to optimize a function given f by equation (3.1). The various modifiable parameters are x_1, x_2, x_3, $\ldots x_n$ whose values can be modified to minimize the value of f. Here each of the parameter x_i lies between some minimum and maximum value given by equation (3.2)

$$V = \min\{f(x_1, x_2, x_3 \ldots x_n)\} \tag{3.1}$$

$$L_i \le x_i \le U_i \tag{3.2}$$

Here we may generate numerous solutions to the problem by setting different values of x_is within the allowable range. $f(x_1, x_2, x_3 \ldots x_n)$ represents the fitness function whose value determines the overall quality of the solutions. Solutions or individuals having lower values of f are said to be more fit as compared to the ones that have larger values of f. The basic objective of the algorithm is to minimize the value of f and find the values of parameters x_i that give this minimal value. We hence plot the fitness landscape or the search space which is a surface plot of f for all possible values of x_is. This plot would contain $n+ 1$ dimension with one axis for each of the x_is and the vertical axis for the function value. Each of the axes of the x_is is bound between their minimum and maximum values given in equation (3.2). This is the search space of the EA and the task is to compute the point such that the value of the function is the least. Since the search space may be very complex with numerous hills and valleys, it may not be possible to use any other technique to compute the global minima. Also the high number of axis or the problem dimensionality imposes a big restriction for the EA. The search space for a problem with 2 dimensions is shown in figure 3.1.

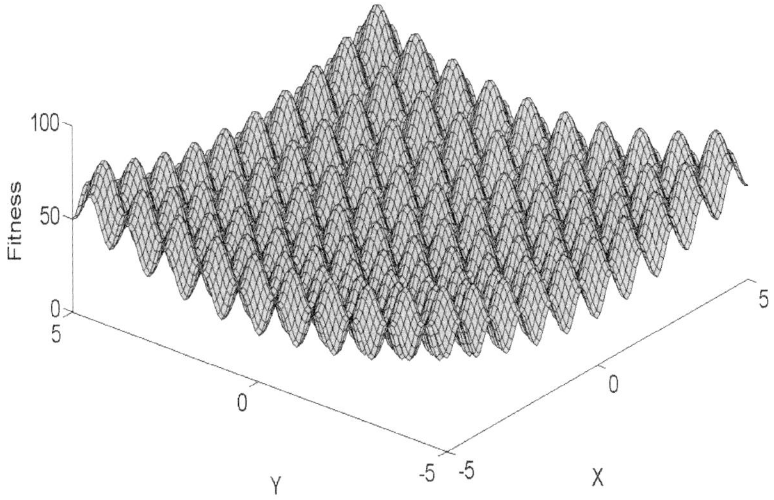

Fig. 3.1 Search Space of Genetic Algorithm

EA is now a well studied disciple that has been modified, customized and designed in numerous ways and means. This algorithm is an excellent optimization agent. In this chapter we study one of the most commonly used forms of using this algorithm which is the Genetic Algorithm (GA). This uses the analogy of the natural evolution of any species where parents in one generation mate to give the offspring that makes an individual of the next generation. Numerous other topics would be discussed in this book in chapters 9 to 12. The various evolutionary algorithms are highly related in terms of problem solving approach which is attributed to their evolutionary nature. These were the results of the pioneering works of a number of researchers that would be discussed in this chapter or in chapters 9 to 12. The notable works include Bäck (1996), Bäck and Hoffmeister (1991), Deb and Goldberg (1989), Fogel (1995), Koza (1992), Michalewicz (1992), and Schwefel (1995).

3.1 Simple Genetic Algorithm

The Simple Genetic Algorithm (SGA) is the most primitive and simple form of evolutionary algorithm that are used for the optimization purposes (Holland 1975; Goldberg 1989). A population in the human context consists of a variety of individuals that have different characteristics and behavior. Each of them has a different level of fitness that is decided as per its capability to live in the present environment. As per the evolutionary concepts the population grows in generations with the individuals of higher generations more fit than the individuals of the previous generation. For the creation of a new generation from an old generation, sexual reproduction is carried out. Here two individuals or parents create a new individual or child of the higher generation. The child carries the selected characteristics of the two parents. Some of the characteristics come from the first parent and the other come from the other parent. A child is usually fitter than both the parents if it takes the good characteristics of both the parents and rejects their bad characteristics. Also many times there is an error in recombination or passage of characteristics. This results in the making of new characteristics in the new individual that evolves from the parents. In this manner different parents yield different children for the next generation. The participation of an individual in this process of generation of children is directly proportional to its reproductive capacity or fitness. More fit individuals are more participative in reproduction as compared to the other individuals. Only some individuals go from one generation to the other or survive the process. This happens as per the Darwin's theory of survival of the fittest. The fitter individuals are more likely to survive and go through the next generation as compared to the weak individuals.

The SGA or simply the GA is analogous to the same concept. The individual is in the form of a solution of the problem. Multiple solutions make the population pool. The fitness of any individual is its optimality in solving the problem. This may be given by a pre-defined function called as the fitness function. Initially a set of random individuals are generated to make a random population pool. This random set of individuals forms the initial population on which the GA runs. We keep iterating over generations till a predefined stopping criterion is met. This may be on the basis of number of generations, time of execution, best fitness value or anything else as per choice. As the generations increase the optimality or fitness must increase. The generation of any population results in the generation of the next population. For this we first select the individuals that survive to the next population. These individuals then participate for the generation of the next population by a set of genetic operators. The most important of these are crossover where two individuals or parents combine for the generation of newer individuals or parents; mutation where new characteristics are added to the individual;

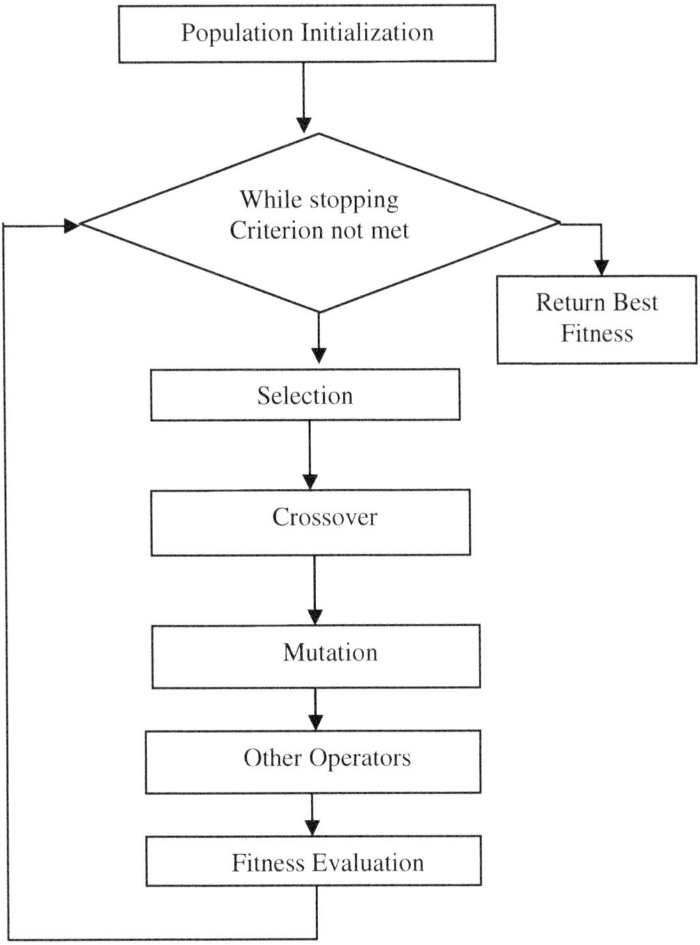

Fig. 3.2 Simple Genetic Algorithm

and elite where the best individual of a population is directly passed to the next generation. This gives us the next generation of population. The fitness evaluation of the new population is done. This may be summarized by figure 3.2.

The various steps of the algorithm are specialized operations that are formulated by an analysis over the biological counterparts. We discuss each of these concepts in the coming sections. We however first study the algorithm from its performance in the search space. Each of the sections would further illustrate the role of the operator in regard to the search space.

3.1.1 Conceptualization

The initial random solutions may be regarded as the random points in the highly dimensional search space or the fitness landscape of the GA. Since they are randomly generated, it is likely that these points or individuals would cover most of the search space leaving no major segment of the search space as blank. The fitness of these individuals is the level in which they are located in the fitness landscape with the fitter individuals lying at the lower levels, considering the problem as a minimization problem. The fitness value gives a decent idea regarding the locality of the area in which the individual is found. We expect the surface of fitness to be smooth and hence the neighboring points must possess similar fitness values. Also we do not expect that the surface would have too many hills and valleys on its surface. This gives us enough idea regarding the shape of the fitness landscape based on the fitness and position values of the different individuals. The different individuals may be regarded as different search agents, which make this problem as a multi-agent system for collaborative search for the global minima. Now the task is to move the agents or the individuals in such a manner that they reach the global minima. A fitter individual or an individual with lower objective value is likely to be found at a place near the global minima. This factor attracts the other individuals towards this individual. As a result the different individuals jump towards the other more fit individuals at the next generation. The weaker individuals may die in the process and stronger ones may produce more individuals in their vicinity. Also the various individuals move by some amount on their own in random directions to look for the possibility of global minima in their vicinity. Again the fitness landscape may be guessed based on the position and fitness values of the agents. Each of these denotes a new possibility of global minima. This process goes on and on. It is highly likely that while traveling the individuals meet the global minima. Towards the end, all individuals converge to some point that is regarded as the final solution of the algorithm.

3.2 Individual Representation

One of the fundamental design issues is the manner in which a solution is represented to solve the problem in a GA. This is also referred by the term of problem encoding. The basic solution of a problem in its native form is called as the *phenotype* representation. This may be specific to the problem and presented in a manner that the algorithm working over the problem understands. This needs to be converted in a manner that the GA can work over. This representation is known as the *genotype* representation of the individual. The most commonly used representations are in form of bit string, numeral vectors or a tree based representation.

The bit string representation is inspired from the biological genes. Here the solution is in the form of a continuous string of bits consisting of *0*s and *1*s of fixed length. Each of these bits represents some characteristic of the individual or the solution to the problem. The individual in its native or phenotype representation needs to be cast into this bit string using some mechanism. The most primitive method of doing this may be the decimal to binary conversion of the problem parameters limited to a fixed precision. Consider the problem of functional minimization where the objective was to minimize f. Now we may convert each of the x_is into binary number. The conversion however needs to be restricted to a fixed number of bits, say t. This would limit the maximum numeral precision that the GA meets. Now we append the binary representation of the entire x_is one after the other. This gives us a very long string of length *n x t* consisting of bits that are either *0* or *1*. This is the genotype representation of the individual. We may easily imagine a mapping between the genotype and the phenotype representation. In most of the other problems the conversion between the genotype and the phenotype may not be that trivial in nature (Lobo *et al* 2007). In such a case we may have to work out a strategy for the effective conversion to take place such that the GA can have a bit string to optimize. A sample bit string representation is given in figure 3.3(a).

1	1	0	0	0	0	1	1	1	1

Fig. 3.3(a) Bit String Representation of individual

The other form of an individual representation may be in the form of string of real numbers. The GA may directly be applied into this string. The easy way of doing this is to append the various numbers one after the other in a continuous manner. This forms an entire string or vector of numbers that may be changed for carting out the optimization. This form of representation is given in figure 3.3(b).

1.123	4.223	9.23	10.112	3.33

Fig. 3.3(b) Double Vector Representation of individual

The other commonly used form of individual representation is in the form of a tree. This is rooted across some parameters and the other parameters are further arranged in a tree-like manner. This forms a very special representation in which the various parameters need to be placed in such a way that the tree-structure if formed. The various genetic operators are applied to this type of structure. This representation is shown in figure 3.3(c).

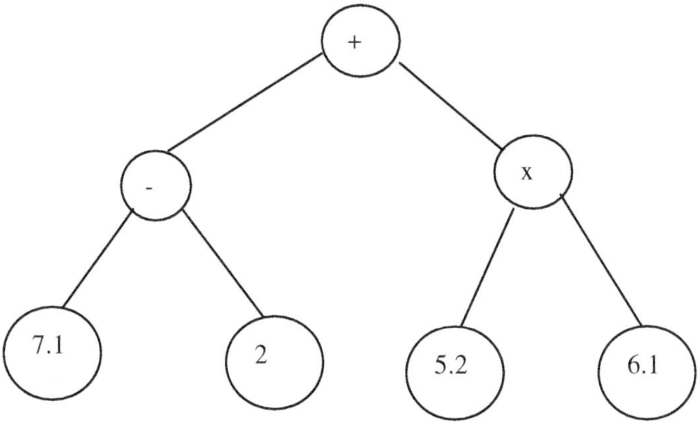

Fig. 3.3(c) Tree Representation of Individual

3.2.1 Role of Good Encoding Technique

The type of representation greatly determines the performance of the GA in problem solving. For most of the problems the problem representation is trivial and hence the role of representation might not always be realizable. For many other problems there might be a characteristic manner of representation. There are a few important aspects in regard to the representation that we discuss in this section.

The problem representation has a direct consequence on the fitness landscape of the GA. A complex fitness landscape has numerous hills and valleys on its surface. This makes it very difficult for the GA to solve. A good representation technique must hence ensure simplicity of the surface. Also many times the representation may have steep walls. This is especially in cases where the problem loses feasibility exceeding some value of parameter and is optimal somewhere just inside the feasibility region boundary. Such sharp changes are further a problem for the GA. They make the fitness value too large or small on minor changes of parameters. The existence of these sharp changes at the ends of the space might not matter, but their presence in the core of the search space around interesting areas usually causes problems. These must hence be avoided. The choice of the fitness function has a role to play here for similar reasons.

The other major factor is the mapping of the phenotype to the genotype representation. In most of the cases a single phenotype may represent a single genotype, but this may not always be true in all forms of representations. Many times the representation is done in a manner that two or more genotype representations map to the same phenotype. This may be easily done by the addition of redundancy in the problem representation. In many other problems this may be unavoidable consequence of representation. In such a context the fitness landscape has multiple global minima that ultimately lead to the same phenotype representation. This helps in the genetic search as the individuals have the option to find any of the minima out of the available redundant minima options. But many other times this may make the fitness landscape complex or result in adding the overall problem complexity, thus yielding problems in search.

The shorter representation techniques are usually better than the longer ones. This is because of the fact that the search space complexity or dimensionality depends upon the length of the genetic individual. Each gene marks an axis of its own. This makes the search space grow and complicates the entire process. This is hence undesirable. The smaller individuals are easier to optimize. The genetic length however mostly depends upon the problem. It is normally not possible to shorten the genetic length and we have to accept the representation in its trivial form.

3.3 Scaling

The individuals in a population pool of the GA have some chance of going to the next generation and surviving in the genetic competitions (Whitley 1989). This depends upon their fitness value, with the fitter individuals having better possibility of going to the next generation. The possibility of surviving and being selected for the next generation is termed as the *expectation value* of the individual. This value denotes the expectation that an individual may have for survival. A higher expectation value denotes a fitter individual. The task of assigning the expectation value to the various individuals in a population is termed as scaling. The scaling is an important operation in the GA since a change in scaling methodology might result in the algorithm being more or less biased towards the fitter individuals. Based on these concepts, there are three commonly used scaling mechanisms. These are fitness based scaling or proportional scaling; rank based scaling and top scaling.

3.3.1 Proportional Scaling

The most basic type of scaling mechanism is when the individuals are directly scaled by their fitness values. The fitter individuals have higher fitness values and hence naturally have a better chance of being selected for the next generation which is the basic working philosophy. The expectation value of the individuals for selection may hence be given by equation (3.3).

$$\mathrm{Exp}(I_i)=\alpha\ \mathrm{fit}(I_i) \tag{3.3}$$

Here $Exp(I_i)$ is the expectation of the ith individual denoted by I_i. $Fit()$ is the fitness function and α is a constant.

The trends observed in this type of scaling are the same as the trends in the fitness function. Hence if an individual has an exceptionally large fitness value as compared to the other individuals, it would have an exceptionally large expectation value as well. Due to this it is highly likely that this individual gets selected again and again in the selection process. This would make the entire next generation dominated by a single individual. This is the case many times where a few individuals occupy prominent position in the genetic landscape at any generations. If this type of scaling is followed, they would naturally dominate and the other individuals would be inclined to suddenly jump towards these solutions. This may be conceived as a limitation in proportional scaling that is removed by the rank based scaling.

3.3.2 Rank Scaling

The other type of scaling is the rank based scaling. Here the individuals get an expectation value based on their rank in the population pool. This scaling solves the problem of domination of the highly fit individuals that was prevalent in the proportional scaling. Ranking of the individuals is done on a continuous scale. Hence in this technique the major factor is the relative score of the individuals and not the value of the score. The expectation value in this method of scaling for any individual I_i may be given by the equation (3.4)

$$\mathrm{Exp}(I_i) = \min + (\max - \min)\frac{\mathrm{rank}(I_i) - 1}{N - 1} \tag{3.4}$$

Here N is the total number of individuals in the population, min and max are the minimum and maximum values of expectations possible.

3.3.3 Top Scaling

Proportional and Rank Based scaling carry the task of scaling in a fitness biased manner where the individuals with higher fitness are likely to get selected numerable number of times, on the other hand the lower fitness individuals are expected not be selected at all or lower number of times. The expectation value of fitter individuals is higher as compared to the individuals with smaller fitness values. This type of scaling gives equal expectation values to all the individuals. Each one of them hence has equal chances of selection and going to the next generation. This mechanism results in an equal treatment to the individuals and hence the fitter individuals are not allowed to dominate. Even in rank based scaling it is likely that the fitter individual over-dominates the solution which can lead to local minima. However in order to make the next generation fitter, it is important to make the fitter individuals more likely than the lesser fitter ones. For this we select only the top few individuals and consider them for scaling, the others are assigned with an expectation value of *0*. Let us say that the total $\alpha\%$ of the individuals is selected. In this manner we stop the domination of a particular solution as well as favor the fitter set of individuals. The expectation value of any individual I_i may be given by equation (3.5)

$$Exp(I_i) = \begin{cases} \frac{1}{N} & \text{if } I_i \text{ is in top } \alpha\% \\ 0 & \text{otherwise} \end{cases} \tag{3.5}$$

Here N is the total number of individuals in the population

3.4 Selection

The expected values calculated using the scaling mechanism gives a fair idea of the possibility of the individual to go to the next generation (Goldberg and Deb 1991; Chakraborthy 1996). The major task now is to actually select individuals based on the expected values. This is done by the operation of selection. Selection takes into account the expectation values and selects the required number of individuals that participate in the genetic process. The selection follows the Darwin's theory of survival of the fittest. Here only the fittest individuals survive and the others are eliminated from the population pool, or do not survive in the subsequent generations.

We discussed the role of the GAs as a search problem in the fitness landscape. Now we know that the different individuals or search agents are located at different locations with different fitness values. Our assumption states the fitness landscape to be relatively simple with not many hills and valleys. In such a context we may assume the agents at locations with high fitness values to be located at un-interesting locations. There would be no use in continuing the search of these agents since they are likely quite away from the global minima. The other agents that have better fitness values are likely to be more close to the global minima. Hence it would be judicious to shift the attention from the agents at poor fitness values to the agents at high fitness values. This means that we kill the agents at the poor fitness value locations and in return produce them at the locations with high fitness value. This job is done by the selection operation.

Numerous methods may be used for the selection of the individuals based on the expected values. Some of the commonly used techniques include Roulette Wheel Selection, Stochastic Uniform Sampling, and Tournament Selection.

3.4.1 Roulette Wheel Selection

Roulette Wheel Selection does the task on selection based on the mechanism of working of a roulette wheel. Here the various individuals are given their own segment on the roulette wheel with the circumference of each segment directly proportional to their expectation values. The roulette wheel contains a pointer that points towards the selected individual. The wheel is rotated and the selected individual is noted. It is natural that the individuals having higher expectation values are more likely to be selected than the individuals with smaller expectation values. The possibility of selection here depends upon the circumference which again depends upon the expectation values. In all we select the roulette wheel N number of times where N is the total number of individuals that are to be selected. This method however may result in the individuals with high expectation values to be selected a very large number of times, since they have a reasonably large circumference on the roulette wheel and the turns of the wheels are all independent.

3.4.2 Stochastic Uniform Sampling

This is another method of selection that tries to do away with the limitation of the roulette wheel selection regarding the possibility of excessive selection of the individuals with a high fitness value. This implementation is similar to the roulette wheel selection. The only difference maintained is that there are N numbers of selection pointers in the wheel in place of *1*. The wheel is now turned only once. The various selectors of the roulette wheel now point to N number of elements. All the items to which these selectors point are selected and the rest are eliminated. Again it is quite possible that the individuals with higher fitness value get selected number of times. This would happen with multiple selectors being fit into the same segment that has a larger circumference. But the excessive selection of the fitter individuals would naturally be avoided. This mechanism plays a big role in the preservation of the diversity by not letting one individual dominate. Multiple selections of different types of individual results in algorithm not converge very fast and maintain diversity.

3.4.3 Tournament Selection

This is another type of selection mechanism. Here the selection is done by a series of tournaments. We randomly pick up two individuals from the population pool. These individuals compete with each other in terms of expectation value. There is a predefined number k chosen that denotes the possibility of the individual with higher expectations winning the tournament. Every time a competition happens, we select a small random number r. If r is found to be smaller than k, the individual with higher expectation is said to have won and vice versa. The winning individual of the tournament gets selected in the outcome. In all there are N numbers of tournaments that are played and the winner is selected. The tournament in this type of selection scheme can also be played with multiple numbers of individuals. The number of individuals that participate in the tournament is known as the tournament size. In a multi-player tournament the individual with the highest expectation values wins with a definite probability that is predefined. In this manner all the ranks of expectation values have some probability of winning. The choice of the winner may be done by a random number as was the case with 2 player tournament. Since the initial selection is random, the individual with higher expectation does not get selected a large number of times. Hence this mechanism is also resistant to domination by individuals with high expectation values.

3.5 Crossover

Crossover does the task of recombination of two individuals to generate individuals of the new generation. The new individual carries some of the characteristics of the first parent and the other characteristics from the other parent. The new individual generated in the process may be fitter than the parents or may be weaker. This depends upon the selection of the characteristics by the individual. The selection of good characteristics from both the parents that result in high fitness results in better individuals and vice versa. The fitness of the individual is computed by the overall performance as a combined representation of the entire individual chromosome. Hence nothing certain can be assured regarding the fitness of individual unless it is measured. Crossover results in constant exchange of characteristics in the individual. GA derives a lot of computational optimality as a result of this continuous exchange of characteristics in between the individuals. It may sometimes be viewed as the phenomenon that tries to recombine two individuals hoping that the resulting individual is better in terms of fitness than both parents. The final result of the GA is the fittest individual. It is hence necessary to generate fitter individuals than parents so that the fitness of the best individual of the population pool keeps getting optimized and can finally be returned by the system.

In terms of the fitness landscape search, this operation has a major role to play in the convergence of the individuals towards some point. The various individuals lie at various locations in the fitness landscape. The crossover of two individuals results in the generation of an individual at a point in-between the parents in the fitness landscape. Hence after a complete cycle of crossover operations, the resulting individuals are found somewhere in-between to where the parents were found. This contracts the entire search space of the GA. Again after the next iteration, the new individuals would be generated between the parents. This would further result in contraction of the search space. Hence at each iteration, we cut off a large part of the search space where the global minima cannot lie as per the present fitness values. This contraction of the search space continues and towards the end all individuals lie at almost the same place. Every time the individuals are likely to be attracted more towards the individuals with high fitness values. As the algorithm runs, new areas in the fitness landscape may be explored by the various individuals in their search and the system may escape from the local minima and get to the global minima.

Crossover rate is defined as the total proportion of individuals that are contributed by the crossover operation to the next generation. The other individuals are contributed by the other genetic operators to the new population. This fraction is mostly kept high to avoid randomness and achieve a high convergence. We study the crossover separately for the bit string and the vector string representation.

3.5.1 Bit String Representation Individuals

Here the two parent chromosomes are represented in the form of strings of bits. The crossover operator mixes the bits of the parents to form the bit sequence of the children. Based on this three types of crossover operators are usually used. These are one point crossover, two point crossover and scattered crossover.

One point crossover chooses a random position in the entire chromosomal length of the individual. This point is called as the point of crossover. The crossover simply mixes the chromosomes of the parents around this point. The bits that lie towards the left of the crossover point are taken from the first parent and the others are taken from the second parent. Similarly the other individual is generated that contains the bits from left of the crossover point from the second parent and the others from the first parent. In this manner two parents yield two children of the next generation. This mechanism of crossover is shown in figure 3.4(a).

Similar to the one point crossover is the two point crossover. Here we select two points in the entire chromosome around which the crossover operation takes place. The missing of the bits is carried over around both these points. For the first individual the values of the bits for the two extremes of the chromosomes are taken from the first parent and the middle section is taken from the second parent. This is opposite for the second child that takes the extremes from the second parent and the middle part from the first. This type of crossover is shown in figure 3.4(b).

The one-point and two-point crossover have a major disadvantage that there is a positional dependency in the use of these the crossover. The combinational manner in which the chromosome is represented has an impact on the algorithm. It is likely that similar characteristics found adjacent in the chromosomal representation of some individual remain adjacent for a long time. The separation of these would require time and generations. If the adjacent placement of these is causing the sub-optimality, then the algorithm would suffer. This positional dependence is removed in the other crossover called the scattered crossover.

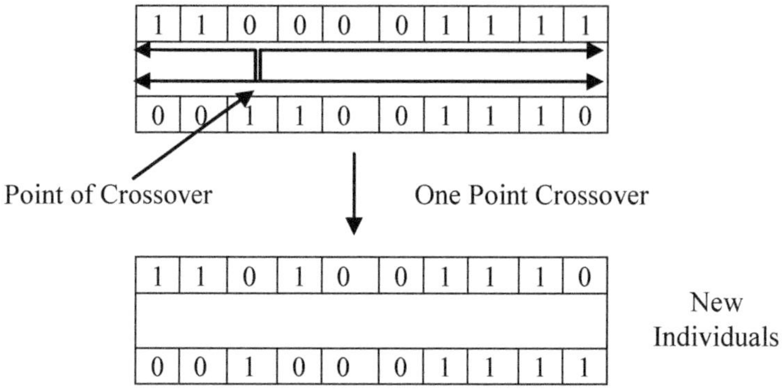

Fig. 3.4(a) One Point Crossover

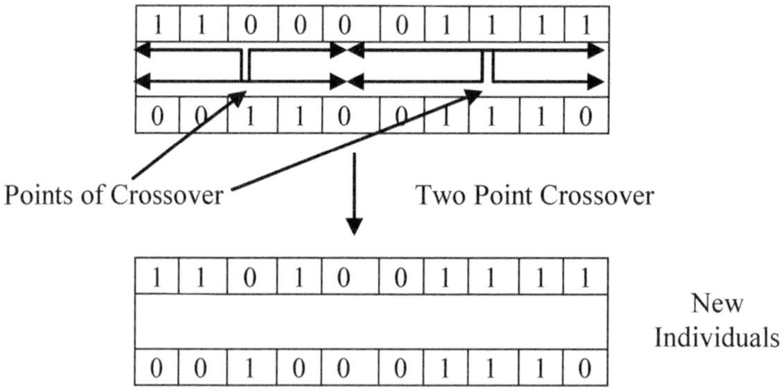

Fig. 3.4(b) Two Point Crossover

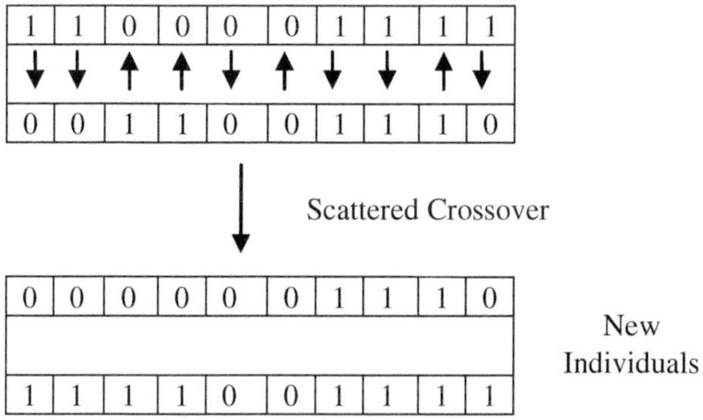

Fig. 3.4(c) Scattered Crossover

In scattered crossover, the selection of which bits come from which of the parents is done randomly at the time of crossover. Some of the bit positions are randomly selected and these bits come from the first parent and the other come from the second parent. The other individual is the converse selection of the bits. This makes two new individuals for the next generation. The inclusion of the randomness in the algorithm to decide the position breaks the positional dependence and this greatly improves the performance. This type of crossover is shown in figure 3.4(c).

3.5.2 Numeral Vector Representation

The other type of representation is in the form of a continuous vector of real numbers. In this representation crossover may be applied in a slightly different manner. We iterate through each and every numeral in the vector and calculate the value of the child chromosome based on the values of the two parents. In this manner the child gets its complete vector assigned. One of the common methods is the *immediate crossover* technique. Here suppose that the ith gene of both the parents is p^1_i and p^2_i, then the corresponding i^{th} gene of the new individual c_i may be given by equation (3.6).

$$c_i = p_i^1 + r\alpha(p_i^2 - p_i^1) \tag{3.6}$$

Here r is a small random number

3.6 Mutation

Mutation is responsible for the addition of new characteristics into the individual. Crossover alone largely does the continuous exchange of characteristics into the individuals in a population. However the optimality cannot be achieved unless the individuals have new characteristics added to them. This operation is performed by the mutation operator. In this operation we randomly change the characteristics of individuals by some amount governed by the mutation rate. This changes the individual and the new individual that emerges may be fitter or weaker than the parent individual. If the changes applied were good, the new individual is fitter and vice versa. The fitter individuals survive in the genetic process and the weaker ones are eliminated as per the Dawin's survival of the fittest theory. Hence if the added characteristics were good, the other individuals would copy them and these would get replicated in the next generations by the crossover operations. If the characteristics were not good the individual might die in a few generations and the characteristics might not spread to as many individuals in the population.

In terms of the fitness landscape, the mutation carries forward the task of expansion of the search space and hinders convergence. The various individuals are the search agents spread around the fitness landscape. Now crossover continuously attracts the individuals towards areas of high fitness. If only crossover is used, the individuals would readily jump towards the other individuals resulting in a fast convergence or contraction of the search space. In such a context it is highly likely that the global minimum gets avoided and the algorithm converges at some other point as reflected by the fitness of the individuals. For the optimal search for the global minima we follow another operation of mutation as well. This enables individuals search in their surroundings as well to look for the global minima. This operation essentially throws off the individuals in the search space to random places away from the present location by certain distance. The distance of deviation of individual from the present location is determined by the mutation rate. Mostly the deviation is small and the individual is thrown to nearby areas. Now the fitness of the individual may be more or less depending upon its location. Since the fitness landscape is assumed to be reasonably simple, we do not expect a very radical change in fitness value for most of the small changes. Both the small and large movement of the individuals has different relevance. The small movements enable the individuals seek the global minima in the surroundings. This is especially important for the fitter individuals that might be very near the global minima but not exactly onto it. Even for other individuals, they might not be enjoying a good fitness value due to their present location as compared to the other individuals, but the minima might lie somewhere very close. This can be rectified by mutation.

For larger movements of the individuals, they are usually thrown at some completely new locations in the search space whose fitness cannot be easily guessed. Many times excessive convergence leads to the individual near the global minima dying in the genetic process or jumping towards the other individuals by large margins that presently possess better fitness values. Such an action leads to the algorithm converge towards local minima. The large mutations are a hope to throw some individual again to such areas to uncover the global minima. In other cases, when the individuals early converge to some point, we might feel the need to again expand the search space to best make use of the computational or search time. This is done by large mutation movements.

We discuss two mutation techniques namely uniform mutation and Gaussian mutation.

3.6.1 Uniform Mutation

The first mutation technique is the uniform mutation. Mutation rate denotes the impact of the mutation. In bit string representation this number represents the probability that a bit in the bit string would be flipped. We flip every bit of the individual represented with this probability. Suppose that m was the mutation rate, then we may expect $m\%$ of the bits are flipped from the present state to the opposite state. If there was a 1 in the gene, it is converted to 0 and vice versa. This is shown in figure 3.5(a). In case of a numeral vector representation of the individual the method of application would be slightly different. Here m or the mutation rate would denote the level by which the individual may be moved in the fitness landscape from its present location. For the application of mutation, we may move the individual by an amount within this limit by a random amount within all the axes. In this manner the values of all the genes are modified by some amount. This is shown in figure 3.5(b).

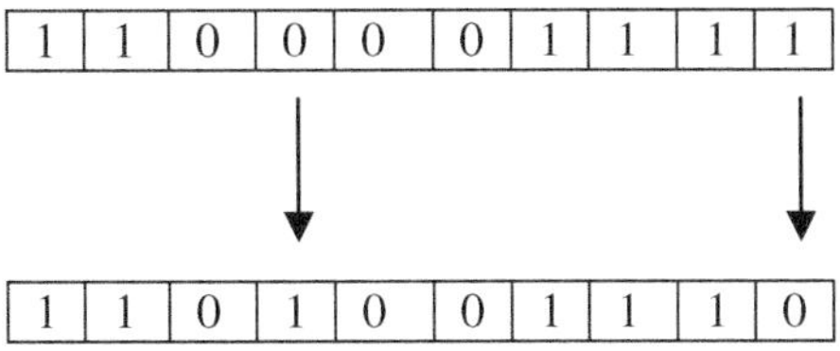

Fig. 3.5(a) Mutation with a bit string representation

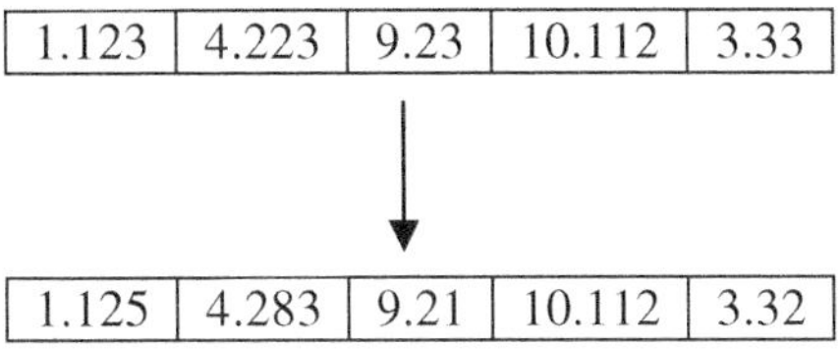

Fig. 3.5(b) Mutation with a numeral vector representation

3.6.2 Gaussian Mutation

The problem with the uniform mutation is its constant mutation rate. Hence the individual is always limited to move within a small predefined amount. This fixed mutation disregards any big changes in the individual. Many times these changes are required in the individual to control early convergence as discussed. This problem is solved with the help of Gaussian Mutation. In this type of mutation we

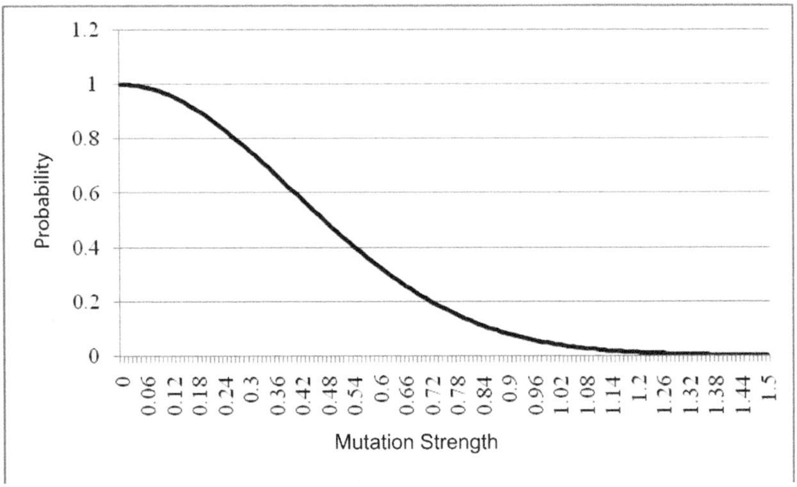

Fig. 3.6 Gaussian Mutation

make use of Gaussian random numbers for the change of any gene value. These numbers obey a Gaussian distribution. Hence usually these numbers are low in magnitude. There are only very few cases where these numbers are of large magnitude. Hence usually the mutation carried is small and occasionally large mutation occurs in the individual. The Gaussian distribution may be controlled by varying the scale and shrink parameters of the distribution. Scale is a measure of the variance at the initial generation of the population. The shrink parameter varies mutation rate so that the variation in population shrinks as per the specified value along with generations. These parameters signify the shape of the distribution. The curve for the Gaussian distribution is given in figure 3.6. They naturally represent a good technique for mutation as the probability of small mutation is high and the probability of large mutation is fairly low.

3.7 Other Genetic Operators

The GA is not only restricted to the use of mutation and crossover as the two genetic operators. Considering the problem and the algorithm behavior we may design our own operators. The basic motivation is to create fitter individuals for the next generation and to aid in the creation of fit individuals. Based on these many operators have been classically used apart from mutation and crossover. We discuss these one by one.

3.7.1 Elite

This operator selects the best few individuals from one generation and passes them to the next generation. The selected individuals are passes as it is to the next generation without any kind of modification. The number of individuals passed in-between the generations by using this operator is called as the elite count. This parameter is usually kept *1* or just greater than *1*. Elite individuals play no role in the optimization process. In the evolutionary process it is highly likely that the best individuals are deformed and sometimes also deleted. The deformation would be a result of the crossover or mutation operators. The deformation may be good or bad depending upon the fitness of the generated individuals. In case the deformation is bad a good individual is completely replaced by a bad one. This is a big loss to the system as the ultimate aim is to return the best individual in the population pool. Hence the elite operator ensures that the best individuals are always passed unaffected. Even if the optimization process results in no optimization, the best individual keeps passing in-between generations and is preserved in the process. In case the optimization results in a fitter individual, the preservation happens with this individual.

3.7.2 Insert and Delete

These are other operators that are not commonly used in practice, but used in some specific applications of GA optimization. Insert adds new individuals to the population pool. These may be completely random individuals that constitute the population pool. The delete operation deletes individuals with very poor fitness from the population pool. Suppose that the problem of consideration is very sensitive with a high probability of generation of infeasible solutions or individuals. These are the solutions or individuals that are in-eligible to be returned as solutions to the problem as they disobey some vital constraint. In such problems one of the solutions is to assign them the most poor fitness value. It is assumed that these solutions would get killed in the process. If the proportion of these individuals in the population pool is very high, then this strategy might face problem. This is due to the affect that these individuals have in the evolutionary process because of their large magnitude. We may consider the operation of delete operators for such individuals. Also many times we add new individuals to the population pool. This is done to add diversity in the population pool that might get killed in the evolutionary process. Many times this may be used to externally add required characteristics in the population pool. These may be good characteristics as per the present state of the problem. The insertion of these characteristics may be by adding new individuals having these characteristics.

3.7.3 Repair

Many times the individuals generated in the evolutionary process are infeasible. These individuals may get into the population pool and affect the other individuals as well. In other cases these individuals are a waste of a search agent, as they cannot be used as possible solutions. Repair is an operator that tries to repair and infeasible solution and converts it into a feasible solution. As soon as an infeasible solution is produced, it may be repaired by this operator before adding to the population pool. The repaired individual may not be very fit, but represents a set of possible characteristics that may be optimized in the evolutionary process.

3.7.4 Hard and Soft Mutation

While discussing mutation we emphasized on the need of big and small changes in the individual in the mutation operation. This was done by a large and small mutation rate. The big changes greatly add diversity and help in excessive search space convergence, giving better chance of exploring the global minima. The small changes were good in exploring the vicinity of the individuals in search of minima. These also restricted the search space convergence. We may sometimes maintain two different mutation operators for carrying both these tasks. Hard mutation is applied very rarely but has a deep impact on the individuals as a result of a high mutation rate. The soft mutation has a small mutation rate and is occasionally applied to the individuals.

3.8 Number of Individuals

The number of individuals is another major parameter of the algorithm. These denote the total number of search agents that aid in the search of the global minima

(Goldberg 1989). Having a large number of individuals would mean that a large number of search agents in search. This would mean a lot of computation for the various genetic operations of scaling, selection, crossover, mutation, etc. In such a case the algorithm would become more computationally expensive and we would be able to continue the search for a smaller number of generations. The search operation would hence have smaller convergence per time step as a result of large number of individuals as well as the computationally expensive nature. It may again be argued that this would make the entire algorithm randomized in nature. However keeping large individuals is effective in case the search space is very complex with large hills and valleys. In such cases having small number of individuals may make the algorithm converge towards local minima. This may be on the basis of the initial fitness values as well as the values that the individuals record in the process of exploration. A large number of individuals mean a large parallel exploration. This is likely to uncover the global minima. It may however be noted that finding the global minima for complex search spaces is a very computationally expensive and difficult task.

Having fewer individuals involves less computation per generation. Hence the algorithm would be able to meet higher number of generations in the limited time. This would make the algorithm converge effectively to some point. The algorithm may further easily explore the areas of the vicinity by the mutation operation. This methodology may make the algorithm converge to local minima, due to lesser overall exploration. However if the search space is simple, there would be existence of only a global minima or very few minima. In such a case this methodology may be useful.

Based on these comments the number of individuals may be judiciously kept which again depends upon the problem and the nature of the fitness landscape.

3.9 Stopping Criterion

A stopping criterion denotes till what limit the algorithm is to be executed. The algorithm continues to execute by generating the higher generations from the lower generations till the stopping criterion is not met (Chakroborthy and Dastidar 1993; Forrest and Mitchell 1993). Once the stopping criterion is met, the best individual is selected and returned as the final system output. The stopping criterion may be kept as any one or more of the following depending upon the problem, its nature and the requirements.

- **Number of Generations:** Here the maximum number of generations is fixed. This criterion finds application in general problems where a limit is kept on the computations that the algorithm performs.
- **Time Duration:** Here the maximum time of execution of the algorithm is specified. This is used in applications that are real time in nature. The algorithm has a specified time of execution. Best individual after this specified time is regarded as the final solution.
- **Stall Generations:** Stall generations are the number of generations in which the fitness of the best individual does not change. This may be seen as a straight

line parallel to the generation axis in a plot of fitness v/s generation for best individual. This criterion specifies the maximum number of stall generations. This may be either due to the convergence of the individuals at some minima where the fitness does not improve, or due to the characteristic location of the best individual at minima with the other individuals unable to explore other better minima in the specified generations. This usually happens in complex landscapes.

- **Stall Time:** This criterion limits the total stall time in place of stall generation as the stopping criterion.
- **Best Fitness:** For many problems the need is optimization only up to a certain limit. Further optimization may not be advantageous for them For these problem we specify the fitness of the best individual as a stopping criterion.

3.10 Diversity

We have used the term diversity multiple times in the text. Here we give it a formal discussion along with some problems and methods of diversity preservation. Diversity is the average distance between the individuals. This distance may be taken as any norm between the attributes of the individual. Hence widespread individuals in the fitness landscape have a high degree of diversity as compared to the individuals that have converged to some point in the algorithm execution. Conventionally the initial solution represents a population of high diversity in the genetic space. The crossover operator does the search space contraction by attracting the individuals towards it. This reduces diversity. This is contradicted by the mutation operation, but crossover is kept dominant to mutation. This makes the algorithm converge or lose diversity along with generations. Towards the end the various individuals lie at almost the same place. This represents a solution of very small diversity. The algorithm further continues dominated by the mutation operation that increases diversity to some value. This phenomenon results in the effective search of the algorithm for global minima in the fitness landscape. It may be commented here that although we make frequent use of the term global minima, any minima can never be guaranteed to be globally minimal in the fitness landscape. There is always a possibility of a better minima lying somewhere at some other part of the search space. We can effectively only design mechanisms for escaping the local minima in search of global minima by formulating better exploration techniques.

The discussed mechanisms have a major side effect. Suppose that an individual was located at some point the fitness landscape with relatively poor fitness values. Now suppose this individual is selected for crossover with another individual that is quite distant and enjoys a good fitness value. The result would be a major movement of this individual towards the other individual. This may lead to complete absence of individuals from the region where the individual was initially located. The search space would contract leaving out the region. Now it is highly likely that the region possessed global minima. This was not brought to light in the working of the GA as the individuals deputed at that region were at some distant locations and thus had poor fitness values. We saw the way in which the individual got shifted by the crossover operator. This affects the algorithm by large means as the algorithm overlooks the global minima. This is due to the intrinsic nature of the GA to attract individuals towards the places with better fitness values. This is the problem with the GA that kills diversity.

In order to overcome this problem, a diversity preserving GA is used. This algorithm tries to control the algorithm diversity by disallowing individuals that lie very far apart to crossover. In this manner an individual can now only crossover with another individual that lies in its locality. Now if we consider the above scenario, the problem would be different. The individual near the global minima would crossover with nearby individuals to generate individuals in the same region. This would be complemented by the mutation operation. Now the individuals have a better chance of exploring the global minima that got overlooked in the former methodology. For crossover to occur between two individuals, they must be separated by a certain threshold distance η. This is given by equation (3.7).

$$\| I_1 - I_2 \| \geq \eta \tag{3.7}$$

This localization of the crossover operator has a special relevance. We see the way in which the various societies exist. An individual of a society normally mates with individuals of same society. This results in the overall specialization of the same society with the growth of better individuals possessing specific traits. In this way every society excels. The migration of individuals from one society to other society takes place where the better of the individuals get transferred. This results in the exchange of characteristics among the society at the global level. In this manner there are global interactions to attain optimality. Here the various societies may be conceived as local minima and the development of societies as convergence to local minima. The global interactions aid in the finding of global minima.

3.11 Chapter Summary

This chapter was the first step towards evolutionary algorithms. The chapter gave an exploration over Simple Genetic Algorithm. The algorithm was chiefly presented as a functional optimizing agent in lieu of the discussions and usage that would take place in the rest of the book. The algorithm involved a variety of steps and usage scenarios. The first major task associated is the individual representation. Here we discussed the simple bit string representation where the complete individual is coded as a sequence of bits, double vector representation where the individual is a sequence of real numbers and tree based representation. The entire algorithm is iterative in nature where the solutions keep improving along with time and generations. A number of operators are used for the generation of higher generation population from the lower generation. We discussed each of these in the chapter. The chapter discussed rank based, top and other scaling measures. Further selection operators were discussed that carry the task of selection of the better individuals for the genetic processes. Here stochastic uniform selection, tournament selection and other schemes were discussed. Crossover was used for the mixing of two parents and generation of the individual. We saw how this operator could be used for different types of representations. One-point, two-point and scattered crossover operators were illustrated. This was followed by mutation which involved modification of the some values of the individual to inject new characters. The other operator of discussion included elite, repair, insert and delete. Fitness function was used for measuring the goodness of a solution or individual. The other topic of discussion was diversity that denotes the average distance between the individuals of the population and stopping criterion that denotes the condition when the evolutionary process needs to be stopped.

References

[1] Bäck, T.: Evolutionary Algorithms in Theory and Practice: Evolution Strategies, Evolutionary Programming, Genetic Algorithms. Oxford (1996)

[2] Bäck, T., Hoffmeister, F.: Extended selection mechanisms in genetic algorithms. In: Belew, R.K., Booker, L.B. (eds.) Proc. of the Fourth Intl. Conf. Genetic Algorithms. Morgan Kaufmann, San Francisco (1991)

[3] Baker, J.E.: Adaptive selection methods for genetic algorithms. In: Grefenstette, J.J. (ed.) Proc. of the First Intl. Conf. on Genetic Algorithms and Their Appl. Erlbaum, Mahwah (1985)

[4] Chakraborty, U.K., Dastidar, D.G.: Using reliability analysis to estimate the number of generations to convergence in genetic algorithm. Inf. Process. Lett. 46, 199–209 (1993)

[5] Chakraborty, U.K., Deb, K., Chakraborty, M.: Analysis of selection algorithms: A Markov chain approach. Evol. Comput. 4(2), 133–167 (1996)

[6] Davis, L.: Handbook of Genetic Algorithms. Van Nostrand Reinhold, New York (1987)

[7] Deb, K., Goldberg, D.E.: An investigation of niche and species formation in genetic function optimization. In: Schaffer, J.D. (ed.) Proc. of the Third Intl. Conf. on Genetic Algorithms. Morgan Kaufmann, San Francisco (1989)

[8] Fogel, D.B.: Evolutionary Computation: Toward a New Philosophy of Machine Intelligence. IEEE Press, Los Alamitos (1995)

[9] Forrest, S., Mitchell, M.: What makes a problem hard for a genetic algorithm? Some anomalous results and their explanation. Mach. Learn. 13, 285–319 (1993)

[10] Goldberg, D.E.: Genetic Algorithms in Search, Optimization, and Machine Learning. Addison-Wesley, Reading (1989)

[11] Goldberg, D.E.: Sizing populations for serial and parallel genetic algorithms. In: Schaffer, J.D. (ed.) Proc. of the Third Intl. Conf. on Genetic Algorithms. Morgan Kaufmann, San Francisco (1989)

[12] Goldberg, D.E., Deb, K.: A comparative analysis of selection schemes used in genetic algorithms. In: Rawlins, G. (ed.) Foundations of Genetic Algorithms. Morgan Kaufmann, San Francisco (1991)

[13] Holland, J.H.: Adaptation in Natural and Artificial Systems. University of Michigan Press, Ann Arbor (1975)

[14] Koza, J.R.: Genetic Programming: On the Programming of Computers by Means of Natural Selection. MIT Press, Cambridge (1992)
[15] Lobo, F.G., Lima, C.F., Michalewicz, Z.: Parameter Setting in Evolutionary Algorithms. Springer, Heidelberg (2007)
[16] Michalewicz, Z.: Genetic Algorithms + Data Structures = Evolution Programs. Springer, Berlin (1992)
[17] Mitchell, M.: An Introduction to Genetic Algorithms. MIT Press, Cambridge (1999)
[18] Schwefel, H.P.: Evolution and Optimum Seeking. Wiley, Chichester (1995)
[19] Whitley, L.D.: The Genitor algorithm and selection pressure: Why rank–based allocation of reproductive trials is best. In: Schaffer, J.D. (ed.) Proc. of the Third Intl. Conf. on Genetic Algorithms. Morgan Kaufmann, San Francisco (1989)

List of Figures

Fig. 3.1. Search Space of Genetic Algorithm
Fig. 3.2. Simple Genetic Algorithm
Fig. 3.3(a). Bit String Representation of individual
Fig. 3.3(b). Double Vector Representation of individual
Fig. 3.3(c). Tree Representation of Individual
Fig. 3.4(a). One Point Crossover
Fig. 3.4(b). Two Point Crossover
Fig. 3.4(c). Scattered Crossover
Fig. 3.5(a). Mutation with a bit string representation
Fig. 3.5(b). Mutation with a numeral vector representation
Fig. 3.6. Gaussian Mutation

Chapter 4
Fuzzy Logic

Abstract. Logic forms a fundamental concept behind the manner in which a variety of works are performed. One tries to process the inputs based on reasoning or logic that helps in the generation of the outputs. In this chapter we first study fuzzy logic that is driven by fuzzy sets and fuzzy rules. Then we present the Fuzzy Inference Systems that use Fuzzy Logic for the generation of the outputs from the inputs. The logic base of these systems gives a clear understanding of the manner in which the system is operating. The chapter would present the various concepts of problem solving using Fuzzy Logic covering membership functions, fuzzy arithmetic and operators and finally the fuzzy inference systems. We also present the analogy of these systems with the Artificial Neural Networks and the design methodology in fuzzy systems. At the end the chapter focuses on the problems and limitations of the fuzzy systems that motivate their hybridization in form of hybrid computing systems that give a better performance.

We are all well used to logic and rules that drive our daily lives. We frequently come across rules like '*If weather is cold, illness is likely*' or '*If rain is high, temperature is moderate.*' Many of our daily actions are based on these forms of rules that are inbuilt into us or learnt along with time. In this manner logic is the driver of decision making in everyday life. One of the aspects of this form of decision making is that they are very discrete or crisp in nature. Anything here can either be true or false. Consider the statement '*If weather is cold.*' One would either expect a true or false as an answer. These types of sets are known as crisp sets. An entity either completely belongs to a set or does not at all belong to the set. Rules formed on these crisp sets further have the same annotation. Since they are based upon fuzzy sets, they would either be completely true or they would be completely false. In case the rules are true, the corresponding action may be taken. However, in case the rules are false, no action is taken. These rules determine the behavior of the systems and are referred as the production rules. Now whenever the inputs are applied, the system simply does the task of evaluating the conditions over which the decision needs to be made. In case a condition is true, the following action is taken.

The work over logic driven systems and production rules started long back with Aristotle who used similar logical systems for decision making. The production rules govern the manner in which the system works or behaves. The innovation lies in the formulation of the correct rules to control the system behavior. As the

A. Shukla et al.: Towards Hybrid and Adaptive Computing, SCI 307, pp. 83–108.
springerlink.com

rules are simple to read and comprehend, the entire system is easy to understand and develop. This forms the basis of the logic based system of mapping the inputs to the outputs. This is unlike the systems we studied so far where the system itself formulated any rule that mapped the input to the output. We were given the entire system as a whole without any understanding of the rules.

A typical production system consists of production rules for its working. Let there be *n* rules in the system. Here any rule R_i may be represented by (4.1).

$$R_i\text{: if } a_1 \wedge a_2 \wedge \ldots .a_n \rightarrow c_1 \wedge c_2 \wedge \ldots .c_n \quad (4.1)$$

Here a_1, a_2, ...a_n are the antecedents and c_1, c_2, ...c_n are the consequents. The ^ is the '*and*' operation. We may easily replace it with the '*or*' (V) operation as well. The '*and*' operation has different meanings in the antecedent side and the consequent side. In the antecedent side it serves as a binary operator whose truth value is given by its truth table. In consequent side it acts like a separator that separates the various actions or the consequents specified.

The prominent features of this production system include the truthiness or falseness of the various antecedents that govern the truthiness of the entire condition. If the condition is found to be true, the rule is fired. If the same is found to be false, the rule is not fired. The firing of rules denotes the truthiness of all the consequents. In this manner the final decision is made and the outputs are computed by the system based on the supplied inputs. These systems can solve a large number of problems. But the intrinsic factor is always that they run on crisp set theory where something needs to completely belong to a set and vice versa.

This forms the basis of the fuzzy sets that we discuss in the next section. The first few sections of the chapter would focus upon the migration from the production rule systems to the systems based on fuzzy logic or the fuzzy inference systems (Kasabov 1995, 1998; Kandel 1991; Yager and Zadeh 1992; Dubois and Prade 1980, 1985).

4.1 Fuzzy Sets

In fuzzy set, it is not mandatory for any entity to completely belong to the set or not. It can always belong to the set by some degree and not belong to it by another degree. Hence the elements of the fuzzy set partially belong to the set (Zadeh 1965, 1968, 1975). The belongingness of an element to the set is known as its membership value and it lies in the range of *0* to *1*. A membership value of *0* means that the element is completely out of the set and does not belong to the set. Similarly a membership value of 1 denotes that the element is completely inside the set and belongs to it to full degree. All other cases are in-between. Suppose the value of membership was 0.75. This denotes a high belongingness of the element to the set and a low non-belongingness to the set.

In the same example consider the set to be '*It weather is cold.*' Now a traditional system would leave it to the human expert to decide whether it is cold or not. In case the temperature is below some threshold level, the expert may say that the statement is true. In case the temperature is above the threshold level, the person may claim the statement to be false. This was the case with crisp numbers. Now consider the fuzzy equivalent of the same statement. Consider the *cold* is measured in terms of temperature. Now if the temperature is 10 degrees, we may say that the level of cold is very high, or that the statement is true to the extent of *0.75*. However if the temperature is 35 degrees, we may say the statement to be true to the extent of *0.25*. In this manner we are more flexible in assessing the degree to which a statement is true or false in fuzzy sets.

In fuzzy sets an element belongs to a set to the extent of its membership value. Let the fuzzy set be denoted by Z. On the lines of the set theory for crisp sets, we define the fuzzy set by (4.2).

$$Z=\{z_i/\mu(z_i)\} \tag{4.2}$$

This defines Z to be a collection of elements z_i such that each z_i has a membership degree of $\mu(z_i)$. Using the above discussions, we may write the membership function *cold* as (4.3)

$$\text{cold}=\{\ldots,\ 10/0.75,\ldots\ 35/0.25\ldots\} \tag{4.3}$$

The function $\mu(z_i)$ is known as the membership function (MF). The membership function takes the value of an element of a set and returns its membership value. In other words it does the task of mapping of the attribute values to the membership values of any set. For most of the problems we define a membership function as an arithmetic equation taking as its input the attribute value. It computes the corresponding membership value. The membership function determines the manner in which the attribute behaves in the belongingness to the set. The MF returns 1 as the membership value for most appropriate set of attribute values. The membership value may be *0* for the highly unlikely value. In the above example one may expect that cold reaches a membership value of 1 around temperatures of 0 degrees. For the temperatures around 50 degrees the value of membership may be zero. The values of membership for the various degrees of temperature may be intermediate for all intermediate cases. We hence plot the membership value of *cold* for various values of the temperature. This is known as the membership plot. This is given in figure 4.1.

Now consider the input *weather* has three MFs associated with it. These are *cold*, *moderate* and *hot*. All three MFs may be plotted in the same membership plot showing the behavior of all three MFs simultaneously. This is shown in figure 4.2. It may be seen that all these three MFs have some key area that they focus upon. The membership value touches its maximum at these regions. The membership value then slowly drifts down in both the sides. These values depict how the association of some attribute towards the membership functions increases or decreases as we change its value. Consider the temperature 15 degrees as an example. Here it belongs to both *cold* and *moderate* with non-zero membership values.

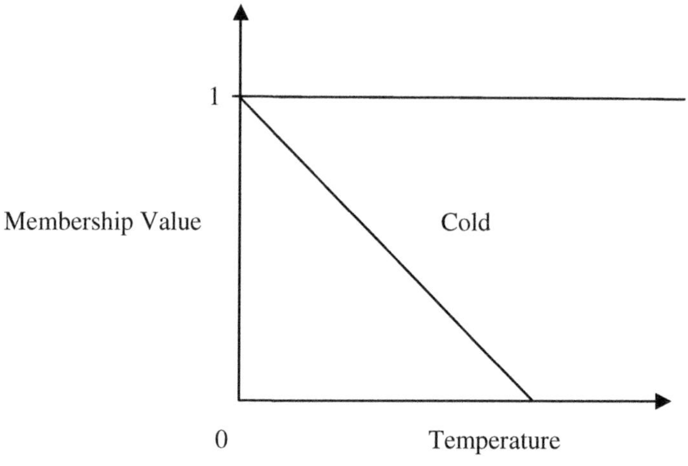

Fig. 4.1 Membership Function

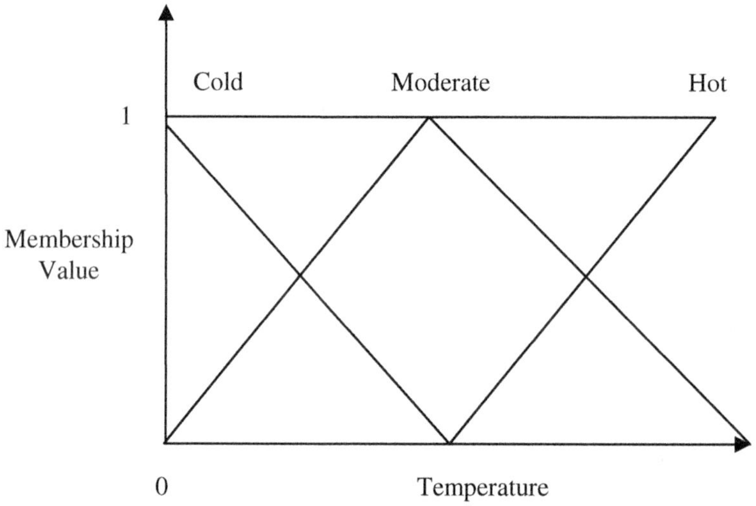

Fig. 4.2 All Membership Function Plots for *Temperature*

As we start decreasing the temperature from its present value, the membership of cold increases and that of moderate decreases.

Various standard MFs are used to represent various daily scenarios. One of the most commonly used MF is the Gaussian membership function that denotes the increase or decrease in the membership value as per the Gauss equation. Various other standard MFs may be used. The various MFs are shown in figure 4.3.

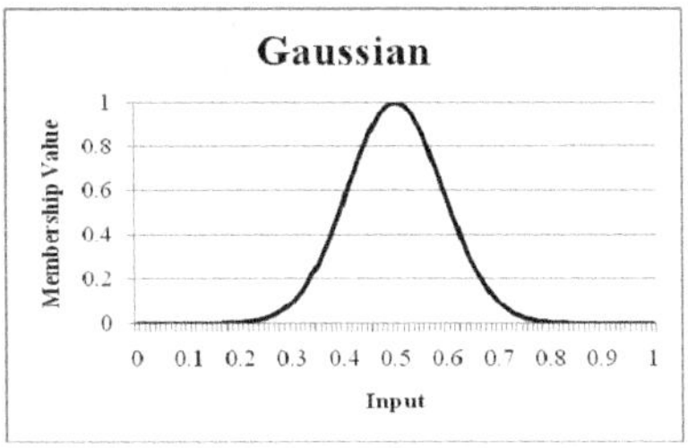

(a) Gaussian Membership Function

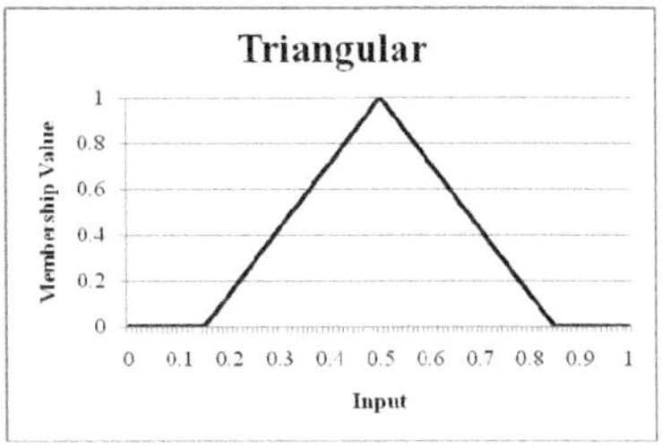

(b) Triangular Membership Function

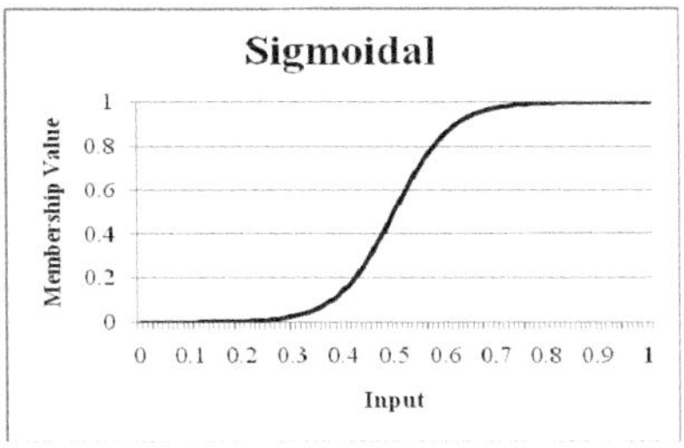

(c) Sigmoidal Membership Function

Fig. 4.3 Different Types of Membership Functions

4.2 Fuzzy Logical Operators

The production systems consist of rules that make extensive use of logical operators for the representation of the condition or the joining of the antecedents. Since the aim is to migrate the production system into fuzzy logic, the first job that we need to consider is the modification of logical operators. Logical operators include the *AND*, *OR*, *NOT* and *Implication*. We consider the basic methodology and lingual meaning of these operators and then design the fuzzy equivalent of these. These operators need to take fuzzy inputs and generate fuzzy outputs (Mamdani 1977; Sugeno 1974, 1985; Tagaki and Sugeno 1985). This means each of the input and output would be indentified by a membership value that denotes the participation of the input in the operator or the belongingness of the input to the membership function for which the rule was framed.

Consider the rule '*if weather is cold and rainfall is low then illness is likely*'. Here temperature (or weather) and rainfall level constitute the system inputs. Suppose that the system is given as input a temperature of T degrees. Now we may easily say that the input T participates in the above rule by a degree of $\mu_{cold}(T)$. Here $\mu_{cold}(T)$ is the membership function of *cold*. This step is known as the fuzzification of the inputs. In this step we compute the membership values of the various inputs to the various membership functions. These values may be used in individual rules as needed. The entire equation may hence be represented in numerical forms after the fuzzification step as (4.4)

$$\mu_{cold}(T) \wedge \mu_{low}(R) \rightarrow \mu_{likely}(I) \quad (4.4)$$

Here the three membership functions are for *cold* (weather), *low* (rainfall) and *likely* (illness). *T*, *R* and *I* denote the inputs for these three variables. $\wedge$ is the '*and*' operator and $\rightarrow$ is the implication operator.

4.2.1 AND Operator

AND is a logical operator that gives a value of *true* if all its inputs are *true* and *false* otherwise. It follows a lingual meaning where a task may only be done when all the prerequisites separated by an 'and' are met. The fuzzy equivalent of the and operator takes two membership values and returns as its output one membership value which denotes the combined participation of the two antecedents in the rule that were separated by the '*and*' operator. Let the two inputs given to the system be *x* and *y* where both lie between *0* and *1*. The output of the system is x AND y that is given by equation (4.5) or equation (4.6). Any of the methods may be used depending upon the problem design.

$$x \text{ AND } y = \min(x,y) \quad (4.5)$$

$$x \text{ AND } y = x.y \quad (4.6)$$

Equation (4.5) performs a minimum operation over the inputs. This method of ANDing the results is known as the *min* method. Equation (4.6) returns the product of the two inputs as the output. This method is known as the *product* method of computing *AND*. Since the inputs lie in the range of *0* to *1*, it may be verified that the outputs also lie in the same valid range. Again, giving various combinations of 0s and 1s gives us the same results in both the methods as given by the crisp or Boolean logic system with *0* representing *false* and *1* representing *true*. Further we may easily note that the equations (4.5) and (4.6) may be easily extended to multi-input system as well with the same operations of *min* and *product*. Consider *AND* operations over any two MFs *X* and *Y*. The corresponding MF resulting after the *AND* operation is shown in figure 4.4.

An important aspect is the realization of these two methods of *AND* operation. Consider that there was a secret treasure that needs to be protected. There are two doors that someone must cross before he can steal the treasure. If P_1 denotes the probability of trespassing the first door and P_2 denotes the probability of trespassing the second door. Now we know from the probability theory that the probability of stealing the treasure would be given by $P_1.P_2$. This is the expression of the *AND* operator where the antecedents may be given by '*if door*$_1$ *is trespassed and door*$_2$ *is trespassed then theft is likely*.' The membership function is naturally analogous to the probability measure. In the same example consider that the ability determined the probability. Now if a person can trespass the more difficult door, he can naturally trespass the other door. This is because the other door is easier to trespass and he possesses a greater ability of trespassing. A person hence needs to possess ability to trespass the harder of the doors which may be given by the minimal probability measure of trespassing. This is analogous to the *min* method.

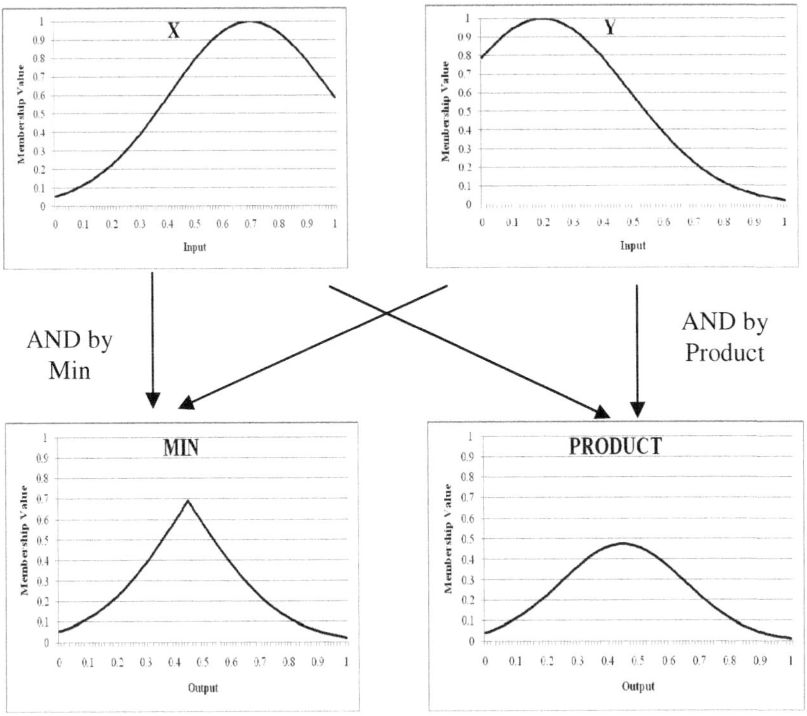

Fig. 4.4 AND Fuzzy Operator

4.2.2 *OR Operator*

OR is another binary operator that takes two inputs and returns one output. A Boolean *OR* operation returns a *true* if any of its input is *true*. In all other cases it returns a *false*. This is also analogous to the lingual meaning of the word '*or*' where any one of the conditions needs to be true for the entire condition to be true and the corresponding action to be performed. Let the two inputs given to the OR operator be x and y. This operator returns the resultant membership value given by x OR y by the two methods given in equation (4.7) and (4.8).

$$\text{x OR y} = \max(x,y) \tag{4.7}$$

$$\text{x OR y} = x+y-x.y \tag{4.8}$$

Equation (4.7) takes the maximum of the presented inputs and is called as the *max* method for computation of *OR*. Equation (4.8) is known as the *probabilistic OR* method for the computation of *OR*. Both result in generation of membership values in the valid range of *0* and *1*. Further both obey the Boolean *OR* truth table with *0* representing *false* and *1* representing *true*. They may be suitably extended to multi input systems as well. We plot the MF *X*, *Y* and *X OR Y* in figure 4.5.

For the realization of these two operators we use the same scenario as discussed in the *AND* operation. Consider that the two doors are placed parallel to each other. Now we need to only trespass the door which can be easily trespassed. This is the one that has a greater probability of trespassing. This means that we chose a maximum of P_1 and P_2 which denotes the probability of trespassing. The rule may now be framed as '*if door$_1$ is trespassed or door$_2$ is trespassed then theft is likely.*' The other method is the *probabilistic OR*. We know that the probability that an event happens out of two events possible with probabilities x and y respectively is $x+y-x.y$. This equation may be used as it is in our case with the two events denoting the trespassing of the two doors. This explains the *probabilistic or* method.

4.2.3 NOT Operator

NOT is a unary operator that takes only one input and generates one output. In Boolean terms the operator returns the converse of the input a *true* is returned as *false* and a *false* is returned as *true*. This is the case with its lingual meaning as well where the *NOT* reverses the usual meaning of any clause. Suppose that the input given is *x*. The application of this operator results in the output as *NOT X* given by equation (4.9).

$$\text{NOT x} = 1 - \text{x} \tag{4.9}$$

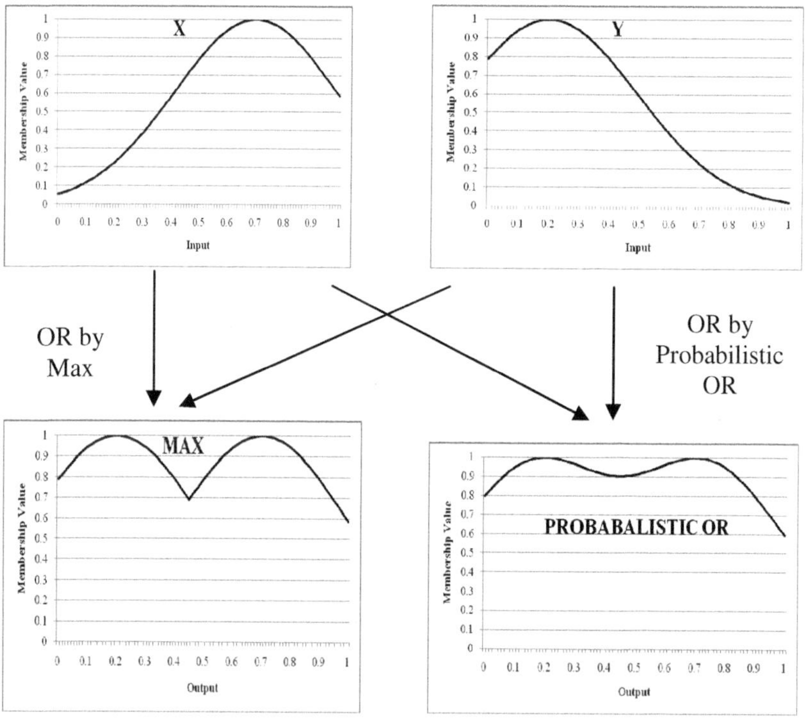

Fig. 4.5 OR Fuzzy Operator

Since *x* lies in the range of *0* to *1*, *NOT x* also lies in the range of *0* to *1*. It behaves in the expected manner for Boolean inputs of *0* and *1*. The plots of x and NOT x are given in figure 4.6.

The operator has a very simple way of understanding if we consider the probability theory. Say the probability of happening of any event is *P*. The probability of the event not happening is given by *1-P*. This is similar in nature to the *NOT* operation of equation (4.9).

4.2.4 Implication Operator

The operators discussed so far enable us to solve the antecedents of any general rule using fuzzy arithmetic. This is because any simple or complex antecedent would employ use of multiple *AND*, *OR* and *NOT* operations. The next task is the computation of the implication operator that maps the antecedents to the consequents. While representing the general rule in (4.1) we stated that the use of *AND* in antecedents and consequents are two different aspects. The *AND* of the antecedents denotes the *AND* logical operator. On the other hand the *AND* in the consequents denotes a separator that separates the various consequents. The implication operation is applied independently for all the consequents that appear in a rule. All these consequents have the same antecedent.

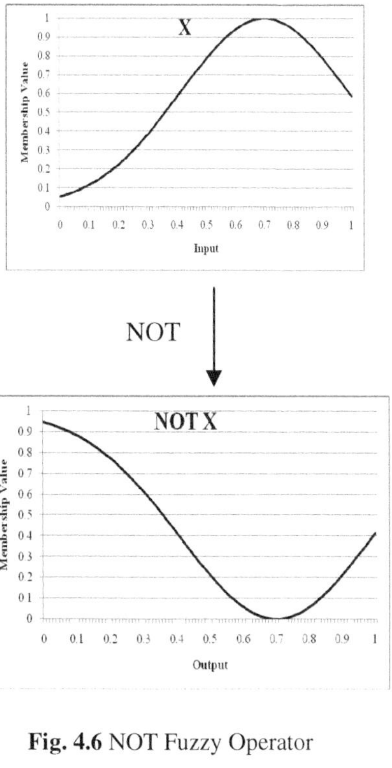

Fig. 4.6 NOT Fuzzy Operator

The implication operator also takes two inputs. The first input is a fuzzy number that represents the computed antecedent. The second input is the entire fuzzy membership function consisting of various attribute values along with their membership degrees. The operator applies the implication of the first input to the entire MF given in the consequent. Let us suppose that x was the antecedent or the first input to the operator. Let the second input be y that is an element of the entire fuzzy MF Y. The output of the operator for every y would be given by equation (4.10) or (4.11). This is similar to the *AND* operator.

$$x \rightarrow y = \min(x,y) \tag{4.10}$$

$$x \rightarrow y = x.y \tag{4.11}$$

Equation (4.10) is the *min* operation that takes the minimum of the input values. This operator slices the membership function that lies above the value of x. This is hence also known as the slicing operator. This effect of the operator is shown in figure 4.7 that denotes the input x and membership function Y with the min operation. The other method represented in equation (4.11) is the *product* operation that takes the product of the input with the MF. This results in a compression of the shape of the MF along the membership axis. This effect of the input x over the MF Y is given by figure 4.7.

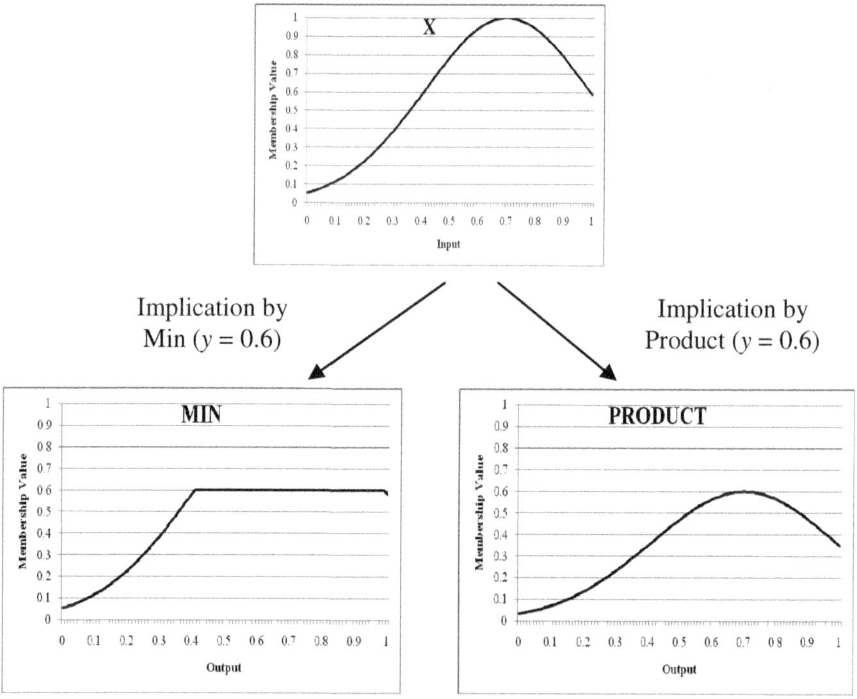

Fig. 4.7 Implication Fuzzy Operator

4.3 Aggregation

The other operator that aids the migration of a conventional logic based system into a fuzzy logic based system is aggregation. The system may have multiple rules and multiple outputs. Individual rules may be conceived as single decision affecting agents. The complete decision of the system regarding the output of each and every output attribute may be done only after consulting all the rules of the system. There must hence be means to calculate the cumulative effect of the various rules available in the system for every output variable. This task is performed by the aggregation operator.

The output of each rule is a MF that denotes the probability of the various possible outputs within the range along with their membership values. The output corresponding to higher membership value is more likely to be the output of the rule in crisp terms than the output with less membership value. The completely irrelevant outputs within the working range have a membership value of zero. The work of the aggregation is to aggregate the various membership functions, each representing the outputs along with the associated membership values, and form a consolidated membership function of the entire rule base. This membership function also gives the output values along with the membership values. This may be regarded as the final output of the systems in fuzzy terms.

We know that the various rules try to affect the various output variables of the system. Consider only one of such output variable, since these are independently handled. The various rules may denote the various circumstances in which there is a characteristic behavior of the system that produces a characteristic output of the variable being considered. The problem has different rules were fired to various degrees, depending upon the membership value of the antecedents. Each rule influences the decision in favor of the values it computed with the associated membership value. The aggregation does the summarization of the various rules and produces a final consolidated membership function that is in agreement to the recordings or the outputs of all the individual membership functions.

The aggregation operator takes two inputs each of them is a membership function and returns a membership function. This may be suitably extended to the multi-input, single output systems as well. The aggregation of two inputs x and y, that belong to the MFs X and Y is given by z that belongs to the MF Z as per equations (4.12) and (4.13).

$$z = x + y \tag{4.12}$$

$$z = \max(x, y) \tag{4.13}$$

Equation (4.12) is known as the *sum* method of aggregation and equation (4.13) is known as the *max* method of aggregation. Both these operations do the required task in a very logical manner. It may be noted that in the *sum* method, all values greater than *1* are trimmed to *1*. This is because the range of the MF needs to be in-between *0* and *1*. The graphs of the MFs using both these operations are given by figure 4.8.

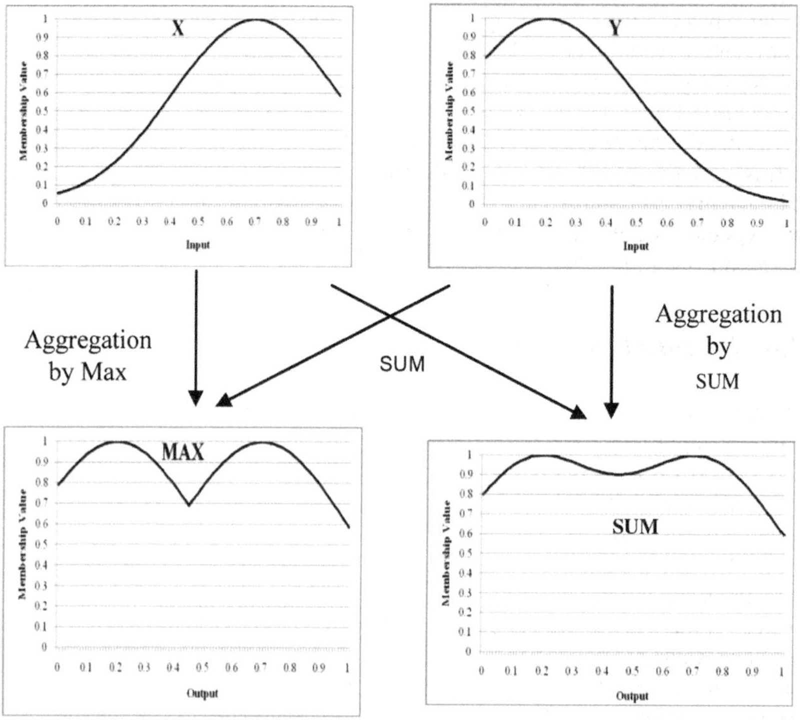

Fig. 4.8 Aggregation Fuzzy Operator

The manner of using these operators has a lot of practical relevance. The sum may be visualized as a mechanism of voting in between the different rules to affect the output. The sum of all the responses of the rules is taken. The largest magnitude is fixed to be 1. The maximum is a similar behavior where we assign the most promising membership value to an output. This is like consulting only the best rule for decision making which would be influential due to its large impact.

4.4 Defuzzification

The last operator that we study is defuzzification. The output so far is a membership function with different output values, each associated with a membership value. This is the fuzzy output of the system. But the requirement with the system would be a crisp number that the problem being modeled demanded. We must hence have a method of defuzzification of the output membership function to generate a crisp output which is a single number. This method is called as the defuzzification operation. This operation considers all the output values along with the associated membership costs and returns a single output number that is the crisp output of the system.

Defuzzification takes as input an entire fuzzy MF and produces a crisp value as the output. The various outputs have an associated membership degree that denotes the probability of the output being declared as the final system output. In this manner we have a series of output, each claiming to the system output by some membership degree. The task is to consider all these values and compute the final system output of the output variable considered.

There are numerous means by which the defuzzification operation might be carried out. One of the most common of these approaches is the centroid method. This method takes the weighted mean of all the outputs, weighted by their membership values. The basic motive behind this approach is that the various membership functions already denote the probability of them being the system output or the crisp output with the outputs having higher membership values being more likely. These membership functions may hence be treated as weights in making the final decision regarding the system output. Suppose that the MF is given by μ(x). Now the defuzzified value may be given by equation (4.14).

$$Centroid = \frac{\int_x x\mu(x)dx}{\int_x \mu(x)dx} \tag{4.14}$$

The plot for a MF and the computed deffuzified output using centroid method is shown in figure 4.9. The other methods of defuzzification include bisector, mean of maximum (MOM), smallest of maximum (SOM), etc.

4.5 Fuzzy Inference Systems

So far we have studied the migration of the rule based production systems into fuzzy driven systems or the fuzzy inference systems. In this section we look at the design of these systems from a greater level. Fuzzy Inference Systems (FIS) are systems that map the inputs to the outputs using fuzzy based rules (Furuhashi *et al* 1993; Mizumoto and Zimmermann 1982; Plonka and Mrozek 1995). These systems are extensively used for the making of the logic systems due to their ease in modeling and accuracy of performance. These systems are driven by rules and are thus referred as the logic based systems.

Logic or rule being the driver is intrinsic factor in many of the systems. For these systems some idea is known how the various types of inputs produce an output. We may hence write simple English rules to describe the system. This description however needs to be completely and well-known by the designer. Now the system is capable of using these English based rules to produce outputs whenever it is given any input. Here the FIS takes advantage by making an easy conversion of these rules to a set of working segments of systems for computing the output. The simplicity in their use and design makes it easy to design systems, as if it was just describing them.

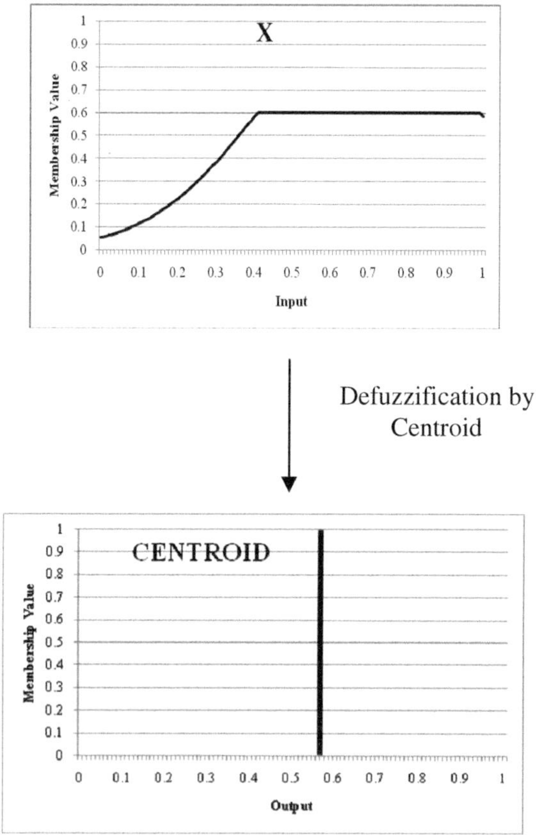

Fig. 4.9 Defuzzification Operator

FIS has many components that must be designed before the entire system can be made to execute. This includes the determination of the inputs and outputs, specifying the membership functions to all the inputs and outputs in the system, formulating the rules, determining the fuzzy operators and testing the system. The formulation of the inputs and outputs is a conventional task of system design that enables the formulation and selection of the inputs and outputs such that there is an easy mapping that relates them. Most of the fuzzy operators we have discussed in the previous sections and any of these may be selected as per the problem and design requirements. We discuss the other concepts one by one. The entire methodology of design of FIS may be given by figure 4.10.

4.5.1 Number of Membership Functions

The foremost job in the design of the fuzzy systems is to specify the number and type of fuzzy MF for each of the inputs and the outputs. This denotes the division

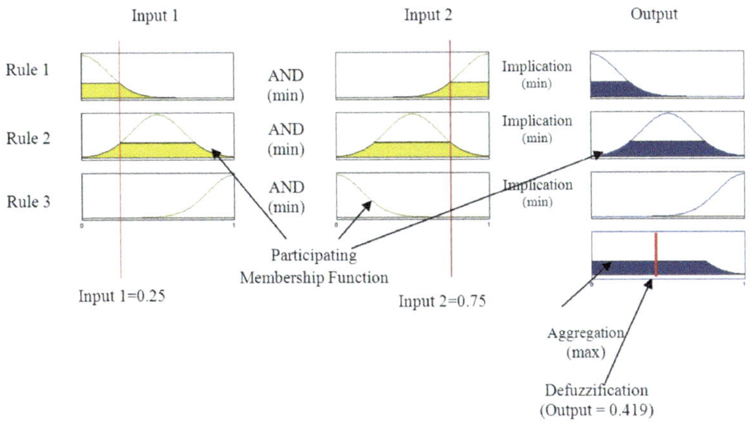

Fig. 4.10 Fuzzy Inference System

of the input in various ranges of inputs. We naturally have a fuzzy meaning of the division here. This means the entire range of value of the inputs is divided into fuzzy sets where the membership of each of the input or output attributes to any set can be in-between 0 to 1. The number of MFs denotes the number of regions in which the input range may be distributed. The number of MFs may be decided as per the problem logic or the rule base. While designing the system, however this decision would always be a difficult one to make. We hence discuss the various aspects of it and considerations that must be kept in mind while deciding the number.

A large number of membership functions contribute a lot towards the making of the problem localized in nature and a drop of the overall generality of the solution. Readers may recall our discussions on these topics from chapter 1. A large number of MFs mean that we divide the input space into more number of segments. Each of these segments has a fuzzy nature. Now there are a large number of segments that will require an equally large number of rules for the mapping of the inputs to the outputs. While the formulation of rules, we would have to consider all kinds of combinations of the various MFs of all the inputs and outputs. This would make the task tedious requiring a more complex mapping. The positive aspects of the system would be a greater control in which the inputs behave. A very large number of MFs can naturally be thought to be analogous of making a lookup table to solve the problem. This however is much difficult to understand and comprehend. Most of the problems however may be simple in nature that can be effectively mapped by using simple MFs which are less in number. The generalization power or the ability to give correct outputs to the unknown inputs is much larger in these systems.

The number of MFs is one of the major factors that drive the generalizing power of the FIS. Many times we are unable to design an effective system that can solve the problem using the set criterions. In such a case, addition of MFs can be of help which gives a greater flexibility in output control as per demands. Again recollect that the problem itself determines the number of MFs needed that give the system enough complexity to give the correct output. The task of designer is to determine this number, below which the performance is not good.

4.5.2 Type of Membership Functions

Membership Functions may be of numerous types as discussed in section 4.1. Many of them are conventional and are being extensively used for problem solving as a standard. One of such standard MFs under use is Gaussian. The others include triangular, bell shaped, etc. FIS also gives the flexibility to design custom MFs that behave as per the human requirements. The type of MF associated with any of the input or output also largely depends upon the problem being considered. We discuss this notion in two separate heads of type of MF and the extent of its domination in input or output range.

The shape of the MF denotes the manner in which the input and output are varying in their behavior in the system. Many systems are such that small changes in the input values make a deep impact on the output at some particular input value. At some other stages the impact may not be that high. This is a common scenario for the use of Gaussian MFs or other non-linear MFs. Smooth changes in outputs may be modeled by triangular MFs. The slope of this determines the change in domination of the MF. This holds equally true for the MF of the output variable as well. Here the MF may be Gaussian in nature if some region is sensitive to inputs and reports sharp changes in output value across these regions. The changes in the shape of the MF is usually not much experimented or changed during the process of experimentation. We usually stick to some shape and try to change the other factors for better output mapping as per requirements.

The other part of the discussion is the length of the ranges in which each MF is dominant. We discussed that the number of MFs are like dividing the entire input or output range into segments. Each segment has some length which cannot be measured in a crisp sense, as the segments are fuzzy. The most trivial manner of dealing with the length of segments is to make them equal in length. However this might not be true for all types of problems. For many of the problems some segments may be very large and others may be very small. This specially is a characteristic of the problem and the manner in which the inputs and outputs behave. If the output nearly remains constant for a large part of the input, the MF may have a long range with a membership value of 1. If the output changes rapidly, we may wish to insert a large number of MFs in the region to better control the sensitive region. This would mean smaller MFs in the region.

Consider the problem of telling a robot the direction in which it has to turn to get to the goal while traveling towards it (Kala *et al* 2010). This is the steering mechanism of the robot. A typical view of the MF for the input of current angle may be as given in figure 4.11. This is formed with the concept that anything more than 90 degrees of angle would mean the same thing, since the robot is expected to turn to the largest extent possible. This accounts for the uneven distribution in terms of shape and length of the MFs in the entire input range.

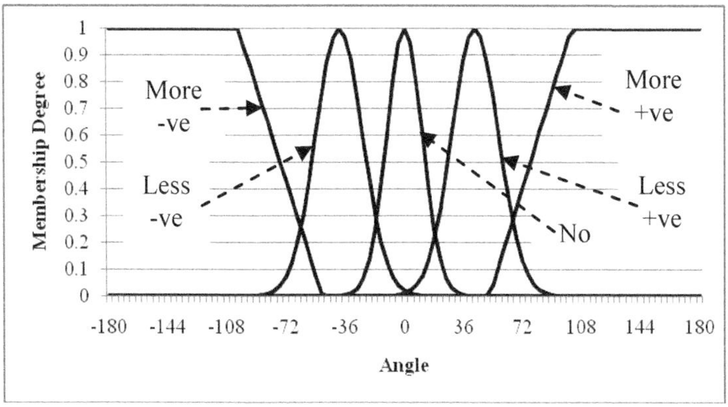

Fig. 4.11 Uneven Membership Function Distribution

4.5.3 Rules

The core part of the FIS is the rules that drive the system. The rules denote the logic that maps the inputs to the outputs. These are in the form of lingual sentences that can be easily framed as fuzzy rules for the functioning of the systems, or for the proper mapping of the inputs to the outputs. The rules are the description of the manner in which the system works. They are like explanations fed in the system that guides its working. We saw how these rules are computed, interpreted and executed using the concepts of fuzzy logic. The only way to manually frame these rules is to understand the system and to comprehend its working. We must understand how the different inputs work in affecting the outputs. The mapping of the various combinations of inputs and outputs can be established by the problem logic known prior.

The number of rules also affects the generalizing ability of the FIS. The larger number of rules makes the problem more localized in nature, similar in the manner as was the case with the number of membership functions. A complex system requires too large number of rules to enable it carry out an effective mapping of the inputs to the outputs. This makes the problem lose generality because of it's over complex nature. Most of the common problems can be easily solved by a small set of rules that is a result of effective input and output selection or effective problem modeling that can be solved using FIS.

4.6 Comparisons between Fuzzy Systems and Neural Nets

We studied the Artificial Neural Networks (ANN) in the previous chapter. In this chapter we presented the Fuzzy Inference Systems (FIS). Both these systems are used to solve the problems where some inputs are given and the determination of the output needs to be carried out. Both the systems were very different in the means by which they modeled the problem and solved it. In this section we compare and contrast few of the major points of these systems.

4.6.1 Differences

The FIS approach to problem solving made use of a logic base where logic was the driver of the complete system. This however was not the case with ANN where the systems were run by means of machine learning or finding out the trends in the historical data and generalizing them to the new data. This further made these systems behave as black boxes where the logic that resulted in the production of outputs was unknown.

The FIS running on the basis of rules made it a very rigid architecture where it could only perform as the rules drove it. The validity of the rules may change with time. Many times some data may not follow a rule. The fuzzy system is very rigid to these scenarios and hence can be effectively used only in cases where the logic base is strong and well-known and verified. The ANN on the other hand is very flexible and can assume any kind of shape from basic to complex. This system is completely driven by the weights that are stored in the connections between the neurons. As the system gets trained, in an online or offline manner, the various weights are changed and the entire system changes its shape as a result. The flexibility enables these systems to attain any shape as needed by the problem without the boundation of underlying rules.

FIS does not have any in-built mechanism to train itself or optimize its parameters for optimal working. This is done in the ANN by the training algorithm that is responsible for the setting of the various weights of the connections in between neurons. These enables the ANN extract out the trends and patterns from the historical database and use it for the functioning of the system.

4.6.2 Similarities

Both ANN and FIS are tools of problem solving with some inputs and outputs. Both of these systems make use of a knowledge base that contains some knowledge for carrying the computation of the output from the inputs. In FIS this knowledge base is in the form of rules. In case of the ANN this knowledge base is in form of inter-neuron connections. These in their own way represent rules in analogy with the fuzzy systems.

The rules in the fuzzy systems are quite analogous to the neurons in the ANN. We saw in chapter 2 that the addition of neurons causes the system to be more flexible to large number of complex output surfaces in the input space. This made the system lose the properties of generalization and made the system much localized in its operation. The similar discussion was carried out in case of number of rules for the FIS as well. The larger number of rules made it possible for the system to specify the output behavior in all the contexts of inputs. This made the problem specialized towards the class of inputs and localized in its operation. In this regard the number of rules in FIS largely plays the role of the number of rules in FIS.

Similar comments can be made between the number of layers in the ANN and the number of MFs in the FIS. The addition of even a single layer in the ANN Multi-Layer Perceptron model marked a great rise in the complexity of the problem which enabled complex output surfaces. This added a lot of connections from the preceding and the next layers which were responsible for the complexity hike and ultimately loss of generality. The similar is the case with the number of MFs in the outputs and inputs in the FIS. The addition of a single MF marks a bulk of hike in the possibilities of rules. The number of rules that can be formed with the addition of a MF in any of the inputs or outputs is quite large. This greatly increases the system complexity. From the discussions of the previous paragraph, we can hence say that the number of MFs is analogous to the number of hidden layers.

4.6.3 Comparisons with RBFNs

Radial Basis Function Networks (RBFNs) were discussed in chapter 2. These are three layer ANNs with an input layer, hidden layer and output layer. An interesting correlation of these networks may be seen when compared with the FIS, besides the already discussed points above.

The various neurons in the hidden layers were seen as the various sub-systems, each of which are deputed at some point in the input space and carry forward the task of making its predictions of the outputs. Various such neurons computed their outputs. The final result was the weighted sum of all the computations by these neurons that made the final system output. The addition of nodes in this system means the deployment of more sub-systems at various parts of the input space that made the problem more localized. The association of any input to any neuron is measured with the help of the activation function which again depends upon the norm of the input to the neuron. Further, the spread or the influential area of these neurons was proposed to be as high as possible to enable different systems compute and mix their outputs. This greatly contributed to the overall system generalizing ability.

FIS may also be visualized in similar terms with the rules being analogous to the neurons in the hidden layer of the FIS. Each rule determines the behavior of some input combination. This may be easily seen as the rule catering to the need of some of the inputs lying at some regions of the input space. In case of the RBFN, norm of the input to the neuron was the driver for the activation of the neuron. Similar phenomenon happens with these systems with the rules being more active for the inputs that they were designed for, or the inputs which show a membership value of 1. As we go away from this set of inputs, the rules become less effective with the decrease in effect decided by the increase in distance or norm. The activation function of the neuron in this case is similar to the MF of the FIS. Another implementation similarity in these two systems is the use of Gauss as the activation function or the membership function.

The various neurons are activated by the input to various degrees. Mostly only a small set of the neurons that lie in the input vicinity are activated to a remarkably good extent. Similar is the behavior of the FIS. Here also a small set of rules that are most appropriate for the input are activated out of all the available rules. The RBFN uses a weighted sum, whereas the FIS uses the aggregation and the defuzzification operators which may be argued to be analogous on the same lines.

4.6.4 Remarks

In this section we discussed the similarities and comparisons between the ANNs and the FIS. This does motivate the creation of a unified model, especially after noting the comparisons with RBFN. It may be seen that all the models do perform similar when they reduce to a simple lookup table which simply stores all the inputs and the associated outputs. This has no generalizing ability and a perfect learning. In ANN with MLP and BPA this would happen for excessively large neurons with each storing an input element. In other neural models it would happen with each neuron representing some input. In FIS the same thing would happen with each rule depicting an input and output.

The motivation here is not the hybridization of these systems based on their similarities and differences in the manner of working. We rather hybridize them considering their advantages and disadvantages and hoping the design of a system that maximizes the advantages and diminishes the disadvantages.

4.7 Limitations of the Fuzzy Systems

While we have discussed most of the associated problems with the FIS in the subsequent sections, we formally write them down in this section. This would be a motivation behind the hybridization of these systems with the other soft computing systems that is the purpose of the book.

One of the foremost problems of these systems is that they are unable to learn. Suppose that the problem we have has a bulk of instances already available from the historical database. In such a context it would be good to have a system that can adapt itself to this dataset. In its native form the human designer is supposed to do the same. The basic approach is to build a system using the available information and test it against the available data sets. The anomalies are noted and the design is modified using the discussed principles. This calls for a lot of work over and over again by the designer to adapt the system to give a decent performance in the scenario given.

The other problem of these systems is of a fixed architecture. The number and type of MFs, their parameters, rules, etc. have to be specified beforehand. This needs to be judiciously designed by the designer of the system. This affects the performance as the designer may make a sub-optimal design of the complete system.

4.8 Type-2 Fuzzy Systems

The fuzzy systems we discussed so far are called as the Type 1 Fuzzy systems or simply T1 fuzzy systems. The other fuzzy systems are the Type 2 fuzzy systems or T2 Fuzzy systems (Kamik and Mendel 1998; Mendel 2003; Mendel and John 2002; Zadeh 1997). These systems make use of T2 fuzzy sets and T2 fuzzy arithmetic for the computation of the output based on the inputs and the rules specified. The basic methodology of problem solving with these systems is similar to the manner we did for the T1 Fuzzy Systems. The basic difference happens in the use of T2 Fuzzy Sets and associated arithmetic. Most of our discussions, experimentations and simulations would be confined to the use of T1 Fuzzy Systems. However we discuss the T2 fuzzy systems for the sake of completeness of the text as well as their increasing use in problem solving with fuzzy systems.

T1 Fuzzy sets denote a number along with its membership or belongingness. T2 fuzzy sets are advancement over these sets. They enable the representation of the fuzziness of a membership function. In other terms they represent the fuzziness of a fuzzy system. They are hence also called as the fuzzy fuzzy systems.

Consider any MF of the fuzzy sets. As per the T2 Fuzzy System theory, this MF is fuzzy. This would hence have fuzzy boundaries and it cannot be stated with certainty regarding the membership value of any attribute by this MF. We may hence denote the element x has a membership value of $\mu'(x)$. Here $\mu'(x)$ is itself fuzzy in nature. Hence this must possess some membership value. Hence a membership function of the T2 fuzzy set is completely represented by $\mu(x,u)$. Here x is the attribute whose membership degree is to be computed and u is the point over which the membership value must be calculated. This represents the membership as a function of x and u and may be represented in a three dimensional plot.

The stated method of dealing with fuzzy systems is extremely complex and required a large amount of computation. We hence place a small constraint that the membership value can only be either *0* or *1*. This plays a big role in simplification of the entire system and the associated computational costs. The realized system is solvable in small times and easy to model and comprehend. These are called as the Interval Type 2 Fuzzy Sets. These are the most extensively used fuzzy sets.

Now the three dimensional plot may be compressed and viewed in two dimensions. The shaded regions may be used to denote the membership value of 1. The non-shaded regions may denote a membership value of 0. The plot that results in this two dimensional graph is called as the Footprint of Uncertainty (FOU). This is shown in figure 4.12.

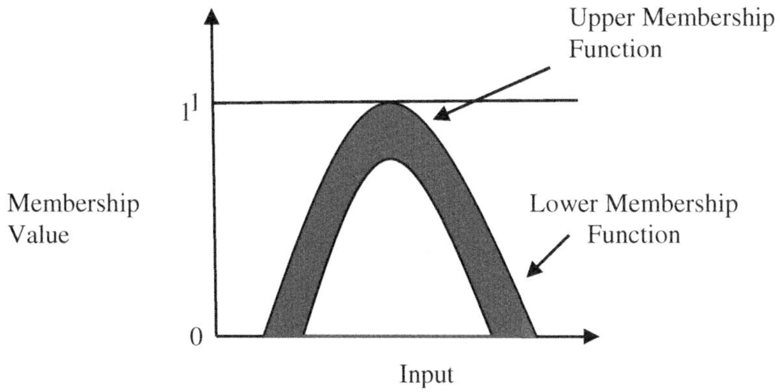

Fig. 4.12 Type II Fuzzy Set

The T2 fuzzy sets are represented using Vertical Slice Representation or Wavy Slice Representation. Suppose we slice the three dimensional membership plot at any value of x, we get a two-dimensional figure with axis as $\mu(x,u)$ and u. Here x is a constant across which the plot was cut. This plot is called as the vertical slice at any particular x. This is a simple and computationally less expensive method of representation. Here we monitor two membership functions. These are the Upper Membership Function (UMF) and the Lower membership Function (LMF). The UMF contains the maximum u for any fixed x. The LMF contains the least u for any fixed x. Wavy-Style representation is commonly used for the theoretical purposes. This makes a lot of use of embedded MF in its representation. The embedded fuzzy system is a general curve that is within the least and the maximum values. Union of all such curves gives us the footprint of uncertainty.

The solution of a T2 FIS is done using the same methodology as the T1 FIS. This makes use of the discussed steps of fuzzification, rule operation, aggregation, implication and defuzzification. Each step is conveniently changed as per the T2 fuzzy arithmetic and ideology.

The first major change in these systems is the manner in which the rules are handled. These differ from the T1 fuzzy sets as the inputs and outputs are supposed to work with are both MFs. We follow a simple all combination strategy to work with this situation. Every combination of MF for every rule to be fired is computed separately. The resulting rules are T1 fuzzy rules that may be trivially computed. The outputs are later combined. Consider any rule '*if x if I then y is O*'. Here I and O are in turn made up of MFs. Let us say that I is made up of membership functions I_1, I_2, I_3, … I_m and O is made up of membership functions O_1, O_2, O_3, … O_n . Now we compute every combination of I_i and O_j. Every combination is worked separately as a T1 fuzzy rule and outputs are finally computed.

Similar is the case with the implication operator. Implication is applied separately for the UMF and the LMF.

The defuzzification in T2 fuzzy systems is similar except for the need of an extra step of Type Reduction. This converts a T2 fuzzy set into a T1 fuzzy set. The type reduction may be performed by the Center-of-Sets (COS) mechanism. This makes use of the Karnik Mendel (KM) algorithm. After the TR, the defuzzification operation may be easily carried out. We may alternatively use Center-of-Sets (COS) defuzzification.

4.9 Other Sets

There are numerous other sets besides the discussed fuzzy sets. There is an extensive work going on in these sets largely due to their higher modeling capabilities. The underlying logic is the same that we need to represent any number along with its degree of association or membership values. These are briefly discussed one after the other.

4.9.1 Rough Sets

Rough sets are the sets that account into the fact that many times there might be no knowledge of the data at all. A common example is the experiments where the data is not recorded properly or is lost. Such scenarios cannot be modeled by the conventional sets that are used for modeling the impreciseness or the uncertainties. The rough set theory defines a boundary which consists of such elements whose existence in set is not precisely known (Pawlak 1991; Polkowski and Skowron 1996). If the boundary is of zero width or does not contain any element, the rough set is called as a crisp set and it becomes a traditional mathematical set with no impreciseness.

4.9.2 Vague Sets

Vague Sets are the sets in which every element has a degree of trueness as well as a degree of falseness (Chen 1955; Gau and Buehrer 1993). Trueness is a measure of the degree to which element is a member of the set. Similarly the degree of falseness measures the degree to which the element is not a member of the set. The degrees of membership always lie between 0 and 1. The difference between these sets and the fuzzy sets is that the sum of trueness and falseness may not necessarily be 1. Their sum is always less than or equal to 1. The membership of any element x may hence be represented as $<\alpha(x), 1-\beta(x)>$. Here $\alpha(x)$ denotes the degree of trueness and $\beta(x)$ denotes the degree of falseness. These obey the relation given by equation (4.15).

$$\alpha(x) + \beta(x) \leq 1 \tag{4.15}$$

The plot of the membership function of these sets is given in figure 4.13.

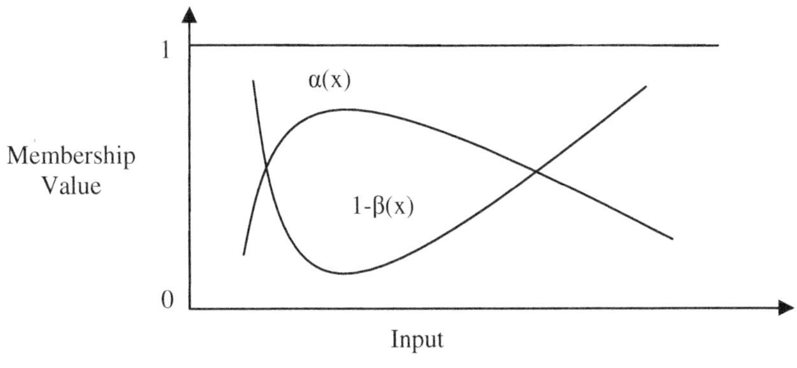

Fig. 4.13 Vague Sets

4.9.3 Intuitionistic Fuzzy Sets

The other kind of sets used is the Intuitionistic Fuzzy Sets (Atanassov 1986, 1994). These are quite similar in concept to Vague sets with the use of two degree of membership associated with any element of the set. In these sets the first degree of membership is used to measure belongingness or the association of the element. The second membership function measures the non-membership or the non-belongingness of the element. It is denoted by <$\mu(x)$, $V(x)$>. Here $\mu(x)$ is the degree of membership and $V(x)$ is the degree of non-membership. These sets may be plotted as shown in figure 4.14.

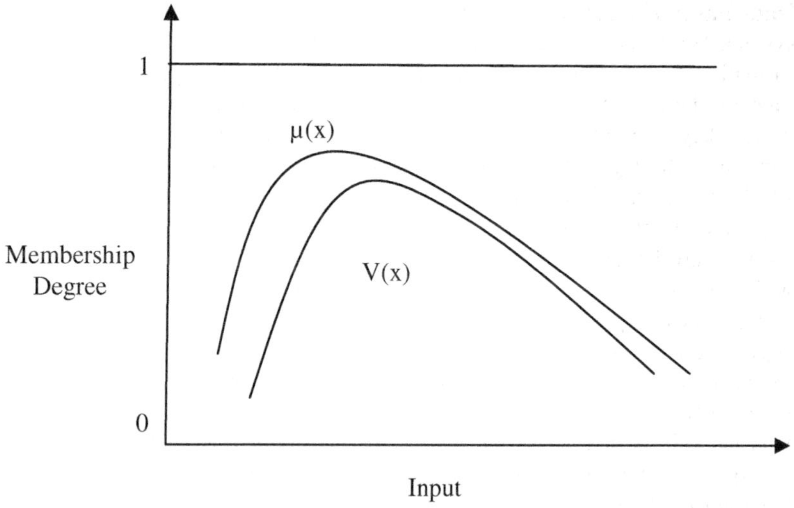

Fig. 4.14 Intuitionistic Fuzzy Sets

4.10 Chapter Summary

In this chapter we presented fuzzy logic which made use of fuzzy sets for information processing. The fuzzy sets are sets where every element is associated with some degree of membership. This differs from the crisp sets where the elements either belonged to the set or do not belong. In other words the degree of membership can be either 0 or 1. The basic motive behind the chapter was to convert the conventional rule based production systems using crisp inputs into fuzzy systems. For this we discussed the fuzzy arithmetic that took fuzzy inputs and generated fuzzy outputs. The Fuzzy Inference Systems first carry out the fuzzification of the inputs where the inputs are fuzzified using different membership sets. The fuzzy inputs are then operated by the fuzzy rules. The rules consist of antecedents and consequents. The AND, OR and NOT fuzzy operators are used for antecedents. The variety of ways in which these operators could be used was discussed in the chapter. The chapter further discussed the implication operator that was used for the mapping of inputs to outputs. The next operator was the aggregation operator. This carried out the aggregation of the outputs of the various rules. This resulted in the aggregated fuzzy output that was converted back into the crisp output by the defuzzification operator. The other part of the chapter dealt with type 2 fuzzy logic. Here fuzziness was introduced into the fuzziness of the inputs. At the end the other types of sets namely Vague Sets, Rough Sets, and Intuitionistic Fuzzy Sets were presented.

References

[1] Atanassov, K.: Intuitionistic fuzzy sets. Fuzzy Sets Syst. 20(1), 87–96 (1986)
[2] Atanassov, K.: New operations defined over the intuitionistic fuzzy sets. Fuzzy Sets Syst. 61, 137–142 (1994)
[3] Chen, S.M.: Measures of similarity between vague sets. Fuzzy Sets Syst. 74(2), 217–223 (1995)
[4] Dubois, D., Prade, H.: Fuzzy Sets and Systems: Theory and Applications. Academic Press, New York (1980)
[5] Dubois, D., Prade, H.: A review of fuzzy sets and aggregation connectives. Inf. Sci. 36, 85–121 (1985)
[6] Furuhashi, T., Hasegawa, T., Horikawa, S., et al.: An adaptive fuzzy controller using fuzzy neural networks. In: Proc. of IEEE Fifth Intl. Fuzzy Syst. Assoc. World Congress, pp. 769–772 (1993)
[7] Gau, W.L., Buehrer, D.J.: Vague sets. IEEE Trans. Syst., Man, Cybern. 23(2), 610–614 (1993)
[8] Kala, R., Shukla, A., Tiwari, R.: Fusion of Probabilistic A* Algorithm and Fuzzy Inference System for Robotic Path Planning, Artificial Intelligence Review (2010) doi:10.1007/s10462-010-9157-y
[9] Kandel, A. (ed.): Fuzzy Expert Systems. CRC Press, Boca Raton (1991)
[10] Kasabov, N.K.: Hybrid connectionist fuzzy production systems: Toward building comprehensive AI. Intell. Autom. Soft Comput. 1(4), 351–360 (1995)
[11] Kasabov, N.K.: Foundations of Neural networks. In: Fuzzy Systems, and Knowledge Engineering. MIT Press, Cambridge (1998)
[12] Mamdani, E.: Application of fuzzy logic to approximate reasoning using linguistic synthesis. IEEE Tranms. Comput. 26, 1182–1191 (1977)
[13] Mendel, J.M.: Type-2 Fuzzy Sets: Some Questions and Answers. IEEE Neural Netw Soc. 1, 10–13 (2003)
[14] Mendel, J.M., John, R.I.B.: Type-2 fuzzy sets made simple. IEEE Trans. Fuzzy Syst. 10(2), 117–127 (2002)
[15] Mizumoto, M., Zimmermann, H.: Comparison of fuzzy reasoning methods. Fuzzy Sets Syst. 18, 253–283 (1982)
[16] Pawlak, Z.: Rough Sets – Theoretical Aspects of Reasoning about Data. Kluwer, Dordrecht (1991)
[17] Plonka, L., Mrozek, A.: Rule based stabilization of the inverted pendulum. Comput Intell: An Intl. J. 11(2), 348–356 (1995)
[18] Polkowski, L., Skowron, A.: Rough mereology: A new paradigm for approximate reasoning. Intl. J. Approx. Reason. 15(4), 333–365 (1996)
[19] Sugeno, M.: Theory of Fuzzy Integral and Its Applications. PhD thesis, Tokyo Institute of Technology, Tokyo (1974)
[20] Sugeno, M.: An introductory survey of fuzzy control. Inf. Sci. 36, 59–83 (1985)

[21] Takagi, T., Sugeno, M.: Fuzzy identification of systems and its applications to modelling and control. IEEE Trans. Syst., Man and Cybern. 15, 116–132 (1985)
[22] Yager, R., Zadeh, L. (eds.): An Introduction to Fuzzy Logic Applications in Intelligent Systems. Kluwer Academic, Boston (1992)
[23] Zadeh, L.: Fuzzy sets. Information and Control 8, 338–353 (1965)
[24] Zadeh, L.: Probability measures of fuzzy events. J. Math. Anal. Appl. 22, 421–427 (1968)
[25] Zadeh, L.: A theory of approximate reasoning. In: Hayes, M.M. (ed.) Machine Intelligence, vol. 9, pp. 149–194. Elsevier, New York (1979)
[26] Zadeh, L.A.: Toward a theory of fuzzy information granulation and its centrality in human reasoning and fuzzy logic. Fuzzy Sets Syst. 90, 111–137 (1997)

List of Figures

Fig. 4.1. Membership Function
Fig. 4.2. All Membership Function Plots for Temperature
Fig. 4.3(a). Different Types of Membership Functions: Gaussian Membership Function
Fig. 4.3(b). Different Types of Membership Functions: Triangular Membership Function
Fig. 4.3(c). Different Types of Membership Functions: Sigmoidal Membership Function
Fig. 4.4. AND Fuzzy Operator
Fig. 4.5. OR Fuzzy Operator
Fig. 4.6. NOT Fuzzy Operator
Fig. 4.7. Implication Fuzzy Operator
Fig. 4.8. Aggregation Fuzzy Operator
Fig. 4.9. Defuzzification Operator
Fig. 4.10. Fuzzy Inference System
Fig. 4.11. Uneven Membership Function Distribution
Fig. 4.12. Type II Fuzzy Set
Fig. 4.13. Vague Sets
Fig. 4.14. Intuitionistic Fuzzy Sets

Part II
Simple Intelliegent Systems

Chapter 5
Speech Signal Analysis

Abstract. Intelligent systems possess the capability to model and solve many problems of practical importance. The best way to understand these systems is do design and develop such systems which exposes their various advantages and disadvantages. This chapter presents the basic analysis technique of speech signals that would further help us in using speech as a medium of developing intelligent systems. In this chapter we study the manner in which we may highlight and extract useful features out of a given speech signals. We discuss the analysis techniques in two heads. The first head consist of the bank of filters approach. Here we present the Fourier, Short Time Fourier and Wavelet Analysis which extract interesting features. Here we would stress the importance of time and frequency domain. In the other head we would discuss the Linear Predictive Coding (LPC) methods. We discuss the manner in which the linear coding helps in analysis. We even discuss the general speech parameters that facilitate good recognition in these intelligent systems.

Speech is one of the most fascinating things that we encounter in everyday life. We hear many kinds of sounds in our everyday life that influence the way we think and act. Many decisions we take are based on the inputs that we receive by speech. Many decisions are communicated by speech for further action. Speech hence forms an integral part of our everyday life. We get an understanding of our surroundings, its happenings, largely by speech. Consider the railway announcements, enquiry or the horn of train. They all enable us to do many things. Again consider the quarreling, horns and other unpleasant noises. These distract our attention and hinder our working. We lose our concentration and ultimately the working efficiency reduces. Disturbances might so many times prohibit two people to exchange their ideas using speech. This all forms the ill effects of speech. The place that speech occupies in the natural world is prominent.

Speech is a signal that changes its amplitude and frequency along with time. As a matter of fact, speech is a highly complex signal that has too many variations in a very small magnitude of time. Even a simple signal might be actually very complex when seen in the domains of time and frequency. Consider the example of heart beat which sound very uniform and simple. This signal as well shows signs of very complex nature. This makes the entire task of working with speech signals very complex and a specialized art.

The extensive use of speech in natural system is a big motivator behind its use in the artificial systems as well. Speech based artificial systems have been dreamt

A. Shukla et al.: Towards Hybrid and Adaptive Computing, SCI 307, pp. 111–128.
springerlink.com

of since a very long time (Atal 1976; Campbell 1997; Naik 1990; Wan and Campbell 2000). The rapid growth in research along with increasing computation power along with time has now created possibilities for these systems to become a reality. Speech based systems may be easily seen in speech based security checks, speech to text convertors, IVRS systems, etc. Many of these systems may be encountered today in everyday life. Speech based intelligent systems further hold a lot of potential of future where they find applications in security, multilingual speech to speech translations, and speech based interface etc. This forms a very exciting area of research and development that has a lot of fruits to offer in the future to make everyday living easy and comfortable.

5.1 Feature Extraction

The highly complex signal of speech can never be worked out as a whole by the intelligent systems we have been discussing so far. Speech continuously changes in terms of amplitude along with time. The frequency as well shows highly complex and different trends along with time (Reynolds 1994, 2002; Shukla and Tiwari 2008b). Now we know that the entire signal is needed for the recognition purposes as all information is important. This is too much of information for the system to handle. Hence we need some means to reduce this vast amount of information into a reasonably small amount that the system can work with. We need to further ensure that the loss in information is minimal with the largest amount of reduction in the dimensionality reduction.

The reduction in problem dimensions is known as dimensionality reduction. One of the best ways for dimensionality reduction is through feature extraction. Features may easily be understood as real attributes that summarize some interesting part of the data and carry a special relevance (Rosenberg 1976; Rosenberg and Soong 1992; Bunke and Kandel 2002). The set of features extracted out of the signal are given as inputs to the system rather than the complete signal. The various features contribute in their own manner towards the identification. Each feature informs the system something that enables the system to make its decision regarding the output.

5.1.1 Good Features in Classification

The various features or attributes that are extracted for the system must be good as per the problem demands. We assume the problem is of classification. Now the various features must be such that the intra class distances are low and the interclass distances are high. This means that the features must remain almost constant for the various members for the same class. The change must however be as large as possible when the same are measured for the members of different classes. This may again be easily visualized in the input space where the various attributes serve as axes and the classes are marked by the labels. The problem is expected to draw decision boundaries that demarcate the various classes.

Consider the problem is of identification of speaker. Now good features would be such that these features remain almost constant for any number of times the same person speaks. This needs to account for the fact that it would be impossible for the person to speak as it is every time. There might be a lot of variations in the various utterances of him. The speed, expressions, etc. are bound to change in the various recordings, even if these recordings are done one immediately after the other. However when another person speaks, this features must change drastically. The vast change in these features would make the task of recognition very simple. This determines the goodness of the features. The change must be large even if somebody else tries to mimic the other person's voice.

5.1.2 Need of Features

We have seen that we carry out extraction of features to reduce the amount of data to a reasonable limit that we can handle (Shukla *et al* 2010). We hence study the major advantages that we get by the application of the feature extraction. The problems associated with speech that necessitate feature extraction may be classified into following broad heads.

- **Computational Limitations:** Computation is always finite in nature. We cannot make very complex learning methods that require a very large training time. The total time required by the system must always be such that the system can be trained and tested within tolerable time. The limitation of computation hence places a threshold on the volume of data that the system can handle. This further necessitates that we limit the total volume of data.
- **Space Limitations:** Storing big audio files requires a large amount of space. The multiple instances of the various recordings that drive the system may also be very expensive in terms of memory. It might not always be possible to store such a large database for constant access. Space is hence another limitation in the usage of the signal as it is.
- **Data Limitations:** The large dimensionality of the problem usually has bigger networks. There is an additional amount of processing required for the system to compute the relations between the various attributes and figure out how they affect the final system. These large networks have a bulk of parameters or weights that must be tuned. This further requires a large number of instances of the training data size. The data might not always be easy to record for every problem. Especially speech signals have this problem where the activity of recording, preprocessing, analyzing and extracting the features is highly difficult to carry out. The data is hence very expensive and hence limited in terms of the number of instances. This prohibits the training of large networks and hence poses a limitation in the entire problem that necessitates the extraction of features.
- **Feature Enhancement:** In its native form the various parts of the recorded signal might not give valuable information. The system would be forced to find out all the relations and the manner in which the various parts of the input contribute to decide the output. This is a very difficult task. If given a signal or a part of the signal, anybody would find it extremely difficult to identify the

word, speaker or anything else. This hence puts a lot of burden on the system in terms of computation, data and space. The better technique would be if someone can figure out these relations or a part of these relations and let the system do the rest. This is something that most of the features do. They enhance the good or the interesting parts in the entire signal and give in the form that is easy for the system to comprehend. This feature enhancement is done manually or by any other automatic technique and not by the machine learning stage.

5.1.3 Who Decides the Features?

In the previous section we have simultaneously presented two contrasting ideologies. In the first we said that the system can be given the entire speech vector. It would decide the relations, interestingness, features, etc. on its own. In this manner it would try to correlate the inputs to the outputs. The other ideology that we presented was in form of features that were extracted by a human and presented before the system as inputs that it might use for learning or classification. These two are quite opposite to each other.

System trying to find out and forming these rules in which the inputs relate to the outputs is a very flexible system. We give no clue to the system for this. It can possess and assume any relation as per its own understanding of the data. However manual feature extraction holds well only when it is known that the features extracted are good. In case the features extracted are not good, the system would naturally fail. The features are always decided by the humans based on his prior knowledge. This knowledge and its validity as per the practical life input that the system might face largely determine the system performance. Many times the prior knowledge of features might be theoretically correct, but the system might not work when given the practical inputs. This is because of noise, wrong assumptions, etc.

5.1.4 Analogies with the Human Brain

Most of the inspiration behind all these systems and approaches has been the human brain. The human brain is a marvelous piece of creation that is able to do fine pieces of work (Konar 1999). Most of the machines that we build cannot even compete the working and precision of a small part of the brain. A large part of the manner in which the brain works is still a matter of mystery for researchers. We talked about the feature extraction and analysis in this chapter so far. An interesting fact would be think over the fact if the brain also performs feature extraction in its working.

The human brain has 10^{11} neurons. These 10^{11} neurons connect to each other. This makes a total of 10^{22} connections between the various neurons. Each of these neurons does computation in parallel and independent of all other neurons. The results are regularly passed in between the various neurons by these connections. This represents a lot of computation that takes place in a unit time in the brain. Also the brain has a lot of continuous learning in an online mode. It keeps learning by constant and regular reinforcement. This creates a possibility of a lot of learning that accumulates over the span of time of the individual. Besides the connections are the form of storage that are always plentiful in number. Hence most of the limitations that we deal are not necessarily present in the human brain. This makes it work as a very flexible and adaptive system that most of our systems cannot imitate.

However most of the processing of the brain is done in hierarchal manner. There is a layered processing where information is constantly processed. The processing makes the information perceived by the sensory organs range from very abstract form to very crude form layer by layer. This operation may have its analogy with the feature extraction where the near layers supply inputs to the later layers that try to cite small features. The still later layers make their own complex features. This goes on and on. Finally the decision may be made based on the features.

5.1.5 Number of Attributes and Classification

The number of attributes is an important factor that we give to any system. We have already discussed that the networks or systems with large number of attributes are slower to learn and more time consuming. Further it may many times be little difficult to train these networks due to the large amount of computations.

The other factor of interest is the how does increasing the attributes affect the classification. We assume that the added attributes are good that add some good characteristics to the system. In other words they possess a high inter class separation and a low intra class separation.

Imagine the input space with two attributes for separation. Now imagine a few points in this input space that you are asked to classify as shown in figure 5.1. It is naturally very difficult to draw a decision line as the classes intermingle with each other. Any system when given this problem would not succeed in getting 100% accuracy due to the highly complex shape of the decision boundary if plotted. Now imagine that we add a new attribute to this recognition system. This would now mean that the input space has a new axis. The figure hence becomes 3-dimensional in nature now. Due to this it is highly possible that the classes can be

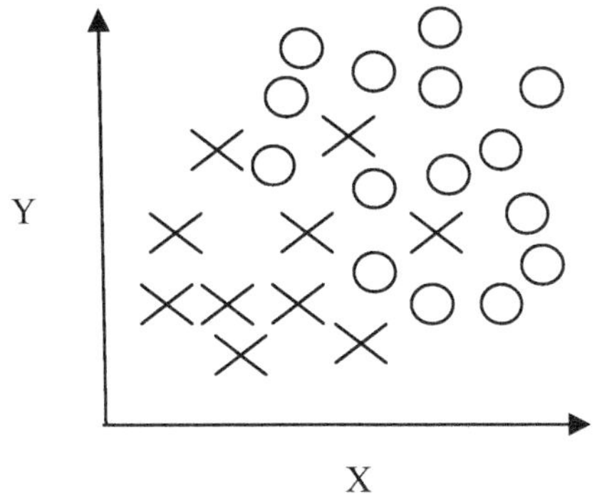

Fig. 5.1 Input Space for classification with less attributes

separated by some decision boundary which is relatively simple. If the added attribute was very good, it might be very simple to do so.

Hence the addition of every attribute adds a new axis to the input space that makes the task of classification easier. If the added attribute is good, it would result in an increase of the interclass separation in the input space. Hence classification would be facilitated. This however may not be necessarily true when the added attribute is not good enough. This is what happens when many times we try to insert too many attributes into the system without prior understanding.

Hence it may be easily seen that the addition of attributes to the system has both positive and negative effects. The positive effect is the ease of classification. The negative ones are the increase in computation time. The attributes for any system may hence be judiciously computed based on their goodness.

5.2 Spectral Analysis

The speech signal may be easily plotted as a conventional signal. Consider any such signal plotted in time domain (Rabiner and Gold 1975; Shukla and Tiwari 2007, 2008). A typical signal might look like the one shown in figure 5.2. Now the task is to analyze this signal and to formulate a set of interesting features out of this. Based on this we may first of all like to apply the common signal processing tools and try to figure out some of the statistics that this signal holds. This gives us the first set of features that may be potentially used for the signal. We know that any signal can be broken down and denoted by a set of commonly used attributes like pitch, frequency, etc. This may effectively represent the entire signal or a large part of it. We may be only interested in the representing the signal to some extent that can be used for the task of recognition rather than trying to represent a very large part of it. It may be noted that whatever we may not be able to represent would be completely lost in the system and hence the ideology is to take everything that might significantly contribute towards the recognition. Throughout the text when we mention any of the features, its contribution in speech, speaker or other recognition systems may be visualized and appreciated.

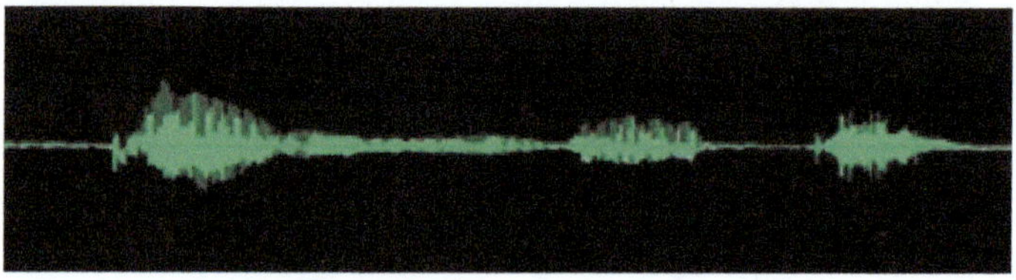

Fig. 5.2 Typical Speech Signal

Now we first study and discuss some of the commonly used features in this section. This would be further highlighted by two heads of features that we would discuss in the coming sections. These are Bank of Filters and Linear Predictive Coding (LPC) coefficients.

5.2.1 *Number of Zero Crossings*

The number of Zero Crossings (NZC) is the first factor that we consider. This denotes the total number of times the amplitude crosses the time axis either from a positive value to a negative value or vice versa. This factor is highly prone to noise and needs to be applied on a noiseless signal. Noise may be present as external signals that interfere with the speech signal. Now this becomes a parameter for many recognition systems. It may be easily seen that the change in NZC would not be very high for small changes in the signal. Hence this alone can differentiate between basic kinds of signals.

Many times we may choose a level crossing in place of zero crossing. Here we fix a level and count the number of times the signal crosses this level. This may be seen as a generalized version of the NZC where the level is not fixed to 0.

5.2.2 *Duration of Utterance*

The duration is the total time of utterance. This forms another factor of interest of most systems. This is the total length of the signal under analysis. The length can also be a factor that can carry a large part of the classification or recognition. It may be easily seen that by length we can differentiate between the long and the small utterances which contributes towards the overall recognition.

An interesting point to be mentioned for this as well as the other factors that we study and consider is the start and the end of the utterance. In most of the cases the speech signal is continuous in nature. We do not know where does the needed region start and where does it end. This is solved by warping over time and making the recognition. We discuss this in chapter 6.

5.2.3 *Pitch*

The pitch is the frequency of the signal. We are used to identifying people or sounds by their frequency. It largely forms the basis of our everyday recognition of people. The use of frequency for differentiating between males and females to some degree is largely observable. Any speech signal over a time period may be broken down into many frequencies that are dominant in that signal. The most fundamental frequency is the most dominant one in the signal under consideration. The pitch forms an entirely separate domain that we look at the signal apart from the time domain that we had been doing. We would study this in the bank of filters in the next section.

5.2.4 *Amplitude*

The amplitude is the signal in its time domain as well. The amplitude may be needed to be sampled along with time to get a finite and limited number of values that can be studied or analyzed over time. This domain of signal may be easily visualized and the effects of change of amplitude with time may be easily seen for any signal. The amplitude and the manner in which it changes represent the information needed for the recognition purposes.

5.2.5 Spectrogram Analysis

We have presented two major concepts here of time and frequency. Both these form two domains of study where the signal is analyzed. The combination of time and frequency is important for any recognition system that optimally performs in both domains.

The spectrogram is a special diagram where we plot the amplitude or the energy along with time and frequency. Hence this represents a 3-dimensional graph with one axis each of time, frequency and amplitude. The time is kept on the x-axis, frequency is kept on the y-axis and the amplitude on the z-axis. One such spectrograph for a dummy speech signal is shown in figure 5.3. It is not possible to easily plot or easily study the 3-dimensional graph. Hence we use 2-dimensional graph with time and frequency axis. The amplitude is plotted with the intensity of the color.

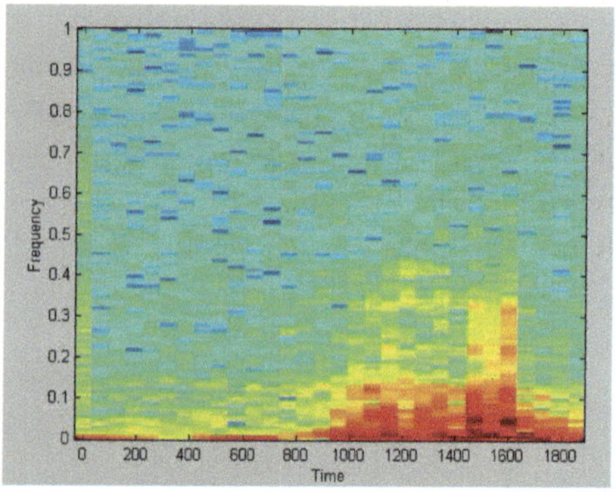

Fig. 5.3 Typical Speech Spectrogram

The spectrogram looks as a beautiful picture representing all the details of the speech signal. It may be taken as an image of the entire signal. It is possible to carry out any recognition task by looking at the speech spectrogram. A manual survey over the spectrogram would illustrate the various regions of interest at the various times and frequencies that are differentiated by the intensity.

5.2.6 Power Spectral Density

The Power Spectral Density (PSD) is another major factor that conveys good information for the recognition purposes. The PSD measures the energy of the

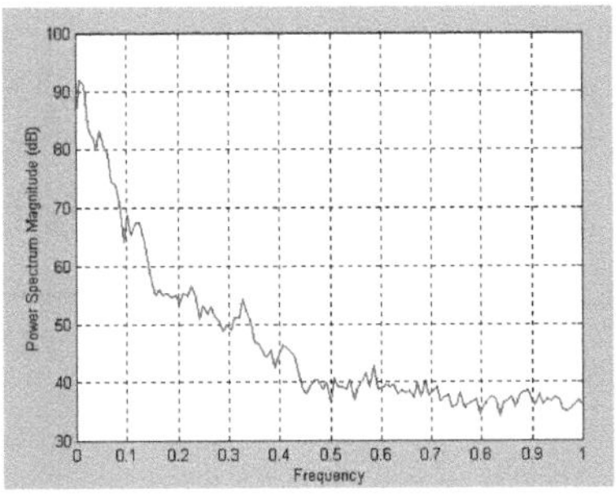

Fig. 5.4 Graph for Power Spectral Density v/s Frequency

signal or its power as a function of frequency. The idea is to study the power distribution over the frequency spectrum of the signal. Here we plot the power against the frequency to give a graph as given by figure 5.4.

The interest for recognition systems is in average PSD and peak PSD. These two factors are of extensive use in systems using speech and enable effective classification.

5.3 Bank of Filters

In this section we present a fundamental class of techniques for the analysis of the speech signal called as the Bank of Filters (Dautrich *et al* 1983; Rabiner and Juang 1993). As the name suggests this approach makes use of a collection or bank of filters for the analysis purpose. The sound signal comes and is passed through each of these filters. Each filter works independent of each other in the processing of the given signal. This may be easily represented in figure 5.5 which shows the signal being divided among the various filter banks. Each bank processes and gives its own output that makes the complete output of the analysis of the signal.

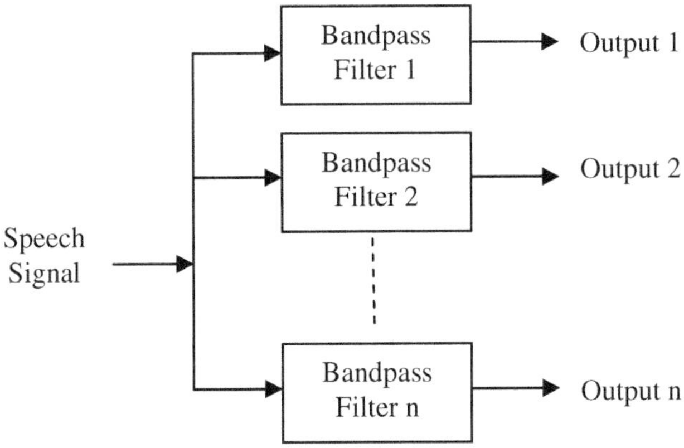

Fig. 5.5 Distribution of signal in Band of Filters approach

Each filter in the bank of approach has a characteristic range of frequencies that it works upon. The various ranges of the filters overlap with each other by some amount that further facilitates two filters acting upon a single frequency. This contributes towards the overall analysis of the signal. The overlap of the frequencies in the frequency domain is shown in figure 5.6.

Let the signal be represented by *s(n)*. Here *n* is the time over which the signal amplitude was measured. Further let us say that the signals are sampled with a sampling frequency of F_s. Here the sampling frequency is decided by the Nyquist-Shannon sampling criterion. As per this theorem the signals must be sampled with the sampling frequency that is at least twice the maximum frequency represented by the signal. Hence is F_{max} represents the maximum frequency, the sampling may be done at time intervals of $1/(2F_{max})$ for no loss of information.

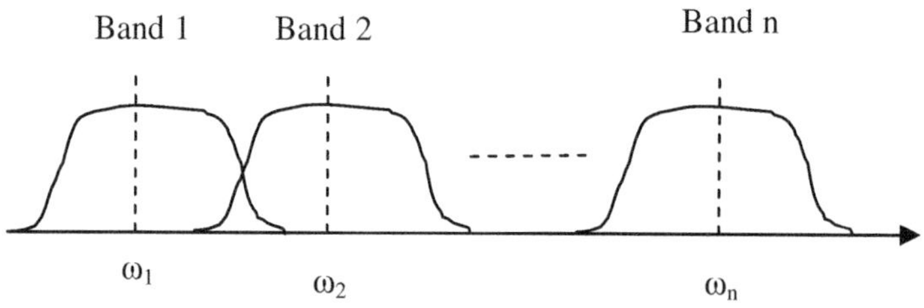

Fig. 5.6 Frequency Distribution in Bands

Every filter *i* has a frequency F_i. The normalized frequency (ω_i) for the filter is given by equation (5.1). The final output of the system (o_i) by the ith filter may be given by equation (5.2).

$$\omega_i = \frac{2\pi}{f_i} \tag{5.1}$$

$$o_i = X_i(e^{\omega i}) \tag{5.2}$$

The final representation of the speech signal may be computed by the individual outputs of the various filters that are placed in the bank. This forms a good means to analyze the given signal.

Again consider the figure 5.1. We have the same signal being applied to a variety of filters or the filter bank. Let the ith bandpass filter possess a response of $h_i(m)$ and a duration of M_i. The resultant output of this filter may be easily given by the convolution operator '*' as given in equation (5.3). This may be further expanded into equation (5.4) as per the definition of the convolution operator.

$$s_i(n) = s(n) * h_i(n) \tag{5.3}$$

$$s_i(n) = \sum_{i=0}^{M_i-1} h_i(m).s(n-m) \tag{5.4}$$

The signals generated after the operation of the filters are passed through a non-linear filter for the addition of non-linearity and then passes through a low pass filter to filter out lower frequencies. This undergoes a sampling rate reduction and then amplitude compensation. The resultant signal is the representation of the original signal after the application of the filter.

We leave a bulk of discussion over the various processing steps and their related issues in this book. We would rather focus on a key element of the implementation of this approach that we would use in the design of speech based intelligent systems.

We now discuss the representation of speech in the frequency domain. In this category we would be discussion three most used transforms that is Fourier Transforms, Short-Time Fourier Transform and Wavelet Transform (Sonmez *et al* 1998; Shukla *et al* 2008, 2009a. 2009b). All these transforms carry the task of extraction of some frequencies out of the signals given for the task of analysis that is later used as inputs in the intelligent systems. The basic methodology in all these is to extract the key frequencies in the signal, the importance of which has been already discussed. At the same time these methods present a tradeoff between frequencies and time domains.

5.3.1 Fourier Transform

The basic analysis technique is the Fourier analysis. Here we transform the signal in the time domain completely into the frequency domain. Into the frequency domain we may easily identify the frequencies present and select the top few frequencies for the recognition purposes. The Fourier analysis does this task of extracting the frequencies out of the given frequency signal.

The Fourier Transform is a transformation in the representation of the signal from the time domain into the frequency domain. Let the signal in time domain be represented by *s(x)*. The conversion of the signal into the frequency domain *S(n)* may be easily carried out using the Fourier transform equation given by equation (5.5). Here we assume the signal to be continuous with time. In case the signal is discrete, Discrete Fourier Transform may be used instead similar to the Fourier Transform.

$$S(n) = \frac{1}{2\pi} \int_{0}^{2\pi} s(x) e^{-inx} dx \tag{5.5}$$

This gives the various frequencies present in the signal in the decreasing order. The first few frequencies are very important and dominant in the system. The last few frequencies are not very important. The breakup would result in infinite number of frequencies, but we can effectively work with only a few of these. We hence select the top few frequencies for our work. The signal may be again represented in the time domain by an additive expansion of these frequencies in the time domain. We would observe that the regenerated signal closely follows the original signal but not exactly. Hence there is a loss of information in neglecting the lower terms or lower frequencies obtained in the Fourier Analysis. However the loss is negligible. Also we have compressed the signal by a high amount. This again clarifies us in the point of selection of the number of attributes to accept after the Fourier Analysis. We must select sufficient number of attributes such that the various classes can be differentiated from each other based on these attributes. The differentiating features must be extracted.

The reverse transform of the signal into the time domain is given by equation (5.6)

$$s(x) = \sum_{n=-\infty}^{\infty} S(n)e^{jnx} \tag{5.6}$$

A major factor in this is that we are completely working over the frequency domain. All information of time is completely lost. Say there was a small utterance which was quite brisk in nature. This existed somewhere in the complete signal which is being analyzed. Now the utterance would not appreciably affect the extracted frequencies since it has a small effect on the overall signals in the frequency domain. This means that this utterance would be largely neglected. But this otherwise is an important factor that can affect the decision regarding the recognition. Hence working entirely in the frequency domain is not optimal. This problem is solved by the Short-Time Fourier Transform and the Wavelet Transform that we discuss next.

5.3.2 Short-Time Fourier Transform

The other transform that follows our discussion of transformation from time domain into frequency domain is Short-Time Frequency Transform (STFT).The major problem with the Fourier transform was the complete absence of information regarding the time. In this domain we hence add time dependency in the Fourier transform to enable simultaneous work in both time and frequency domains. The Fourier transform breaks the entire signal in time domain into continuous time fragments. The entire signal is the collection of these fragments placed one after the other. Then Fourier transform is applied into these entire fragments one after the other. This gives us the frequency components or the top selected frequency components from all these fragments. The result is the combined list of all the frequency components in all the fragments that are extracted out of the signal. This may be understood with the help of figure 5.7.

The STFT is hence betterment towards Fourier transform in terms of the mentioned criterion. The STFT is visualized as the process of moving a window that applies Fourier transform in the entire signal from one extreme position to the other extreme position. The length of the time frame occupied by a single fragment is called as the window length. This is an important parameter in this transformation. If the length of the window is too small, the STFT would be dominated by information on time domain. There would be virtually no information about the frequency. Similarly a very large window length would make the algorithm be dominated by frequency. There would be almost no information about time. This would behave similar to the Fourier transform. Based on these concepts the length of the window may be fixed.

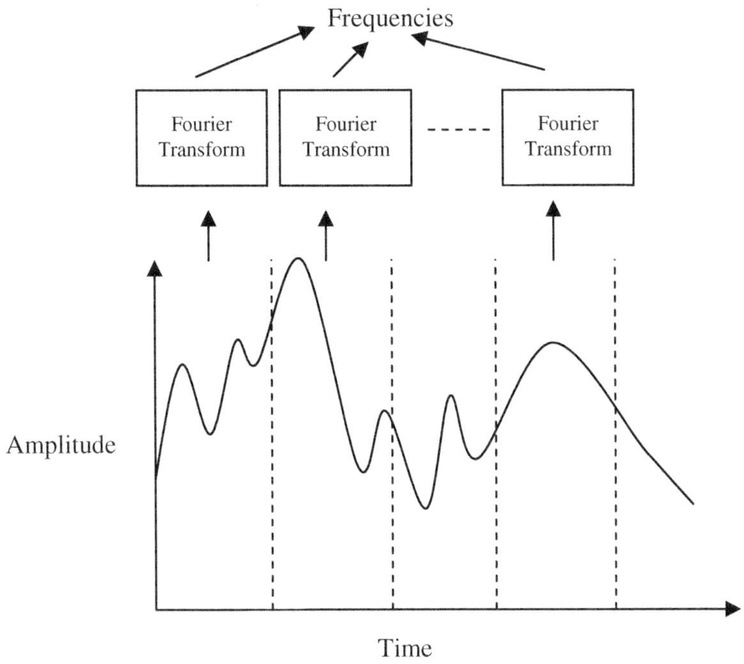

Fig. 5.7 Short Time Fourier Transform

Consider any signal *s(x)* that has to be transformed. The application of STFT on this signal to generate the frequency domain signal *S(n)* for the fragment starting at *i* may be given by equation (5.7).

$$S_i(n) = \int_{-\infty}^{\infty} s(x).w(x-i)e^{-jnx}dx \tag{5.7}$$

It may be easily seen that the major factor here is that the signal s(x) is multiplied by the time window *w(x)* that has been shifted by *i*.

Here *w(x)* is the window function that isolates a small part of the signal that is of interest.

5.3.3 Wavelet Transform

The last transform that we study in this text is the wavelet transform. We saw the addition of time dependent features in Fourier transformation that made the STFT. In STFT however the size of the window is fixed. This creates a problem in practical usage of this technique in signal analysis. This problem is solved by the use of Wavelet transform that represent an even more generalized transformation. In this technique the size of the window is not fixed. Rather it can take a more variable size as per requirements that may be stated.

The wavelet analysis technique breaks down the signal into approximations and detail. That approximation may be further broken down resulting in approximation and detail. This forms the second level of the system. We may continue indefinitely in breaking the approximation of the previous level into approximation and detail. For n-level decomposition, there are n + 1 possible ways to decompose, or encode, the signal. The interestingness or the information content in the details keeps reducing along with levels. Only the top few levels carry relevant information that may be required for the identification purposes. We hence extract the details only to some levels that may be required by the system, or that capture the needed information that may be useful for recognition.

This signal is applied in a similar manner. Consider the signal $s(x)$ that is to be analyzed by wavelet analysis. Let the transformed signal be $S_{a,b}(n)$. The transformation may be given by equation (5.8).

$$S_{a,b}(n) = \int_{-\infty}^{\infty} s(n)\phi_{a,b}(n)dn \tag{5.8}$$

Here $\phi_{a,b}(n)$ is called as the mother wavelet. a is called as the scale of the wavelet. b defines the shift of the mother wavelet for the transformation to be computed.

5.4 Linear Predictive Coding

The other model that we study for the purpose of signal analysis is the Linear Predictive Coding (LPC). This method assumes a linear relation between the current value of the signal and its past values (Atal and Hanaver 1971; Kumar *et al* 2009). Consider that the signal to be analyzed is represented by $s(n)$ where n is the time. In the LPC model we assume that the signal represents a linear equation between the past p values of the signal. Hence $s(n)$ may be easily represented as a weighted equation between $s(n-1)$, $s(n-2)$ … $s(n-p)$. This may be represented by equation (5.9).

$$s(n) \approx a_1\, s(n-1) + a_2\, s(n-2) + \ldots\ a_p\, s(n-p) \tag{5.9}$$

Here we have used the symbol $\approx$ in place of equality since there is an approximation involved. Using this equation we may not be able to exactly represent the signal equation, but we may be able to represent a good part of it. p is the order of the LPC. Sometimes an additional excitation may also be added to the signal. This gives us equation (5.10).

$$s(n) = \sum_{i=1}^{p} a_i s(n-i) + Gu(n) \tag{5.10}$$

Here $u(n)$ is known as the normalized excitation and G is known as the gain that is applied over the excitation.

Equation (5.10) may also be represented in the z-domain by applying the z-transform over the signal. This is given by equation (5.11).

$$S(z) = \sum_{i=1}^{p} a_i z^{-i} S(z) + GU(z) \tag{5.11}$$

The transfer function of this equation is given in figure (5.12).

$$H(z) = \frac{S(z)}{GU(z)} = \frac{1}{1 - \sum_{i=1}^{p} a_i z^{-i}} \tag{5.12}$$

This model or system for the analysis of the signal can be easily represented by figure 5.8. This represents an excitation source of $u(n)$ that gets magnified by the gain G. This after the application of the filter leads to the output $s(n)$.

Based on equations (5.9) and (5.10) we may easily calculate the error that is induced by the representation of the signal as a liner sum of the past p values of the signal. This is given by equation (5.13).

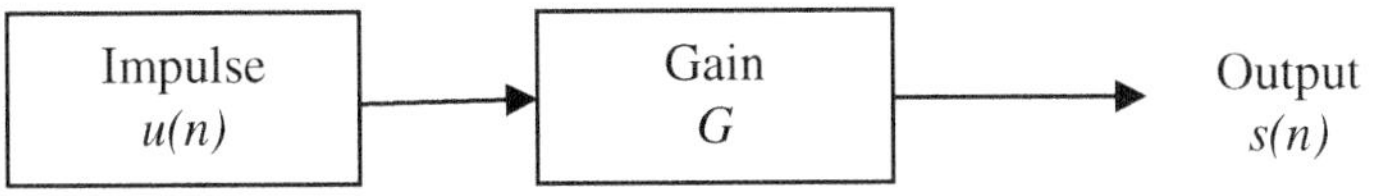

Fig. 5.8 Linear Predictive Coding model

$$e(n) = s(n) - \sum_{i=1}^{p} a_i s(n-i) \tag{5.13}$$

The transfer function of the error may be given by equation (5.14).

$$A(z) = \frac{E(z)}{S(z)} = 1 - \sum_{i=1}^{p} a_i z^{-i} \tag{5.14}$$

The problem is the determination of the coefficients a_i based on the speech signal being analyzed. The error is computed at every point of a small range of signal over a definite time period. Mean square error serves as the resultant error over the region. This is given by equation (5.15).

$$E(m) = \sum_{n} (s(n) - \sum_{i=1}^{p} a_i s(n-k))^2 \tag{5.15}$$

This error is the net error that needs to be minimized. Based on this principle the coefficients may be computed. It may be done by the application of the Yule-Walker equations that can be efficiently solved using Levinson-Durbin recursion. The side-products of the Levinson-Durbin recursion are the Reflection Coefficients (RC). These may be denoted by k_1, k_2, k_3, …k_p. From these reflection coefficients we also get the Arcus Sin Coefficients (ARCSIN) which are more stable than reflection coefficient.

Line Spectral Frequency (LSF) are closely related to formant frequencies and they have some desirable properties which make them attractive to represent the Linear Predictive Coding (LPC) filter. Given the LPC coefficients a_i, the LPCC coefficients C_i, may easily be computed using the recursion given in equation (5.16).

$$c_n = \begin{cases} a_n + \sum_{k=1}^{n-1} \frac{k}{n} c_k a_{n-k}, 1 \le n \le p \\ \sum_{k=n-p}^{n-1} \frac{k}{n} c_k a_{n-k}, n > p \end{cases} \quad (5.16)$$

The log area ratio (LAR) coefficients L_i are derived from the linear prediction (LPC) coefficients. The LAR may be computed from the LPC by using equation (5.17) and (5.18).

$$L_i = \log(\frac{1+k_i}{1-k_i}) \quad (5.17)$$

$$L_i = \sum_{j=0}^{m} L_{ij} s_j(n) \quad (5.18)$$

The relation between the various discussed features is shown in figure 5.9.

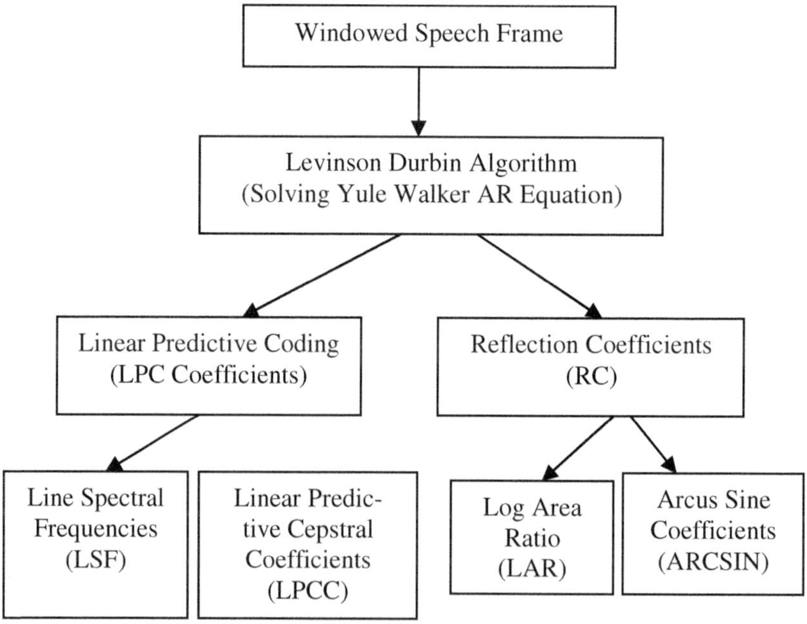

Fig. 5.9 Relation between different LPC coefficients

5.5 Chapter Summary

This chapter presented the mechanism of analysis of the speech signal. The underlying motivation is to have good features that may aid effective intelligent system design for recognition and verification. The chapter explored the various concepts of features from their goodness to their numbers and properties. The chapter further presented spectral analysis and bank of filters approach to signal analysis. The spectral analysis approach discussed the features of number of zero crossings where the number of times the signal crosses zero line is calculated. Other features of discussion were pitch, amplitude, spectrogram analysis, and power spectral density. The chapter further presented the Bank of Filter approach as an analysis technique. Fourier transform was presented that carried the task of conversion of the signal from time domain to the frequency domain. This decomposed a signal into constituent frequencies. We discussed how the top few frequencies could be used as features for the intelligent systems. The other technique was Short Time Fourier Transform where a time window of fixed length was used. This induced time dependence into the Fourier transform and hence results in good analysis. This was further generalized in Wavelet Analysis where the time window could be variable in length. This was followed by the Linear Predictive Coding analysis. Here we tried to predict the future values of the signal as a linear function of the past values. We further presented a variety of such parameters like Line Spectral Frequencies (LSF), Linear Predictive Cepstral Coefficients (LPCC). Log Area Ratio (LAR) and Arcus Sine Coefficients (ARCSIN)

References

[1] Atal, B.S.: Automatic Recognition of Speakers from their Voices. Proc. IEEE. 64(1), 460–475 (1976)

[2] Atal, B.S., Hanaver, S.L.: Speech Analysis and Synthesis by Linear Prediction of the Speech Wave. J. Acoust. Soc. Am. 50(2), 637–655 (1971)

[3] Bunke, H., Kandel, A.: Hybrid Methods in Pattern Recognition. World Scientific, UK (2002)

[4] Campbell, J.P.: Speaker Recognition: A Tutorial. Proc. IEEE 85(9), 1437–1462 (1997)

[5] Dautrich, B.A., Labiner, L.R., Martin, T.B.: The effect of selective signal processing techniques on the performance of a filter bank based isolated word recognizer. Bell Syst. Tech. J. 62(5), 1311–1336 (1983)

[6] Konar, A.: Artificial Intelligence and Soft Computing: Behavioral and Cognitive Modeling of the Human Brain. CRC Press, Boca Raton (1999)

[7] Kumar, R., Ranjan, R., Singh, S.K., Kala, R., Shukla, A., Tiwari, R.: Multilingual Speaker Recognition Using Neural Network. In: Proc. of the Frontiers of Res. on Speech and Music, FRSM 2009, Gwalior, India, pp. 1–8 (2009)

[8] Naik, J.M.: Speaker Verification: A Review. IEEE Comm. Mag. 28(1), 42–48 (1990)

[9] Rabiner, L.R., Gold, B.: Theory and Application of Digital Signal Processing. Prentice-Hall, Englewood Cliffs (1975)

[10] Rabiner, L.R., Juang, B.H.: Fundamentals of Speech Recognition. Prentice Hall International, Englewood Cliffs (1993)

[11] Reynolds, D.A.: Experimental Evaluation of Features for Robust Speaker Identification. IEEE Trans. Speech Audio Process. 2(4), 639–643 (1994)
[12] Reynolds, D.A.: An Overview of Automatic Speaker Recognition Technology. In: Proc. of the IEEE Intl. Conf. Acoustics, Speech, and Signal Process (2002)
[13] Rosenberg, A.: Automatic Speaker Verification: A Review. Proc. IEEE 64(4), 475–487 (1976)
[14] Rosenberg, A.E., Soong, F.K.: Recent research in automatic speaker recognition. In: Furui, S., Sondhi, M.M. (eds.) Advances in Speech Signal Processing, pp. 701–738. Marcel Dekker, New York (1992)
[15] Shukla, A., Tiwari, R.: Fusion of face and speech features with ANN for speaker authentication. IETE Tech. Rev. 4(5), 359–368 (2007)
[16] Shukla, A., Tiwari, R.: A Novel Approach Of Speaker authentication By Fusion of Speech and Image Feature using ANN. Intl. J. Info. Comm. Technol. 1(2), 159–170 (2008)
[17] Shukla, A., Tiwari, R.: Intelligent Biometric system: A case study. Intl. J. Info. Technol. Res. 1(3), 41–56 (2008b)
[18] Shukla, A., Tiwari, R., Kala, R.: Real Life Applications of Soft Computing. CRC Press, Boca Raton (2010)
[19] Shukla, A., Tiwari, R., Meena, H.K., Kala, R.: Speaker Identification using Wavelet Analysis and Modular Neural Networks. In: Proc. of the Natl. Symp. Acoust. 2008, Vishakhapatnam, India, pp. 125–130 (2008)
[20] Shukla, A., Tiwari, R., Meena, H.K., Kala, R.: Speaker Identification using Wavelet Analysis and Modular Neural Networks. J. Acoust. Soc. India 36(1), 14–19 (2009a)
[21] Shukla, A., Tiwari, R., Meena, H.K., Kala, R.: Speaker Identification using Artificial Neural Networks. J. Acoust. Soc. India 36(1), 20–25 (2009b)
[22] Sonmez, K., Shriberg, E., Heck, L., Weintraub, M.: Modeling Dynamic Prosodic Variation for Speaker Verification. In: Proc. of the Intl. Conf. on Spoken Lang. Process., vol. 7, pp. 3189–3192 (1998)
[23] Wan, V., Campbell, W.M.: Support Vector Machines for Speaker Verification and Identification. In: Proc. Of the 2000 IEEE Neural networks for signal process. workshop, vol. 2, pp. 775–784 (2000)

List of Figures

Fig. 5.1. Input Space for classification with less attributes
Fig. 5.2. Typical Speech Signal
Fig. 5.3. Typical Speech Spectrogram
Fig. 5.4. Graph for Power Spectral Density v/s Frequency
Fig. 5.5. Distribution of signal in Band of Filters approach
Fig. 5.6. Frequency Distribution in Bands
Fig. 5.7. Short Time Fourier Transform
Fig. 5.9. Relation between different LPC coefficients

Chapter 6
Intelligent System Design in Speech

Abstract. Speech based systems may be used for a variety of tasks that we use in everyday life. The applications using speech may be broadly classified into speaker recognition and speaker verification. The recognition deals with the identification of the person who is speaking. The task of verification systems is to verify the claimed identity of the person. Here we make use of speech as a biometric modality. Besides, speech is extensively used for word identification whose application includes the speech to text conversion. The variety of applications has a large number of issues associated with them. In this chapter we give a broad overview of these systems, their working and design issues. Some of the issues addressed in the chapter include the variation of speech, effect of noise, text dependent and independent recognition, start and end of signal detection, etc. Most of the discussions revolve around the pattern recognition approach for the recognition systems. This would give an application base of the discussed approaches that we have studied so far or would be studying later in this book.

Speech is a valuable tool for application design that provides a completely new paradigm in terms of interface, authentication, identification or other tasks as required by the system. Speech based systems have been put into a variety of use that we can see all around. The common examples would be the speech based authentication systems or the speech driven IVRS systems that we commonly use. Speech provides a lot of easy functionality from the user's perspective since a lot of effort is not required from his end. At many places speech becomes the only option as the use of other techniques may not be viable. Consider a person driving a car or a person operating a machine at distance at some other end as an example.

The use of speech as a biometric is equally important where we try to identify people by their voice. Speech comes under the category of behavioral biometrics where the values of the features change along with time. The task is the identification or the verification of the person based on whatever he speaks. Speech has further been extensively used in a fused mechanism where they are mixed with other biometric modalities that result in the generation of the bi-modal or multi-modal biometric systems. These systems have largely been the result of the failure of one technique or biometric modality to completely solve the problem.

Speech forms a highly dimensional input as we saw in chapter 5. Both amplitude and frequency changes numerable times per second even in the simplest speech signal. Every characteristic is important that contributes towards the recognition. Speech in nature is continuous in nature or into analog form. The

A. Shukla et al.: Towards Hybrid and Adaptive Computing, SCI 307, pp. 129–151.
springerlink.com

digitalization of speech into digital bits itself leads to some loss of information which is negligible to detect and does not prohibit any recognition. This is done by sampling. The signal still represents a very high dimensionality that cannot be handled by the systems that we engineer.

Mostly while working with soft computing systems we may take a black box approach. Here we specify all the inputs with the possible outputs without necessarily specifying the underlying logic. The system itself learns and figures out the logic that maps the inputs to outputs or the system itself finds the patterns in the inputs. This depends whether the learning was supervised, unsupervised or reinforcement. Based on the discussions of previous chapters we know that the designed intelligent systems work over the highly dimensional input space where each input attribute has its own axis. The addition of even a single input attribute may hence add a lot of computational load on the system which would even make the system fail to perform. Hence the speech itself cannot form an input to these intelligent systems that we build.

The reduction of data without resulting in a much loss of information is done by feature extraction. In the context of speech these features come from the speech analysis techniques that we discussed and presented in chapter 5. The speech may be used as an input that may be analyzed by any of the techniques presented. These features may be used as inputs for the task of recognition. This solves the problem of high dimensionality of speech input by dimensionality reduction on speech. This even enhances the features that are good for the recognition purposes. These enhanced features may be easily memorized and generalized by the system that we build.

6.1 Recognition Systems Using Speech

In chapter 1 we discussed the basic architecture of any recognition system. Speech also represents a recognition system that has all the various modules that we discussed there. These modules are input, pre-processing, segmentation, feature extraction, pattern recognition, and historical database. We discuss them briefly in context of speech signals in this chapter. This may be represented as given in figure 6.1.

6.1.1 Input

Speech is normally recorded with the help of a microphone. The recording forms the input to the system. The signal must be as less noisy as possible for the system to perform well. This ideally requires a recording environment that is noiseless. This environment is usually hard to maintain as many wanted and unwanted noises creep into the recording. Further the microphone needs to be of a high quality to fully capture the sound and deliver it to the connected systems. Many microphones have an in-built noise reduction mechanism where they are able to filter out the noise from the speech being given as an input.

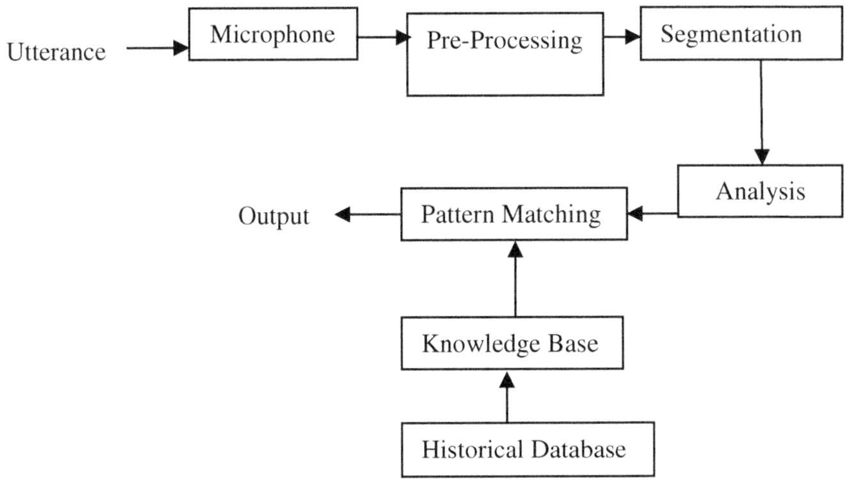

Fig. 6.1 General Architecture of Recognition System

The speech spoken as analogue form is coded internally by the system software being used into the digital format. These recordings are usually stored into a "*wav*" format. The other sound formats may also be used for storing speech utterances or signals. The speech is sampled before the software works upon the signal for the computation of the features in the given signal. This sampling reduces the amount of data in the signal and makes the data workable for the software in order to make it give the results in a finite time period. It is natural that the recording may be of a very high bit rate. This is done for the clarity of voice in the recorded signal. A more bit rate normally has a better quality and occupies more space. It carries more information about the signal per second of its utterance. While working with the signal analysis we may not want a very high bit rate. This is done by sampling. Sampling follows the Nyquist-Shannon principle. Here the sampling rate is fixed at twice the largest frequency represented in the signal.

6.1.2 Preprocessing

The input signal normally has noise in it. The removal of noise needs to be carried out before the signal can be used for the recognition purposes. This is important from the context of speech especially because most of the real life applications with speech have noise in the surroundings. This may be background noise of people speaking, noise of electro-mechanical signals, moving vehicles close to recording, etc. Hence many of the systems are recorded in a clean environment and testing is done in real life scenarios with some noise. It may usually be the case that the noisy recorded signals after preprocessing are clean enough for the purpose of recognition. In such a case the preprocessing has its own importance in cleaning up of the given signal.

The preprocessing forms a vital task as the signal is always prone to noise. The noise removal may be facilitated if we have some idea regarding the noise and its behavior. The removal takes place by the construction of a noise removal filter that performs an operation opposite to that of the noise. In this manner the effect of noise is reduced to an appreciably good extent. One of the commonly found noise in these signals is the impulsive noise where these is a sudden occurrence of noise at some part of the signal.

In any of the case, the removal of noise takes place by passing the signal through a noise removal filter. The use of low pass filters, median filters, extended Kalman filters, or the specially designed adaptive filters is common.

6.1.3 Segmentation

The other task associated in the recognition system involving speech is segmentation. Segmentation extracts out the relevant part of the speech out of the complete signal that is given to the system for the purpose of recognition. Say the task was the identification of word. The input may usually be a continuous speech representing some sentence. Now one may wish to extract out the exact word and use it for the recognition system. This extracted or segmented part of the speech may then be used for the purpose of recognition by the system.

The segmentation is surprisingly not an easy task in speech signals. This is not the case with most of the other systems where we may be able to get the Region of Interest by simple surveys. In case of visual information it is many times easy to extract an object due to its non-connectivity with the other objects. Any simple edge detection system may hence do bulk of the task with little left for the segmentation. We definitely assume no occultation in between the various objects that is difficult to handle.

In case of speech it is very difficult to recognize where does the needed speech signal start from and where does it end in the complete speech signal that is given to the system as an input. This is the problem usually addressed as the problem of continuous speech recognition. Many other recognition systems may either assume the input to be only the needed part or that the word is manually segmented.

In order to better understand the problem in continuous speech, consider a sentence being spoken normally by a speaker. It may sometimes be thought that silence in between words may be common to separate the words. But in reality this would not be the case with most of the spoken sentences that we normally speak in a good and continuous flow.

The problem of inability of isolation of the desired part of the signal is not only prominent in the speech recognition system, rather this is a problem faced in the speaker recognition systems as well. A major reason for this is the fact that most of the systems employed are text dependent where the text that the speaker is supposed to speak is fixed. Another major reason for the same is the fact that speaker recognition systems have to capture interesting features in the complete signal or the complete sentences. We discussed that feature enhancement is an important step that gives the recognition system good inputs based on which the task of recognition may be easily done. The same applies in the segmentation where good parts of the speech may be needed to be segmented and not the noise and silence that do not carry any information that may aid in the recognition.

Hence we may not be able to clearly isolate the utterance needed out of the complete signal that is given to the system. This problem is solved by using a technique called as Dynamic Time Warping (DTW) that we discuss next.

6.1.4 Dynamic Time Warping

Dynamic Time Warping (DTW) is a technique used to solve the above mentioned problem of segmentation of the needed signal out of the complete signal that is given to the system (Campbell 1997; Rabiner and Juang 1993; Pelecanos and Sridharan 2001). The basic problem is that we are to match two signals to find out if they are same or not. It is natural that they would not match perfectly. Hence the problem is to match the recognition score between these two signals. However it is unknown where does the signal lie in the master signal in which matching is to be performed.

For this we make use of a dynamic window that iterates through the signal. It tries to generate every possible combination regarding the start and duration between the signals. All the possible combinations are checked for their matching score. The best matching corresponds to the most likely start and duration. In this manner the segmentation may be carried out. The recognition score decides the level of matching based on which the decision regarding the acceptance or rejection of the match may be made. The recognition score may be made using any statistical or intelligent techniques as we shall see later in this chapter. The algorithm may make use of the principles of the dynamic programming for this computation. Using DTW is naturally time consuming as compared to the simple segmentation methods and requires a very different modeling scenario. Hence this step is usually assumed to be mixed with the pattern recognition in the model or architecture of a speech based recognition system.

6.1.5 Feature Extraction

The next task associated with recognition of speech is feature extraction (Mammone 1996; Sonmez *et al* 1998; Reynolds 1994; Kumar *et al* 2009). This is done by signal analysis techniques discussed in chapter 5. This limits the data size such that the input features can be handled by the recognition system. This also results in reducing the learning and testing time and effort as well as the volume of the training data size required for the system to be effectively trained. The extracted features are much simpler to learn than the entire signal. The extraction further enhances the features which is important for the optimal performance of any intelligent recognition system.

The extracted features represent the signal in their own manner. All the needed information is represented in them and the unnecessary or redundant information is left out. In many of the analysis techniques it would be possible to re-construct the complete signal with very little deviation from the original signal or very little loss of information. Since we are limiting the information by the analysis technique by only choosing a limited number of features, it may not be possible to get the complete original signal back.

6.1.6 Pattern Recognition

The last task is pattern recognition. This is done using any of the intelligent methods or other methods (Farrell 1990; Kala *et al* 2009). The class of neural networks that we presented in chapter 2 forms good and efficient methods for pattern recognition. These systems are easily able to classify or cluster the inputs into known classes. The other methods include Hidden Markov Models (HMM), Reinforcement Learning, Fuzzy Systems, or Template Based Matching. We briefly summarize the methodology of the neural networks for these types of problems. We do not deal with the other methods in this text.

The Artificial Neural Networks (ANN) are provided with a historical database that records the previously known inputs. If the learning is supervised the outputs are also known. In case the learning is unsupervised, the outputs are not known. In either of the cases the ANN is supposed to learn the inputs and form rules, trends or patterns in them. This summarizes the whole data into smaller and more compact representation. In implementation this representation is in the form of the weights of the connections in the ANNs and the various other parameters that they hold. The learnt ANNs are then ready to take unknown data. It is assumed that the learning would form generalized rules that must hold good for the unknown data as well. In such a case the unknown inputs given to the system would give correct outputs. This is the output of the pattern recognition phase that forms the final output of the system.

6.2 Types

In the above sections we have briefly mentioned the entire paradigm of intelligent systems as recognition systems. In this section we present an overview of the various kinds of systems and classify the entire domain. The basic type of systems from the point of view of intelligent systems that we are discussing is the recognition systems that more or less have a general structure discussed in section 6.1. These systems using speech as a means may be broadly classified into 4 types. These are speech recognition, language identification, speaker recognition and speaker verification (Campbell 1997).

We may easily see that there are two notions being presented. The first notion mainly deals with the text of the signal that is being spoken. Here it is not important who is speaking or how is the signal being produced, rather the content of the speech or its phonetics is important. The other notion being presented is that of the speaker based systems. Here the text or phonetics is not important, rather the source is important which represents how the signal is produced. This carries the information about the speaker.

These systems may be understood by figure 6.2. Each of these is discussed in the next sections.

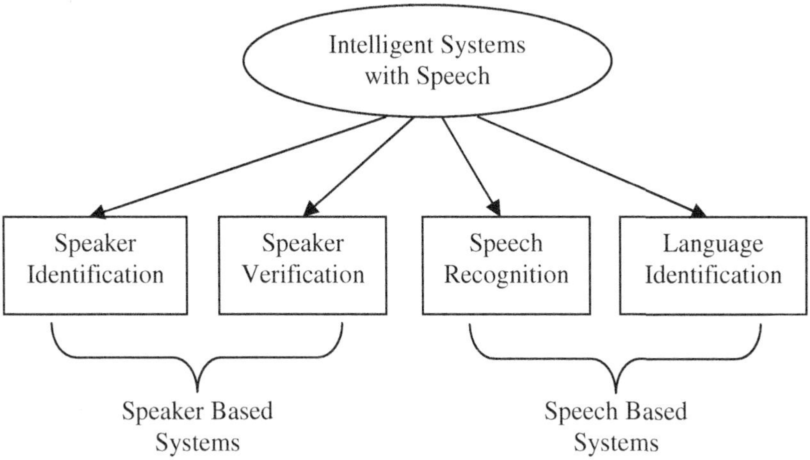

Fig. 6.2 Types of intelligent systems in speech

6.2.1 Speech Recognition

The first kind of system is the speech recognition systems. These systems try to understand the text of the spoken word or sentence (Gish *et al* 1991). One of the major applications of these systems is in speech to text conversion where these systems extract out the text spoken by the person. The other common example includes the making up of intelligent interfaces of machines where the machines are operated based on the commands that are given by the human speakers. This forms a hands-free interface to any system which has its own importance in interface design.

These systems in some or the other manner try to exploit the information of the phonetics in speech. Every word is a collection of phonetic sounds associated with it. This sound largely depends upon the language, but currently we focus our attention only on the single language systems. Multilingual speech recognition is possible but much difficult than the single language speech recognition. The phonetic sound associated with the word may be visualized as a collection of nodes that contribute a sound in the total utterance of the word. The features we discussed in chapter 5 are good in capturing these features or the phonetic sounds. By capturing these sounds we may easily make up a speech recognition system.

The lower formant frequencies F1 and F2 are typically very helpful in recording the phonetic sounds. It is known that these frequencies largely carry the information regarding the phonetics and very little regarding the source. Hence these are stable or constant when the same word or phonetic sound is spoken by various speakers and change drastically when the word or the phonetic sound changes. The identification of the phonetic sound or the complete word may be done using many techniques available in literature.

6.2.2 Speaker Recognition

The other broad type of problem that we deal is the speaker recognition (Campbell 1997; Martin and Przybocki 2000; Doddington 2001; Shukla *et al* 2009a; Shukla and Tiwari 2007). This problem deals with the identification of the speaker who spoke the text. The speaker recognition system tries to identify the speaker based on the personal characteristics as well as the characteristic manner in which a person speaks a text. We are usually able to identify people by their voice. This happens numerous times in everyday life when someone calls us or speaks to us. We may be able to recognize them even without seeing them. This forms the motivation for the speaker recognition systems. Here we engineer machines that are able to identify people by their voices that are given as input to the system. These systems also make use of the personal characteristics of the people to carry out the identification.

The analysis of speech signals that we studied in chapter 5 is again capable of extracting out the features out of the speech signal that convey information about the speaker. This is a source based approach where the information about how the signal was produced is of a large importance. This helps in figuring out the speaker who spoke the lines. It is important for the features to change largely with the differences in speakers and to remain as constant as possible with the same speaker. The features must remain constant no matter how many times the speech is recorded.

The higher formant frequencies F4 and F5 are two characteristics that convey good information regarding the source or the speaker. These features largely remain constant with the same speaker and change largely along with a change in speakers. They convey very less or practically no information regarding the phonetics which makes them good features with limited dependence on text.

6.2.3 Speaker Verification

The other system of common interest is the speaker verification system. In these systems a person claims some identity and the system is supposed to verify whether the claimed identity is correct or fake. These systems verify or authenticate the identity of the speaker and are hence widely used for security purposes. These systems may be easily deployed at the entrance gates which presently make use of deputed staff for checking. The automated systems have two major advantages that are not found in the manual system.

The first major advantage is a higher level of security. These systems are automated and use speech that is a characteristic of a person. It is hence very difficult to befool these systems. The same may not be necessary for the manual systems where a little slag in the procedure for checking can be hazardous. Besides, these systems save a lot of human capital which again is advantageous and cost saving.

The other major advantage is of speed and ease. The automated systems are able to do a bulk of authentication very rapidly. These systems can process large volumes of queues for authentication with good level of security in a comparatively short time. The same procedure being done by the manual labor is time consuming. Besides it is a discomfort for the person to undergo large stages of non-automated security checks which is not the case with automated systems. The person may simply speak and pass on.

The principles that we discuss with the speaker detection system regarding the source based identification hold good for this system as well. The same features and methodology for use drives these systems to carry out the authentication or verification of the individual.

6.2.4 Language Identification

These systems carry forward the task of identification of the language based on whatever is spoken by the speaker. This forms a very different paradigm of application into speech based systems that are seldom studied as a completely different category. We highlighted the language that plays a role in the speech identification. We further highlighted the point that different languages have different phonetics because of which it may not be genuinely easy to unite the various languages for a unanimous speech recognition system. These factors differentiate the language identification system from the other speech recognition system.

The language identification is also largely a source based approach where the phonetic based parameters are more important in the determination of the final decision regarding the language that forms the output of these systems.

6.3 Further System Classifications

The classifications presented in section 6.2 form the major classifications of the entire domain of speech based intelligent systems. We may further be able to classify the systems on the other factors on which the system performance and use depends upon. The system to be build and deployed into the real life scenario highly depends upon its purpose, environment and the prospective users. Further the system constraints, user demands and other factors also depend upon the practical real life scenario where and how the system is being deployed and the underlying usage. Every classification has some advantages and disadvantages. The more generalized systems are usually harder to make and have a limited accuracy. This is the converse in the more specific systems that we study in these classifications.

6.3.1 Text Dependence

Based on the text the speaker recognition and verification systems may be broadly classified into two major heads. These are text independent systems and the text dependent systems (Wan and Campbell 2000).

The text dependent systems have a specified text that the speaker has to speak for the recognition or identification purposes. If the speaker tries to speak another word or another text, the recognition would fail. This system is very easy to imaging for the verification system. Here every speaker has a particular word or a combination of words. The identification happens only when the speaker speaks this particular word. The actual system would be such that the person comes, says this word and is granted the access or rejected based on the system judgment. For a speaker recognition system a library of words is usually maintained. The system is able to identify the speaker only when they say any word in this library. If a new word is spoken the system would not be able to carry forward the task of identification and may give an erroneous output.

The other type of systems is the text independent systems. These systems are much generalized in the sense that they do not depend at all on the text. These systems are able to identify the person or authenticate him based on whatever he speaks. In case of verification a person may come, say anything and based on this the person may be given access or restricted depending upon the system judgment. In identification as well, the system would identify the person irrespective of whatever he speaks.

6.3.2 Speaker Cooperation

The other kind of classification that we may perform is the speaker cooperation. Based on this we classify the systems as speaker cooperative systems and the speaker non-cooperative systems.

The speaker cooperative systems are the ones where the speaker is supposed to speak in a definite manner to carry out the recognition or verification. These systems are very much prone to the manner in which the person speaks. A variation may not result in the correct identification or verification. The restrictions that are usually put upon the speaker are the time of speech, duration of speech and other speech variations. In these systems the speaker is supposed to only speak when he is asked to do so. A lot of variations in time durations may also not be desirable. In case the system fails to identify the word, speaker or carry forward the verification, these systems would demand the speaker to repeat the speech. These restrictions make it reasonably simple to engineer systems as many of the issues relating to the speech recording and analysis and left at the user end.

Besides the practical design and ease of identification, a major factor behind these restrictions is the variations in speech. Even if the same speaker speaks a sentence twice, there might be variations in the various features recorded by the system. This creates a problem in the design of these systems and their usage. The reasons of these variations are the mood of the speaker, context in which the speech is spoken, speaker consciousness to system, etc. All these restrict the performance. The problem with the speech recognition is even more difficult due to the variations in the different languages and dialects. The same word in the same language may be spoken with some variations by people from different areas or nationality. The dialects differ the manner in which the same word is spoken.

6.3.3 System Robustness

Speech largely deviates from its ideal behavior that is usually a problem in most of the recognition systems. The real life applications making use of speech are hence much challenging in terms of design and implementation (Rose *et al* 1994; Mammone *et al* 1996; Kala *et al* 2009). This gives rise to a completely new domain in speech that is called as robust speech recognition. This domain takes into account all the various issues in effective use of speech that come in practical life and tries to handle all of these. The task of working with these real life signals requires the systems being much more robust in nature as compared to the simple speech based systems that operate in a noiseless environment with a lot of support from the user in terms of how and what he speaks.

The quality of speech may be very poor based on which the robust speech recognition systems are designed and operated with. Many times there is a lot of addition of noise into the signal that makes the original signal very poor and hard to understand even for a human. Various kinds of background noise exist that is very difficult to make out or filter. In many scenarios the environment may not be very noisy, but some noise may appear at the time the recording was made. This may be in the form of a person speaking, an airplane landing, sudden lightning, etc. All these are unexpected noises that make the recognition system highly difficult to engineer. In many other scenarios, noise is embedded right into the recording environment. The common examples may be of a market place where the background noise is always present or of a place with a noisy fan, etc.

Noise may not only be present at the source. Many times noise is induced in the transmission of the signals as well. The channel has its own noise forms that are induced over the signal. The various operations at the source and the destination may also result in noises of various kinds that all adds to the total noise in the system.

Speech signal carries various features that may be destroyed by these noises. Now any recognition system uses these features for recognition which cannot be afforded to be destroyed. Working with robust speech recognition system hence deals with the making up of models that overcome the destruction of noise and cleverly reconstruct the original signal. The feature extraction hence extracts the good features.

6.4 Template Based Matching

We discussed a variety of approaches, especially the neural approaches in chapter 2 that enable design of intelligent systems for the recognition and verification purposes. We further saw in section 6.1 how the features are extracted and facilitate recognition by these intelligent systems. The features form the input to the intelligent systems that work in the feature space with each feature as an axis of a highly dimensional space. The output class may be represented as a label as discussed in chapter 1. This is shown for a two feature system by figure 6.3. The entire task of the intelligent system is to make the decision boundaries that separate the classed from one another. The use of the discussed neural approaches is good as these approaches are able to make very flexible decision boundaries that separate the various classes from one another.

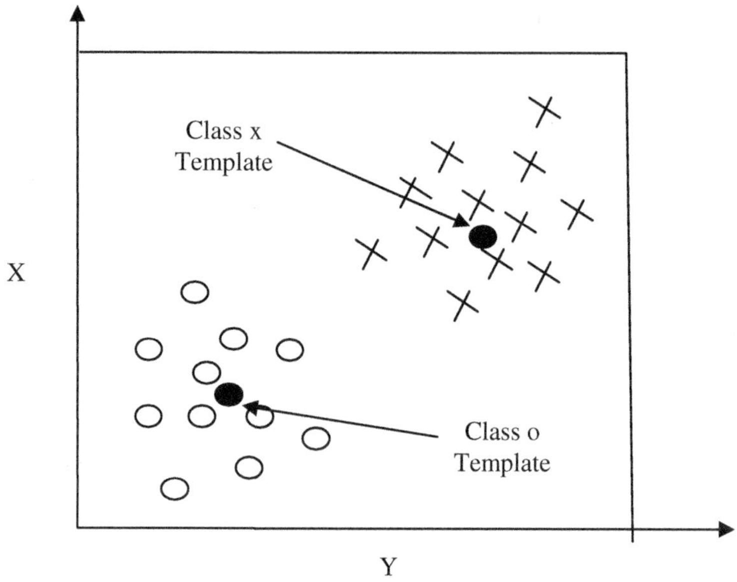

Fig. 6.3 Template based approach

The ease of separation of the classes largely depends upon the quality of the attributes or features chosen. Good attributes have a very high inter class separation and a very low intra class separation. In the feature space this means that the various instances of the same class are very close to each other and the other classes are far apart from each other. In such a case we may easily replace the instances of a single class by a template at the average location of the instances of that class. This forms a feature space with reduced storage of points. The various points represented per class are known as the template of the class, since it depicts the average behavior of the class. This is shown in figure 6.3. Now any unknown input is more likely to lie close to one of the templates. Based on the distances we may return the template closest in the feature space to the location of the unknown input. This forms the output of the system. Such systems are known as the template based matching systems.

Now an interesting observation would be the comparison of this approach to the neural approaches discussed in chapter 2. The neural approaches have a better flexibility in the modeling of the classes and decision boundary that usually has a better recognition score, especially when the input attributes are not very ideal. However, as we shall see later in the chapter, recognition systems need constant addition, deletion and updating for real life usage. The neural approaches may require a long training phase for learning which makes these operations slow and expensive. This is not the case with the template based systems where only the required template is to be changed, added or deleted.

6.5 Speaker Verification System Design

We talked about the meaning, behavior and working of the verification systems in section 6.2.3. In this section we look at the design and some of the underlying issues to be taken care of in a verification system (Naik 1990; Rosenbrg 1976). The verification system follows the general template of a recognition system that we discussed in section 6.1. However this has some additional properties associated with it that makes this system complete. The added functionalities include the enrolment, verification and system update. This makes the pattern recognition system represented in figure 6.1 to expand to the additional steps. These are shown in figure 6.4. We discuss the various added steps in the next sub-sections. This would be followed by the discussion over the other issues of the verification systems for design purposes.

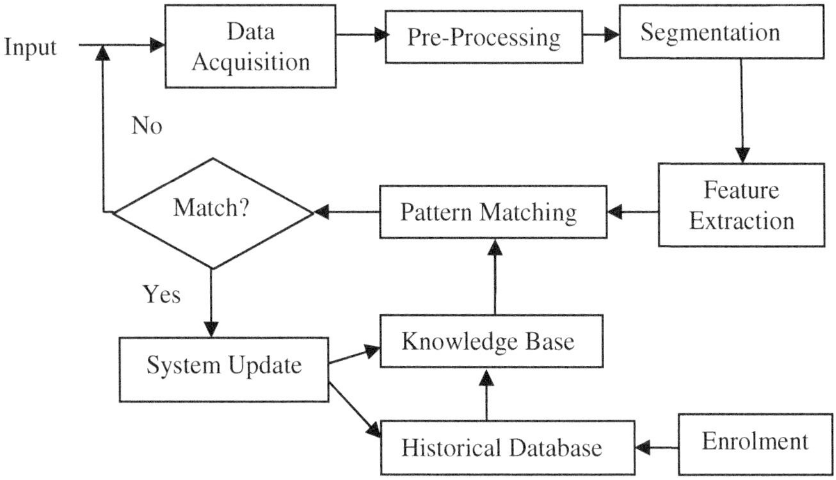

Fig. 6.4 Biometric Verification System Design

6.5.1 Enrolment

The first major addition in these systems is that of enrolment. This is the step in which a new user is added to the system. Enrolment begins with the recording of the credentials of the user. This goes through preprocessing and segmentation and ultimately features extraction. The features extracted by the system for the added individual is the unique identity of the individual which is possessed by himself alone and cannot be easily imitated. This feature set is added to the database storing the features of the system.

If the system used for pattern recognition is a simple template based matching, the new system is ready. The recorded features are the template for this speaker which is stored in the simple lookup table of features. However a neural system may require re-learning to learn about the new added feature set. In this manner the newly added feature or speaker becomes a part of the entire learnt database which has its rules in the form of connection weights of the ANN and determine the generation of outputs from the inputs.

6.5.2 Verification

The next task to be performed is the verification. In this step we judge whether the person is same as the identity he claims. Verification systems are usually used for security (Doddington 1965). At many of the places security has to be very strict to avoid even the smallest possibility of infringement. For these reasons the verification systems are made to work such that they disallow all possible cases which may be possessing chances of fraudulent persons claiming identity. We have seen that speech may greatly vary due to the various internal and external factors like noise, mood of the speaker, etc. Hence it is largely possible that the voice of a genuine person is identified as fraudulent due to a very strict matching. For this reason a person may many times be asked to repeat the speech to again try the identification. It is possible that this recording might not have so large variations from the ideal case and may be correctly identified. This possibility of repetition may hence be used to correct the decision of the system. Now it is natural that again and again speaking the word to get authentication is discomfortable from the user's perspective who could have better been served by the manual system. However genuine cases being identified as fraudulent is not very common in nature. Besides, the boost that these systems bring in security is much more important that the personal comfort of a few people who may be wrongly identified at first instance.

6.5.3 System Update

The last unique addition of these systems to the recognition scheme is that the system may be optionally updated based on every correct recording. If the person is correctly identified, his feature vector may be updated in the system. We know that the various biometric modalities for the same person change along with time and age. We may hence want that the system itself adapts itself to these changes in the course of time. This is done by the system update on successful identification of the system. The old value may be updated to the new value. If the system is based on learning, the learning algorithm may be used immediately or at a certain frequency. Hence as the biometric value of the speaker changes with time, the system adapts to these gradual changes.

6.5.4 Verification Ratios

The accuracy or the performance of the verification system may be assessed by two fundamental ratios. These are False Acceptance Rate (FAR) and false Rejection Rate (FRR).

FAR is a relative measure of the total number of imposters that the system grants access. An imposter is granted the access when he is identified to be the person with the claimed identity which he is actually not. It is a relative measure of the security infringement cases that occur in the system. This rate must hence be as low as possible. Normally an erroneous decision by the system may not be made when the features of the person claiming identity are quite different from the features of the actual person. This means that these two are located at reasonably distant locations in the feature space. However the decision might be erroneous in case they possess similar feature or lie in close vicinity in the feature space. It may be noted that most of the biometric systems including speech represent characteristic features rather than the ones that may be visual or auditory in nature to the common man. Hence the similarity here does not necessarily mean that that these are similar to hear or see. Two distinctly sounding speech signals may actually lie quite close in the feature space. In general the best way to judge the working of a verification system is to try to try to mimic a genuine person to challenge the system to make correct decision. Same as the general human checker, mimics may be difficult for the system to identify.

The other ratio of interest is the False Rejection Rate (FRR). This rate measures the total relative number of cases that the system correctly identifies wrong and rejects the access. This ratio must be as high as possible for the ideal working of the system. It is a relative count of cases where the infringement was prevented. This is normally very simple when the genuine person's features and the claimed person's features lie far apart in the feature space.

Another common manner of representation of the ratio is as True-speaker Rejection Rate and Impostor Acceptance Rate.

6.5.5 Matching Threshold

Another concept we place here is of the match threshold. It is natural that at any instance the feature vector of the person claiming identity would never match exactly with the genuine person's feature vector. It would be matching to a certain degree only. If the matching degree is too high, we can say that probably the person claiming the identity is genuine as it closely matches the genuine person. Similarly if the matching degree is very poor, then we can be almost assured that these two are two different people. In verification system we must make concrete decisions regarding the matching. The verification algorithm is supposed to give an output of true or false depicting a match or a no match between the persons. In such a case we need to fix a certain threshold of matching. If the match is above this threshold, we give the output that the claimed identity is genuine. However if the match is below this threshold, we say that the identity claimed is fake.

This threshold decides the level of security which again governs the FRR and FAR. Suppose the matching threshold is fixed to be too low. In such a case the system would give a positive match even if the matching is not very high. This would make many imposters match with the genuine claimed identity. This would make the FAR large and the FRR low. Similarly if the FAR is fixed to be too high, the

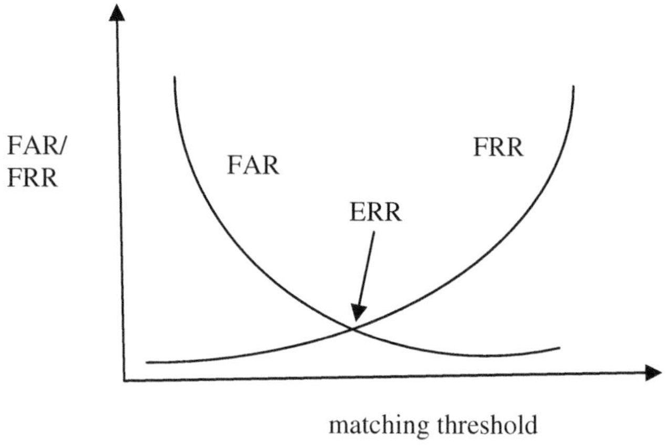

Fig. 6.5 False Acceptance Rate and False Rejection Rate

system would return a negative match for most of the cases. In this manner most of the genuine cases would also be rejected and the false cases would be strictly rejected. This would make the FAR very low and the FRR very high. The change in FRR and FAR for the various values of threshold is plotted in figure 6.5.

The intersection point of the two rates is known as the Error Equivalence Rate (ERR) or the Crossover Error Rate (CER). This point is desirable for the system as it balances the two opposing effects of accepting a fake person on denying a genuine person which is what happens at the two extremes of the threshold. This serves as an ideal threshold level for many systems.

The threshold largely depends upon the system where the biometric is being used. Suppose that the system is being employed at a regular office for authentication where the chances of infringement are low. Here the system may have a poor threshold set. This is because we cannot afford wastage of time in multiple recordings of the same person till a perfect match is found. Here time and ease of use may be given a priority over the security. In the same context for many of the high-security areas the threshold may be fixed very high. A person may not mind speaking again to get an access. The high security is more important than time and the ease of the person.

This may also be visualized in the feature landscape where the features are plotted against the governing classes. In case the threshold is too low, points lying far away from the region where the genuine person is found in the feature space are acceptable. On the other hand if the threshold set is very high, the claimed person needs to be reasonably close to the region where the genuine person is found in the feature space.

6.5.6 Mimic Resistance

Another factor we study in the category of the verification systems is the mimic resistance (Reynolds 2002). This is an important consideration to keep in mind while designing of any system. This refers to the minimal distance in the feature space where a positive match is returned. This is naturally analogous to the concept of threshold discussed above. Here we place a major factor that determines the design and setting of this parameter. The effective testing of a verification system can only be done when we ourselves try to exploit its security by trying the various inputs that may or may not be faced in the real life deployment. In the case of speech systems, the basic methodology of testing might be to try to mimic the original speaker. The intelligent system must be able to demarcate the difference between the genuine and the fake speaker. This would enable it maintain a high FRR and low FAR.

In this measure of mimic resistance we compute this boundary radius that demarcates the true person and the mimic. We draw this around the template of the feature vector of the genuine person. This boundary must be large enough to accommodate the variations of speech in different utterances of the genuine person. At the same time this boundary must be short enough to separate even the best mimic.

6.6 Recognition with Closed and Open Data Sets

Recognition is one of the major categories of systems where speech finds application as discussed in section 6.2. The performance of the recognition systems may be easily reported in terms of the percentage of correct recognitions. An important criterion in the testing of these systems that comes from the practical relevance is the closed and open data sets.

In a closed data set the system is trained with various instances of a set of classes. The testing involves giving other instances of the same classes. Here we do not experiment the system by giving any new class to the system. Hence the task of the system is to simply find the best possible output based on the input given. This is one of the classes that the system was informed while training.

The other kind of testing is a much generalized version. Here the system is trained with various instances of some set of classes. However at the time of testing an entirely new class may also be given. The system hence has to first make a decision whether the given inputs would correspond to any of the classes or not. Then the system is supposed to return the correct class to which the input belongs. The system may hence not give any output if it finds that this is an entirely new class which it has not been trained with. In this type of training verification stage is involved along with the recognition.

Consider the problem of identification of speaker. Here a closed data set would correspond to the problem where at the time testing no new speaker comes and speaks. However if the testing involves the speech of some person completely new whose speech wasn't a part of the training data set, the system would be an open data set system.

6.7 Making of a Good Database

Making of a good database is an important activity in the speech based systems. This is important because of the variations in speech that come in various situations that we have been discussing in the chapter. A good database must be diverse in nature to ensure that the application that results is robust against most of the scenarios. The underlying notion in machine learning is that the system learns the historical database. Hence if any of the input is repeated, it would be able to give the correct output based on its learning. If the learning is generalized enough, we may extend the learning to new inputs as well. But these inputs give the correct output only in case the inputs are like some of the earlier presented inputs. Completely new inputs may result in wrong outputs.

We can easily see that from the model of classification in feature space as well. The learning results in summarizing the bulk of points given while training into a finite set of points. The testing with unknown inputs results in the correct outputs only when it lies somewhere close to a previously known input in the feature space. The completely unknown inputs are like random points in the search space that may lie far from the known points. Though a decision may be made depending upon the closest class found, it is likely to be erroneous.

The major point behind this discussion is that the system must be trained with a very diverse set of inputs that the system may ever face in real life application. Then only an optimal performance may be expected in all the cases. Else the system might only give correct outputs to similar inputs to ones given in training. This means that we have a diverse training that necessitates the making of a very diverse data set.

The dataset made for the speech based systems must have diversities in terms of age group of speakers, gender of speakers, different words, phonetics, dialects, accents, etc. Multiple recordings may be done for every speaker to capture all the variations in the same speaker in constant situations as well.

6.8 Example of Speaker Recognition

After studying the speech based intelligent system, we now take a procedural approach to make a speaker recognition system. In this section we discuss the basic methodology to carry out simple speaker recognition. In this section we briefly discuss the work of Shukla and Tiwari (2008) and Kala *et al* (2010).

6.8.1 Making up of a Database

The first task that needs to be performed is the making up of a good database. The database consists of the utterances of 20 speakers that primarily belong to the same age group. The text of speech was a set of 36 Hindi words that were spoken in continuation by the various speakers. Some of these words are "*ab is baar aap uske*" The complete details of these words may be found in Shukla and Tiwari (2008). These words consisted of a set of Hindi consonants that were interwoven between the Hindi vowels. A large range of combinations of consonants and vowels were covered for a robust recognition. Each recording was saved in a separate *wav* file.

6.8.2 Speech Processing

Each of the recordings was processed with the help of GOLDWAVE Software. A sampling frequency of 44 kHz was used for the purpose. Each of the word was extracted from the utterances of the speakers in the consonant-vowel-consonant (CVC) context. The individual extracted speech utterances were given to the MATLAB signal processing toolbox. The various speech features were extracted using this toolbox. For this experiment the extracted features were Time Duration, Number of Zero Crossing (NZC), Maximum Cepstral, Average Power Spectral Density (Average PSD), Pitch Amplitude, Pitch Frequency, Peak Power Spectral Density (Peak PSD), Formant Frequencies F1, F2, F3 and F4.

6.8.3 Inputs and Outputs

The input for the system was the various features that were extracted by signal processing. The system output is the speaker who spoke the particular input. The output needs to be digitalized so that it is used in the training and testing. For every input we represent a set of n outputs where n is the total number of speakers in the system. This may be represented by $<o_1, o_2, o_3, \ldots, o_n>$. Here every o_i represents the chances of the ith person being the speaker of the applied input. Here the chance is measured on a scale of -1 to 1. A -1 denotes the impossibility of the input belonging to the ith class or the ith speaker. A 1 denotes the assurance of the input belonging to the ith class or the ith speaker. In case of the supervised learning the outputs are to be specified for the various inputs. Using this method the output for any speaker i may be given by a vector with a 1 at the ith location and -1 at all other locations. For $i=2$ this is given by $<-1, 1, 1, -1, \ldots -1>$. At the time of determination of the output of any applied input, we may simply return the class i that has the maximum value of o_i in the set of outputs $<o_1, o_2, o_3, \ldots, o_n>$. This corresponds to the class that has the largest value of chance of being the output. Details of the method can be found at Shukla *et al* (2008, 2009b).

The complete data set obtained by the extraction of the features was collected. Normalization of the data was carried out by division of every attribute by using a normalization equation to make all the data lie between 0 and 1. This usually results in a better training of the ANN using Back Propagation Algorithm. This data was divided into two sets of training and testing. These were used for the training and testing of the system that we discuss next.

The complete methodology of the training and testing may be summarized by figure 6.6.

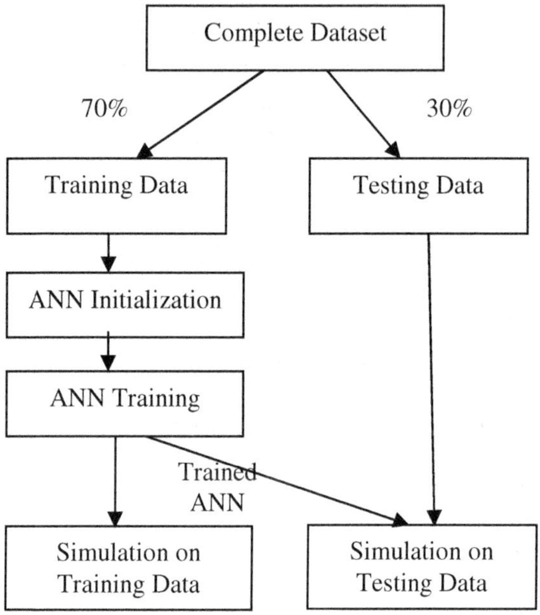

Fig. 6.6 Working Methodology of Speaker Identification

6.8.4 Training

The next task performed is of training. This made use of the training data set that made about 70% of the entire dataset. MATLAB Neural Network toolbox was used for this purpose. The ANN Multi-Layer Perceptron model was used. The model was initialized by setting the number of layers, neurons, activation function and training algorithm. Later the training algorithm attributes were fixed like momentum, learning rate, epochs and goal. The system was then trained using the training algorithm. This was a supervised learning technique adopted. The training curve of the system is given in figure 6.7.

6.8.5 Testing

The last stage involved the testing of the trained neural network. In this step unknown inputs reserved for the testing purpose were given to the system. This made approximately 30% of the database. The corresponding outputs to these inputs were computed using the trained neural network. These outputs were compared with the standard outputs of the systems that were known from the testing database. This was the final efficiency of the system.

After trying numerous cycles of training and testing with different network architectures and training parameters, the best testing accuracy was determined. This is the final accuracy of the system. The accuracy that this system gave using the above approach was 90% in the recognition of speaker.

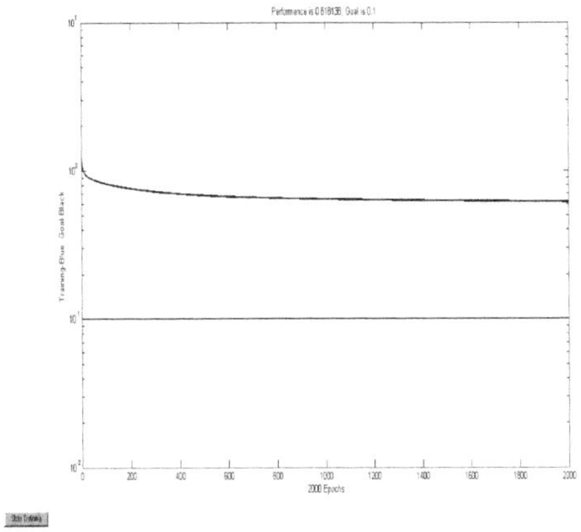

Fig. 6.7 Training curve for the problem

6.9 Chapter Summary

This chapter explored the various designs, concepts and applications of intelligent systems based on speech. The chapter stared with the basic recognition scheme used by the speech applications. Here we discussed about the various steps including data collection, pre-processing, segmentation, feature extraction and pattern matching. We additionally discussed about time warping which is an important requirement in speech since we do not know which part of the signal needs to be matched from the ones available in database. The chapter further explored the various types of speech based intelligent systems. Here we discussed about speaker recognition, speaker verification, speech recognition and language identification. Speaker recognition made use of speech as a biometric modality where the system is supposed to identify the person. Speaker verification deals with the decision making whether the person is same as per the claimed identity. Speech recognition deals with the recognition of the words spoken by the person. In language identification we try to identify the language being spoken. The systems were further classified based upon text dependence, speaker cooperation, and system robustness. All of these made an impact on the features and pattern matching model of use. These were all dependent on the kind of application where the systems are to be used. The next topic of discussion was the verification systems, where we stated the overall methodology of working in verification. The new steps of enrolment and system update were introduced. The chapter also presented the concepts of verification ratios of False Acceptance Rate and False Rejection Rate. The other topics of interest were Open and Closed data sets. The entire chapter was summarized by the making of a system for speaker recognition. The complete step by step procedure was stated. It started with the making of the database to pre-processing, segmentation, and feature extraction. Based on the extracted features the training and testing data sets were made. The training data set was used for network learning. The complete system was tested by using the testing dataset. In this manner the system was used for solving the problem of speaker recognition.

References

[1] Campbell, J.P.: Speaker Recognition: A Tutorial. Proc. IEEE 85(9), 1437–1462 (1997)

[2] Doddington, G.: Speaker Recognition based on Idiolectal Differences between Speakers. Eurospeech 4, 2521–2524 (2001)

[3] Doddington, G.R.: Speaker Recognition – Identifying People by their Voices. Proc. IEEE 73(11), 1651–1665 (1965)

[4] Farrell, K.R., Mammone, R.J., Assaleh, K.T.: Speaker Recognition using Neural Networks and Conventional Classifiers. IEEE Trans. Speech Audio. Process. 2(1), 194–205 (1994)

[5] Gish, H., Siu, M.H., Rohlicek, R.: Segregation of Speakers for Speech Recognition and Speaker Identification. In: Proc. of the IEEE Intl. Conf. Acoustics, Speech, and Signal Process, Toronto, Ont., Canada, pp. 873–876 (1991)

[6] Kala, R., Shukla, A., Tiwari, R.: Fuzzy Neuro Systems for Machine Learning for Large Data Sets. In: Proc. of the IEEE Intl. Advance Comput. Conf., Patiala, India, pp. 541–545 (2009)

[7] Kala, R., Vazirani, H., Shukla, A., Tiwari, R.: Fusion of Speech and Face by Enhanced Modular Neural Network. In: Proc. of the Intl. Conf. on Info. Syst., Technol. and Manag., ICISTM 2010, Bankok, Thailand, pp. 363–372 (2010)

[8] Kumar, R., Ranjan, R., Singh, S.K., Kala, R., Shukla, A.: Tiwari R, Multilingual Speaker Recognition Using Neural Network. In: Proc. of the Frontiers of Res. on Speech and Music, FRSM 2009, Gwalior, India, pp. 1–8 (2009)

[9] Mammone, R.J., Zhang, X., Ramachandran, R.P.: Robust Speaker Recognition. IEEE Signal Process. Mag. 13, 58–71 (1996)

[10] Martin, A., Przybocki, M.: The NIST 1999 Speaker Recognition Evaluation - An Overview. Digital Signal Process 10(1), 1–18 (2000)

[11] Naik, J.M.: Speaker Verification: A Review. IEEE Comm. Mag. 28(1), 42–48 (1990)

[12] Pelecanos, J., Sridharan, S.: Feature Warping for Robust Speaker Verification. In: Proc. A Speaker Odyssey, The Speaker Recognition Workshop, pp. 213–218 (2001)

[13] Rabiner, L., Juang, B.H.: Fundamentals of Speech Recognition. Prentice Hall International, Englewood Cliffs (1993)

[14] Reynolds, D.A.: Experimental Evaluation of Features for Robust Speaker Identification. IEEE Trans. Speech Audio Process. 2(4), 639–643 (1994)

[15] Reynolds, D.A.: An Overview of Automatic Speaker Recognition Technology. In: Proc. of the IEEE Intl. Conf. Acoustics, Speech, and Signal Process (2002)

[16] Rose, R.C., Hofstetter, E.M., Reynolds, D.A.: Integrated Models of Signal and Background with Applications to Speaker Authentication in Noise. IEEE Trans. Speech Audio Process. 2(2), 245–257 (1994)

[17] Rosenberg, A.: Automatic Speaker Verification: A Review. Proc. IEEE 64(4), 475–487 (1976)

[18] Shukla, A., Tiwari, R.: Fusion of face and speech features with ANN for speaker authentication. IETE Tech. Rev. 4(5), 359–368 (2007)

[19] Shukla, A., Tiwari, R.: A Novel Approach Of Speaker authentication By Fusion of Speech and Image Feature using ANN. Intl. J. Info. Comm. Technol. 1(2), 159–170 (2008)

[20] Shukla, A., Tiwari, R.: Intelligent Biometric system: A case study. Intl. J. Info. Technol. Res. 1(3), 41–56 (2008)
[21] Shukla, A., Tiwari, R., Meena, H.K., Kala, R.: Speaker Identification using Wavelet Analysis and Modular Neural Networks. In: Proc. of the Natj. Symp Acoust. 2008, Vishakhapatnam, India, pp. 125–130 (2008)
[22] Shukla, A., Tiwari, R., Meena, H.K., Kala, R.: Speaker Identification using Wavelet Analysis and Modular Neural Networks. J. Acoust. Soc. India 36(1), 14–19 (2009a)
[23] Shukla, A., Tiwari, R., Meena, H.K., Kala, R.: Speaker Identification using Artificial Neural Networks. J. Acoust. Soc. India 36(1), 20–25 (2009b)
[24] Sonmez, K., Shriberg, E., Heck, L., Weintraub, M.: Modeling Dynamic Prosodic Variation for Speaker Verification. In: Proc. of the, Intl. Conf. on Spoken Lang. Process., vol. 7, pp. 3189–3192 (1998)
[25] Wan, V., Campbell, W.M.: Support Vector Machines for Speaker Verification and Identification. In: Proc. of the IEEE Neural networks for signal process. workshop, vol. 2, pp. 775–784 (2000)

List of Figures

Fig. 6.1. General Architecture of Recognition System
Fig. 6.2. Types of intelligent systems in speech
Fig. 6.3. Template based approach
Fig. 6.4. Biometric Verification System Design
Fig. 6.5. False Acceptance Rate and False Rejection Rate
Fig. 6.6. Working Methodology of Speaker Identification
Fig. 6.7. Training curve for the problem

Chapter 7
Intelligent Systems Design in Music

Abstract. Music is an exciting area where the computational intelligence has cast a deep impact. In this chapter we study a few intelligent systems in the domain of music. The vast volume of music available in various formats has necessitated the need for their automated classification. Here we discuss the system for the identification of genres as well as the recognition of artists. These systems have a variety of application in playlist generation, music suggestion, music fetching etc. The other part of the chapter would focus upon the composition of music. Here also we discuss a variety of methods using Genetic Algorithms and Neural Networks. The manual assistive design of Genetic Algorithms enables the automated composition of music as per human demand. The neural approach uses a series prediction phenomenon to compose music when some part of it is known. These systems enable good composition techniques which may be employed to assist human composers in their task.

Music is a fascinating field which is now assisted by the intelligent systems that carry out many operations to automate or semi-automate most of the tasks of the field. The effect of music is prominent at each and every place. The wonderful natural sounds of rains, chirping of birds, etc. represent music in their own way. The music instruments with their melodious sounds have always caught the attention of people. Unlike speech which moves in language dependent phonetics, the music is a jingling of the musical nodes. The effect of music to convey spirit, energy and other emotions can be easily felt in the manner it is played by the various instruments.

Music forms a very interdisciplinary field of research from a scientific point of view. The automated intelligent systems for musical classification, synthesis and recognition especially need valuable inputs from the domains of physics, acoustics, AI and soft computing, cognitive science etc. This has hence emerged a completely new field of its known that is occasionally addressed as computational musicology or cognitive musicology.

The intelligent techniques have greatly extended their fruits to the domain of music. It is hence now possible to intelligently process, analyze, compose and classify music. These further extend their applications in numerous areas. The synthesis of music artificially forms another promising field which enables the production of sounds artificially similar to those generated by the natural musical instruments. This has again given a completely new paradigm to the musical world where the musicians can now play their instruments by their artificial

A. Shukla et al.: Towards Hybrid and Adaptive Computing, SCI 307, pp. 153–173.
springerlink.com

counterparts. The sudden burst in the volumes of music has been another major motivator for the automated processing of the music which cannot be now done manually because of the large amount of time needed and the increasing consumers' demands. This opens gateway for a lot of research in these allied areas where the challenge is to efficiently analyze and classify music.

Like speech, music is also a very complex signal when viewed in the time or frequency domain. The musical instruments, despite their simple architecture and look are easily able to generate a rich class of music that displays complex behavior. This makes the analysis difficult mainly due to the highly dimensional nature and the complex relation of the signal for identification of genres and artists.

The processing and analysis of these signals follows the same norms as that of speech. Most of the features that we use in the various recognition schemes of music have already been discussed in chapter 5. Here also the main function of analysis is to enhance the features and carry out dimensionality reduction.

In this chapter we would focus our attention on two major problems. The first problem is the classification of music. Here the motivation is to identify the various user desired classifications and classify the given music samples accordingly. The other problem is that of automated music composition. In this problem we develop automated systems that can intelligently generate good music on their own.

The first task that we carry out in this chapter is the musical classification. Here we would classify the music on the basis of genre and artist. This would be done using a variety of models and features that facilitate easy classification. From the point of view of intelligent system design this is a simple classificatory problem where the task would be do construct or predict the decision boundaries in the feature space. From the context of music, the focus of the section would be to judge the good characteristics or features and extract them using the signal analysis techniques.

The other task of concern in this chapter is music composition. Music composition deals with the arrangement and selection of the correct genres, harmonies, melodies and timbre in music in such a manner that gives a soothing sound to the human as well as matches his taste. In music composition we would see how the artificial intelligent systems are able to construct or compose simple music as per the human demands. Here we make an extensive use of Genetic Algorithms for the problem solving. The problem is hence an optimization problem where the optimal combination of nodes is to be computed.

We broadly divide this into four categories that form the next four sections. Section 7.1 would discuss the work of Kosina (2002a) in musical genre recognition. In section 7.2 we present the work of Scott (2001) for genre and artist classification. Section 7.3 is a discussion over the work of Tokui and Iba (2001). Here a human assistive genetic algorithm has been used to carry out composition of music. Finally in section 7.4 we present the work of Horner and Goldberg (1991) for music composition using computer assisted genetic algorithms.

7.1 Genre Classification

The first model of study is the genre classification. Whenever we hear any music we are easily able to classify it as classical, rock, pop, metallic, etc (Tzanetakis and Cook 2002; Li and Ogihara 2005; Li et al 2005; Burred and Lerch 2003). These classifications enable us select and play songs as per our interests and mood. Selecting the song from a particular genre and developing liking for a particular genre is a common phenomenon with people. The purpose of an automated genre classification system is to automatically classify the music into its genre or type of music.

Genre types do not have any formal and concise definitions. These have more or less been formed by the various people on their knowledge or understanding. The same music may be interpreted to be of different genres by two different people. This irregularity is further a problem in the automated genre recognition systems. The machine learning approaches present promising solutions for these types of systems where the underlying knowledge is not to be known in advance. The system is itself able to extract out knowledge or rules based on the supplied inputs and outputs. The general scheme for recognition of this system is same as that discussed in chapter 6. We hence focus our attention only to two major points that affect the system performance and working. These are feature extraction and machine learning. The system developed by the authors is called as MUGRAT (Music Genre Recognition by Analysis of Texture) (Kosina 2002b).

7.1.1 Feature Extraction

The feature extraction analyzes the signal and extracts out good features from it that are able to separate the classes or the genres in the feature space (Rabiner and Juang 1993). These are broadly classified into two types. These are speech surface features and musical beat features.

7.1.1.1 Musical Surface Features

The musical surface features are the ones obtained from the direct representation of the musical signal. These features include the following:

- **Spectral Centroid:** This feature computes the centroid of the spectrum measured by the magnitude of the Fourier Transform. It is also called as Spectral Brightness. This factor is given by equation (7.1)

$$C = \frac{\sum_{i=1}^{N} S_t(i).i}{\sum_{i=1}^{N} S_t(i)} \tag{7.1}$$

  Here $S_t(i)$ is the magnitude of the Fourier transform at frame *t* and frequency *i*.
- **Rolloff:** It is defined as the frequency *R* corresponding to *r%* of the magnitude distribution. It is given by equation (7.2). r was fixed to be 50% for the experimental purposes by the authors.

$$\sum_{i=1}^{R} S_t(i) = r\sum_{i=1}^{N} S_t(i) \tag{7.2}$$

- **Flux:** This measures the local spectral change. This is given by equation (7.3).

$$F = \sum_{i=1}^{N} (N_t(i) - N_{t-1}(i))^2 \tag{7.3}$$

Here $N_t(i)$ is the normalized magnitude of the Fourier transform at frame t.

- **Zero Crossing Rate:** This is a measure of the total number of times the signal crosses the zero line.

These features were not used directly in the experiment. Rather the mean and variance of these values were used for the experimental purposes. The mean and variances were computed for 40 windows by the authors. This captured good features for spectral surface.

- **Low Energy:** This is the percentage of windows that have less energy than the average energy of all 40 windows.

7.1.1.2 Beat Features

The other category if features are the beat related features that try to capture the musical beat for the purpose of genre identification. The extraction of the features from the musical signal involves a series of steps that are performed one after the other.

The first step is the decomposition of the signal by the wavelet analysis. We studied that this technique extracts the features out of the signal by time flexible windows to capture more information regarding time and frequency.

The next step to be performed is the envelope extraction. Here a series of steps are involved that help in the extraction of the envelope of the musical signal. In this stage the first step involved is full wave rectification. This converts the time domain signal into the temporal envelope. The simple manner to apply this rectification is to simply take the magnitude of the signal that transforms the negative values into the positive direction.

Then a low pass filtering is applied to the resultant envelope. This results in the smoothening of the envelope. In this step we compute the value of the signal by balancing its historical and current values that induces a smoothing effect. Weighted mean over the historical and current values is taken. This is given by equation (7.4).

$$s'(n) = \alpha\, s'(n-1) + (1-\alpha)\, s(n) \tag{7.4}$$

After this the signal passes through the stage of downsampling. This reduces the total data stored in the signal which generally results in a loss of information. This however makes the algorithm computationally very inexpensive as it has to handle large volumes of data.

Towards the end comes the stage of mean removal. Here we take the difference between the current and the predicted value of the signal (E). This is given by equation (7.5). This centers the signal to zero for autocorrelation.

$$y(n) = s(n) - E(s(n)) \tag{7.5}$$

After the envelope has been extracted using this procedure, autocorrelation is taken. The autocorrelation is a strong statistical tool used to measure the similarity of a signal with itself in its delayed form. This forms a valuable tool in signal analysis. This is given by equation (7.6).

$$y(k) = \frac{1}{N}\sum_{i=1}^{N} s(n)s(n-k) \tag{7.6}$$

The function *y(k)* given by equation (7.6) is the autocorrelation function that gives the correlation measure against any delay. On plotting *y(k)* against *k* we get an autocorrelation graph. This graph has peaks corresponding to the lags after which where the largest similarity was found in the given input signal. The next step involved is beat histogram calculation. The amplitude of the first three peaks of the autocorrelation function is added to a beat histogram. This is done for all the analysis windows. This makes the complete histogram of beats. The beat with the highest beat in this histogram would be the main or the dominating beat. The next highest beat is the next most important or dominating beat in the music.

At the end feature computation takes place that uses the beat histogram for the computation of the various features that may be used for the task of recognition. The various features that are computed from here are as follows:

- Relative Amplitude: This is done for the first two peaks. The relative measure here means that the amplitude is divided by the sum of all other amplitudes.
- Amplitude Ratio: This measures the ratio of amplitude of second peak to the first peak.
- Period of beat: The period in Beats Per Minute (BPM) is measured for the first two beats.
- Sum of Amplitudes: All the amplitudes of the histogram are added together as a feature. This measures the strength of the beat in the music.

7.1.2 Pattern Recognition

The other major task of interest here is the pattern recognition once the features have been identified and extracted. The entire data set is divided into the training and the testing data sets (Kala *et al* 2009). The basic methodology would be to learn the rules represented in the training pattern. This would form a learned system whose results may be generalized to the new inputs as well. The testing input is now applied to measure the recognition accuracy or the performance of the system.

In this method a k-Nearest Neighbor Classifier has been used for the experimental purposes. Consider the feature space where the classification is to be performed. This method uses the distance between the applied input and the other classes in the feature space as a means for classification. The attempt is to classify the input to the class whose inputs lie close enough to this input. This method uses the k-nearest neighbors of the applied input in the feature space. These then compete for the decision regarding the status of the final output class to that input. The relative frequencies are considered and based on them the best class is chosen out of these neighbors.

7.1.3 n-Fold Cross Validation

Another technique not discussed in the book that is used here is of n-fold cross validation. We hence comment on it in this section.

The choice of the training and testing data set carries a lot of relevance in any machine learning problem. The same problem and same data may give a high accuracy for some combination of training and testing data and some other for another combination. Hence if we get a high testing accuracy for some problem for some combination of training and testing data set, the result can ideally not be generalized for other combinations as well. This may be easily visualized as well as verified from the feature space as well. Different training data sets mean different training points. In classification especially one of the major problem is of class dependency. If some training set contains maximum entries of a particular class and very little of the other class, it may become more oriented towards the class. This may hence give biased outputs. The distribution may not always be possible to control always.

Hence to solve this problem we divide the complete data set into n parts. Each part holds some data items from the entire data set. Each part is given the same number of elements. The system is trained with n-1 parts and tested with the part left out with which it was not trained. This is repeated for all the parts. Every time one part is used for testing and the others for training. The average of all the recordings is taken and is reported as the final system accuracy. This manner we largely go away with the problems of the recordings being too much specific on the distribution of the training and the testing data sets.

7.1.4 Experimentation

The last section that we discuss is to experiment this algorithm and compute the recognition score of efficiency of the system to carry out classification. The experimentation from the work of Kosina (2002a) has been reported here. 189 songs were taken from 3 genres namely metal that had 63 songs, dance tracks that had 65 songs and classical that had 61 songs. Sampling rate was 44.1 kHz and bit rate of 128 or 160 kbps. Random 3 seconds samples were taken from the music files. The accuracy of classification reported keeping the value of k in k-Nearest Neighbor approach as constant was 88.3598%.

7.2 Musical Classifications

Here we discuss another model for carrying out the classification of music. Here we would classify the genre along with the artist (Whitman *et al* 2001; Mandel and Ellis 2005; Shukla *et al* 2008). Again we use the same methodology of discussion. Our focus would be two areas of feature extraction and recognition that highlight the importance of music and machine intelligence respectively.

We saw in the previous section that musical samples contain information that may be used for the purpose of genre classification. The capturing of good features results in supplying valuable information to the intelligent systems that are able to carry out the recognition by themselves. Another important aspect of the system is that we did not supply any information to the system in forms of rules, logic or prior understanding of the system. It was able to figure them all out by itself without any external inputs. The system just used the training inputs for figuring out these rules and logic. This forms the black box working with the neural networks, where the inputs are supplied without any underlying logic. In this section also we would focus upon the extraction of good features and then using them for learning and testing in similar manner.

Two problems have been considered here. The first problem is that of the genre recognition. The second problem is the identification of the artist behind the music. Both use same features for the recognition system.

7.2.1 Feature Extraction

The first major job to be carried out in the design of this system is the identification and the extraction of the features that aid classification or recognition. In this model a total of 124 features are used. These are as follows:

- **LPC Coefficients:** These are the set of parameters attained from the LPC signal analysis. A total of 34 coefficients are extracted and used in the system.
- **DFT Amplitudes:** The other class of parameters is the coefficients from the Fourier Analyses. The Discrete Fourier Transform is used for the extraction of 34 features for representation of the signal.
- **Log of DFT Amplitudes:** The DFT values attained in the previous step may still be difficult for the system to work with. These values are also used in the logarithmic scale by taking the log of DFT amplitudes. This forms 34 new values to be used as features.
- **IDFT of log of amplitudes:** The values obtained from the previous step are passed through an Inverse Discrete Fourier Transform (IDFT). This gives a new set of 34 values to be used as features.
- **MFCC Features:** This forms another set of features used by the system. A total of 15 coefficients from the Mel Frequency Cepstral Coefficients are taken. These coefficients are computed from windowing the data with Hamming window, taking of the amplitude values, converting the output values to filter bank outputs, calculating the log of base 10 and then the cosine transform.
- **Volume:** This is the last attribute that are given to the intelligent system as feature. This is a single attribute feature that represents the variance of the samples.

The entire feature vector consisting of all the features hence has a total of 124 attributes as discussed. This is a very large input that any system can handle. By feature extraction we had reduced the highly dimensional input of music signal into a much more compact form, but the dimensionality of 124 is still high. A solution may be taken to be taking much less attributes. But this may naturally spoil the recognition. Consider the feature space. Taking lesser attributes meads deciding the class to which the input belongs with much lesser axes. The closely lying class would naturally be difficult to separate without having the information of the other dimensions that is the case with taking more dimensional inputs.

The problem here is that we want to give the system much lesser inputs so that it is able to perform well without sacrificing the input attributes. This is done by using a technique of dimensionality reduction. In this problem Principle Component Analysis (PCA) is used s a means of dimensionality reduction.

7.2.2 Dimensionality Reduction

The dimensionality reduction is a technique to reduce the total number of dimensions or inputs to the systems. Feature Extraction also carries out the same task and may hence be regarded as a type of dimensionality reduction technique. This limits the size of the input to the intelligent systems to a limit that they can effectively handle. In feature extraction we had a fair idea regarding the interestingness of the various features that enhance the predictive capability of the ANNs. Take the formant frequencies as an example. We knew that there was a fair chance of identifying people by their formant frequencies that made it an interesting feature. This made it very simple for the intelligent recognition system to carry out the recognition.

In this section we consider a completely different case. Here we do not have an idea regarding the interestingness of the various attributes. We need to mine out a smaller set of interesting attributes from this pool of attributes. This is done by Principle Component Analysis (PCA) (Jollife 2002; Smith 2002). It is natural that many attributes would be better or more interesting for carrying out recognition from the others. In many other cases some characteristic mathematical combination of attributes would be better in recognition.

Principle Component Analysis does this mining. It takes as its input the various dimensions intended to be given to the system. It takes the various data recordings or data sets carrying these attributes as well. It analyzes the differences and changes in these attributes or their combinations across the various data recordings. It is natural that the attribute that changes a lot in between the data set is the best for the purpose of recognition. The ones that remain almost constant may not be of any use for the system. This analysis results in a set of new attributes that are the output of this analysis technique. These attributes completely identify all the features originally given to the system. Hence no information is lost. Further these are arranged in descending order of interestingness as per the PCA analysis. We may naturally take the top few attributes and leave the rest. This results in loss of information as we have rejected some information. But the overall goal to extract the best features out of the entire lot of features is met. Further we did not have to supply any information about the various features in the problem context. All this was figured out by the PCA itself.

7.2.3 Pattern Recognition

The last job to be done in this system is to train and test the features and use them for the task of recognition or classification. This is done by using a Multi Layer Perceptron (MLP) trained by using Back Propagation Algorithm (BPA). Like any technique, here we have two distinct phases of training and testing. In the training phase the system is shown the inputs and the outputs. The input to this network in training is the list of features for the training data that were extracted in the previous step along with the outputs. The trained network is used for the testing purpose where the unknown or the testing inputs are applied and the system is asked to do the recognition. The system generated outputs are matched with the actual outputs to give the system efficiency.

While the inputs generated from the feature extraction and dimensionality reduction module may be used directly as inputs after normalization, an interesting point here is the coding of the outputs. The outputs need to be digitalized and represented in the format that the system can work upon.

In chapter 6 we introduced a manner of coding the system outputs for such classificatory problems where the total number of system outputs is same as the total number of classes, with each output measuring the probability of that class being the final output class. This type of coding has a problem that the total number of outputs becomes very large in nature that again becomes a problem for the ANN. This is especially the case with systems that have too many output classes.

The other technique that one may think of is to give simple numerical equivalents to the various classes. Say the first type of genre is 0.1, the second type is 0.2 and so on. Decimals are used to make the outputs lie within 0 to 1. This system also has a problem. Since we are using an ANN the training would never be perfect and no output may be perfectly determined. Consider that the final output class of the system had an equivalent of 0.5. Now the system would naturally produce as the training output a number close to 0.5. Now consider that a new input is applied that largely belongs to this class with numerical equivalent of 0.5 but has some characteristics of the class having the numerical equivalent of 0.9. This may produce as its output a number largely close to 0.5 but with some orientation to 0.9. Say the output is 0.6 which represents a completely different class. The input may not have any characteristics of the inputs of this class. Giving an output of 0.5 would have been ideal and 0.9 would have been understandable, but not 0.6. This is the problem that this encoding faces.

The same problem can be easily understood in the feature space as well. Now the problem is that of predicting the function that gives the numeral equivalents of the classes. This may be plotted as a functional prediction problem plot with the value of the function as one of the axis or the vertical axis and the other input attributes as the other inputs. Consider the case where the various classes have large inter class separation. This plot may be seen as a functional surface with discrete levels at various points in the functional landscape. These discrete levels correspond to the numerical equivalents of classes. The inter class discrete value transitions would be by smooth values to give a smooth overall look to the functional landscape curve. For a single input attribute this may be as shown in figure 7.1.

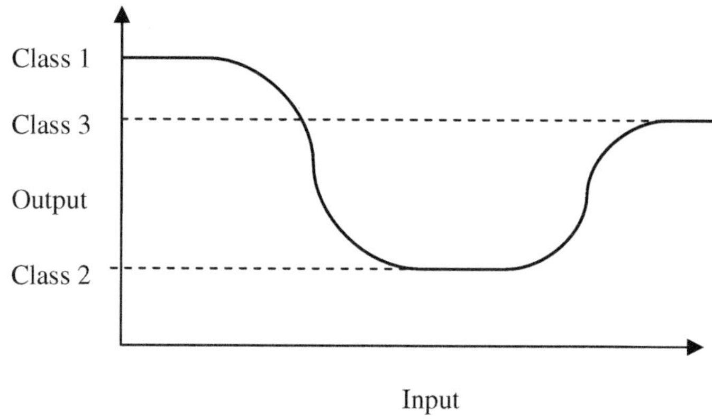

Fig. 7.1 Function to be predicted with numeral equivalents and high inter class separation

This plot may still be memorized by the ANN considering its curve fitting approach with the training inputs. But in practical problems we know that the inter class separation may not always be high. Further the system may have some noise in it. In such a case there would be far too many sharp changes in the discrete levels in the system. This is especially the case when two classes lying close in the feature space occupy numerical equivalents with large differences. In very less or probably no feature space the system has to make a large transition in values. Again consider that a class in the training input is wrongly identified. This would try to influence the system by its own numerical input which might drive the system by a big force in the wrong direction to allow for curve fitting with this noisy value. This requires far too many neurons and further makes the entire system very sensitive. This may be seen by figure 7.2.

The point that we take from this discussion is that the numeral equivalents of class must be as far apart as possible. Secondly any combination of classes must have the same separation in terms of values in their numeral equivalents. This is naturally not possible in the single numerical value equivalent per class. We hence need to add integers in the representation of every class. This makes the numerical equivalents a set of integers.

With this we introduce another space that is known as the output space. Every output in the case of a classification problem is a class (Shukla *et al* 2008, 2009). Every class c_i may be represented as a set of integers $<c_{i1}, c_{i2}, c_{i3}, \ldots c_{in}>$. Here n is the total number of output attributes that must be as small as possible for representation. This may be easily plotted as a space with n dimensions. Every output representation or output class is a point in this output space. From the above discussions the motivation is to be able to represent every class such that the distance between any two classes is the same and is as large as possible. The minimal and the maximal values of the axis are -1 and 1 respectively as per the normalization of inputs and the outputs follow in neural networks. In other words we are expected to find points in this output space that lie very far from each other and are at some definite distance from each other.

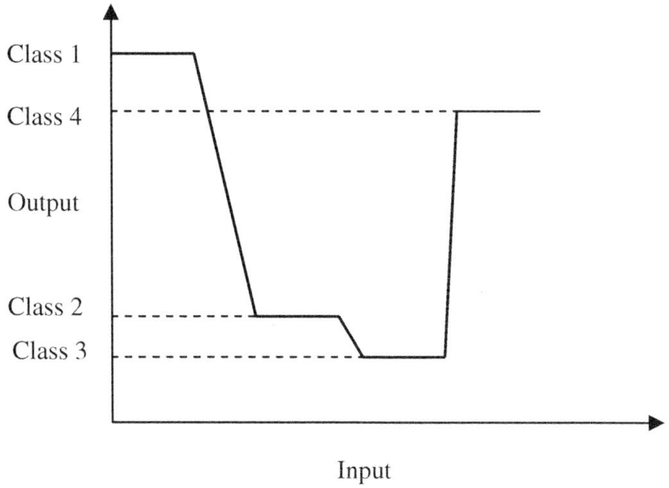

Fig. 7.2 Function to be predicted with numeral equivalents and low inter class separation

In this problem of music classification we are expected to find the genre as well as the artist. Both the problems are solved independent of each other. Genre has four types these are Rock, Classical, Soul/R&B and Country & Western. These are the four classes that must be considered in this first classificatory problem. Here we may consider a three-dimensional output space. The representation of Rock may be taken as <1, 1, 1>, Classical as <-1, -1, 1>, Soul/R&B as <1, -1, -1> and Country & Western as <-1, 1, -1>. The distance between any two classes is 2. Further this was not possible using any combination in a lower or two-dimensional space. The output space for this combination is given in figure 7.3.

Similarly for the other problem of artist classification, the output classes possible are R01, R02, R03 and R04. Here we may use a similar representation i.e. R01 as <1, 1, 1>, R02 as <-1, -1, 1>, R03 as <1, -1, -1> and R04 as <-1, 1, -1>.

7.2.4 Experimentation

The experimentation of the system reported by Kosina (2002a) is presented. In this system the authors used data from 8 Compact Disks. There were 2 disks per genre. The total data set consisted of 4425 which were divided into training data set, testing data set and validation data set.

The purpose of the validation data set is to stop leaning or curve over-fitting. As the network keeps training itself, it is natural that the errors would reduce. This is due to the curve fitting mechanism of the ANN where it tries to imitate a curve that fits the training data elements. As the network keeps training it, the errors reduce and fitting is better. To further reduce the error the network may rigorously try to fit the points which may spoil its overall shape. This is the problem of over-fitting. Validation data is a complete separate data set. The network is trained only

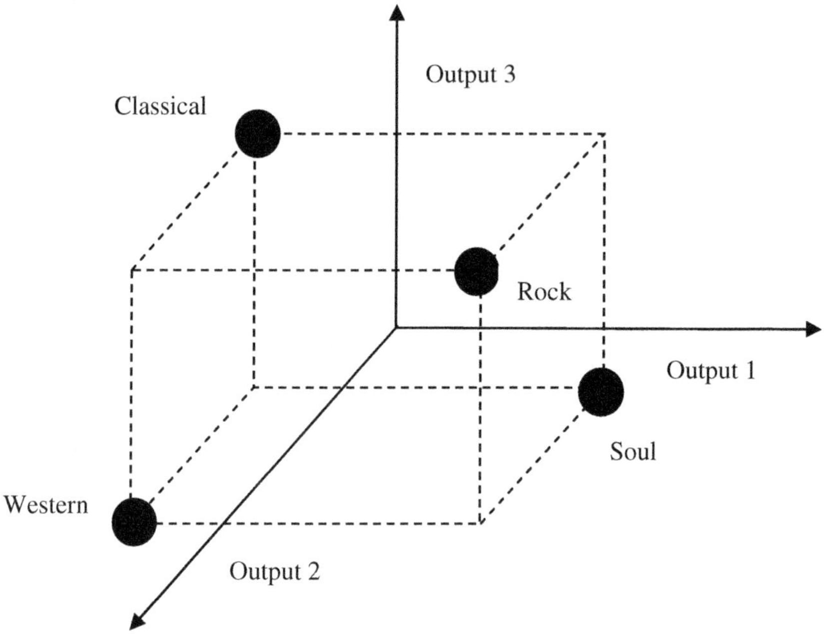

Fig. 7.3 Output Space for the problem

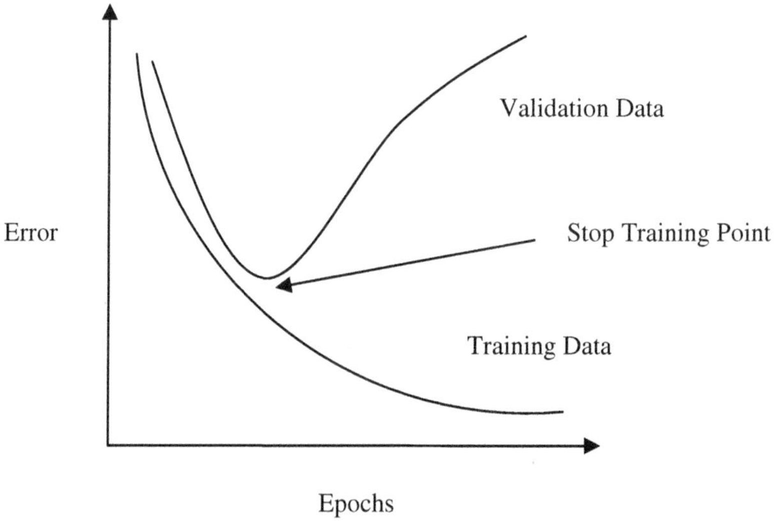

Fig. 7.4 Early Stopping of neural network with validation data

till the performance of the network on validation data is reduced. When overfitting takes place, the error on training data reduces and the validation data increases. The network training is stopped at this point of time. This is known as early stopping. It may be represented as shown in figure 7.4.

The net performance of the system in this manner for genre was found to be 94.8%.

For the experimentation of the artist a total of 4 compact disks were used out of which 2187 data instances were extracted. Similar procedure was followed in this classification as well. The final system had a recognition rate of 92.4% in recognizing artist as reported by the authors.

7.3 Music Composition by Interactive Evolutionary Computation

We now discuss another wonderful application of the intelligent systems in music. This is of composition of music. Music composition has always been a work of an art where a lot of expertise is required to generate the correct rhythm that is enjoyed by the listeners (Mozer 1994; Taube 1991; McCormack J 1996). All the fine pieces of music that we regularly listen and admire have been finely crafted by composers. Further there is a great disparity in music. Some people like a specific type of music while the others like some other type of music. A piece of music that is of interest to one person may be completely disliked by another person and vice versa. This adds a dependence of audience to the music that is well taken care of in professional music composition.

Music may ultimately be visualized as a collection of nodes, timbres, melodies and harmonies one after the other. All these produce some specific sound that is a very small part of the entire music. The collective sound produced by these entire nodes one after the other results in a good piece of music that people hear. The task of music composition may hence be seen as a combinatorial problem where the task is to find the most optimal combination of nodes that gives a good hear to the listeners.

Genetic Algorithms (GA) have been a very valuable tool for optimization problems. These algorithms are able to effectively search the search domain and figure out the optimal point within the time constraints. These algorithms further find a lot of application in these combinatorial problems as well where the task is to optimize the combination of values. Every combination may be rated good or bad by some means which forms the fitness function of these algorithms. The Traveling Salesman Problem is a classic example of the application of Genetic Algorithms for combinatorial problems. In all these problems it is important that the search space is so large that an exhaustive search cannot be performed.

We have mentioned many times the complexity of search domain largely determines its performance. The number of modifiable parameters that affect its fitness values constitute the dimensionality of the algorithm. Each of these adds a dimension into the search space of the GA. The higher the dimensionality, more difficult would be the optimization. Common music that we hear is very good to listen but in terms of GA it is highly complex. This is because of the huge number of nodes that may be combined in numerous ways. The total length of the music is usually high. This adds a lot of dimensionality to the problem. It is natural that the GA would fail in such complex problems. But the motivation here is the composition of simple music or parts of music or a simple rhythm that is a problem of much less dimensionality and may be composed with the help of Genetic Algorithm.

In this section we present Interactive Evolutionary Computation (IEC) Technique for solving the problem of music composition. The basic methodology involved would be the same as was the case with Genetic Algorithms. We would mainly stress on the manner in which the music is expressed as a genetic individual as well as the IEC that is a unique feature of the use of GA in music composition. Rest most of the concepts are the same that come from basic GA.

7.3.1 Interactive Evolutionary Computation

The Interactive Evolutionary Computation (IEC) tries to use the optimization powers of the evolutionary algorithm along with the expertise of the humans in judging the solution, to create a unique interface between machines and humans for problem solving (Takagi 1998; Kamalian *et al* 2004). There are systems where the computer cannot judge the quality of a solution. Music is one such system. It would be hard to make the system evaluate music that may easily be performed by humans. This puts a restriction on the use of automated systems for the generation of music. The intervention of human as a judge is mandatory for these types of systems. The other problem that comes is that music is highly specific to humans. The likings change from person to person and play an important role at the time of composition. Any music composer has some idea regarding the targeted audience and their likings before he composes something.

Let us look at the problem of music composition from the point of view of genetic optimization in the fitness landscape. Here we are given a fitness landscape that is a measure of the fitness value at all combinations of the system parameters. The objective is to find the most optimal point. In music composition there are numerous problems when we operate in this manner. The first is that the landscape is too large, dimensional and complex. The other problem is that it is constructed by humans. Due to this factor there are very few computations that may be performed. Further the humans may not be consistent in their evaluations. This is naturally a problem in terms of genetic evaluations. The same position may give two different fitness values when measured two different times.

The other factor here is that there might be too many optima in the entire landscape. Same person may like a large variety of music. These may greatly differ from each other in terms of the positions in the fitness landscape. This again greatly increases the complexity of the fitness landscape that is usually a big problem to solve. The complex landscapes are very difficult to mine for optima. Further the multiple optima may many times confuse the GA and may result in driving it in wrong directions. This necessitates effective diversity preservation techniques to respect the different optima that are explored in the search process. Further the person making the fitness evaluations must know the kind of music that is to be composed. This would greatly help in guiding the evolution process in the correct direction.

7.3.2 Genetic Architecture

The entire problem may conceptually be divided into two separate parts. Both of these are solved separately in an interchangeable manner in the method being discussed. The first is the generation of good musical phrases that may sound good for the user. This step deals with the optimization of the sound of the individual short tunes that contribute in the music.

Once a collection of good short musical phrases has been done, the other task is to generate music from these phrases. In this problem the task is to have the correct combination of the phrases for a good overall musical melody. The various good phrases may not always combine to produce good music as per the listeners' requirements. The music may sound good only when the phrases are combined in some characteristic manner.

These two problems are highly dependent on each other. Without good phrases a good music cannot be generated. Only good phrases may also not lead to good music. A unique combination is to be sought out. Hence the user is given a choice to switch between the two optimization processes. If the user feels a need for better phrases, he can revoke the phrase optimization and if he feels the need of better combinatorial optimization, he can revoke the other optimization.

In the discussed approach the optimization of the phrases has been done using Genetic Algorithms (GA). For the computation of the correct ordering or combination Genetic Programming (GP) has been used. Genetic Programming is not dealt here. Interested readers may refer chapter 10 that gives an extensive coverage over this topic.

7.3.3 Individual Representation

One of the major tasks here is the representation of the individual for the evolutionary process. This is different for both the problems of musical phrase synthesis that is performed by GA as well as the optimization of the phrase combination that is done by GP. Each of these is discussed separately.

In the GA, the individual is represented as a two-dimensional array. The size of the array depends upon the parameters that are set by the user. The various factors considered here are the number of timbres, the length of the phrase represented by an individual and the unit time resolution (eighth or sixteenth note). Each of the cell of this array stores the strength of the signal or the velocity which is an integer between 0 and 127 and occupies 4 bits.

The other algorithm is the GP. Here the representation is in form of a tree. In this representation the leaves of the tree represent the GA individuals. Each of these individuals has an ID number that is stored as the leaf node and it points to an individual of the GA. The non-leaf nodes store the genetic functions. These functions may take a number of parameters as GA individuals that are supplied by the leaf node as per the GP methodology. The length of the rhythm is kept close to the value specified by the human by imposing length constraints.

The various GP functions used are as follows. All these have some relevance from the point of view of music composition. All these in some manner are likely to contribute to the generation of better music.

- **Sequence:** This operation takes two nodes. Both these nodes are played consecutively one after the other.
- **Repetition:** This operation takes two nodes. The first node is played numerable number of times, till the playing time is equal to the length of the second node.
- **Concatenation:** This takes two node arguments. It plays the first half of the first node and the second half of the second node.
- **Reverse:** This plays the node given as argument in the reverse manner.
- **Random:** This plays any random node. No argument is supplied in this operation

The GP representation of any individual is a collection of these operations along with the GA individuals in a tree-like structure. One such individual is represented in figure 7.5.

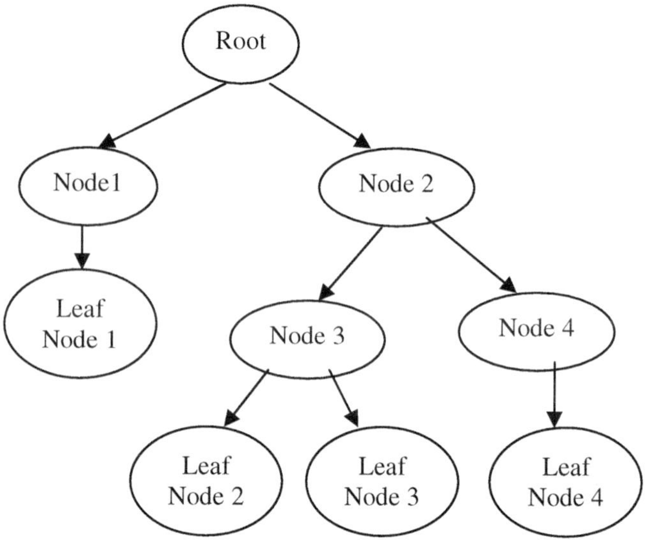

Fig. 7.5 Genetic Programming Individual

7.3.4 Genetic Operations

The other factor that is needed in design is the genetic operations that result in the generation of higher generation of individuals from the lower generations. This is also different for both the GA and the GP used in the algorithm.

In GA the major operations used are as follows:

- **One Point Crossover:** This is a standard one point crossover that results in mixing of two individuals for the generation of a higher generation individual.

- **Part Exchange:** In this technique we exchange parts of timbres in between two individuals.
- **Random:** This mutates the values of the individuals in a random manner.
- **Rotation:** This operation rotates the loci.
- **Reverse:** This operation reverses the loci.
- **Timbre Exchange:** This operation selects an individual and exchanges timbre within it.

The GP uses the standard operators for the task of generation of higher generation individuals from the lower generation. These are sub-tree crossover and mutation. Besides, both approaches use elitism as an operation.

7.3.5 Evaluations

We have already discussed the fact that the humans would be supposed to evaluate and assign fitness values in the mentioned approach. A problem was identified as too much of dependency on humans that restricts the scope of large number of evaluations. This problem is taken care of in the approach. The individuals are partially judged for fitness by the use of neural network. Only the individuals that possess a fine fitness values as per the neural assessment may be given to the humans for assignment of the actual fitness values.

The neural network needs to be trained beforehand. This is done by recording the initial assignments of the fitness values by the humans. A number of these values are initially collected that serves as the training data for the neural network. After training it may be used for the testing purposes where new GP individuals are judged for their fitness.

7.4 Computer Assisted Music Composition

The other model that we study in this chapter is of computer assisted music composition (Cope 1987; Assayag et al 1999; Holtzman 1980; Vaggione 2001). Here again we would be making use of the GA for the task of composition of music. The method presented here adopts a different mechanism of individual representation of the individual and the related operation. Here we try to compose music that best matches a predefined pattern. Hence the fitness function is automated in this method unlike the previous method where the fitness evaluation was assisted by human. The basic methodology used in solving this problem is called as thematic bridging which is useful in transformation of some musical pattern over some other musical pattern over some specified duration. We discuss the various salient features of the algorithm and its working in the coming section.

7.4.1 Thematic Bridging

The basic methodology used is of thematic bridging. Basically this method attempts to carry out the transformation into the initial pattern by using a set of transforms one after the other. This makes the final pattern. Here pattern is a collection of nodes that represent the basic structural part of the music. The initial, final as well as all the intermediate patterns are constituents of the final sequence generated as a result of the application of this bridging. This forms a novel method of generating a sequence when the initial pattern and a list of operations is given.

Consider any arbitrary initial sequence (i) <*a, b, c, d*>. We consider that this sequence is subjected to four operations one after the other. Let these operations be *rotate, delete first node, rotate, mutate first node*. Here all words have their usual meaning. The resultant sequence after the first operation would give sequence (ii) <*d, c, b, a*>. This serves as the input to the next operation that after its application results in sequence (iii) <*c, b, a*>. Similarly after the application of the third operation would give sequence (iv) <*a, b, c*>. Towards the end the last operation would result in sequence (v) <*e, b, c*>. The complete sequence may be obtained by the concatenation of the sequences (i)-(v). This is given by <*a, b, c, d, d, c, b, a, c, b, a, a, b, c, e, b, c*>. In this manner the complete sequence is generated by thematic bridging.

Many times the generated sequence may not be desirable due to non matching of the initial and the final patterns in the limited set of operations. In such a case the operations may be modified and the entire procedure may be re-run.

7.4.2 Individual Representation

The individual representation largely determines the performance of the entire optimization done by the GA. Hence a good representational strategy is of importance, especially when dealing with a highly dimensional problem like music composition that has a very complex fitness landscape in the conventional representation techniques.

In this problem the individual is a set of operations that are applied one after the other. We saw earlier how easily the complete sequence can be generated once the sequence of operations is known. We may simply apply the operations ne after the other. The various operations can be coded to be represented as a gene of the individual. Many operations may further require a parameter for their functioning. These operations are given a place in the individual for the specification of the parameter. This parameter is specified as a real number in between 0 and 1.

The various operations that are possible in this algorithm as per the implementation is:

- **No-Op:** This operation carries no effect on the individual and leaves it unchanged.
- **Add:** This operation is used to add an element in the pattern.
- **Delete:** This operation deletes an element
- **Mutate:** This modifies the value of an element.
- **Rotate:** This is used for rotating the entire individual.
- **Exchange:** This operation exchanges elements.
- **Rotate Incorrect:** It rotates the incorrectly placed elements.
- **Exchange Incorrect:** This operation exchanged incorrectly placed elements.

7.4.3 *Fitness Evaluations*

The last concept we study in this system is of fitness evaluation. The fitness function operates in two hierarchies. The first hierarchy used in this method tries to match the generated individual against a final pattern. The higher fitness is awarded for a better match. The similarity between the patterns is measured both on the basis of the content of the patterns as well as the relative ordering of these. When an exact matching is found in the first hierarchy the system moves into the second hierarchy of fitness evaluation. In this hierarchy the fitness evaluation takes place on the basis of duration. An exact duration as desired by the algorithm is given the highest fitness. A deviation from this value is suitably penalized.

7.5 Chapter Summary

This chapter dealt with the design, issues and applications of intelligent systems using music. Music forms an integral part of one's life and the variety of music in large numbers that we come across is the motivation behind these systems. The chapter discussed specific intelligent systems for the problems. A total of four systems were discussed. The first system under discussion was musical genre recognition. Here the system was asked to classify the given music into one of the predetermined genes. This was done on a variety of features. Next a system for genre and artist classification was presented. This system could simultaneously determine the artist as well as the genre behind the music. These systems were based upon neural learning of the training data and performance analysis over the testing data. We later moved into the genetic domain. Here the system being designed was for music composition. The system worked in a human assistive mode for the fitness evaluation. The system further made use of neural networks for segregating the poor sounding nodes from the better sounding nodes as per the specified user inputs during music composition. The last system for discussion was again for the task of music composition. This system tried to generate music as per the specified objectives. The system made use the evolutionary paradigm for computation.

References

[1] Assayag, G., et al.: Computer Assisted Composition at Ircam: PatchWork and OpenMusic. The Comput. Music J. 23(3), 59–72 (1999)
[2] Burred, J., Lerch, A.: A hierarchical approach to automatic musical genre classification. In: Proc. of the Sixth Intl. Conf. on Digital Audio Effects, DAFx-03, London, UK (2003)
[3] Cope, D.: An expert system for computer–assisted composition. Comput. Music J. 11(4), 30–46 (1987)
[4] Holtzman, S.R.: A Generative Grammar Definition Language for Music. Interface 9(2), 1–48 (1980)
[5] Horner, A., Goldberg, D.E.: Genetic Algorithms and Computer Aided Music Composition. In: Intl. Comput. Music Conf., ICMC 1991, Montréal, Québec, Canada, pp. 479–482 (1991)

[6] Jollife, I.T.: Principle Component Analysis. Springer, USA (2002)

[7] Kala, R., Shukla, A., Tiwari, R.: Fuzzy Neuro Systems for Machine Learning for Large Data Sets. In: Proc. of the IEEE Intl. Advance Comput. Conf., Patiala, India, pp. 541–545 (2009)

[8] Kamalian, R., Takagi, H., Agogino, A.M.: Optimized Design of MEMS by Evolutionary Multi-objective Optimization with Interactive Evolutionary Computation. In: Deb, K., et al. (eds.) GECCO 2004. LNCS, vol. 3103, pp. 1030–1041. Springer, Heidelberg (2004)

[9] Kosina, K.: Music Genre Recognition. MSc Thesis, Technical College of Hagenberg (2002a)

[10] Kosina, K.: MUGRAT (2002b), `http://kyrah.net/mugrat/`

[11] Li, T., Ogihara, M.: Music genre classification with taxonomy. In: Intl. Conf. on Acoust., Speech, and Signal Process., Philadelphia, USA (2005)

[12] Li, T., Ogihara, M., Li, Q.: A comparative study on content-based music genre classification. In: Proc. of the Intl. Conf. on Res. and Development in Inf. Retrieval, Toronto, Canada, pp. 282–289 (2003)

[13] Mandel, M.I., Ellis, D.P.: Song-level features and support vector machines for music classification. In: Proc. of the 6th Intl. Conf. on Music Inf. Retrieval, London, UK (2005)

[14] McCormack, J.: Grammar based music composition. In: Stocker, R., et al. (eds.) Complex Systems 96: From Local Interactions to Global Phenomena. ISO Press, Amsterdam (1996)

[15] Mozer, M.C.: Neural network music composition by prediction: Exploring the benefits of psychoacoustic constraints and multi-scale processing. Connection Science 6(2-3), 247–280 (1994)

[16] Rabiner, L., Juang, B.H.: Fundamentals of Speech Recognition. Prentice Hall International, Englewood Cliffs (1993)

[17] Scott, P.: Music Classification using Neural Networks (2001), `http://www.stanford.edu/class/ee373a/musicclassification.pdf`

[18] Shukla, A., Tiwari, R., Meena, H.K., Kala, R.: Speaker Identification using Wavelet Analysis and Modular Neural Networks. In: Proc. of the Natl. Symp Acoust., Vishakhapatnam, India, pp. 125–130.

[19] Shukla, A., Tiwari, R., Meena, H.K., Kala, R.: Speaker Identification using Wavelet Analysis and Modular Neural Networks. J. Acoust. Soc. India 36(1), 14–19 (2009a)

[20] Li, S.: A Tutorial on Principle Component Analysis (2002), `http://www.cs.otago.ac.nz/cosc453/student_tutorials/principal_components.pdf`

[21] Takagi, H.: Interactive evolutionary computation. In: Proc. Int. Conf. Soft Comput. Inf. Intell. Syst., pp. 41–50 (1998)

[22] Taube, H.: Common Music: A Music Composition Language in Common Lisp and CLOS. Comput. Music J. 15(2), 21–32 (1991)

[23] Tokui, N., Iba, H.: Music Composition with Interactive Evolutionary Computation. In: Proc. of IEEE Syst., Man and Cybern. (2001)

[24] Tzanetakis, G., Cook, P.: Musical genre classification of audio signals. IEEE Trans. Speech Audio Process 10(5), 293–302 (2002)

[25] Vaggione, H.: Some Ontological Remarks about Music Composition Processes. Comput. Music J. 25(1), 54–61 (2001)

[26] Whitman, B., Flake, G., Lawrence, S.: Artist detection in music with minnowmatch. In: IEEE Workshop on Neural Networks for Signal Process, Falmouth, Massachusetts, pp. 559–568 (2001)

List of Figures

Fig. 7.1. Function to be predicted with numeral equivalents and high inter class separation
Fig. 7.2. Function to be predicted with numeral equivalents and low inter class separation
Fig. 7.3. Output Space for the problem
Fig. 7.4. Early Stopping of neural network with validation data
Fig. 7.5. Genetic Programming Individual

Chapter 8
Intelligent System Design in Face

Abstract. Speech forms a major biometric that makes it possible to reap the numerous fruits of automation and verification. The later of the book focus on the shortcomings of speech that motivate the development of hybrid systems using face along with speech. Hence it is important to have a brief idea about face as a biometric that uniquely couples with speech. This chapter would present various techniques of using face as a biometric. We deal with two kinds of systems. The first class of systems uses dimensionality reduction techniques over the visual information that face represents. Here we would discuss the Principle Component Analysis (PCA) and Regularized Linear Discriminant Analysis (R-LDA). All these methods have their own mechanisms for dimensionality reduction. The other class of algorithm identifies landmarks in face and uses these for the recognition purpose. We also analyze the various advantages and shortcomings of these systems.

Face is the most common biometric systems used in everyday life. We may not be able to recognize a person by his voice over telephone, but it would be relatively easy to identify him when he comes before us by his face. The face of every human is believed to be quite different from each other. The faces of identical twins may also differ by some magnitude. This makes face a well thought and studied biometric system that is extensively used. The facial input is given in the form of an image to the recognition system. The task of the recognition system is to process the image and figure out the person whose face it is. In this manner a person may be identified or authenticated in an easy manner.

The face especially forms a very exciting biometric that holds a lot of information in it. Every part of the face has many characteristics that aid in the entire recognition process. The rich set of features in the face is both a boon and a curse. At the positive side the wide variety of features give rise to a lot of possibilities for the extraction of good and robust features. At the other side the use of face may require intense processing due to the high amount of information content. We would have to throw away a large part of the given data while extraction of features before the input can be given to the intelligent systems for recognition. This throwing away of data can sometimes mean the loss of valuable information that may not be represented in the features extracted in the problem. This further may harm the recognition purpose.

Speech has been extensively used as a biometric and for other intelligent systems. We saw some good applications of intelligent system design using speech in

A. Shukla et al.: Towards Hybrid and Adaptive Computing, SCI 307, pp. 175–184.
springerlink.com

chapter 6 and 7. The use of speech as a biometric deserves a special mention due to the possibilities of various systems and applications stated in chapter 6. The extensive use of speech has caught the interest of a wide variety of people. This has further resulted in a lot of development in various areas of speech. The constant urge to achieve higher accuracies and development of more robust systems is natural. In any system, speech itself has a threshold that may not be easy to cross. This brings us towards the notion of fused systems where we mix speech with the other possible biometric modalities like face, lips, ears, etc. These fused systems have a much higher recognition score and are able to give a much better performance as compared to the single modality systems.

8.1 Face as a Biometric

Face is one of the biometric modalities that closely go with speech (Shukla and Tiwari 2007, 2008; Kala *et al* 2010). This is in terms of the ease to the user as well as the ease to design. We would study these issues of fusion and their application in chapter 17. This chapter is mainly intended to give a brief introduction of biometric recognition using face that can act as an aid or a prerequisite for the discussions in chapter 17. This would further enable us to engineer the fused biometric systems in totality.

Face in this context may also be taken as a representative of the visual biometrics (Turk and Pentland 1991; Er et al 1999; Zhao *et al* 2000; Phillips *et al* 2005). We saw the manner in which we handle speech in chapter 5 where the emphasis was upon the use of processing techniques. The input to these systems was in the form of speech signals. In facial biometric systems we would take the input in the form of an image. The image processing is somewhat different from that of signal processing. Using face as a specific topic of study, we may also be able to appreciate the use of other biometrics like ear, palm, lips, etc. All of them would have a similar manner in which the various processes of identification are carried out.

The stages involved in the recognition using face are similar to the general recognition scheme that we presented in chapter 1. This comprises of data collection, pre-processing, segmentation, feature extraction and pattern recognition. All these steps have the same role in the problem of face recognition as well. The input is taken in the form of an image and hence the various steps involved are in context to visual information processing.

The main focus of this chapter would be on two steps. These are dimensionality reduction and pattern reduction. In dimensionality reduction we would be concerned with the methods which may be adopted to reduce a high dimensional face input into a smaller dimension input that can be given as input to the recognition system. We purposefully do not use the term feature extraction here. This is because dimensionality reduction may be taken as a much broader term that has other means to carry out the reduction task besides feature extraction. We would see in this chapter how the use of various dimensionality reduction techniques carries these tasks.

The other important step is of pattern recognition. Once the high dimensional face input has been reduced to a lower dimensional input, the next task is to build an intelligent system that can use this lower dimensional input for the purpose of pattern recognition or identification of the person.

In this work we would make use of the experimentation of Shukla *et al* (2009a, 2009b, 2009c) who build numerous models for dimensionality reduction and recognition. The models made use of Principle Component Analysis (PCA), Regularized Linear Discriminant Analysis (R-LDA) and Morphological Analysis (MA) for the dimensionality reduction. The recognition systems consisted of the use of Multi-Layer Perceptron with Back Propagation Algorithm and Radial Basis Function Networks.

The first step to be carried out in the recognition system is dimensionality reduction for the purpose of limiting the input size given to the recognition system. An important factor in dimensionality recognition is to strike a balance between the equality of the features that are extracted and aid in recognition with their numbers. We know that extracting too many dimensions may be good in terms of the ease of classification in the feature space. Using large information given by the various features may make it easy to decide the class of an input. In other words the large amount of axes in the feature landscape makes it easy to construct decision boundaries for class separation. This however may greatly increase the training and testing time and associated storage. Smaller number of dimensions may not be able to carry forward recognition. Every attribute conveys some contribution in the ease of recognition. This may be an important factor to consider while deciding the number of dimensions. Adding good dimensions, without which the recognition would be difficult, may sometimes be acceptable even at the cost of time and associated problems of dimensionality.

Here we would discuss about different methodologies of dealing with the problem of dimensionality reduction. Each of these would be discussed in the next sections.

8.2 Visual Dimensionality Reduction

The first class of dimensionality reduction technique we use is the visual reduction techniques. In this technique we use statistical methods for the dimensionality reduction. These tools find out the variances in between the various attributes and try to use the best possible attributes as a whole or after suitable operations. The selection of attributes to be used as inputs to the recognition system is the task of a statistical tool. We give no information to this tool regarding the meaning and the interestingness of the different attributes. The statistical tool is supposed to find out everything on its own without any aid.

The basic intention in these systems is that we have some data and want to use this entire data for recognition. We naturally cannot do so because of the large sizes of the data. This means the reduction of dimensions. Up to now we have propagated the use of human intelligence to figure out features in this data. This is done by an analysis of the problem and figure out what may be good for recognition. We saw how the various formant frequencies, wavelet analysis coefficients and other factors are motivating reasons for their use in the speech and speaker recognition problems. In the visual dimensionality reduction systems we however leave all these computations to the system. The various statistical tools hence largely act as black boxes which are assumed to give good features.

In case of face recognition problem we are given the photograph of the face of the person. We are supposed to reduce the dimensionality of this photograph to enable good recognition. In place of figuring out good features, we give the entire facial photograph to be analyzed by a statistical system and the necessary variances are taken into account to compute the features. As we give the entire image vector to the system, this technique is called visual dimensionality reduction technique.

Now it would be wise to analyze the pros and cons of this technique over the use of conventional style of figuring out the possibly interesting features and extracting the same. This technique figures out all the various possible features on its own. Hence we may expect the extraction of features that are supported by the training dataset rather than those which the designer feels are good. This further means that features extracted would be flexible to any combination of any attributes or values. This is not the case when the humans try to figure out the parameters. Humans would usually extract only features that are simple. They might not consider the fact how various factors may unite in a characteristic manner to give good features. This may make the system sub-optimal in working. The visual systems as discussed may make any simple or complex mapping to represent the physical data into features as computed by the statistical analysis.

These systems however may suffer from a major problem. Ideally a large amount of data may be required by the system to compute the interestingness of the various features. This may not be possible in a practical scenario where the data is limited in nature. Further too much of data would be very hard for PCA to analyze. This may lead to a poor decision regarding features. For these reasons these systems may give poor performance. The intelligent recognition system may work assuming the features to be enough to classify the data which may be wrong. The other factor is that many times the humans may be able to figure out actual parameters that affect the recognition based on their understanding or recognition. This may be in the form of well-known logic or rules known to the humans. This further may not be possible for any method to automatically figure out even with the greatest of computations. In such systems the conventional feature extraction may help. Suppose a person can be correctly identified by the dimensions of his eyes. In such a case we may only be interested in using some image processing for computing these features. The real life data and real life applications has its own complexities, issues and behavior. The variety of methods needs to be experimented in order to get an idea about the best method and parameters that may solve the problem.

We would briefly mention the two different systems used for experimentation. These are Principle Component Analysis (PCA) and Regularized Linear Discriminant Analysis (R-LDA). Each of these is discussed one by one.

8.2.1 Principle Component Analysis

The first system we present is Principle Component Analysis (PCA). In chapter 5 we had used an analysis technique called Fourier Analysis for the extraction of features from the speech signal. This is a similar approach applied in the context of images. We do not discuss the mathematical details of this analysis in this book, rather give an overview of the general working of this method (Jollife 2002; Smith 2002).

The PCA aims in the extraction of components. The task is to reduce the high dimensional vector into the lower dimensional vector. The component represents an interesting feature that helps in this reduction in dimensions. It is natural that the high dimensional vector would have a large amount of redundancy in it. This is because of a large amount of storage of information in between the various attributes of the high dimensional vector. The key is to exploit this redundancy in representing the same data in smaller dimensions.

The entire process of dimensionality reduction in PCA happens with the extraction of components from the given data. We try to find the most interesting components by an analysis over the data that has a large amount of change in value. This naturally represents an interesting feature to be used for the recognition tasks. This is the most predominating component for the recognition. This may be correlated with the fundamental frequency in the Fourier domain. This was the frequency that was most dominant in the entire speech signal and was hence of maximum interest in the Fourier analysis.

As we move, we keep extracting the components from the dimensional vector. At any time the analysis technique tries to formulate a component that carries the best recognition or has the maximal change in the possible components from the data. This component is the one that has the highest degree of correlation in analysis from components left to be extracted and the least degree of correlation from the components that have already been extracted from the data analysis.

As we keep adding components to the pool of extracted components, the interestingness keeps getting fade. The newly extracted components have a very poor recognition ability that makes the addition of these components not very useful. We observed a similar thing while Fourier analysis where the extraction of higher degree Fourier components hardly had any impact in the information of the original signal.

Hence the best components are selected and the rest are not selected. It is natural that the original image may not be re-constructed without taking all the components. But the top few components may give a fair amount of idea regarding the image. We saw the same concept in the Fourier analysis where we studied that having the top few coefficients, we can almost reconstruct the original signal back. In this manner there is a little loss of information with a great dimensionality reduction.

8.2.2 R-LDA

The other form that we study for dimensionality reduction is R-LDA (Plataniotis and Venetsanopoulos 2005; Huang and Hwang 2005; Liu et al 2004). This is the regularized form of Fisher's discriminant, type of linear discriminant analysis (LDA) (Ye and Janardan 2005;Lu *et al* 2003; Yu and Yang 2001).One of the major disadvantages in the use of PCA is that it does not consider the outputs while performing the analysis. This means that the data of the same class and different classes are dealt in the same manner while performing the analysis. This is especially poor for the classificatory problem where we have floated the notion numerous times that the inter class separation needs to be high and the intra class separation needs to be reasonably low. Hence the outputs have a role to play in the dimensionality reduction as well.

As an example, consider the feature space with two data instances *A* and *B*. Suppose that both these instances lay very close to each other and are members of two different classes. Now an analysis by the PCA would reflect that these two instances are redundant. This is however not true. In reality these two instances are very different as they belong to different classes. Further the low separation necessitates an interesting problem where the low inter class separation may be a problem for any recognition system.

This problem is solved by the R-LDA which takes into account the class as well while performing the data analysis. In this method the basic approach is to extract the components in such a manner that the disparity between the different class elements is high and between the same class elements is low. In this manner the analysis technique considers the same classes and different class separately. This is ideal for the performance in case of classificatory problems.

8.3 Morphological Analysis

The other technique of interest is where we extract the features based on our understanding of the problem (Card et al 1991; Maragos and Schafer 1986). We call such systems or analysis technique as Morphological Analysis (MA). This technique makes use of an understanding of the basic formulation mechanism of any shape, object or picture. In this we analyze the basic constituents of the shape and the manner in which they are related with each other. The overall aim is to formulate some features from the understanding of the object or image that can be used as direct features by the recognition system for the purpose of identification as per the problem demands. Since these represent features, it is important that these remain constant is the different recordings of the same class elements and change drastically in between classes.

In face these methods find an interesting application. The face is filled with so many interesting parts. The eyes, nose, mouth, etc. may be easily located. Morphological technique tries to use these features for the task of recognition. In other words we try to formulate features sing this knowledge of the various parts in the face. Many features may be easily jotted down based on these parts. The distance between the eyes, distance between eyes and mouth, distance between mouth and nose are some of the features. Now it is known that these are the features that do not change in a person. No matter how many times the recording be made, these features would remain relatively constant. Further a pool of some of these features makes an interesting identification system. There would be very few or rather no couple of people that would possess the same values for all of these values or very slightly deviation of these values. In this case it would be wise enough to use these as the features for the recognition systems.

Another important factor we consider here is that of variations in the recording conditions. The foremost factor is taken as light. We know that in visual dimensionality reduction techniques each and every pixel of the picture and its stored value has a role to play in the final recognition. The light changes cause a change in the gray level of the pixel as well. This causes a change in the analysis of both the discussed techniques. Hence these techniques are not very robust against the changing light conditions.

The scenario is quite different in case of MA. These methods carry out the identification or the location of the various landmarks in the facial image. The detection of the landmarks is quite robust against the changing lightning conditions. Unless the lightning is too bright or too dark or highly un-uniform, the detection can be easily carried out. Once the correct identification of the landmarks is done, it is very easy to find out the other metrics or features that do not depend upon the lightning conditions. This makes these algorithms very stable against the changing conditions. The other variations that are occasionally placed forth the face recognition is the presence of props, styles, etc. The MA is surely much more robust towards these as compared to the visual dimensionality reduction techniques.

The process of extraction of features using this approach follows a sequence of simple steps. The first task to be carried out is to convert the image from the color image to a grayscale image. The conversion to grayscale makes it easy for the algorithm to work with a single value that is the gray value rather than a set of values as per the color coding of the picture. Further color seldom has any unique property that cannot be depicted in gray scale.

The next task is to get the binary gradient image. For this we need to get the threshold value above which would be treated as white and below which would be treated as black. This value depends upon the changes in contrast around every pixel. This threshold value is computed using the edge and sobel operator.

The next task is to remove the linear gaps in the image. This is done by the application of linear structuring elements, such as dilating the binary gradient image. After this operation the holes are filled using the region filling algorithm. Next Extraction of eight-connected sets of pixel components is then done to suppress light structures connected to the image border. Next steps include filtering, thinning, and pruning. This image is superimposed with the gray image resulting in the outlined initial image.

Using this image the various features in the image are identified. Normal Centre of Gravity (NCG) is used for each of the extracted features. This may be used to compute the distances between the various landmarks detected in the face which forms the features to be used in recognition.

8.4 Pattern Recognition

Once the patterns have been detected and extracted, the task is to make an intelligent system that would carry the task of pattern recognition. This system does the task of person identification using the extracted features. For this purpose neural networks have been used. The system hence uses the methodology of learning. The data is divided into training and testing sets. The training set is used for the tuning of the various neural networks parameters. The training may be performed by using the training algorithm specific to the model under use. It is important to train the neural network such that the network so formed may be generalized to the other data as well in the testing phase. After the neural network learning is over, the testing data is used for the task of finding the accuracy of the finally evolved system.

The system may use any of the approaches discussed in the book or other approaches to carry out the task of pattern recognition. Each network may have its own methodology and way of working. It may not be easy to make any prediction regarding which network or which training mechanism would lead to the optimal performance. This is due to the fact that the neural behavior may highly depend upon the data and problem. The different problems and associated data may showcase different trends and behavior when trained on different neural networks. The networks for our interest in this chapter are the ones used in experimentation by Shukla *et al* (2009a, 2009b and 2009c). These are Multi-Layer Perceptron with Back Propagation Algorithm and Radial Basis Function Network.

8.4.1 Multi-layer Perceptron

The Multi-Layer Perceptron with Back Propagation Algorithm is the first model with which the data may be trained for the task of identification of the person by face. These networks are very generalized in their approach and try to solve the problem using a curve fitting approach where they try to find a curve based on whose values the decision may be made. Classification is usually considered a localized problem where the output at a place in the fitness space is highly dependent upon the output at the nearby places. This makes these networks less likely for the classification problems.

8.4.2 Radial Basis Function Networks

The other networks considered for the experimentation are the Radial Basis Function Networks. These networks also try to adopt a curve fitting mechanism. However these networks are much localized in their approach. These apply various neurons at various places of the feature space to look after the outputs of the nearby places in the fitness landscape. Hence these networks form a more localized approach for solving the problem of classification. The localized approach may be better that the curve fitting method where the entire network is supposed to fit the entire fitness landscape in a generalized manner without the division of the fitness landscape. The spread of the neurons largely controls the generalizing capability of these neurons where the far neurons carry global trends.

8.5 Experimentation

Here we discuss the experimental framework used in the work of Shukla *et al* (2009a, 2009b and 2009c). The database used for the purpose was Grimace

Database (Spacek 2010). The database contains 360 colored face images of 18 individuals. There are 20 images present for each subject. Database images vary in expression and position. The size of each image is 200x180. Half of the images were used for training data set and the other half for testing.

In the first experiment Principle Component Analysis was used with Back Propagation Algorithm. The authors reported a recognition rate of 92.77% in the use of Multi-Layer Perceptron (MLP) neural network with Back Propagation Algorithm (BPA). The recognition score was 94.44% in the use of Radial Basis Function Network (RBFN).

The same experiment was repeated using R-LDA as dimensionality reduction tool. The recognition score in the use of MLP with BPA was found out to be 93.88%. The recognition score in the use of RBFN was 96.11%.

The next experiment was done using Morphological Analysis and the comparisons were made with the other approaches used. MLP with BPA was used for the task of pattern recognition. Here Morphological Analysis methods gave a recognition score of 93.33%. The recognition score was 92.88% when R-LDA was used for the dimensionality reduction purpose. The application of PCA as a dimensionality reduction tool gave a recognition score of 92.77%.

Apart from the differences in magnitudes of recognition scores which may or may not follow the theoretical speculations, all the networks with all the dimensionality reduction techniques showcased a reasonable recognition score.

8.6 Chapter Summary

This chapter presented the case of face as a biometric modality for carrying out biometric recognition as well as verification. The chapter presented the notion of the various interesting features in face for use as biometric. The chapter further formed the basis of construction of multi-modal biometric system. The problem of face recognition was dealt with. For solving the problem a dimensionality reduction technique and a pattern recognition technique was formulated. The dimensionality reduction techniques were studied under two headings. The first heading dealt with the visual dimensionality reduction techniques where the complete set of images was given to the system for finding the features that can better perform classification. Here we discussed Principle Component Analysis and Regularized Linear Discriminant Analysis. The other part of the chapter focused upon pattern recognition techniques. Here two types of neural networks were discussed, multi layer perceptron with back propagation algorithm and radial basis function networks. Experimental and simulation results showed that these techniques could carry effective recognition.

References

[1] Card, S.K., Mackinlay, J.D., Robertson, G.G.: A morphological analysis of the design space of input devices. ACM Trans. Inform. Syst. 9(2), 99–122 (1991)

[2] Er, M.J., Wu, S., Lu, J., Toh, H.L.: Face Recognition with Radial Basis Function(RBF) Neural Networks. IEEE Trans. on Neural Nets 13(3), 697–710 (1999)

[3] Huang, S.Y., Hwang, C.R. (2005) Kernel fisher's discriminant analysis in gaussian reproducing kernel hilbert space. Technical report, Institute of Statistical Science, Academia Sinica, Taiwan (2005)

[4] Jollife, I.T.: Principle Component Analysis. Springer, USA (2002)

[5] Kala, R., Vazirani, H., Shukla, A., Tiwari, R.: Fusion of Speech and Face by Enhanced Modular Neural Network. In: Proc. of the Intl. Conf. on Info. Syst., Technol. and Manag., ICISTM 2010, Bankok, Thailand, pp. 363–372 (2010)

[6] Liu, W., Wang, Y., Li, S.Z., Tan, T.N.: Null Space Approach of Fisher Discriminant Analysis for Face Recognition. In: Maltoni, D., Jain, A.K. (eds.) BioAW 2004. LNCS, vol. 3087, pp. 32–44. Springer, Heidelberg (2004)

[7] Lu, J., Plataniotis, K.N., Venetsanopoulos, A.N.: Face Recognition Using LDA-Based Algorithms. IEEE Trans. Neural Networks 14(1), 195–200 (2003)

[8] Lu, J., Plataniotis, K.N., Venetsanopoulos, A.N.: Regularization studies of linear discriminant analysis in small sample size scenarios with application to face recognition. Pattern Recognition Lett. 26(2), 181–191 (2005)

[9] Maragos, P., Schafer, R.W.: Applications of morphological filtering to image processing and analysis. In: Proc. IEEE Intl. Conf. Acoust. Speech, Signal Process., Tokyo, Japan, pp. 2067–2070 (1986)

[10] Phillips, P.J., et al.: Overview of the Face Recognition Grand Challenge. In: Proc. of the, IEEE Conf. on Compu. Vision and Pattern Recog., CVPR 2005, vol. 1, pp. 947–954 (2005)

[11] Shukla, A., Tiwari, R.: Fusion of face and speech features with ANN for speaker authentication. IETE Tech. Rev. 4(5), 359–368 (2007)

[12] Shukla, A., Tiwari, R.: A Novel Approach Of Speaker authentication By Fusion of Speech and Image Feature using ANN. Intl. J. Info. Comm. Technol. 1(2), 159–170 (2008)

[13] Shukla, A., et al.: Intelligent Biometric System using PCA and R-LDA. In: Proc. IEEE Intl. Global Cong. on Intel. Syst., Xiamen, China (2009a)

[14] Shukla, A., et al.: Face Recognition using morphological method. In: Proc. of the IEEE Intel. Advance Comput. Conf., Patiala, India (2009b)

[15] Shukla, A., et al.: Face Recognition based on PCA, R-LDA and Supervised Neural networks. Intl. J. of Engg. Res. Ind. Appl. 2(3), 243–255 (2009c)

[16] Li, S. (2002), A Tutorial on Principle Component Analysis, `http://www.cs.otago.ac.nz/cosc453/student_tutorials/principal_components.pdf`

[17] Spacek, L.: Grimice Database (2010), `http://cswww.essex.ac.uk/mv/allfaces/grimace.html`

[18] Turk, M.A., Pentland, A.P.: Face Recognition Using Eigenfaces. In: IEEE Conf. on Comput. Vision and Pattern Recog., pp. 586–591 (1991)

[19] Ye, J., Janardan, R., Li, Q.: Two-Dimensional Linear Discriminant Analysis. Adv. Neural Inform. Process. Syst. 17, 1569–1576 (2005)

[20] Yu, H., Yang, J.: A Direct LDA Algorithm for High-Dimensional Data with Application to Face Recognition. Pattern Recog. 34(10), 2067–2070 (2001)

[21] Zhao, W., Chellappa, R., Rosenfeld, A., Phillips, P.J.: Face Recognition: A Literature Survey. CVL Technical Report, Center for Automation Research, Univ. of Maryland, College Park (2000)

Part III
Evolutionary Computing

Chapter 9
Swarm Intelligence

Abstract. The manner in which the swarms behave and survive is a great inspiration to build evolutionary algorithms. The flocking of birds, foraging of ants, etc. has been a big inspiration for the optimization and search algorithms that we study in this chapter. We first present the Particle Swarm Optimization which is an optimization tool inspired by the flocking of birds. We present the manner in which multiple particles walk on the fitness landscape to search the optima. We then present the Ant Colony Algorithm which is a search based algorithm inspired from the working of the ant colonies. We would present the manner in which the ants proceed to search the most optimal solution and help the others by the deposition of the pheromone. The other algorithm for discussion is the bee algorithm inspired by the honey bees. We also present Stochastic Diffusion Search, Probability Based Incremental Learning and Biogeography Based Optimization in this chapter.

Swarm Intelligence is a broad area that tries to imitate the natural swarms and their working. These swarms possess great intelligence that is exhibited by the manner in which they work and are able to solve problems. These swarms have been both a mystery as well as an inspiration for the researchers. This gives a new paradigm to problem solving. The very simple nature and behavior of the swarms is able to solve some very big problems very effectively. The swarms make use of multiple agents for the task of problem solving. All these exhibit similar and related behavior while functioning. This enables the problem solving for most of the major problems where the traditional methods fail to give optimal solution within the timeframe.

Consider the common example of ant colonies. It can be commonly seen that the ants are easily able to collect at the place where something interesting is found. This behavior where they all search, coordinate and ultimately collect is a matter of great surprise. The very simple behavior of the individual ants adds up to make a fantastic exhibition of the collective behavior that they showcase. The other examples may be easily seen from the nature.

This class of algorithms imitates the swarm behavior of the birds, fishes, ants, etc. We study and imitate the collective performance, coordination and search performs by the swarms. These are hence great optimization agents just like GA. These are much simpler in their operation and behavior.

The various natural counterparts have resulted in a large amount of variety of these algorithms. All these algorithms have some common principles based on which we regard them as swarm algorithms. Apart from the collective effort to

A. Shukla et al.: Towards Hybrid and Adaptive Computing, SCI 307, pp. 187–207.
springerlink.com

solve problem, all these algorithms may be characterized by an exchange of information and interaction amongst them that helps in coordinating them and directing them. This further makes these algorithms better than the completely random algorithms. Further the use of these algorithms is done only when it is not possible to exhaustively solve the entire problem due to time, computation or other limitations. This is rather true with all the soft computing paradigms.

9.1 Particle Swam Optimization

The first of the algorithms in the class of swarm algorithms that we present is Particle Swarm Optimization (PSO). Particle swam optimization is an inspiration from the flocking birds. Here also we have multiple search agents that try to search for the global minima in the fitness landscape, just like the GA. This operator however hoes not use expensive operators like mutation and crossover for the generation of new generation from the old generation. Here the various individuals move along with iterations, which are analogous to the term generations in the GA (Kennedy and Eberhart 1995; Shi 2004; Clerc 2002; Ventor and Sobieski 2002; Eberhar and Shi 2001). The individuals, in each iteration, move in the direction and magnitude given by the current particle velocity. Hence every individual or particle has two components assigned to it. These are its position and velocity. The position denotes the value of the various parameters of the individual in the individual representation, or the location in the fitness landscape. The velocity denotes the speed and direction with which the individual is moving in this fitness landscape in search of the global minima. In this manner the individuals keep walking in the fitness landscape. The velocity and positions are constantly updated along with iterations. In this manner the algorithms keeps executing. Like the GA all the individuals finally converge towards the center at the end.

Most of the aspects of the algorithm are similar to the ones already discussed in GA. We quickly refer to these terms and then present the algorithm and its working in the next sub-sections.

9.1.1 Terms and Concepts

The individual or the search agent in this algorithm is referred as the particle. Multiple particles in the search space constitute the population. Unlike the GA, the population always remains constant with time. Hence there are no reproductions and killings of the weak individuals. The individuals however change their positions with time. These individuals may be easily represented by the vector of numerals. In addition to position, an individual has an extra term called as velocity. This denotes the speed with which the individual is moving in the fitness landscape for the search operation. The velocity is measured in every direction or every axis of the search space. Hence there are as many components in the velocity vector as there are in the position vector. Each component of velocity is assigned to a particular attribute or axis. This doubles the total space required for the representation of the entire individual.

Each attribute of the individual constitutes an axis of the search space or fitness landscape with the vertical axis denoting the fitness value. The fitness function is used to designate fitness to the individual which is a measure of its goodness to solve the problem. The better individuals have a higher fitness value as compared to the weaker individuals. The ultimate task is to search for the point in the fitness landscape corresponding to the global minima. This search operation largely depends upon the complexity of the search space or the number of hills and valleys that it possesses. We assume a not very complex search space for most of our discussions.

The entire algorithm is iterative in nature. This means that the search keeps searching for individuals in every iteration. We may terminate the algorithm at any stage and return the fittest individual in the search so far. The best individual keeps improving along with time and iterations.

9.1.2 Algorithm

Here we discuss the algorithm design. The various particles are the search agents deputed at various locations of the search space. Further, each of these particles has a certain velocity with which it is moving. The particles take a unit step in each iteration whose magnitude depends upon its current velocity. As the particles move they record their fitness values. This is the value of the fitness in the fitness landscape. The best fitness is stored permanently by the system that can be returned as solution whenever the algorithm terminates. If a particle in its motion finds a better fitness region, the best fitness value is updated. This forms the overall structure of the algorithm.

Now the major task is the modification of the fitness and the velocity value of the particles based on a strategy that optimizes the search with respect to iterations and promises a convergence to the algorithm execution. The distances may be easily updated by the velocity vector. Consider a particle i whose position vector at iteration t be given by $P_i(t)<p^i_1, p^i_2, p^i_3, \ldots p^i_n>$. Let the velocity vector of this particle be $V_i(t)<v^i_1, v^i_2, v^i_3, \ldots v^i_n>$. Now the position vector at iteration t+1 for any attribute j would be given by equation (9.1).

$$p^i_j(t+1) = p^i_j(t) + v^i_j(t) \tag{9.1}$$

Looking at equation (9.1) we may easily define velocity as the distance moved by a particle in unit iteration.

The particle in its entire motion stores the value of the best position in terms of fitness that it has encountered so far. Hence the various search agents or particles are capable of memorizing their best locations. This is referred as the history of the particle. Each particle stores its own history. Let us assume that the history or the best position encountered by the particle i is given by $Pb_i(t)<pb^i_1, pb^i_2, pb^i_3, \ldots pb^i_n>$. This further necessitates the update of the best position of each particle after every iteration. In this step a particle compares its present fitness value to the best fitness value recorded so far and updates it if necessary.

There needs to be a means of coordination of the various individuals so that they may be moved in a well-structured coordinated manner. This is done by the recording of the global best point for the entire population. Every individual knows the globally best position known to the system so far. For this the individuals constantly exchange the fitness values and compare the same. The best fitness value is broadcasted and communicated to all the individuals. This is known as the globally best position where the best fitness value was recorded. This value is also updated after every iteration. Let this value for the entire population be given by $Pg(t)<pg_1, pg_2, pg_3, \ldots pg_n>$.

Now any particle knows the best position in its path and the globally best position. The motion of the particle or the update in its velocity is carried by two strategies. The first strategy tries to modify the velocity in such a manner that the individual walks towards the point where it recorded its best fitness or P_b^i. The second strategy tries to move the particle in a manner that it moves towards the globally best individual or Pg. The net update in the velocity is the cumulative result of both these intensions. The velocity update for any iteration $t+1$ from iteration t may hence be given by equation (9.2)

$$v_j^i(t+1) = v_j^i(t) + c_1 r[pb_j^i(t) - p_j^i(t)] + c_2 r[pg_j(t) - p_j^i(t)] \tag{9.2}$$

Here r is any random number in the range of 0 to 1. c_1 and c_2 are constants that are usually taken to be 1.

In this equation the second term denotes the update in velocity in the direction of the particle history and third term denotes the update in the direction of the global best position.

The complete algorithm is given by figure 9.1.

9.1.3 Maximum Velocity

An important aspect with velocity is its maximal value. We know that if the velocity is very low, the individuals would walk very slowly in the fitness landscape testing their fitness at minutely small movements. This would result in a large number of fitness evaluations and a very few distance of walk. This would slow down the algorithm performance due to the large time wasted in the fitness computations. The algorithm would take a very long time to converge. We may easily see that the algorithm explores a minutely small part of the fitness landscape in very large time steps. This may give poor performance. The advantage of small velocity however lies in the fact that it is able to behave well in extremely sharp changes in the fitness landscape. Consider the case where the global minimum is at some location surrounded by steep walls. In such a case small velocity would be able to penetrate inside and record it.

Keeping velocity high also has a negative impact. It is very possible that the algorithm completely over-jumps the global minima. The recordings of its entire path may hence be un-appealing, even though the path lies in the location of the global minima. This would especially be the condition when the global minimum

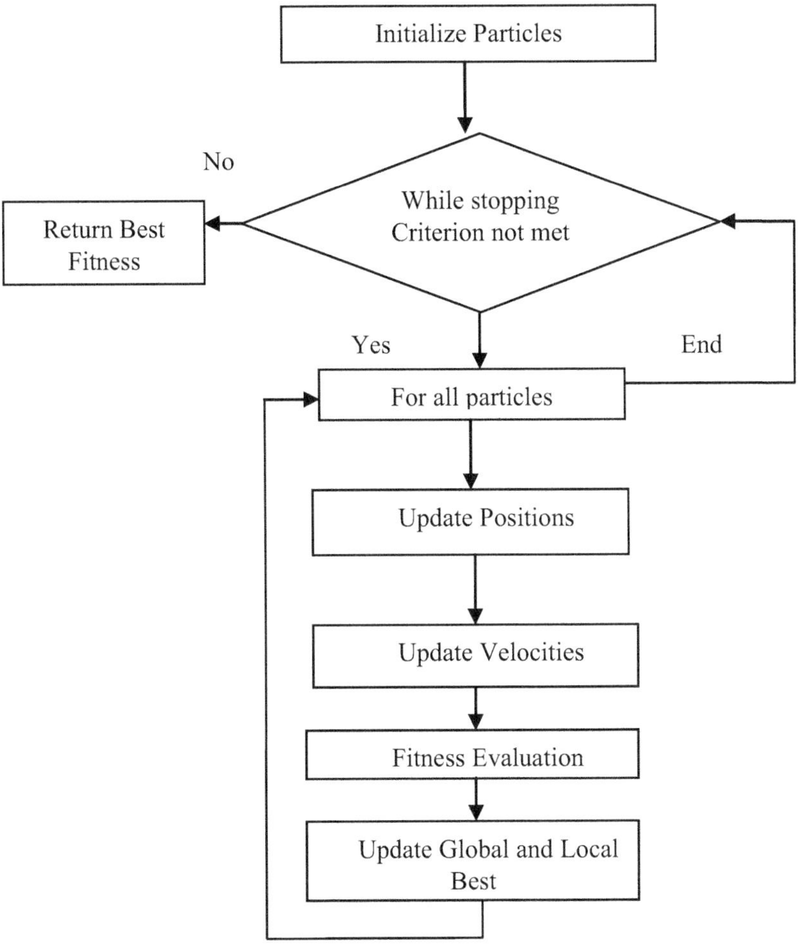

Fig. 9.1 Particle Swarm Optimization

has sharp boundaries. In such a case it is relatively difficult to predict the global minima. Large velocities are however less time consuming.

The loop in the algorithm run keeps modifying the values of velocity to its optimal values. Many times the modification may lead to very large velocity values. This is usually a problem for the PSO and must be avoided. We hence fix a maximum value of the velocity v_m beyond which the velocity cannot increase. In case the algorithm happens to fix a value of velocity greater than v_m, the velocity is trimmed to v_m.

9.2 Ant Colony Optimization

The other algorithm that we study under the head of swarm intelligence is Ant Colony Optimization (ACO). This is another very commonly used algorithm that finds application in the optimization and search tasks. This algorithm is an inspiration from the method of working of the ant in ant colonies and the manner in which they collect and deposit the pheromones (Maniezzo *et al* 2004; Dorigo *et al* 1996; Dorigo and Caro 1999; Stutzle and Hoos 1997; Parpinelli *et al* 2002; Stutzle and Dorigo 1999).

Unlike the earlier discussed approaches, the ACO is more like a graph search problem. This class of algorithm is also called as the meta-heuristic search techniques. For this we shift our focus to the conventional graph search techniques where graphs were defined as a collection of vertices and edges. The edges enable the traversal from one point to the other. There is a fixed source from where the search starts. The objective is to reach the destination as easily as possible. The traversal along the edges involves costs that need to be minimized or optimized. We would first discuss the entire algorithm using this notation of graphs. Afterwards we comment upon the use of these algorithms over the fitness landscape in continuous domain and give it a treatment that we have been doing for the other approached discussed.

9.2.1 Algorithm

The various agents of search in this algorithm are called ants derived from their analogy to the biological counterparts. The ants are supposed to move from the start to the goal and find out the path that connects the source to goal. Each ant independently figures out or makes its own path by walking in its own manner. Suppose that the ant at any point of time is located at node i. Now it may be able to make a variety of moves that make it reach different nodes in the graph. The complete set of nodes that the ant can reach by making any of the possible moves is known as the *tabu list*. At every location the ant is supposed to select one of the possible moves or one of the possible states j from the tabu list that mark its next move.

The ant is guided by two phenomenons that help it to make every move. These are called as the attractiveness and traits. *Attractiveness* are the moves in tabu list that attract the ant in making these moves and are the result of set heuristics. These heuristics depend upon the problem. The moves that make the ant reach nearer to the goal have better heuristic values as compared to the other moves in the tabu list. In this manner attractiveness constantly pushes the ant to choose paths such that the total path length seems to be optimized based on heuristics. These may be denoted by η_{ij}.

The other factor that governs the motion of the ant is traits. *Traits* are the moves that attract the ant because of the historical better performance of these moves. ACO is an iterative algorithm where the solutions are generation at time iterations. It is very important for an ant to learn from the experiences of the previous ants. These experiences are in the form of traits. Traits demarcate the good moves from the bad moves based on the experiences of the ants that made the moves. If the experiences were good that resulted in the generation of an optimal overall path, the traits are attractive for such moves. These may be denoted by τ_{ij}.

The probability of movement of any ant from the node i to the node j is the net effect of these two values. This may be hence given by equation (9.3).

$$p_{ij} = \frac{\tau_{ij}^{\alpha} + \eta_{ij}^{\beta}}{\sum \tau_{kl}^{\alpha} + \eta_{kl}^{\beta}} \tag{9.3}$$

Here α and β are user defined constants to control the contribution of these two factors and lie between the range of *0* and *1*.

As the ants reach the goal, they evaluate the total quality of the path that they had followed. Based on this quality, the ants deposit pheromones at all the edges that they had taken to reach to the destination starting from the source. All the edges are given an equal amount of pheromone that directly depends upon the quality of the path traversed by the ant. If the path is of high quality, the amount of pheromone deposited is high and vice versa. Like this all the ants deputed at the system perform the task of pheromone deposition. This complete task of search and pheromone deposition by all the ants in the system constitutes a single iteration of the algorithm.

Continuous pheromone deposition may usually lead to over deposition of pheromone. The pheromone at some good edges would become too large for the system to handle. Hence there is a concept of pheromone evaporation. This limits the amount of pheromone deposited at any edge. As per this concept the pheromone keeps evaporating from the system, depending upon the present deposited value. This evaporation takes place at every iteration of the algorithm.

The net update in pheromone for any edge *ij* at iteration $t+1$ may hence be given by equation (9.4).

$$\tau_{ij}(t+1) = \rho\tau_{ij}(t) + \Delta\tau_{ij} \tag{9.4}$$

Here ρ is the evaporation constant; $\Delta\tau_{ij}$ is the combined addition of pheromones of all the ants and $\Delta\tau^{k}_{ij}$ is the addition of pheromone by ant *k* given by equation (9.5) and (9.6).

$$\Delta\tau_{ij} = \sum_{k} \Delta\tau_{ij}^{k} \tag{9.5}$$

$$\Delta\tau_{ij}^{k} = \begin{cases} Q_k & \text{if ij is in the path of the k}^{\text{th}} \text{ ant} \\ 0 & \text{otherwise} \end{cases} \tag{9.6}$$

Here Q_k is the quality of the path of the kth ant that may be fixed depending upon the problem.

Another concept used in ACO is of *daemon actions*. The ACO may many times need centralized actions that require the participation of all the ants. The presented algorithm behaves like different ants moving independent of each other without any coordination. Hence the daemon actions are used to perform the tasks that require the participation of all the ants for various types of monitoring or specific needs. These are problem specific and do not usually affect the optimization process. These may be used to run a local optimization search, or to update global information.

This algorithm is given in figure 9.2.

9.3 Artificial Bee Colonies

The Artificial Bee Colonies (ABC) is another novel optimization algorithm that we study in context of the Swarm Intelligence. The ABC uses the natural bees and tries to imitate the same behavior. The natural bees are very good in searching for some food source. Whenever any bee finds the food, it signals the other bees by its dance. This signals the other bees regarding the quantity and the location of the food source. This helps in directing the other bees towards good sources of food in their search for food. These bees are able to attract a large number of bees and carry forward the task. The ABC algorithm tries to use the same concepts and working methodology that is exhibited by these bees (Pham *et al* 2006; Karaboga 2005; Karaboga and Basturk 2007).

The ABC algorithm models every individual solution of the optimization problem as bees. The bees hence represent the points in the fitness landscape. This representation is similar to the problem representation in GA by using vector of numerals. The food and the amount of food in this algorithm refer to the goodness of the location or the fitness of the point in the fitness landscape. This may be taken as the fitness function that returns the fitness value of any individual or solution of the optimization process.

The bees or the solutions of the problem may be classified into three types in this algorithm. These are as explained in the following sections. We then discuss the complete algorithm.

9.3.1 Employed

These are the bees that have visited the food source before and have a complete idea of its location and amount. These bees try to exploit the neighboring locations of the food source and search the best places of food in the surrounding areas of the present value. These bees have memory, so they know the places they have visited before and the quality of food there. This enables them store the best place of food that they encounter in their journey. This information is further shared and collaborated with the other bees in the population.

These solution or bees represent the local search performed by the algorithm. Since these solutions are at places with good fitness values, it is highly likely that the nearby places may have even better fitness values. This necessitates a need to explore the surroundings to reach the minima. The local search performed by these bees enable the algorithm to move or be directed towards the minima in the fitness landscape. This search of the employed bees is hence a valuable tool for the optimization purposes.

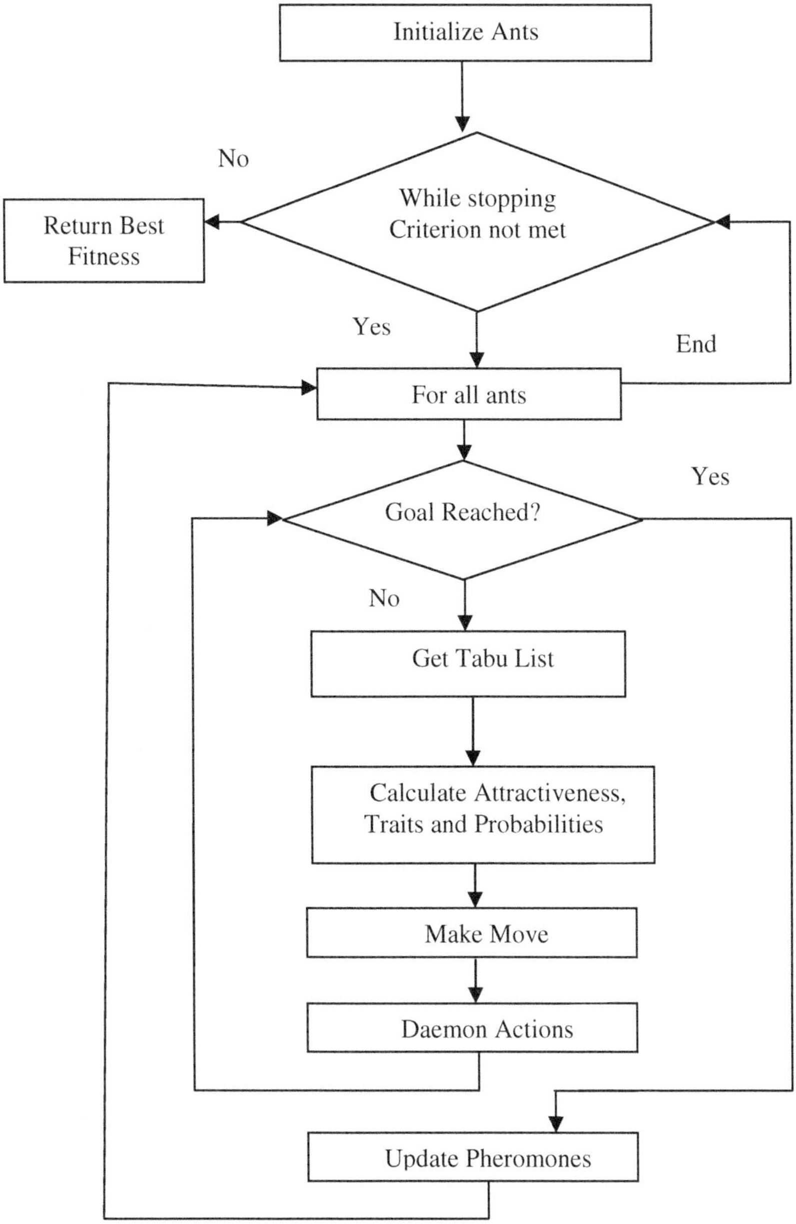

Fig. 9.2 Ant Colony Optimization

These bees while searching store the best searched place or location in the fitness landscape. Hence at every movement the present fitness value or quality of food is compared with the best value known so far by the bee. If this location is better in terms of fitness, the bee memorizes the present location as the best location and forgets the past. In this manner the monitoring of best place is done while the search is on.

9.3.2 Onlooker

These are the bees that are searching for food that try to go to the areas or locations that have already been visited by the employed bees. The employed bees are located at good locations and are hence source of attractions by these bees. These bees get attracted and try to move to the same locations as possessed by the onlooker bees. The quality of food is a point of attraction. Hence the employed bees with better quality of food are more attractive than the bees with poor quality. The onlooker bees continuously try to get to the best places in this manner. As the employed bees, these bees also have memory for storing the best places that they encounter during their search.

The findings of the employed bees in terms of the best locations of the food or better fitness value in the fitness landscape is a great source of information which must be used for directing and coordinating the other bees. This is done by these bees. These solutions or bees move towards the areas where the fitness or the quality of food is good. It is natural that more attention must be given for search to the areas with good fitness values. This is because of the better possibility of finding the global minima in the vicinity. This task is carried out by employing more agents to these areas. This is done in this algorithm by the onlooker bees. These bees travel to the better fitness areas. This is a completely probabilistic function. The attraction or the probability of movement of a bee towards some location is directly proportional to its fitness or the quality of food. These solutions or bees play a major role in guiding the algorithm towards the global minima. The search may usually result in exploration of better fitness points or the global minima in the course of movement of these bees.

9.3.3 Scouts

These are the bees that do a random search for the food. It is important for the bees to explore new areas for possible sources of food there. This plays a major role in uncovering the new areas and new sources of food. This task is carried out by these bees. They are completely random in nature and their operation of search.

The scouts act as randomized search in the fitness landscape in search of the global minima. This plays a role in avoiding the algorithm to get trapped in some local minima. This is especially important when the fitness landscape is complex in nature and the chances of finding the global minima are poor. In such a situation these bees explore new areas in the fitness landscape and uncover vital information. These bees are used in this algorithm in similar nature.

The memory is used in these bees for storing of the information regarding the location and the quality of the food. This helps in possible directing of the other bees.

9.3.4 Algorithm

The entire algorithm uses these three types of bees or solutions for its functioning. The general algorithm framework is given in figure 9.3.

The algorithm starts with the random initialization of the bees. This is common to most of the evolutionary algorithms where random agents are placed onto the fitness landscape. The employed bees are taken as the solutions with known locations as per their definitions. Then the quality or the fitness of these bees are measured using a fitness function. This gives us valuable information regarding

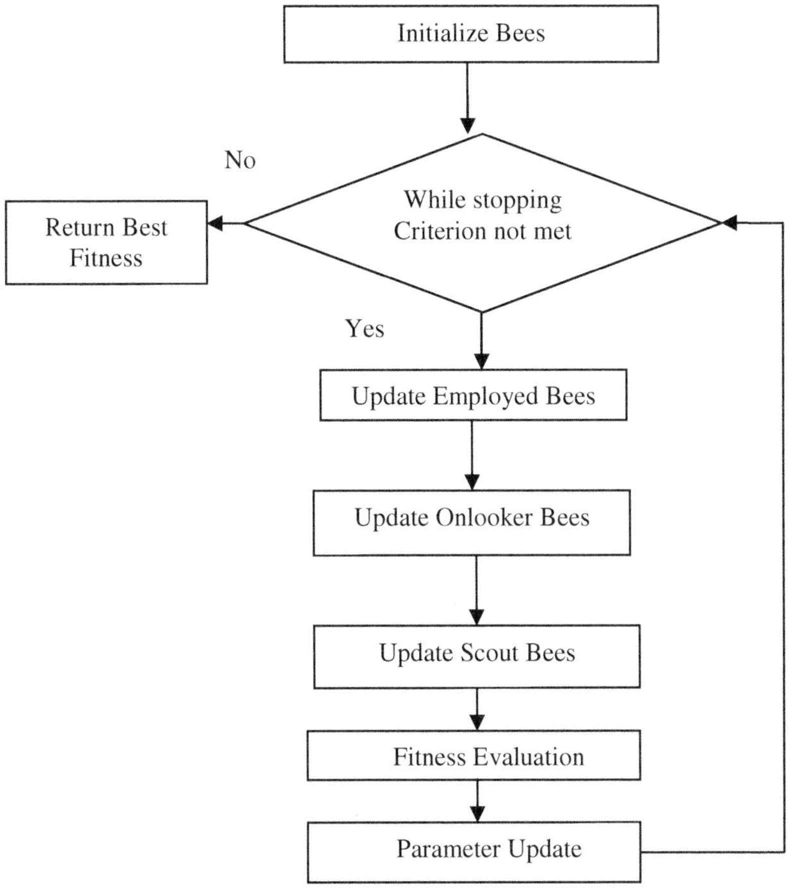

Fig. 9.3 Artificial Bee Colonies Algorithm

the nature of the fitness landscape. The probability associated with each of the food source may hence be measured. At the next stage the onlookers try to use these probabilities to decide their food locations. The higher probability locations are more likely to be selected than the low probability locations. These bees move to the selected food locations. The employed bees explore the neighboring areas in search of better fitness points in the fitness landscape.

It is possible that at some time the algorithm converges to the minima and the improvement with generations is too small. In such a case there might be a transition in the bee operations. Now the employed and the onlookers may be taken as the scouts that facilitate more exploration of the search space.

9.4 Stochastic Diffusion Search

The other algorithm of study is the Stochastic Diffusion Search (SDS). This search technique tries to add good characteristics into the individual representing the solution of the problem. The individual may be represented using any genotype representation as was the case with the GA. The SDS algorithm is a simple algorithm that attempts to make a population pool of multiple individuals and amplify the fitness value of the goodness of any solution by the maximization of the good characteristics. The algorithm operates in generations. The population pool contains multiple individuals with each individual representing a potential solution (de Meyer 2003; Nasuto 1999; de Meyer *et al* 2002, 2006; Nasuto *et al* 1998; Nasuto and Bishop 1999).

Unlike the GA, the solutions or the individual in this technique are classified into active and inactive solutions. This decision is made by some function. The active solutions are the feasible solutions that possess a good fitness value. The inactive solutions are the weak solutions that possess a poor fitness value. In this manner the entire population pool is divided into two separate pools. The attempt is to have as many solutions active as possible with good fitness values. The entire phase where the individuals are divided into these two pools is known as the test phase of the algorithm.

The active individuals are rated good as per the problem definition and the definition of activeness. The inactive individuals however represent poor solutions. We hence need to modify these solutions in some manner to make them active. This is done by establishing an interaction between the individuals. Every inactive individual selects an individual to modify it. Now the selected individual may be active or inactive.

In case the selected individual was active an action is initiated between the inactive and the selected active individual. This causes some exchange of characteristics between the two individuals. Some of the characteristics are taken from the active individual and embedded into the inactive individual. This cases the inactive individual to behave more like an active individual. This may make it active that may be reflected in the next generation of the algorithm.

In case the selected individual was inactive, the chances of an interaction producing good results are vague. Hence the interaction does not take place. Instead the algorithm decides to completely replace the inactive individual with a randomly generated individual. This may again be active or inactive. Nothing can be said for this individual as it was completely randomly generated. Since the original individual was itself inactive, it may be assumed that the results may not be less likely.

This phase is known as the diffusion phase. This initiates collaboration among the various particles where they exchange their information and characteristics. This collaboration has a deep impact in the systems running under the notion of collective intelligence or swarm intelligence. The motivation is to use the others experiences and findings to magnify the performance of the individual in the population pool. This greatly contributes to overall algorithm convergence and exploration of the global minima.

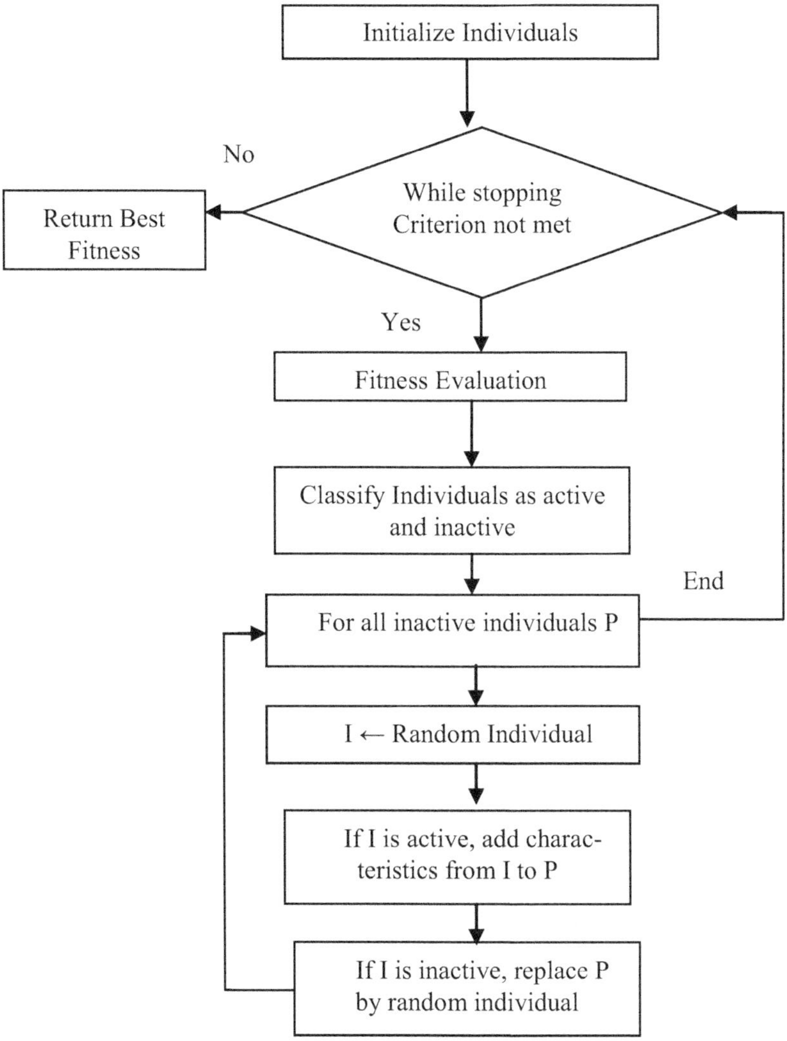

Fig. 9.4 Stochastic Diffusion Search

The function used to designate the activity of the individual is kept probabilistic in nature. This may make different decisions regarding the activity of the same individual when called different times. This adds randomness to the classification of the activity of the algorithm that may further play a role in the optimal algorithm exchange of information or characteristics.

The general algorithm of SDS is given in figure 9.4.

9.5 Probability Based Incremental Learning

The other algorithm that we study in this chapter is the Probability Based Incremental Learning (PBIL). This method of evolution uses a probability vector that aids or guides the generation of individuals and their modification over time or generation. In this manner the operations are applied on this vector that further results in the generation of the individuals. The vector learns over generations where the various probability entries are tuned and finally assume some fixed values. This algorithm hence tries to combine the Competitive Learning and Genetic Algorithms. We deal with the various aspects of this algorithm one by one (Servais and Jangel 1997; Yang and Yao 2003, 2005, 2008; Southey and Karray 1999; Baluja 1994; Yang 2005; Hohfeld and Rudolph 1997).

9.5.1 Representation

Any conventional GA represents a series of individuals that make up the population. Consider any individual to be represented by X_i $<X_i^1, X_i^2, X_i^3 \ldots X_i^n>$. In case the representation is in the form of a bit string vector, every X_i^j would be either 0 or 1. The entire algorithm is based upon the creation of a higher generation of individuals from a lower generation.

In this approach however we do not represent or try to evolve the individuals. We rather try to evolve a probability vector $V<V_1, V_2, V_3, \ldots V_n>$. Here each V_j denotes the probability of occurrence of *1* in the jth location of the final evolved individual considering a bit string genotype representation. This vector measures the likeliness of any position being towards 1 or 0. This may be used as a measure of the convergence of the algorithm as well.

From our understanding of the GA with bit string representation we know that the start generations are random collections of 0s and 1s. As the algorithm continues, the solutions start converging. The positions which are relatively fitter with a presence of 1 get 1 and vice versa in all the individuals. Some locations get the needed bit very early when the other bit is wiped off from all the individuals making the population. This is because of the other bit bringing the unwanted characteristics. Some other bits require a lot of time to do this. We studied the same under the schema theory in chapter 3. Towards the end all the bits converge to one common representation that is fit as per the fitness measure.

The same is true with this algorithm as well. Initially we have no idea of the fitness landscape or the behavior of the fitness unction. In such a case all bits have an equal probability of being 0 or 1. We hence assign a value of 0.5 to all the locations of this vector. Hence at the initialization phase of the algorithm all V_j is given a value of 0.5.

As the algorithm proceeds, we start getting clearer regarding the goodness of 0 or 1 at every location. Hence the vector starts getting a 1 or 0 stored at its various locations. This depends upon the likeliness of 1 or 0 at that location. This task is performed by the learning algorithm as we shall see in the coming sections.

Hence only the vector is carried in between generations using this technique. The entire task is the optimization or the convergence of this vector. There is no need of multiple individuals constituting the population or the complex genetic operators for the generations of the higher generations from the lower generations.

9.5.2 Generation of Samples

We may not pass the population from one generation to the other, but a population is needed for the physical relevance and learning purposes. This may be generated by using the probability vector discussed. Each individual consists of a set of attributes of the form X_i $<X_i^1, X_i^2, X_i^3 \ldots X_i^n>$. In order to generate the term X_i^j we consult the jth attribute of the vector V_j. We generate a random number between 0 and 1. In case this number is less than V_j, the attribute X_i^j is given a value of 0 and vice versa. In this manner we may generate the samples or population from the population for the training purposes.

9.5.3 Training

The major task associated with the algorithm is the training of the probability vector along with generations. This vector is passed between generations to mark the convergence or the exploration of the algorithm. We have seen the generation of a population based on this vector. The best individual of the selected population is used for the training purpose. Let this individual be B $<B_1, B_2, B_3, \ldots B_n>$. The training involves the tuning of the probability vector $V<V_1, V_2, V_3, \ldots V_n>$. At every iteration this vector tries to lean towards the best individual. This would add the probabilities of the best individual into this vector and would further facilitate generation of healthier individuals.

The updating of the jth attribute of the vector V_j from the jth attribute of the best individual B_j is given by equation (9.7)

$$V_j(t+1) = (1-\alpha)\, V_j(t) + \alpha\, B_j(t) \tag{9.7}$$

Here α denotes the learning rate. t is the iteration or the generation of the algorithm.

The complete algorithm is given in figure 9.5.

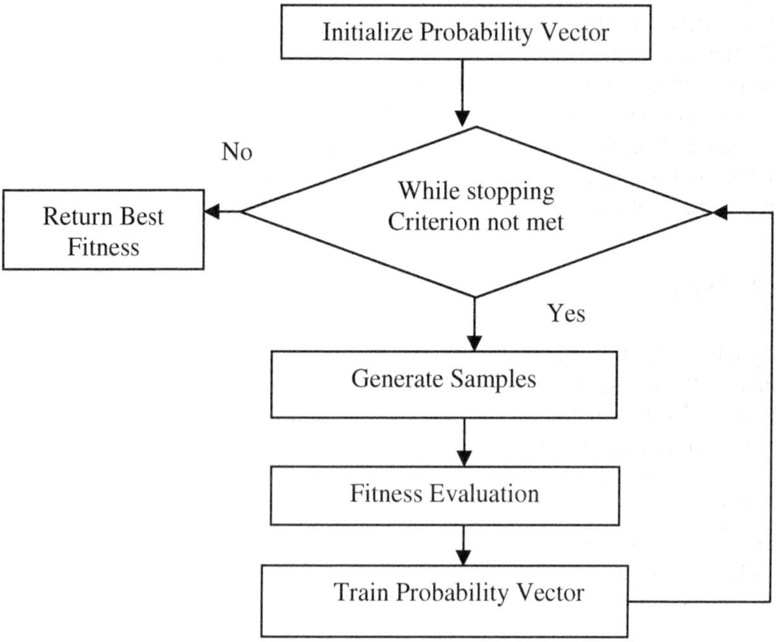

Fig. 9.5 Probabilistic Based Incremental Learning

9.6 Biogeography-Based Optimization

The last algorithm that we study in this chapter is the Biogeography-Based Optimization (BBO). This optimization technique makes use of the concepts of habitats for the optimization purposes (Simon 2008, 2009; Ergezer *et al* 2009; Kumar *et al* 2009; Du *et al* 2009; Rarick *et al* 2009). It maintains different habitats with each habitat with some different characteristics. The habitat has an emigration as well as some immigration rate. This determines the inflow and the outflow of the characteristics from one habitat to another. This way the flow of information or the collaboration is modeled in this algorithm. A habitat with a high fitness value is very likely. Hence its immigration rate is normally very low. This prohibits the transfer of characteristics inside the habitat. Since it has a high fitness value, it is less likely that adding more characteristics from relatively weak habitats would be helpful. On the other hand the emigration rate is quite high. This facilitates a lot of transfer of characteristics from the individual. Since this represents a very fit habitat, it is likely that the outflow of characteristics would be of a very high value for the other habitat. The outflow of characteristics does not mean that the individual loses its characteristics. Rather the copies of the characteristics are flown out of the individual. The entire collection of all habitats constitutes the ecosystem in this algorithm. The constant coordination and flow of characteristics within the habitats causes the optimization to take place. We study the various aspects of this optimization technique one by one.

9.6.1 Representation

In this discussion we carry a few analogies in regard to the similarity of this algorithm with the conventional GA. The habitat may be regarded analogous to the individual of the GA. Each habitat is associated with its own emigration rate and the immigration rate. Hence the complete representation of the habitat includes the genotype representation of the solution along with its immigration rate and the emigration rate. The ith habitat may be denoted by X_i $<X_i^1, X_i^2, X_i^3, \dots X_i^n>$. The ecosystem may be regarded as analogous to the ecosystem population. The generation may be used with the same meaning. This would help in better understanding of the algorithm on the lines of the GA.

The immigration rate is the rate with which the characteristics flow inside any habitat. This is high for the habitats with low fitness value and vice versa. These rates are arranged in a monotonically non-increasing manner. The immigration rate for any habitat X_i may be given by the immigration function denoted by $\lambda(X_i)$.

The emigration rate is the rate with which the characteristics flow outside any habitat into the other habitats. This is high for the habitats with high fitness value and vice versa. These rates are arranged in a monotonically non-decreasing manner. The emigration rate for any habitat X_i may be given by the emigration function denoted by $\mu(X_i)$

Consider the discussion relating to the fitness of the habitat along with its immigration and emigration. It can be easily understood that these both rates are converse of each other.

9.6.2 Habitat Modification

The major task to be carried out at every generation or iteration is the modification of the habitat. At every iteration we make some changes to all the habitats of the ecosystem such that the optimization is enhanced and the latter habitats are better than the earlier ones. This modification of the habitat for every iteration is carried out by the inflow and outflow of characteristics in-between the various habitats. This exchange of information drives the algorithm at every generation.

The modification of the habitat is applied independently for every combination of habitat X_i and X_j. The combination of habitat may undergo modification with an exchange of information or not that depends upon its immigration and emigration rate $\lambda(X_i)$. Here one habitat acts as the donor and the other acts as the receiver. The flow of information from the donor to the receiver only takes place in accordance with the immigration rate of the receiver and the emigration rate of the donor. We generate random numbers to probabilistically select the emigration to happen and the immigration to happen. If both the events agree to happen by this probabilistic selection, then the transfer is agreed to take place. In all other cases no transfer takes place between the habitats.

Let the donor be X_i and the receiver be X_j. In order to carry out a transfer, any one characteristic or gene is randomly selected. Let the selected gene be k. Now the donor transfers the kth characteristic into the receiver. This may be given by equation (9.8).

$$X_j^k \leftarrow X_i^k \tag{9.8}$$

In this way an exchange of information among the habitats largely causes the optimization to take place. This may be taken analogous to the crossover operation of the GA.

The operation may result in deformation of the best habitats. For this reason we use the elite operation to directly transfer the best habitats into the next iteration or generation as was the case in the implementation of GA.

9.6.3 Mutation

The other operation that we need to perform in the habitat is mutation. This is needed for the addition of new characteristics to the ecosystem. The mutation takes place independently for every habitat. The mutation rate is determined from the immigration and the emigration rate. This denotes the probability of the habitat undergoing mutation. If any characteristic or gene is to be mutated in this algorithm, it is replaced by a completely new characteristic or gene. This selection takes place with a rate corresponding to the mutation rate.

The complete algorithm may be given by figure 9.6.

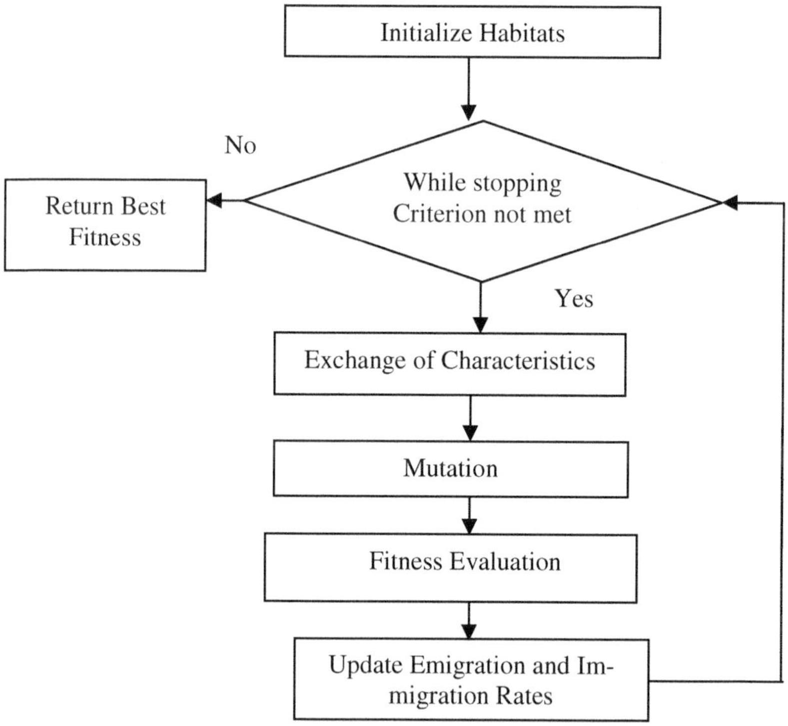

Fig. 9.6 Biogeography-Based Optimization

9.7 Chapter Summary

This chapter presented a variety of algorithms under the general concept of swarm intelligence. Swarm Intelligence is an attempt to imitate the intelligence, coordination and search techniques of the natural swarms that includes honey bees, fish schools, flocking birds, etc. This class of algorithms makes use of multiple individuals for carrying out some search operation. In this chapter we presented a variety of these algorithms by taking an inspiration from their natural counterparts. We first studied the particle swarm optimization. Here the problem consisted of multiple particles moving with some velocity at some part of the search space. Each of these carried forward the task of search of global and local minima in the fitness landscape. The next algorithm was ant colony optimization. Here different ants facilitated the task of graph search. Each ant movement was guided by attraction and traits. The ants deposited pheromones in proportion to their fitness on reaching the goal. Next algorithm of study was the artificial bee algorithm. This algorithm used three types of bees namely employed bees, onlookers, and scouts for the search operation. The next algorithm discussed was stochastic diffusion search. This algorithm classified the individuals into active and passive based on some classifying criterion. The interaction was carried between the inactive and active individuals for increase in number of active individuals. New random individuals were added whenever two inactive individuals tried to interact. Next algorithm discussed was probability based incremental learning. This algorithm had a probability vector that was passed between generations. At each iteration a random population based on this vector is generated and is used for training of the vector. Last algorithm for discussion was biogeography based optimization. In this algorithm different habitats with emigration and immigration rates were used. The exchange of characteristic based on these rates and addition of new characteristics in each habitat caused the optimization.

References

[1] Baluja, S.: Population-Based Incremental Learning: A Method for Integrating Genetic Search Based Function Optimization and Competitive Learning. Technical Report: CS-94-163 (1994)

[2] Clerc, M., Kennedy, J.: The Particle Swarm-Explosion, Stability, and Convergence in a Multidimensional Complex Space. IEEE Trans. Evol. Comput. 6, 58–73 (2002)

[3] Dawei, D., Simon, D., Ergezer, M.: Biogeography-Based Optimization Combined with Evolutionary Strategy and Immigration Refusal. In: IEEE Conf. on Syst., Man, and Cybern., San Antonio, pp. 1023–1028 (2009)

[4] de Meyer, K.: Foundations of Stochastic Diffusion Search. PhD thesis, University of Reading (2003)

[5] Dorigo, M., Caro, G.D.: The Ant Colony Optimization metaheuristic. In: Corne, D., et al. (eds.) New Ideas in Optimization, pp. 11–32. McGraw Hill, London (1999)

[6] Dorigo, M., Maniezzo, V., Colorni, A.: Ant System: Optimization by a colony of cooperating agents. IEEE Trans. on Syst., Man, and Cybern. – Part B 26(1), 29–41 (1996)

[7] Eberhart, R.C., Shi, Y.: Tracking and optimizing dynamic systems with particle swarms. In: Proc. IEEE Congr. Evol. Comput., Seoul, Korea, pp. 94–97 (2001)

[8] Ergezer, M., Simon, D., Du, D.: Oppositional Biogeography-Based Optimization. In: IEEE Conf. on Syst, Man, and Cybern., San Antonio, TX, pp. 1035–1040 (2009)

[9] Hohfeld, M., Rudolph, G.: Towards a Theory of Population-Based Incremental Learning. In: Proc. of the 4th IEEE Conf. on Evol. Comput., pp. 1–5 (1997)

[10] Karaboga, D.: An Idea Based On Honey Bee Swarm For Numerical Optimization, Technical Report-Tr06, Erciyes University, Engineering Faculty, Computer Engineering Department (2005)

[11] Karaboga, D., Basturk, B.: A powerful and efficient algorithm for numerical function optimization: artificial bee colony (ABC) algorithm. J. Global Optimization archive 39(3), 459–471 (2007)

[12] Kennedy, J., Eberhart, R.C.: Particle swarm optimization. In: Proc. IEEE Int. Conf. Neural Networks, Perth, Australia, pp. 1942–1948 (1995)

[13] Kumar, S., Bhalla, P., Singh, A.: Fuzzy Rule base Generation from Numerical Data using Biogeography-based Optimization. Institution of Engineers J. of Electron. and Telecomm. Engg., 8–13 (2009)

[14] Maniezzo, V., Gambardella, L.M., Luigi, F.D.: Ant Colony Optimization. In: Onwubolu, G.C., Babu, B.V. (eds.) Optimization Techniques in Engineering, pp. 101–117. Springer, Heidelberg (2004)

[15] Meyer, K.D., Bishop, J.M., Nasuto, S.J.: Small-World Effects in Lattice Stochastic Diffusion Search. In: Dorronsoro, J.R. (ed.) ICANN 2002. LNCS, vol. 2415, pp. 147–152. Springer, Heidelberg (2002)

[16] Meyer, K.D., Nasuto, S.J., Bishop, M.: Stochastic Diffusion Search: Partial Function Evaluation. In: Swarm Intelligence Dynamic Optimisation. Studies in Computational Intelligence, vol. 31, pp. 185–207 (2006)

[17] Nasuto, S.J.: Resource Allocation Analysis of the Stochastic Diffusion Search. PhD Thesis, University of Reading (1999)

[18] Nasuto, S.J., Bishop, J.M., Lauria, S.: Time Complexity Analysis of the Stochastic Diffusion Search. In: Proc. 1998 Neural Computation Conf., Vienna, Austria (1998)

[19] Nasuto, S., Bishop, M.: Convergence Analysis Of Stochastic Diffusion Search. Intel. J. Parallel, Emergent and Distributed Systems 14(2), 89–107 (1999)

[20] Parpinelli, R.S., Lopes, H.S., Freitas, A.A.: Data mining with an ant colony optimization algorithm. IEEE Trans Evolutionary Comput. 6(4), 321–332 (2002)

[21] Pham, D.T., Eldukhri, E.E., Soroka, A.J.: The Bees Algorithm – A Novel Tool for Complex Optimisation Problems. Intell. Production Machines and Syst., 454–459 (2006)

[22] Rarick, R., Simon, D., Villaseca, F.E., Vyakaranam, B.: Biogeography-Based Optimization and the Solution of the Power Flow Problem. In: IEEE Conf. on Syst., Man, and Cybern., San Antonio, pp. 1029–1034 (2009)

[23] Servais, M.P., De Jager, G., Greene, J.R.: Function Optimisation using Multiple-Base Population Based Incremental Learning. In: Proc. of the 8th Annual South African Workshop on Pattern Recog. (2003)

[24] Shi, Y.: Particle swarm optimization. IEEE Neural Netw. Soc. Mag. 8–13 (2004)

[25] Simon, D.: Biogeography-Based Optimization. IEEE Trans. Evol. Comput. 12(6), 702–713 (2008)

[26] Simon, D., Ergezer, M., Du, D.: Population Distributions in Biogeography-Based Optimization Algorithms with Elitism. In: Proc. of the 2009 IEEE Intl. Conf. on Syst. Man and Cybern., pp. 991–996 (1991)

[27] Southey, F., Karray, F.: Approaching evolutionary robotics through population-based incremental learning. In: Proc. IEEE Intl. Conf. Systems, Man, and Cybern., vol. 2, pp. 710–715 (1999)

[28] Stutzle, T., Dorigo, M.: ACO Algorithms for the traveling salesman Problem. In: Miettinen, K., Neittaanmaki, P. (eds.) Evolutionary Algorithms in Engineering and Computer Science, pp. 160–184 (1999)
[29] Stutzle, T., Hoos, H.: Improvements on the ant system: Introducing MAX–MIN ant system. In: Proc. of the Intl. Conf. on Artif. Neural Netw. and Genetic Algorithms, pp. 245–249. Springer, Wien (1997)
[30] Venter, G., Sobieski, J.: Particle Swarm Optimization. In: Proc. 43rd AIAA/ASME/ASCE/ AHS/ASC Structures, Structural Dynamics, and Materials Conf., Denver, CO (2002)
[31] Yang, S.: Population-based incremental learning with memory scheme for changing environments. In: Proc. of the 2005 Conf. on Genetic and Evol. Comput., pp. 711–718 (2005)
[32] Yang, S., Yao, X.: Population-Based Incremental Learning With Associative Memory for Dynamic Environments. IEEE Trans. Evol. Comput. 12(5), 542–561 (2008)
[33] Yang, S., Yao, X.: Experimental study on population-based incremental learning algorithms for dynamic optimization problems. Soft Comput. 9(11), 815–834 (2005)
[34] Yang, S., Yao, X.: Dual Population-Based Incremental Learning for Problem Optimization in Dynamic Environments. In: Proc. of the 7th Asia Pacific Symp. on Intell. and Evol. Systems, pp. 49–56 (2003)

List of Figures

Fig. 9.1. particle Swarm Optimization
Fig. 9.2. Ant Colony Optimization
Fig. 9.3. Artificial Bee Colonies Algorithm
Fig. 9.4. Stochastic Diffusion Search
Fig. 9.5. Probabilistic Based Incremental Learning
Fig. 9.6. Biogeography-Based Optimization

Chapter 10
Genetic Programming

Abstract. Evolution is becoming a comprehensive tool for automated problem solving that enables the systems to attain their optimal form. Evolution includes the three basic methodologies of Genetic Algorithm, Genetic Programming and Evolutionary Strategies. In this chapter we focus our attention towards the use of Genetic Programming as an evolution tool where programs to solve the problem are evolved along with time. We state the working of these algorithms with a keen eye on their differences with the Genetic Algorithm. Selection, Mutation and Crossover enable the generation of one population to other. A major problem in their use is the unnecessary growth of the code along with the generations which would be addressed in the chapter. Later we present a specialized form of Genetic Programming called the Grammatical Evolution that uses a set of Grammatical Rules running by Backus-Naur Form to evolve systems. This forms another great class of algorithms for optimization or evolution.

Evolution and its immense potential can easily be seen in everyday life from the natural systems. Life evolved with simple organisms that kept attaining complex forms as per requirements. The continuous adaptation to the natural environment is a continuous process that keeps happening with time. As a result the individuals keep getting better with generations as per the changing environment. We discussed in chapter 3 how this led to the making of the evolutionary algorithms. This class of algorithms provided an iterative manner of solving problems where the solutions kept improving with time. This makes these algorithms as excellent optimization tools for system optimization.

EA is widely used for various applications in various domains. One of the fundamental applications includes the optimization where EA is used to find the optimal parameter values to any system or algorithm (Fogel 1992; Forrest 1991; Michalewicz 1993). We studied the various systems in unit-II that made use of various models of Artificial Neural Networks (ANNs) or Fuzzy Inference Systems (FIS) for problem solving. All these systems required the tuning of the parameters that were weights in the case of ANNs and membership function parameters in case of FIS. This was done using some training algorithm by the system. It is natural that the optimization power of the EA can be used to effectively set these parameters. This is done by the use of EA as the training algorithm for the system.

Another problem we highlighted in Unit-I while discussing the various systems was the need to specify architecture for the system. This consisted of the number

A. Shukla et al.: Towards Hybrid and Adaptive Computing, SCI 307, pp. 209–234.
springerlink.com

of neurons in the ANN and the number of type of membership functions in the FIS. The optimization power of the EA can be easily extended to this domain as well where EA can be a tool for optimization of the entire system architecture. This has various advantages that make this a completely different and well-studied discipline of evolutionary Artificial Neural Networks (ENN) (Yao 1999; Angeline *et al* 1994) and Evolutionary Fuzzy Inference Systems (EFIS) (Cordon et al 2001; Pena-Reyes and Sipper 2001).

The use of evolution in the ANNs, FIS or other systems is highly extensive due to the offerings of the EA. We study a lot many interesting application of these algorithms in system evolution and optimization in Unit-IV. The bulk of their use in literature and the increasing advancement of EA in system design and optimization tools makes us devote an entire unit for this class of algorithms. In the subsequent unit whenever we refer to evolution, it may refer to any one of the discussed approach or even their combination.

The EA may be classified into three fundamental types. These are Genetic Algorithms, Genetic Programming and Evolutionary Strategies. All these differ in some or the other terms that affect their overall working. The GA was dealt in chapter 2. Evolutionary Strategy would be discussed in chapter 10. This chapter presents the other technique i.e. Genetic Programming.

Genetic Programming (GP) is a class under Evolutionary Algorithms that is used for problem solving inspired from the natural evolution process. The basic fundamental is similar to the use of Genetic Algorithms (GA). We hence briefly revise our discussion over Genetic Algorithms and then discuss the various operations in GP one by one by comparing with GA (Altenberg 1994; Banzhaf *et al* 1998a, 1998b, 1998c; Langdon 1998; Luke and Spector 1997; Tackett 1993).

10.1 Analogy with Genetic Algorithms

Chapter 2 focused upon Genetic Algorithm (GA) and discussed how these solve problems. Being evolutionary in nature, they take an iterative approach. The initial solution is randomly generated that constitutes the initial population of the problem. The entire algorithm continues in generation where one generation leads to the creation of the higher generation. As the generations increase, the solutions keep getting fitter and solve the problem better as compared to the previous generations. The algorithm makes use of a scaling mechanism to rate the various individuals on the basis of their ability to solve problem. Fitness function is used to assign a score to each solution of the problem or each individual of the population. The selection operator selects the individuals to participate in the genetic operation that creates the higher order generation. This follows a probabilistic selection criterion with the fitter solutions having more probability of selection (Mitchell 1996; Whitley 1994).

The algorithm then uses crossover as the genetic operation to mix two individuals and in turn generate a higher order individual of the next generation of population. The crossover takes some features from the first parent and some features from the second parent and in turn results in the creation of two new individuals. This is supplemented by the mutation genetic operator that randomly modifies some of the values of selected individuals. It adds new features to the population pool. This may be copied in the crossover operation if they result in individuals with high fitness. The other commonly used operator is elite that places the best individual from one generation directly to the next generation without modification. This stops the algorithm from killing of best individuals in the generation of population pool by non-selection or destruction at genetic process.

The major difference between the GA and the GP lies in problem encoding. This further makes the necessity for specialized genetic operators that work with the different encoding technique. The basic block diagram of the GP is given in figure 10.1. We discuss the various aspects of this approach next.

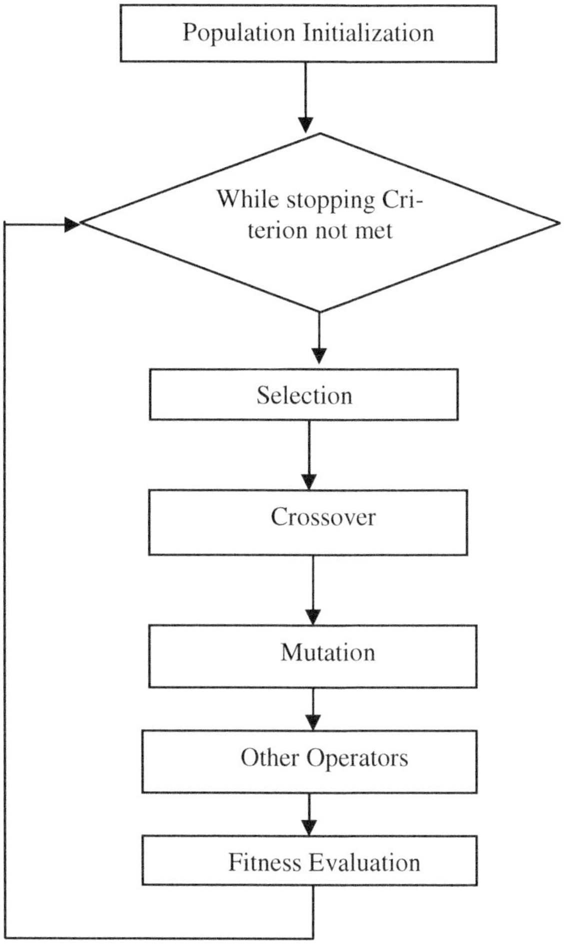

Fig. 10.1 Genetic Programming

10.2 Problem Representation

The first major task as well as the point of difference between the GA and the GP is the manner in which a problem is represented or encoded. In GA we followed a simple phenotype to genotype problem encoding. Here in most of the cases the various parameters were arranged in a linear manner. This may have been encoded in binary for bit string representation or represented as a string of real numbers in numeral vector representation. Both these representations might be realizable in most of the circumstances. This made the task of representation simple as well as easy to visualize. The GA worked over this individual to optimize the various values it represents.

The GP mapping is however not so trivial. In GP the various individuals are programs that solve the problem. Each individual represents a program in itself that is usually represented in a tree-like manner (Koza 1989, 1992, 1994; Stephens and Waelbroeck 1997). The evaluation of the fitness of the individual is done by the execution of the program that it represents. Hence it is necessary for the implementation of GP that the individual be realized as a program that can be executed to compute the individual fitness. The program aspects are casted into a tree-like structure. The tree may consist of non-leaf nodes as well as the leaf nodes. The non-leaf node usually stores the program notations, statements or operands. The leaf nodes are used for storing the various parameters of the system. In this manner the entire program may be realized as a tree.

Consider that the problem was to compute a polynomial whose output values was quite close to the output values of the expression x^2+x+1. Now we need a representation that would allow us to represent polynomials. For this we first define a set of terminal sets that would occupy the leaf node of the GP tree. This set consists of all real values and the variable x and may be given by (10.1).

$$T = \{x, R\} \tag{10.1}$$

Here R is the set of real numbers within the predefined limits.

Similarly we may define a non-terminal set. This set hold the non-terminal values of the program and occupies the non-leaf nodes of the GP tree. This consists of all the various possible operators used in polynomial. This is given by (10.2)

$$NT = \{ +, -, x, / \} \tag{10.2}$$

Here all arithmetic operators have their usual meaning.

Now we may define any polynomial by using these expressions. Some of the examples are given in figure 10.2. These correspond to the expressions x^2+x+1, $x+y+4$, $2x+5$ and $1+x+1+x$ respectively. It may be clearly seen that the non-leaf node is occupied by the NT set and the leaf node with T sets. Again it may be seen that we would be able to represent any polynomial expression by using these trees.

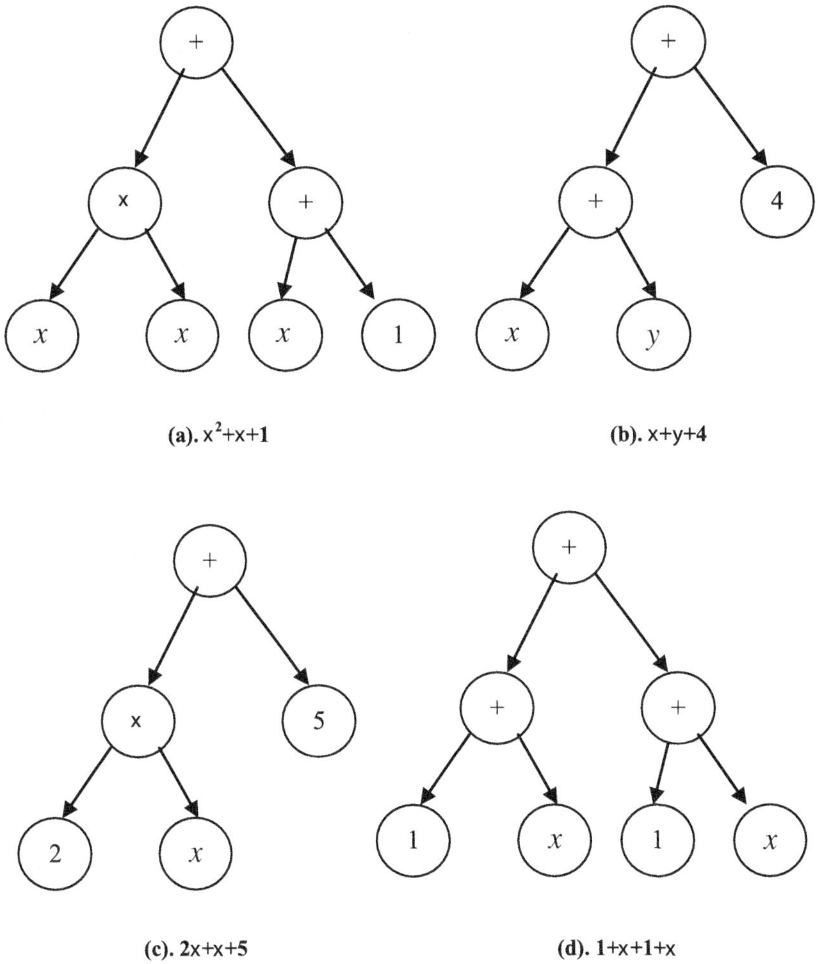

Fig. 10.2 Genetic Programming Individuals

The various trees stand for various polynomials. It can be seen that we can represent a polynomial by various kinds of trees. Hence this representation is a many to one mapping between the genotype and the phenotype where many varieties of genotypes make a single phenotype representations. This makes the fitness landscape complex, but existence of many global minima in the search space may be a plus point. Some of the varieties of writing the same expression 2x+x+5 are given in figure 10.3.

Another major question that comes here is the validity of the program or the individual being represented. Every program has a syntax that must be followed for the correct working of the program. The incorrect syntax program is u-executable. Hence the individual also needs to be syntactically correct as per the program logic being used. The various operations must have the needed number of parameters and the parameters must have the correct matching with the operands or

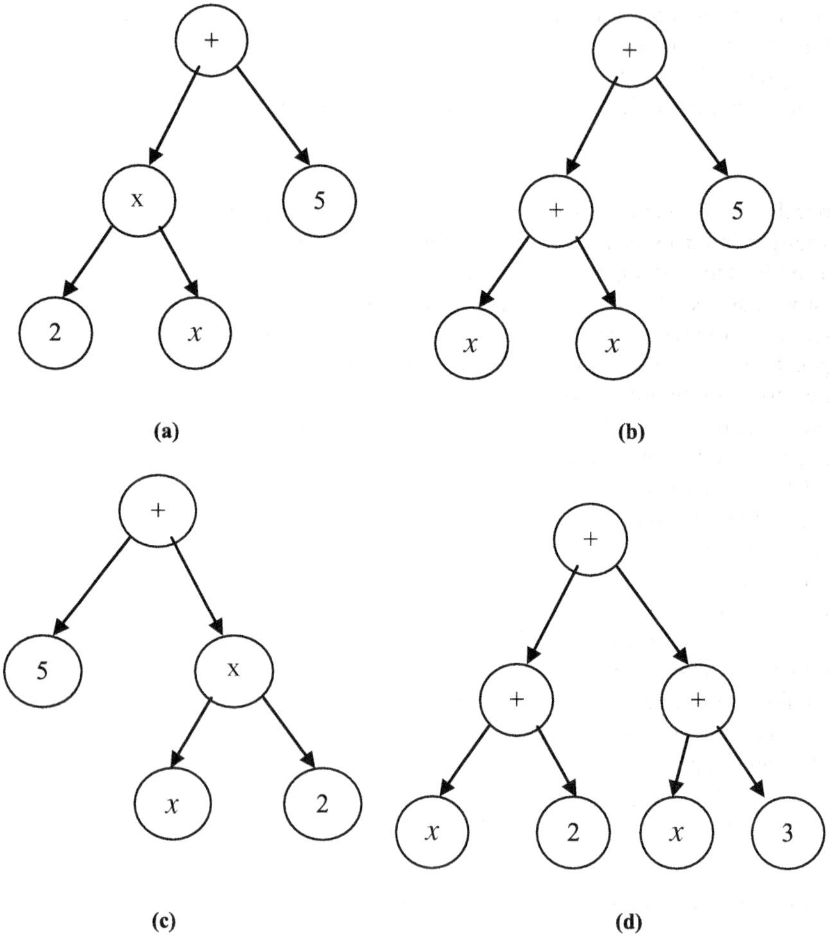

Fig. 10.3 Different Programs representing the same expression

statements. Hence every operation of the algorithm that modifies the individual or the tree needs to explicitly ensure that the tree is syntactically correct. This would give a valid fitness measure to this individual. In case an individual has a wrong syntax, it may be taken as an infeasible solution.

10.3 Fitness Function

The role of the fitness function is to evaluate the fitness of the individual and to assign it some score based on its performance. The ultimate aim of the GP is the optimization of the fitness value. The fitness function does the task of computing the fitness of any individual. It takes the individual as an argument and returns the fitness. The fitness function may be trivially specified by the program as was the case in the use of GA.

The major issue here may be the genotype to phenotype representation that is more complex as compared to the GA. The individual in a GP is a program whose fitness value is found out only when the program is executed. The execution of the program means to give it the inputs that the program is supposed to work with and to execute the individual over these inputs and match the corresponding outputs. This assigns the fitness value to any of the individuals.

In the above stated example of polynomial estimation, the fitness function would measure the deviation between the actual polynomial to be predicted and the predicted polynomial. The polynomial to be predicted was bounded between an upper bound (*L*) and a lower bound (*L*). Let this polynomial be g(x) given by equation (10.3).

$$g(x)=0.5\ x^2 + 0.5\ x+1 \tag{10.3}$$

Here $L \leq x \leq U$

The predicted function can be achieved by decoding the individual of the GP using the tree based notation. Let the decoded function be given by *I(x)* which is some polynomial of *x*. The error at any specific point x between the two measures its deviation. We add all such deviations across discrete set of points in between *L* and *U*. This gives the total error or the deviation in the system. This is given by *f(x)* in equation (10.4).

$$f(I) = \sum_{x=0}^{1} | g(x) - I(x) | \tag{10.4}$$

f(x) denotes the fitness function of the individual denoted by *I(x)*. Suppose that the individual *I(x)* was given by equation (10.5).

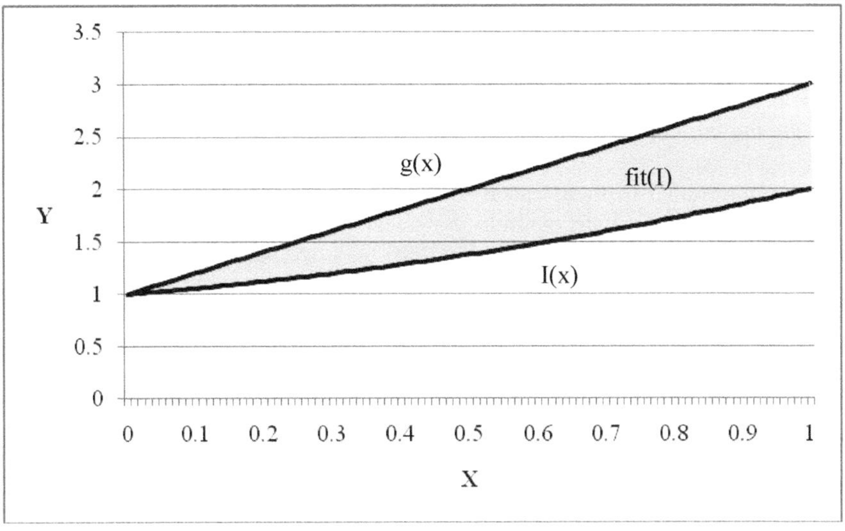

Fig. 10.4 Fitness function for the problem

$$I(x)=2x+1 \tag{10.5}$$

Now we plot *I(x)* and g(x) in a graph. The total fitness is the area between the curves denoted by the shaded region. This graph is shown in figure 10.4. The fitness needs to be as small as possible for the task considered.

10.4 Selection

As per the analogy between the GP and the GA, we may now have an initial population ready with random individuals or random programs that constitute the initial population. This population participates in the evolutionary process of constant modification or improvement as the algorithm runs. The fitness of the best as well as the average individual is expected to increase as we go along with generations in the program. This happens with the application of various genetic operators that create a new generation of individuals from the previous generation. The algorithm continues working in this manner.

The first genetic operator that we consider is selection. Selection is supposed to select the best fitness individuals in a probabilistic selection mechanism with the fitter individuals having larger probability of being selected. This is analogous to the Darwin's theory of survival of the fittest. Multiple selections are made by the algorithm for the participation in the Genetic operations and these go to the next generation. Mostly the fitter individuals may go in multiples to the next generation and the weaker ones may be left out or die in the process.

The selection operator uses an additional step of scaling that designates probabilities or the expectation values to the various individuals to go to the next generation. This may be either by means of fitness value or by means of the rank of the individual out of all the available individuals in the population pool. The scaling based on rank helps in saving over domination of the fitter individuals and their too many selections for the next generations.

There are many kinds of selection operators that are used. The most fundamental selection technique is the roulette wheel selection. Here the various individuals occupy the various positions in a roulette wheel. The wheel is spun and the winning individual is selected. The circumference of the individuals in this case is directly proportional to their fitness values. A similar technique is the Stochastic Uniform Sampling which controls the selection of an individual multiple times that may be a case with the roulette wheel. Here we place *n* selectors in a single wheel and after rotation all the individuals that stand against any of the selector are selected by this operator. The other technique is the tournament selection. This mechanism selects individual by organizing tournaments among them. In every match the fitter individual has a larger probability of winning than the weaker individual. The final winner is selected by this operator. This is repeated multiple times for multiple selections.

This operator is similar in its working to the GA and the same fundamentals that were used in GA may be used here as well. The role of this operator in the overall evolutionary process is also the same as was the case with GA.

10.5 Crossover

This is the reproductive operator that mixes two individuals to generate a new individual that contains the characteristics of both the individuals. The aim is to have fitter overall individuals by selecting the best features out of the parent individuals. In case of GP, the crossover operator carries the same task by mixing two programs for the generation of a new program that carries features of both as well as is syntactically correct. This is done slightly differently from the GA counterpart where there were strings of bits or numerals (Schaffer and Morishima 1987; Shaffer and Eshelman 1991).

In GP we first select a sub-tree out of the two trees representing the program. This may be done randomly out of all the sub-trees possible, each rooted with some non-leaf node. The other task that happens is the exchange of the sub-trees between the selected pair of individuals. The sub-tree of the first individual goes to the second individual and vice versa (Soule 1998). This makes two new individuals for the next generation. This procedure is shown in figure 10.5.

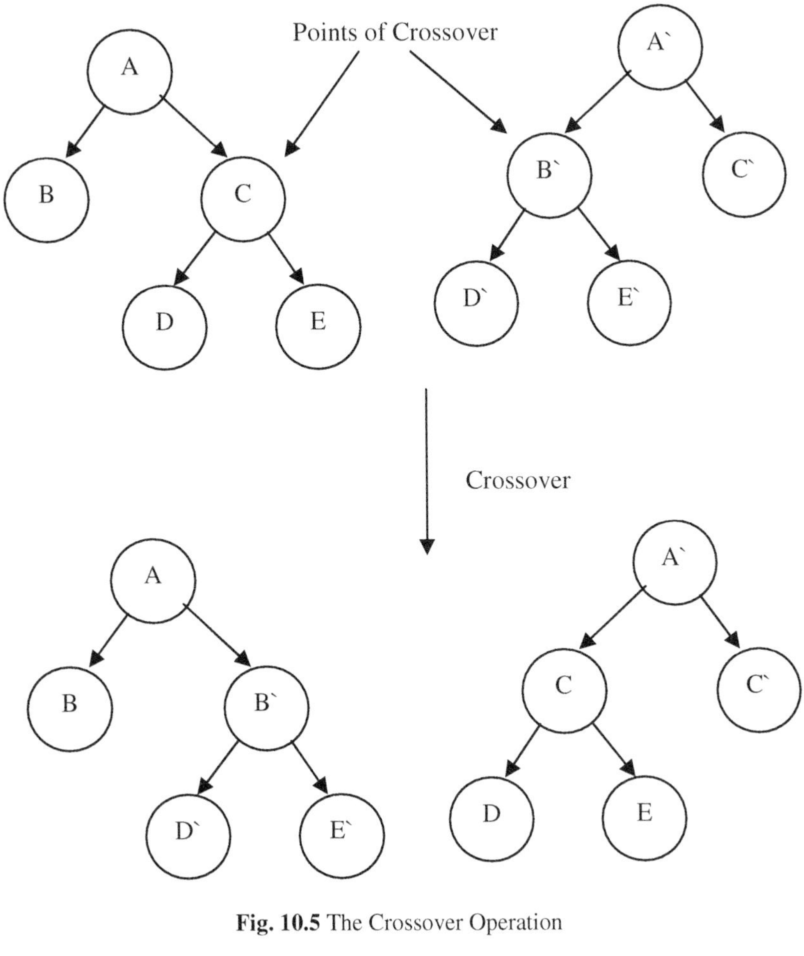

Fig. 10.5 The Crossover Operation

The exchange of sub-trees between the two individuals to generate two new individuals may be visualized as the exchange of two modules between the programs. A program is made up of many modules. Each module does some functionality that contributes to the functionality of the overall code. Each module here represents any non-leaf node. The level of this node denotes the complexity of the functionality it denotes. The nodes very near the leaves may be very simple providing some very basic computation. The nodes very near the tree root may be the nodes providing complex functionalities with large computations. Now a program has good and bad modules. The good modules do what they are expected to do and in turn result in the overall optimal performance of the program. The bad ones might behave unnaturally and would hinder the solution of the problem. Different solutions or programs would have different good and bad modules. The purpose of exchange of modules between programs is done by the crossover. By this operation we hope that the program would be given a better piece of module that performs the work better as compared to the module it currently possessed. With this the individual hopes to maximize its fitness by replacing bad modules with good modules. If the exchange is good, the overall fitness increases and the individual would continue as a part of the population pool. If on the other hand the exchange was bad. The overall fitness would decrease and the program may die out of the population pool in the subsequent generations.

The crossover is a major factor of optimization in GP that itself contributes a bulk of the optimizations. This operator hence produces about *90%* of the individuals needed for the next generation.

10.6 Mutation

The other major factor is mutation. Mutation performs the task of addition of new characteristics in the individual that are random in nature. It tries to overcome any loss of characteristics from the system that may be reinserted by mutation. In GP as well the operator is applied to add random characteristics or sub-trees into the tree based representation of the individual. This operation results in the creation of new sub-trees in the program that may perform better than the original sub-trees in terms of providing optimal functionality that result in a high fitness. If the functionality provided is good, we may expect the individuals to survive the evolutionary process and later share the characteristics or the sub-trees with the other programs or individuals. If the sub-trees added are not good, these might get killed along with time.

The process of the application of this operator is simple. We have an individual or a tree over which the operation is to be performed. We select any non-leaf node of this tree. This is then replaced by a completely new sub-tree in the original tree (Soule 1998). This results in the creation of a new program. The added sub-tree may be entirely random in nature. An important aspect might be the structure and depth of the added mutated sub-tree. This is also randomly decided by the algorithm. In many cases it may be kept somewhat close to the actual tree being replaced. Another important characteristic is that the evolved tree needs to be syntactically correct. This operation is shown in figure 10.6.

We discussed the various tree nodes acting as modules that contribute to the overall program to solve the problem. The mutation is the process of adding a new program code or a program module in the program being developed. Suppose a program that represents an individual of the GP is not functioning properly. Now we may wish to replace a bad part of the code that is responsible for the poor performance with a new code that solves the problem expected out of the module. This is done by the mutation operator. It adds new functionalities to the search space that might be desirable. In this manner this operator makes its contributions. The contribution of this operator is somehow kept reasonably low. This is to due to the low addition of good characteristics as compared to the crossover operation we discussed earlier (Spears 1993).

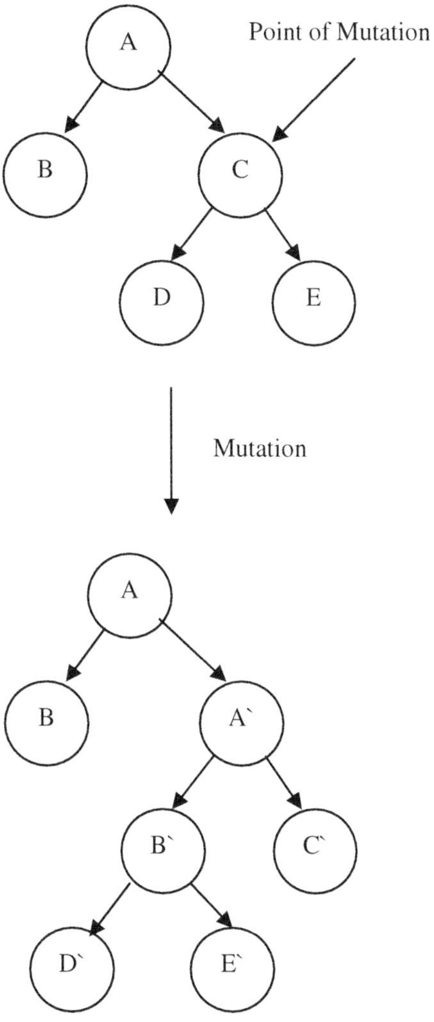

Fig. 10.6 Mutation Operation

10.7 Other Commonly Used Operators

Apart from mutation and crossover, the GP may use many types of operators, each for some specific task or overall control of the working of the algorithm. These operators may be taken as a standard from the fundamentally used operators, or may be completely independently defined and developed by the designer. The self-designed operators are especially useful for the customized problems where some specific requirements may be needed as per the problem logic. Here we discuss some of the commonly used operators in the GP usage.

10.7.1 Elite

Elite is the operation where the best individual from one generation passes directly to the other generation without any kind of modification. This operator ensures that the best individual is not destroyed in the evolutionary process. This is especially important considering the fact the ultimately the best individual is returned as the final solution of the evolutionary process. Hence its preservation is important. In case the evolutionary process generates a fitter solution at any generation, the best individual would get changed. In case no fitter solution is generated, this individual needs to be preserved against modifications or lack of selection that may be a result of the evolutionary process.

10.7.2 Replication

The evolutionary process may seldom result in major changes in the individual by the application of the mutation operation. This greatly adds new diversity to the population by the inclusion of a large amount of new functionality in the form of new codes that are integrated or injected in the individuals. Due to this the original individual is modified and a new individual is generated. The mutation may result in good or bad changes depending upon the modified segment of code. Due to these reasons it may be good to have multiple copies of the fit individuals in the population pool. Later the mutation and crossover may change these multiple copies differently. As a result the multiple copies get differently oriented and try to enhance good characteristics in the other individuals. In this manner the replication doesn't result much in spoiling the diversity of population. This however adds good characteristics to the other programs that enable them attain higher fitness. In this manner in operation is desirable.

10.7.3 Architecture Altering

This is the other genetic operation that tries to modify the architecture of the selected individuals in terms of their architecture. Prolonged evolution of the individuals in the population pool, driven by the methodology of survival of the fittest may have disadvantage. It may result in the generation of a large amount of code that is redundant or poorly framed. This creates individuals or programs with a very poor architecture. Further, due to this there is an unnecessary increase in the individual or the program length. Hence the entire individual is not as desired, when generated by the evolutionary process. This plays a big role in making the evolution slow and sub-optimal. Hence many times we may wish to carry forward the task of improving the architecture of the individual to guide the evolution in the correct direction.

To carry this operation we define a collection of operations that carry forward the task of alteration. All these are framed as per the program growth and the problem logic and are specified at prior. As the algorithm continues, some of the individuals are generated by this operation. We select the individuals and apply any of the alteration operation to control the individual. The altered operation need to make the individual better so that it can impact the other individuals in the population pool and can help in directing the entire evolutionary process.

10.8 Convergence

One of the major issues in any evolutionary technique is the convergence in the problem space or the search space (Grefenstette 1986; Angeline and Pollack 1993). In terms of a representation of the individual as a numeral real valued vector, we discussed the search space as a highly dimensional space consisting of individuals at various places. We further discussed that as generations increase, all individuals jump towards the better fitness points and ultimately converge at the global minima. In this manner the optimization was carried out with the use of GA using the concept of search space.

The bit string representation of the individuals marks another interesting case. Here the various individuals are represented as a string of bits, where every bit can be either 0 or 1. The initial population consists of random 0s and 1s at the various places of the strings. Now a 0 may be highly desirable at some places and 1 at the others. As a result of the survival of the fittest, the population of the next generation contains more individuals with 0s at some specific location and 1s at some other specific locations. This was because of the higher average fitness of the individuals possessing those bits. After a few more iterations, the other individuals completely disappear as a result of non-survival to the evolution process. This makes the population attain a form <00****11****>. Here a 0 represents that all individuals of the population have 0 at that location. A 1 denotes the presence of 1 at all the individuals at that location. A presence of * means that some individuals have an occurrence of 1 and the others of 0. This is called as the schema of the genotype representation. Now it may be seen that the positions marked with 1 and 0 have been completely optimized and no further change in these values may take place. This is the reduction of the dimensions of the search space.

As the algorithm continues, more and more positions get fixed. Hence after a few more generations, the condition of the schema might become <00****11**01>. In this manner the algorithm continues with more and more locations holding definite values. The generations after which the value of defined for any position depends upon the usefulness of that position. A very dominating position or very interesting position that largely determines the fitness value would immediately get some value of either 0 or 1. The other positions might take time to get their final value. At the end all individuals make the algorithm converge to some schema. Here the schema holds all the values.

We address the same concept of search space and schema in the approach of GP as well to better understand how does the algorithm perform in the search space and how does it ultimately converge to some point that is returned or computed as the final output of the system.

The GP consists of trees by which the problem is represented. This tree representation constitutes the search space of the individual in case of GP where every node stands for a complete dimension in the search space that can expand infinitely unless a threshold of expansion is specified by the designer which is usually the case in their use. This represents a very characteristic search space which is unlike the multi-dimensional search space that we dealt with so far. This again represents the possibility of completely flexible, multi-length and multi-complexity individuals by the use of GP that was not the case with GA where the individual was just a set of fixed length parameters that was bounded in all directions.

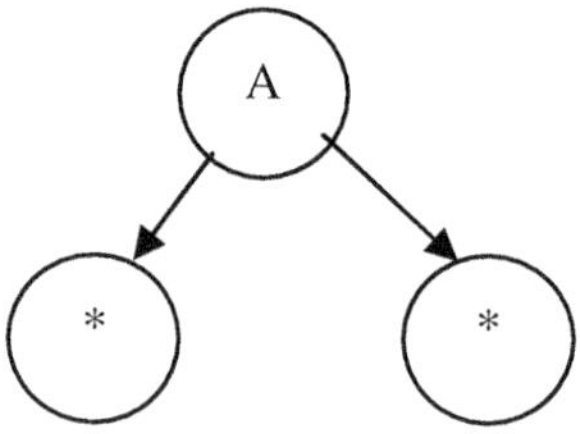

Fig. 10.7(a) Schema at the initial few generations

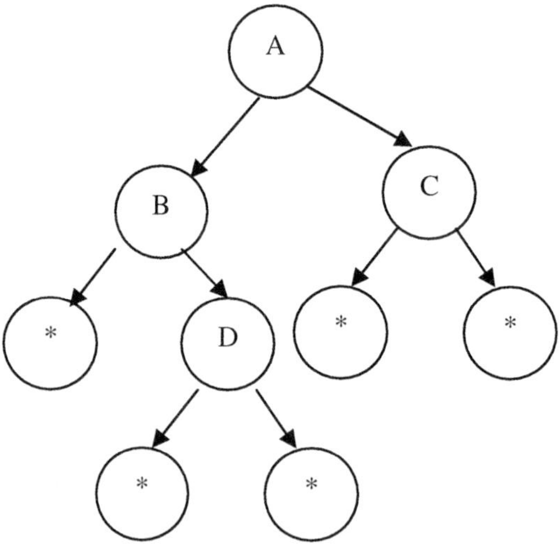

Fig. 10.7(b) Schema at the later few generations

As we start the evolution process, the individual consists of different trees of varying depts. Nothing specific can be said about the functionality of the various parts of the code in such a stage. Now we know that each tree represents a program that may be composed of various modules and sub-modules. As the evolution process grows, the various modules start taking their shape. The well performing segments those are necessary for the problem solving start emerging out. As a result the individual may be represented in a similar manner by the form given in figure 10.7(a). Here the notations have the same meaning as discussed in the schema of GA. This is known as the schema of the GP. Now the task of the algorithm is relatively simple. It needs to optimized the wildcard (*) entries only. The others have taken a decent shape with an optimal performance. This later generates the schema shown in figure 10.7(b). The entire process goes on and on in this manner in the search space of the GP. The entire process involves the encouragement of good modules that grow more than the weaker modules in a particular section of the program and soon occupy the entire population. Once a module or segment has been fixed in this manner, the process continues for the other segments or the other modules also.

10.9 Code Growth

The GP results in effective code that is able to solve the problem and generate programs as the solution. However the complete evolution as per the discussed norms has a major problem of complexity. Many times the code may unnecessarily grow with time. This may add negligible performance boost in the system, but may add a lot of nodes instead. These results in the placement of very large codes in the population pool that further result in large individuals. The large sizes denote a large search space that is always difficult to handle in the evolutionary process. Further the large search space usually ends up in convergence to local minima as it is not possible to explore a large part of it as well as due to its highly complex structure. In such a case the program growth hinders the performance by driving the evolution in the wrong direction and making it too slow and cumbersome. These large programs do give a good fitness value which is marginally larger than the smaller counterparts but they place a lot of additional limitations in the evolutionary process of the GP (Soule 1998).

It is hence important to control the program from taking too large sizes. From our discussions of problem solving we know that many problems can be effectively solved in simple programs. This requires some neat practices and effective styles of programs. Making the code too bulky for problem solving is usually not the solution. In this section we discuss some basic methods of code growth control.

10.9.1 Parsimony

A major technique of code growth control is to penalize the large codes by some penalty that depends upon the problem size. This makes the computation of the overall fitness of the individual to take an additional parameter of the program size. The performance parameter awards the higher performance by the overall code. The penalty parameter penalizes the additional size. Hence the code would normally grow in case there is a large performance boost with the addition of the growth. This is when the reward parameter becomes dominant and the reward is too large than the penalty that is imposed on the system. The initial growth of the network is attributed to this clause. Later however as the algorithm grows, the increase in code size has a negligible improvement in the performance. As a result there is a very less reward and a strong penalty that is imposed by the system. This restricts the code from growing more than a limit. This in turns generates small manageable codes that are used for problem solving. Many times a threshold size may be specified. The solutions larger than this size may be rejected outright by assigning them the poorest fitness value to ensure that they are killed in the evolutionary process.

10.9.2 Code Modification

The other technique that is hard to implement is the code modification. In this step we try to modify the entire program architecture such that the resultant architecture or code is less in size without a reduction in performance. This is especially possible due to the large amount of redundancies in most of the codes generated by the GP where a large part of computation results in a small overall computation that is actually needed by the program. This operation is applied on the inoperative code that is the extra part of a code whose removal does not affect the program working. The step involves the understanding of the program and suitable alteration techniques that need to be build. If applied in the evolutionary process, this operator becomes computationally very expensive and can thus be used to a limited degree.

10.9.3 Parameter Modification

The GP has various parameters that control the algorithm working. The main ones are the mutation rate and crossover rate. The application of the operators may further be controlled to make the algorithm behave as per the demands. The common approaches in this mechanism include the control of the mutation rate, crossover rate, clever selection of the points of crossover and neglecting the destructive crossover events.

10.10 Generating Complex Programs

The other factor of interest is the mechanism by which complex programs may be generated. The problem to be solved may demand a large complexity which needs to be generated by the algorithm for the effective problem solving. This needs the creation of large sized trees because of the inability of the small trees or codes to solve the problem. This requirement of the large size trees makes the problem complex in nature that ideally requires a large computation. It may be easily seen that the entire search space for the problem is very high. A conventional approach in such a large search space may not be useful. This requires specialized search techniques or the specialized operations to be performed while optimizing the problem.

In this section we discuss two fundamental means of handling this problem of generating large programs as the individuals of the evolutionary process. The first method is the Automatically Defined Functions (ADF) where some general architecture is specified that governs the evolution of the entire code. The other mechanism we study is the compression and expansion technique where a library of modules or functions is made and maintained for the generation purposes (Angeline 1994). Both are discussed one by one and then a brief comparison is presented.

10.10.1 Automatically Defined Functions

The first class of solution includes the Automatically Defined Functions (ADF). It is known that the large program to be evolved can be fundamentally broken down into independent modules with each module providing some different functionality. The independent modules are a natural consequence of the manner in which humans designed programs as well. A function is broken down into some functions with each function having its own implementation and providing its own functionality. It has its own local declarations and initializations. This modularity helps a lot for the programmer to make its program. By ADF we try to bring the same functionality to the programs evolved by the GP.

In this approach we define a hierarchical arrangement of functions. Each one of these behaves like an independent module of the entire program and has an independent evolution and optimization. The main complex program may hence be solved as a combination of a predefined hierarchy of solutions. The evolution process involves the optimization of the performance of each of these modules separately as well as the overall program. Initially the various modules are randomized in nature which starts attaining their shapes as the evolutionary process continues to perform.

10.10.2 Emergent Library Builder

The other method we study here is the emergent library builder. In this mechanism a complete library of modules is built as the solutions evolve along with generations. This library stores a number of modules that provide good functionality that might be necessary for the functioning of the program to solve the problem. The programs may easily refer to the functionalities of this library while evolution that helps them import good functionalities into their evolution without an unnecessarily addition of the program code. As a result an optimal functionality is added, that is known to be helpful to generate good results in the program without any hard labor. The library of modules maintained by the system is referred as the *Genetic Library Builder* (GLiB). The library is constantly added with good modules to enable it store vital modules (Angeline 1994).

The entire process may be broken down into two steps. These are *compression* and *expansion*.

10.10.2.1 Compression

The first step of compression deals with the shortening of existing code that becomes very long. Here we take as input a large piece of code that crosses a threshold limit and try to divide it into modules or functions. This may be done in two ways. The first manner of division of program above threshold limit is known as the *depth compression*. Here we define the maximum depth a node can reach. As soon as the tree crosses this limit, a new function or module is made that performs the functionality of this entire tree. This tree is replaced by a suitable call to this function so constructed. The function is stored in the *GLiB*. In this way the code length gets reduced. This mechanism is shown in figure 10.8.

The other type of compression that may be performed is the leaf compression. In this compression the division takes place in the leaves. This is similar to the depth compression with the difference that each unique leaf is replaced with a unique parameter name. This type of compression is shown in figure 10.9.

10.10.2.2 Expansion

The other important aspect of the algorithm is expansion. This is the opposite of the compression mechanism. Here we seek the definition of any function from the GLiB and replace the definition in the actual individual. This expands any program or solution to its actual representation. This operation may be carried out in the application of the mutation process. Here it is not possible to mutate or modify the entry in the GLiB as it would affect all the individuals that refer to this library. Their program and ultimately the fitness value would change as a result of the modifications. We hence carry the operation in the full program.

10.10.3 Comparisons

This approach may be regarded as a more generalized version of the ADF where no hierarchies are pre-specified, but rather they evolve along with time themselves. This generality provides a lot of flexibility to these solutions regarding the evolution of the architecture of the program which can take any kind of shape. This is unlike the ADF where the basic hierarchy was specified to be mentioned that restricted the ADF from evolving many kinds of solutions to the problem. In this method this approach may solve the problem in a better manner. However, the

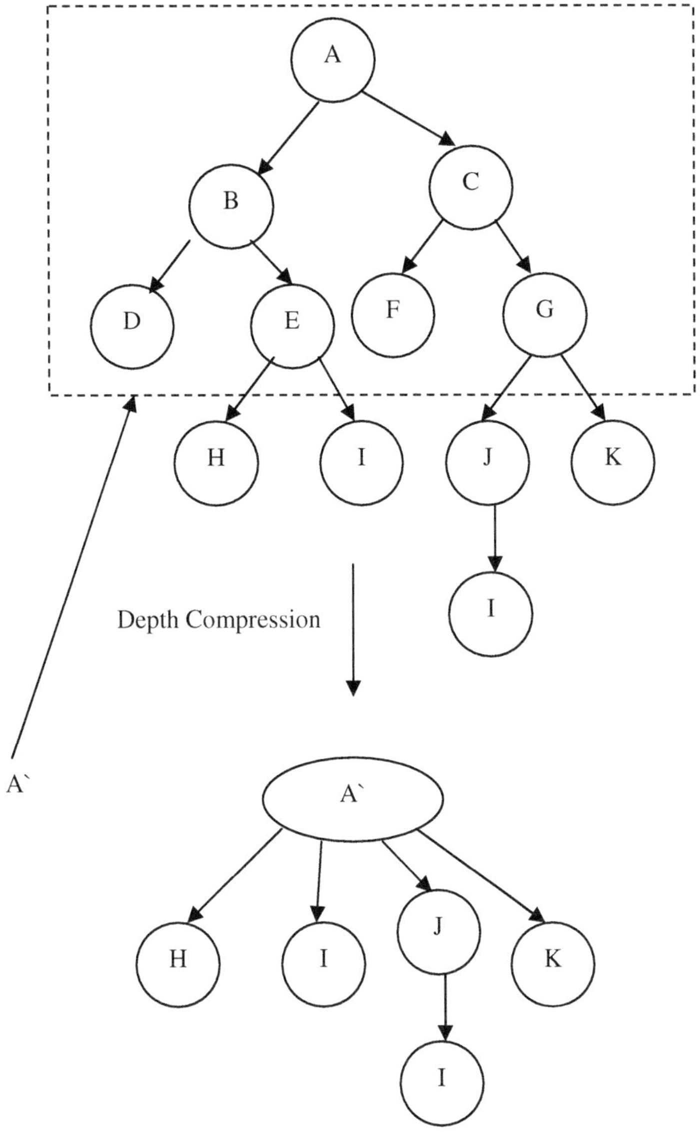

Fig. 10.8 Depth Compression

addition of this generality greatly expands the search space and makes evolution too complex and time consuming. It may be possible that the optimal program may not be able to evolve at all because of the complex structure and large dimensionality of the search space. This is a big hindrance to the use of this mechanism.

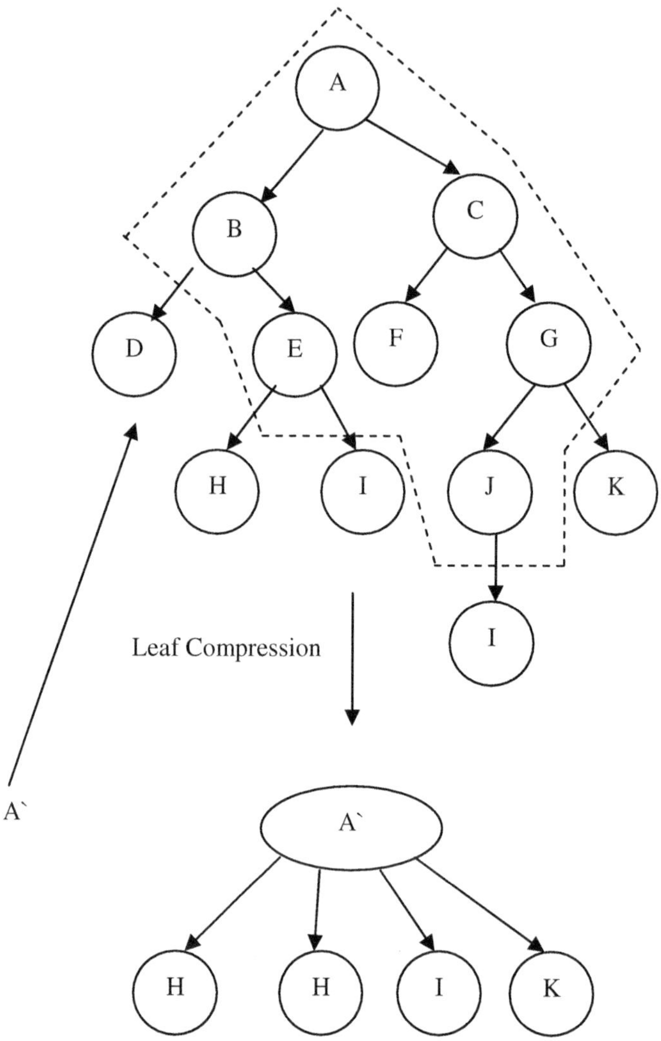

Fig. 10.9 Leaf Compression

10.11 Grammatical Evolution

So far we have studied the GP mechanism of evolving programs or individuals that led to solving the problem of study. In this section we present a very interesting application of GP with the use of Grammatical Evolution (GE). The individual in a GE problem consists of a set of integers or genes. These can be converted into the phenotype representation using a language that is known before hand. The main motive of this technique is to evolve instructions which when operated over a known grammar leads to the phenotype representation or the solution of the problem. Since the grammar is predefined or in-built, small individuals are able to convey a lot of meaning and represent complex forms. As a result of this, the search space is simple and less complex. The dimensionality of the search space is relatively less that makes it possible to easily evolve the optimal solution. The individual is a simple vector of integers which enables the use of conventional operators used in GA. Each of these numbers however conveys a special meaning in the overall making of the phenotype representation. They decide the entire structure and associated parameters of the phenotype (O'Neil and Brabazon 2001; O'Neil and Ryan 2001, 2003; Rosca 1997; Ryan et al 1998, 2002).

Grammar carries a special meaning in the GE as it provides the very base on which GE operates. A grammar is defined as a set of terminating symbols, non-terminating symbols, production rules and start symbol. This may be given by equation (10.6)

$$G = \{T, N, P, S\} \tag{10.6}$$

Here *G* is the grammar that is specific to the solution, *T* is a set of terminating symbols, *N* is the set of non-terminating symbols, *P* is a set of production rules and *S* is the start symbol.

10.11.1 Grammar

The grammar is specified with the help of Backus Naur (BNF) Form. The BNF form is useful for specifying the various states, symbols, syntax and rules of the language and enables the clear understanding and generation of the language or grammar. The non-terminating symbols are the symbols that are not a part of the final language of the system and need to be replaced by proper terminating symbols for the generated sentence to be understood by someone. The terminating symbols are the symbols that are a part of the final statement that the system or user may understand. Production rules are a collection of rules that specify the replacement syntax or rules of any non-terminating symbols with any other terminating or non-terminating symbols. Suppose that the statement at any instance of time contains a large set of non-terminating symbols. Now we may wish to replace these symbols by a variety of symbols. This may be needed to get terminating symbols so that the final understandable statement can be constructed. Alternatively we may need rules to replace the non-terminating symbols with appropriate set of symbols such that the needed architecture of the final statement is evolved. This enables the making of statements as per requirements. Every language must specify a symbol which is the base symbol. The entire construction of the sentence happens by replacing this symbol with some set of symbols allowed by some production rules. This replacement goes on and on until the needed statement is evolved.

We may use the grammar to construct any statement in this manner. The construction always starts with the start symbol. We keep on selecting the rules judiciously to replace the symbols with new set of symbols. This replacement goes on and on. At any point the statement would consist of a set of terminating and non-terminating symbols. We use the desired rules to replace the non-terminating symbols by a set of new symbols. Finally the desired stamen may be evolved that contains only the terminating symbols. It may be easily seen that this is the general manner in which the English sentences are constructed using grammatical rules that guide the entire process of the construction of the sentence, till the final form is constructed.

10.11.2 Individual Representation

The individual in this form of representation is a long list of integers that forms a big string. This is the same manner in which the individual was represented using vector of numerals in case of GA. However this is not a direct coding between the phenotype and the genotype. The mapping of the genotype to the phenotype is done using the underlying grammar. We discuss this mapping that would enable us to compute the fitness of any individual based by converting it into its phenotype representation.

Consider the grammar that evolves Boolean expressions. This may be easily defined in BNF by (10.6) that has N, T, S and P given by equation (10.7)

N = { <expr>, <biop>, <uop>, <bool> } (10.7)

T = { and, or, xor, nand, not, true, false, (,) }

S = { <expr> }

P={
1. <expr> ::= (<expr> <biop> <expr>)
   | <uop> <expr>
   | <bool>

2. <biop> ::= and
   | or
   | xor
   | nand

3. <uop> ::= not

4. <bool> ::= true
   | false
   }

Closely observing the rules we can see that the various rules have various choices associated with them. Each rule has some distinctive non-terminating symbol that they replace. Consider the rule 1 as an example. This rule presents three ways in which the expression *<expr>* may be replaced. These are *(<expr> <biop> <expr>)*, *<uop> <expr>* and *<bool>*. The number of choices that the rules offer is a very important parameter as they aid in the conversion process that we shall see later. The number of choices for the various rules is summarized in table 10.1.

Table 10.1 The number of choices for every non-terminal expression

S.No.	Expression	Choices
1.	<expr>	3
2.	<biop>	4
3.	<uop>	1
4.	<bool>	2

Let the individual be a set of integers given by <I_1, I_2, I_3, I_4, …, I_n>. Now we start the process of conversion of this individual from its genotype representation into its phenotype representation. As per the BNF norms, the grammar always starts with the initial starting symbol *S*. Hence the initial phenotype representation consists of *<expr>*. We extract the first gene out of the individual and read its value. Based on this value we select the appropriate rule that is fired or used for replacement of the *<expr>* that constitutes phenotype. After this step the second gene is extracted. Here also we select the rule that is to be used and replace the non-terminating symbol with the appropriate replacement. This process continues on an on with the genes being extracted and used for the replacement of the non-terminating symbols in the phenotype representation.

At any time the individual in its phenotype representation consists of some terminating and some non-terminating symbols. The strategy followed selects the first non-terminating symbol for replacement. Now the next gene is read. Let the read gene be denoted by *g*. We have a set of options by which the non-terminating symbol may be replaced. This is determined by the set of production rules. For the Boolean expression evolution problem, this was given in table 10.1. Let there be a total of c choices for replacement. Now we select the rule *i* for the replacement of the selected non-terminating symbol, where i is given by equation (10.8). This is repeated for every gene.

$$i = g \bmod c \tag{10.8}$$

Now two types of conditions are common to happen. The first is when the phenotype representation consists of only non-terminating symbols even when there are genes left in the individual. In such a case there is no use of reading the rest of the genes and the phenotype representation is regarded as final. In such a case the genotype representation of the individual carries extra information that is not required. The other common case may be that the entire genotype representation is read, and there are still non-terminating symbols in the phenotype representation. In such a case the strategy may be to again start reading the genotype individual from the first position. This is known as wrapping. We may wrap the individual two or three times. If still the individual contains non-terminating symbols, it may be regarded infeasible. In many other cases, wrapping may not be followed and the individual may directly be regarded as infeasible by the program.

In this manner we are able to map any genotype representation into the corresponding phenotype representation. The other aspects, mechanisms and operations of the GE are exactly same as the GA. In this manner GE may be used for problem solving.

10.12 Chapter Summary

This chapter presented an evolutionary technique called as genetic programming. This technique was similar to the working of the genetic algorithms where different kinds of operators are used for the generation of higher generation population from a lower generation population which generally results in improved fitness. The genetic programming consists of an individual representation in form of a program that is excited to get the real individual or its fitness measure. Here the different individuals correspond to the various types of programs. The operators of selection, crossover, mutation and other customized operators are used for this purpose. Crossover mixes two sub-trees and generates two new sub-trees or programs. Mutation adds random sub-trees in place of a sub-tree. The other operators include elite where the best individuals are transferred directly from one generation to the other, replication which maintains replicated copies of the better individuals, and architecture altering that rectifies the architecture of the individuals or programs. The chapter further moved into the genetic programming invariant of search space where the tree schema was discussed along with the mechanism of convergence. The genetic programming may usually result in excessive codes. A number of methods for code control were presented including parsimony or penalizing large codes, code modification and parameter modification. We then discussed the mechanism of generating complex functions by the genetic programming technique. Here we discussed about automatically defined functions and emergent library builder. Towards the end the chapter presented a technique of linearly representing the program so that the conventional genetic algorithm could be used. Here the concept of grammar was used for individual representations. The individuals had a problem specific common grammar which was used for the individual genotype to phenotype conversion.

References

[1] Altenberg, L.: The Evolution of Evolvability in Genetic Programming (1994). Advances in Genetic Programming, pp. 47–74 (1994)
[2] Angeline, P.J.: Genetic programming and emergent intelligence. In: Complex Adaptive Systems archive Advances in genetic programming, pp. 75–97. MIT Press, Cambridge (1994)
[3] Angeline, P., Pollack, J.: Competitive environments evolve better solutions for complex tasks. In: Proc. of the Fifth Intl. Conf. Genetic Algorithms, pp. 264–270. Morgan Kaufmann, San Francisco (1993)
[4] Angeline, P.J., Sauders, G.M., Pollack, J.B.: An evolutionary algorithm that constructs recurrent neural networks. IEEE Trans. Neural Netw. 5, 54–65 (1994)
[5] Banzhaf, W., Nordin, P., Keller, R.E., Francone, F.D.: Genetic Programming An Introduction On the Automatic Evolution of Computer Programs and Its Applications. Morgan Kaufmann Series in Artificial Intelligence (1998)
[6] Banzhaf, W., Nordin, P., Keller, R.E., Francone, F.D.: Genetic Programming: An Introduction. Morgan Kaufmann, San Mateo (1998b)

[7] Banzhaf, W., Poli, R., Schoenauer, M., Fogarty, T.C.: Genetic Programming. In: Banzhaf, W., Poli, R., Schoenauer, M., Fogarty, T.C. (eds.) EuroGP 1998. LNCS, vol. 1391, p. 97. Springer, Heidelberg (1998c)
[8] Cordón, O., Herrera, F., Hoffmann, F., Magdalena, L.: Genetic Fuzzy Systems: Evolutionary Tunning and Learning of Fuzzy Knowledge Bases. World Scientific, Singapore (2001)
[9] Fogel, D.B.: Evolving Artificial Intelligence. PhD Thesis, University of California, San Diego (1992)
[10] Forrest, S.: Emergent computation: self-organizing, collective, and cooperative phenomena in natural and artificial computing networks. In: Forrest, S. (ed.) Emergent Computation, pp. 1–11. MIT Press, Cambridge (1991)
[11] Grefenstette, J.J.: Optimization of control parameters for genetic algorithms. IEEE Trans. on Syst., Man Cybern. 16(1), 122–128 (1986)
[12] Koza, J.R.: Hierarchical genetic algorithms operating on populations of computer programs. In: Proc. 11th Int. Jt. Conf. on Artif. Intell., vol. 1. Morgan Kaufmann, San Mateo (1989)
[13] Koza, J.R.: Genetic Programming: On the Programming of Computers by Means of Natural Selection. The MIT Press, Cambridge (1992)
[14] Koza, J.R.: Genetic Programming II: Automatic Discovery of Reusable Programs. The MIT Press, Cambridge (1994)
[15] Langdon, W.B.: Genetic Programming and Data Structures: Genetic Programming + Data Structures = Automatic Programming! Kluwer, Dordrecht (1998)
[16] Luke, S., Spector, L.: A Comparison of Crossover and Mutation in Genetic Programming. In: Proc. Second Annual Conf. Genetic Programming, pp. 240–248 (1997)
[17] Michalewicz, Z.: A hierarchy of evolution programs: an experimental study. Evol. Comput. 1, 51–76 (1993)
[18] Mitchell, M.: An Introduction to Genetic Algorithms. MIT Press, Cambridge (1996)
[19] O'Neill, M., Brabazon, A.: Grammatical swarm. In: Deb, K., Poli, R., Banzhaf, W. (eds.) GECCO 2004. LNCS, vol. 3102, pp. 163–174. Springer, Heidelberg (2004)
[20] O'Neill, M., Ryan, C.: Grammatical evolution. IEEE Transactions on Evolutionary Computation 5, 349–358 (2001)
[21] O'Neill, M., Ryan, C.: Grammatical Evolution. Kluwer, Boston (2003)
[22] Pena-Reyes, C.A., Sipper, M.: Fuzzy CoCo: a cooperative coevolutionary approach to fuzzy Modeling. IEEE Trans. on Fuzzy Syst. 9(5), 727–737 (2001)
[23] Rosca, J.P.: Analysis of Complexity Drift in Genetic Programming. In: Proc. Second Annual Conf. on Genetic Programming, pp. 286–294 (1997)
[24] Ryan, C., Collins, J.J., O'Neill, M.: Grammatical Evolution: Evolving Programs for an Arbitrary Language. In: Banzhaf, W., Poli, R., Schoenauer, M., Fogarty, T.C. (eds.) EuroGP 1998. LNCS, vol. 1391, pp. 83–95. Springer, Heidelberg (1998)
[25] Ryan, C., Nicolau, M., O'Neill, M.: Genetic Algorithms Using Grammatical Evolution. In: Foster, J.A., Lutton, E., Miller, J., Ryan, C., Tettamanzi, A.G.B. (eds.) EuroGP 2002. LNCS, vol. 2278, pp. 1–4. Springer, Heidelberg (2002)
[26] Schaffer, J., Morishima, A.: An adaptive crossover distribution mechanism for genetic algorithms. In: Proc. of the 2nd Intl. Conf. on Genetic Algorithms, pp. 36–40, Hillsdale, NJ (1987)
[27] Shaffer, J.D., Eshelman, L.J.: On Crossover as an Evolutionarily Viable Strategy. In: Belew, R.K., Booker, L.B. (eds.) Proc. of the Fourth Inl. Conf. on Genetic Algorithms, pp. 61–68. Morgan Kaufmann, San Francisco (1991)
[28] Soule, T.: Code Growth in Genetic Programming. PhD Thesis, University of Idaho, Moscow, Idaho, USA (1998)

[29] Spears, W.M.: Crossover or Mutation? In: Whitley, L.D. (ed.) Foundations of Genetic Algorithms, vol. 2. Morgan Kaufmann, San Francisco (1993)
[30] Stephens, C.R., Waelbroeck, H.: Effective degrees of freedom in genetic algorithms and the block hypothesis. In: Proc. 7th Int. Conf. Genetic Algorithms, pp. 34–40. Morgan Kaufmann, San Mateo (1997)
[31] Tackett, W.A.: Genetic programming for feature discovery and image discrimination. In: Proc. of the Fifth Intl. Conf. on Genetic Algorithms, pp. 303–309. Morgan Kaufmann, San Mateo (1993)
[32] Whitley, L.D.: A genetic algorithm tutorial. Statist. Comput. 4, 65–85 (1994)
[33] Yao, X.: Evolving artificial neural networks. Proc. IEEE 87(9), 1423–1447 (1999)

List of Figures

Fig. 10.1. Genetic Programming
Fig. 10.2(a). Genetic Programming Individuals: x2+x+1
Fig. 10.2(b). Genetic Programming Individuals: x+y+4
Fig. 10.2(c). Genetic Programming Individuals: 2x+x+5
Fig. 10.2(d). Genetic Programming Individuals: 1+x+1+x
Fig. 10.3. Different Programs representing the same expression
Fig. 10.4. Fitness function for the problem
Fig. 10.5. The Crossover Operation
Fig. 10.6. Mutation Operation
Fig. 10.7(a). Schema at the initial few generations
Fig. 10.7(b). Schema at the later few generations
Fig. 10.8. Depth Compression
Fig. 10.9. Leaf Compression

List of Tables

Table 10.1. The number of choices for every non-terminal expression

Chapter 11
Evolutionary Strategies

Abstract. Genetic Algorithms with their complex genetic operators make a complex system that is time consuming and difficult to comprehend. Most of the systems we encounter in everyday life may not require that complex formulation. In this chapter we study a major constituent of the Evolutionary Algorithm i.e. Evolutionary Strategies. The chapter would focus upon the manner in which this problem solving technique tackles the problem with much less individuals and simple genetic operators. We further discuss the adaptation of the operators that this technique employs. The chapter especially would focus upon the adaptive nature of the Evolutionary Strategies. Here the individuals assess the fitness landscape and make their move accordingly. The general framework of Evolutionary Strategies generates λ individuals from a population of μ individuals with each individual resulting from crossover by ρ parents. Each individual may survive for k generations. The chapter explores the various concepts, issues and applications of this problem solving methodology.

Evolutionary Strategies (ES) is a type of Evolutionary Algorithm (EA) that is a valuable tool for the optimization purposes. The EA are classified as Genetic Algorithms (Mitchell 1996; Goldberg 1989; Whitley 1994), Genetic Programming (Koza 1992; Michalewicz) and Evolutionary Strategies (Back *et al* 1991, 1995). The ES particularly is a simple implementation of the concept of EA that uses simple techniques for the evolution of one generation to the other. The ES in this manner is able to optimize the problem of study and find the set of parameter values for which this optimal solution is obtained. This technique closely follows Genetic Algorithms (GA) in terms of evolutionary concepts and problem representation. Here also we assume that the problem is to be optimized by a set of parameters. These parameters make the phenotype representation of the solution.

As per the evolutionary methodology, ES also take the key concepts from Evolutionary Algorithms. A solution to a problem is a set of parameters. This is known as the individual of the evolutionary process. A pool or collection of individuals is the population pool. The population evolves in iterations known as generations. The goodness of the individual is measured in terms of its performance measured by the fitness function. The fitness function assigns a score to any individual. This is in analogy to the biological fitness of the natural counterpart. The entire algorithm is based on the natural analogy of evolution of the individuals along with time. As the generations increase the optimality or the fitness of the individuals increase in general. In this manner the individuals or solutions keep

A. Shukla et al.: Towards Hybrid and Adaptive Computing, SCI 307, pp. 235–247.
springerlink.com

improving along with time and generation (Angeline 1994; Back et al 1997; Ebin and Smith 2003; Fogel 1992).

The ES uses a similar concept to the GA or the EA in terms of evolution. The major difference between the two approaches lies in terms of the number of individuals. The general methodology of problem solving with GA generally implies deputing individuals in large numbers at various points in the search space. In ES we are generally able to solve the same problems with the use of much lesser individuals than required by the GA. This again makes use of simple concepts rather than bulky operators used by the GA. In ES the computation required per generation is hence much smaller. In other words we are able to achieve a high degree of performance in small times with a little larger generation in many problems by the use of ES (Back 1991, 1995; Beyer 1992; Schwefel 1995; Schwefel and Rudolph 1995; Yao and Lin 1997).

In this chapter we follow the same methodology of discussion that we used in the previous chapter. We mainly cite the differences in working and concept between the GA and the ES. In this manner we present the working of the ES.

11.1 Adaptation

Adaptation is a major issue in system design. Any system requires a set of parameters for its processing. These parameters play a great role in deciding the manner in which the system works and its performance. The examples of parameters include momentum, learning rate, etc. in the neural networks and mutation rate, crossover rate, etc. for the genetic algorithms. Many times a single value of the parameter may lead to sub-optimality. This is when the constant parameter value restricts the system performance and wastes a lot of computation. In other scenario it could lead to driving the system in wrong direction. Hence for these systems a constant value of the parameter may not be desirous. We may need an automated control of these parameters that dynamically sets these parameters at the algorithm run and always tunes them to the most optimal value as per the problem scenario and the system state (Angeline and Pollack 1993; Beyer 2001; Eiben *et al* 1999; Hansen and Ostermeier 1996, 1997; Kita 2001).

This task of fixing of the parameters of the system dynamically while the algorithm is executing is termed as adaptability. It refers to the fact that the system designed is adaptable to multiple scenarios and conditions. It enables the system to adjust its parameter values by an analysis over the present scenario and context. The algorithm may be said to possess some intelligent technique using which it can compute or guess the optimal parameter values. This technique utilized highly depends upon the knowledge of the parameter behavior, problem scenario, type of adaptability desired, etc. The adaptive systems usually perform better than the non-adaptive systems due to the constant tuning of the system parameters at each and every point of time in the algorithm execution.

Consider the training of the neural network by Back Propagation Algorithm (BPA) as an example. Here the learning rate and momentum are the factors that greatly determine the performance of the training of the network. We saw the role played by both these terms in the training in chapter 2. Now suppose we keep these factors constant with time. It is natural that the system may not perform very well. The learning rate would be preferred to be kept high when the system is far away from local or global minima. Similarly the learning rate may be needed to be kept low when the system is near local or global minima. Similar is the case with momentum. A small value of momentum may be desirable when the current state of the system has converged to local minima or global minima. In other scenarios, the momentum may be kept low. Such a system requires an adaptive control of these parameters that can look at the scenario and fix their values dynamically. This is the motivation behind the adaptive systems.

The discussion of adaptive systems in this chapter is to encourage the adaptive nature of the ES. We look at the next few sections how the ES is able to adapt the reproductive parameters to make the most optimized evolution possible. We would give a further discussion over adaptation and adaptive systems in chapter 17 where we discuss the various types and properties of adaptation, along with a deeper motivation for engineering adaptive systems.

11.2 Individual Representation

The individual may be discussed in terms of the representation in phenotype and genotype representation. The phenotype representation is a representation that displays the attribute values in the phenotypic search space of the algorithm. We have been generally referring this by the term of search space in the entire text. This representation is similar in nature to the representation followed in the GA. Here the various attributes are usually mentioned in a convenient representation or technique as needed by the algorithm.

The difference in representation is in the genotype form in which the search is performed. One of the motivations behind the use of ES is to build adaptive systems. We must hence build a mechanism that helps the algorithm tune the various system evolutionary parameters dynamically in the algorithmic run. Based on the working of the GA, mutation is identified as a major parameter that affects the algorithm and must be controlled to optimally make the algorithm execute. We hence embed the information regarding the parameter in the individual itself. As the algorithm runs, the attribute values are only not optimized, but rather the parameter values are also optimized. The initial values may not be desirable, but as the algorithm runs, the values keep getting optimized along with time and generations.

These parameters that are embedded into the ES individual are referred to as the strategy variables. While the normal attribute variables are the basic objective of the optimization process, the role of the strategy variables is to guide the evolution by enabling the evolution across a set strategy. An individual may hence be represented as given in equation (11.1).

$$\hat{I} =< \theta, \hat{x}_1, \hat{x}_2, \hat{x}_3, \ldots, \hat{x}_n, \hat{\sigma}_1, \hat{\sigma}_2, \hat{\sigma}_3 \ldots \hat{\sigma}_{n\sigma}, \hat{\alpha}_1, \hat{\alpha}_2, \hat{\alpha}_3 \ldots \hat{\alpha}_{n\alpha} > \quad (11.1)$$

Here $<x_1, x_2, x_3, \ldots x_n>$ is the phenotypic representation attributes of the individual. $<\sigma_1, \sigma_2, \sigma_3, \ldots \sigma_{n\sigma}>$ and $<\alpha_1, \alpha_2, \alpha_3, \ldots \alpha_{n\alpha}>$ are the strategy parameters. We discuss the role of these parameters in mutation later in the text. n is the dimensionality of the problem. n_σ and n_α are the dimensions of the strategy parameters. Here n_α may be given by equation (11.2).

$$n_\alpha = n_\sigma (n - \frac{n_\sigma}{2})(n_\sigma - 1) \tag{11.2}$$

It may be seen that the additional strategy parameters in the individual representation add dimensions in the genotype search space when viewed from the point of view of conventional GA. But these strategy parameters play a big role in driving the individual in the most optimal manner for every generation.

11.3 Evolution

The other major aspect associated with the ES is the manner in which the individuals participate in the evolutionary process. ES is usually defined by notations that give an idea of the manner in which the population evolves with time. This tells us about the kind of ES used in the evolution process. The various notations would be discussed one by one (Schwefel and Rudolph 1995).

11.3.1 (1+1) ES

The simplest form of ES is the *(1+1)* ES. Here there is a single individual that does the task of search in the search space. Hence there is no difference between the population pool and the individual. This individual at every generation makes another individual by the application of the mutation operator. In this system crossover or recombination is not possible due to the presence of only one individual. Hence mutation always decides the fate of the individual. The selection operator is however applied in this system. The selection operator selects the fitter of the original individual and the individual resultant from the mutation operator. The fitter individual goes to the next generation and the weaker one dies as a result of poor fitness in the evolutionary process. This process repeats at every generation till the stopping criterion is not met.

An interesting fact with this kind of system is that a new population pool comprising of the old as well as the new population is made. The selection operator does the task of selection over this population pool. It may be hence possible that the old generation individual is fitter and gets promoted to the higher generation in place of the newly created individual. If again this individual is fitter than the newly generated individual, it is again promoted to the next generation. In this manner an individual may again and again get promoted to the next generation without the promotion of the newly generated individuals. This especially happens for the individuals at some local or global minima where traveling on any side gives weaker individuals.

The total number of generations for which any individual lives is called as the individual *lifetime* in case of the ES. This corresponds to the total number of generations for which the individual is promoted from one generation to the other.

11.3.2 ($\mu/\rho+\lambda$) and ($\mu/\rho+\lambda$, k) Representations

The other ES that is a generalized version of the *(1+1)* ES is the $(\mu/\rho+\lambda)$ ES. This ES assumes that the population pool contains a total of μ individuals. At every generation these μ parents produce λ individuals as a result of recombination or crossover operation. Here each recombination or crossover operation takes place by a combined operation of ρ individuals. Hence ρ parents produce *1* child in the evolution. The next step is to make a combined population pool consisting of the old and the new population. There are μ parents and λ children totaling $(\mu+\lambda)$ individuals. These individuals then undergo a selection operation to compete for the survival for the next generation. Now a total of μ individuals are selected that survive and go to the next generation. Here also it is possible that the fitter parents would go from one generation to the other without being killed. Similarly many children may not get a chance to go to the next generation. In this manner every individual in this mechanism has a lifetime that denotes the total number of generations for which the individual passes without being killed. Many times this may also be represented by $(\mu+\lambda)$ denoting a default value of the number of parents ρ of 2.

The other ES is the $(\mu/\rho+\lambda, k)$ ES. In this mechanism also a total of μ individuals of the population make λ individuals for the next generation by the application of ρ parents for each individual generation. Out of these a total of μ individuals go to the next generation. However, here any individual can have a maximum lifetime of *k* generations. The selection operator cannot select any individual that crosses the lifetime of *k* generations. For this we maintain a counter with every individual that counts the number of generations for which the individual is alive. The counter increases by one for every generation of the individual. This is initialized to zero for all newly created individuals of any generation.

11.3.3 (μ/ρ, λ) Representation

The last ES is the (μ/ρ, λ) and (μ/ρ, λ, k) ES. This is similar to the ES discussed in the previous sub-section with the only difference that the ',' operator is used in place of the '+' operator. This change in operator causes a little change in the functioning of the ES with the other characteristics remaining unchanged.

In the previously discussed ES we had made a pool of population comprising of the old as well as the new generation and the selection of thc individuals was done from this pool. In this method however, the pool only contains the newly generated individuals by the recombination operation. The older ones are excluded from the operation. This is similar to the selection scheme used by GA. λ is always chosen greater than μ.

The lifetime of any individual in this method of ES can be one 1. This is because every individual dies after the generation of the children for the next generation. Individuals are restricted to be carried forward from one generation to the other.

11.4 Recombination

The next step to be performed in the working of the ES is the recombination operation. This operation is similar to the working of the crossover in GA. As discussed in the above section, this operator makes use of ρ individuals or parents for the generation of children. The attempt is to combine the good characteristics of each of the parents to generate healthy children that have a better fitness value as compared to the parents. In this manner we expect to constantly improve the fitness at every generation. Similar to the GA, this operator results in convergence of the algorithm along with generations.

The recombination may be done in multiple ways depending upon the choice of the designer. Suppose that the ith attribute of the child is given by q_i. Let the same attribute value of the jth parent be given by p^j_i. Since there are a total of ρ parents, j can lie between 1 to ρ. The following are the ways in which the child can be generated by using the ρ parents.

- **No Recombination:** In this recombination technique, no change takes place between the parent and the child. Any one of the ρ parents is selected and copied to make the child. This selection is done randomly. This operation is given by equation (11.3).

$$q_i = p^k_i \tag{11.3}$$

  k is a number randomly chosen every child and fixed for all i.
- **Global Intermediary Recombination:** In this technique we use the information of all the parents to generate the child. This is done by averaging the parent attribute values for the generation of the child attribute value for all attributes. The child is simply the mean of all parents. This operation is given by equation (11.4).

$$q_i = \frac{1}{\rho}\sum_{k=1}^{\rho} p_i^k \tag{11.4}$$

- **Local Intermediary Recombination:** Here we do not use the information or attribute value of all the parents, we rather use the information or attribute value of only a couple of parents. The individual generated is a mix of these two selected parents. The selection of these two parents may be done randomly. This is given by equation (11.5).

$$q_i = u_i\, p^a_i + (1-u_i)\, p^b_i, \tag{11.5}$$

  a, b are random numbers, u_i is a constant between 0 and 1, usually kept as 0.5

- **Uniform Recombination:** This mechanism chooses every attribute value of the child from any of the parents. The sections are random and independent in nature. Any attribute may be contributed totally by any of the parents. This is given by equation (11.6).

$$q_i = p^k_i \tag{11.6}$$

k is a number chosen randomly for all *i*.
- **λ Point Recombination:** In this mechanism we first select λ points in the individual representation. This divides it into $\lambda+1$ part. Now each part is contributed by any of the ρ parents. The number of parts $\lambda+1$ are always kept greater than the number of parents ρ. This enables each parent to contribute at least one part in the making of the child.

In this operator each child is generated independently by using any of the above techniques. We select a total of ρ parents by any selection mechanism. Then these parents recombine to give a child. This is done for all the λ children generated by the parents in this step.

11.5 Mutation

The next major operator involved in this technique is mutation. The role of mutation is to change the values of the various attributes and adjust them such that the individuals are placed at better position with better fitness values. This operation is performed by the addition of some values at the various parameters that moves the individuals in the search space. For this operator we consider the individual as given in equation (11.1). This contains both the attributes as well as the strategy parameters in the individual representation.

The mutation is performed in steps (Schwefel and Rudolph 1995). The first step involves the modification in the σ strategy parameters. This determines the step size for every attribute being mutated. A larger step size corresponds to a more disruption of the attribute value and vice versa. In place of deciding the value of this strategy variable we take the help of normal distribution. These are not kept constant as was the case with the conventional GA. This operation is given by equation (11.7).

$$\tilde{\sigma} = < \hat{\sigma}_1 e^{z_1+z_0}, \hat{\sigma}_2 e^{z_2+z_0}, \hat{\sigma}_3 e^{z_3+z_0} ...., \hat{\sigma}_{n_\sigma} e^{z_{n\sigma}+z_0} > \tag{11.7}$$

Here z_0 ε $N(0,\tau_0^2)$ and z_i ε $N(0,\tau^2)$. $N(\xi, \psi^2)$ is the normal distribution for with mean ξ and variance ψ^2.

By this normal distribution, each individual tries to explore its near surroundings in search for the minima. The attempt for every individual is to jump to the nearest minima that one may encounter in the surroundings (Back *et al* 1991). In this manner the mutation operator plays a big role in the ES to enable every individual walk towards the minima that lies near to it. Since there are multiple individuals in most of the implementation of ES, we may expect each of these to try to find a local minima and ultimately some individual getting to the global minima in this manner. This manner of working with the mutation adds a lot to the adaptability of the algorithm to enable it fix the correct parameter values when compared with the traditional GA. This concept of the individuals working with the normal distribution is shown in figure 11.1.

The choice of the parameters τ_0 *and* τ depends upon the convergence model of use. A convergence model denotes how fast and assuming what type of surface the individual of the evolutionary process tries to go to the minima. The convergence here denotes the movement of the individual with mutations towards the minima. Now we do not have an idea of the behavior of the surface of the fitness function. We cannot make numerous calls to the fitness function that enable us compute the local minima and then advance to it. We hence assume a model that is known as the convergence model. The convergence model governs the convergence of the individual to the minima as per the present scenario or its position in the fitness landscape. This is further an indication towards the adaptability of the algorithm.

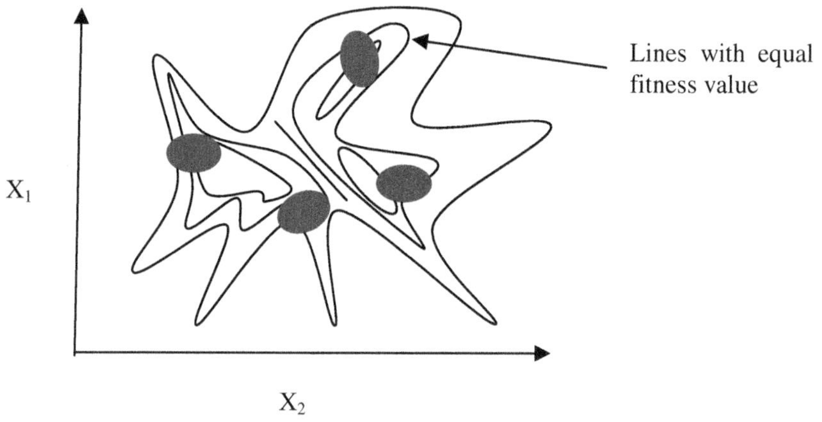

Fig. 11.1 Evolutionary Strategy Individual Exploration

Numerous convergence models may be used. The model for the fastest convergence is sphere for which the values of the parameters τ_0 *and* τ may be given by equation (11.8) and (11.9).

$$\tau_0 = \frac{K\delta}{\sqrt{p_m}\sqrt{2n}} \tag{11.8}$$

$$\tau = \frac{K(1-\delta)}{\sqrt{p_m}\sqrt{\frac{2n}{\sqrt{n_\sigma}}}} \tag{11.9}$$

Here K is a constant that determines the speed at which the convergence takes place. This is specified by the user. δ is a constant that is normally kept as 0.5. p_m is another additional parameter that may be fixed as 1.

The next step to be performed in mutation is the modification of the α parameters. We saw the role of the σ parameters in the search for the optimal points in the fitness landscape. However these parameters can only enable the algorithm to search along the axis of the fitness landscape. This may further sub-optimize the algorithm. We need the algorithm to be able to handle the correlated mutations as well. This adds a rotational factor to the algorithm that enables optimal mutations. This is done by the addition of these parameters. This modification is given by equation (11.10).

$$\tilde{\alpha} = < \hat{\alpha}_1 + z_1, \hat{\alpha}_2 + z_2, \hat{\alpha}_3 + z_3 \ldots \hat{\alpha}_{n_\alpha} + z_{n_\alpha} > \qquad (11.10)$$

Here z_i ε $N(0,\beta^2)$. β may be kept a value of 5^0 (~0.0873).

So far we have mutated the σ and α strategy variables. The next step is to perform the mutation over the attributes that the algorithm was supposed to optimize. This may be given by equation (11.11).

$$\tilde{x} = < x_1 + cor_1(\hat{\sigma},\hat{\alpha}), x_2 + cor_2(\hat{\sigma},\hat{\alpha}), \ldots x_n + cor_n(\hat{\sigma},\hat{\alpha}) > \qquad (11.11)$$

Here $cor_1(\sigma, \alpha)$, $cor_2(\sigma, \alpha)$, … $cor_n(\sigma, \alpha)$ are the elements of the variable $cor_1(\sigma, \alpha)$ which may hence be given by equation (11.12).

$$cor(\sigma, \alpha) = < cor_1(\sigma, \alpha), cor_2(\sigma, \alpha), cor_3(\sigma, \alpha) \ldots cor_n(\sigma, \alpha)> \qquad (11.12)$$

cor(σ, α) is a random vector with normally distributed and correlated components. This vector is calculated by equation (11.13).

$$cor(\sigma, \alpha) = T\, z \qquad (11.13)$$

Here z is given by equation (11.14)

$$z = <z_1, z_2, z_3, \ldots z_{n\sigma}> \qquad (11.14)$$

Here z_i ε $N(0,\sigma_i^2)$.

T is given by equation (11.15), (11.16) and (11.17).

$$T = \prod_{p=1}^{n_\sigma - 1} \prod_{q=p+1}^{n_\sigma} T_{pq}(\tilde{\alpha}_j) \qquad (11.15)$$

$$j = 0.5\,(2n_\sigma - p)\,(p+1) - (2n_\sigma + q) \qquad (11.16)$$

Here $T_{pq}(\alpha)$ is a square matrix whose any element t_{ab} is given by equation (11.17).

$$t_{ab}(\alpha) = \begin{cases} \cos(\tilde{\alpha}_j) \forall a = p, b = p \wedge a = q, b = q \\ -\sin(\tilde{\alpha}_j) \forall a = p, b = q \\ \sin(\tilde{\alpha}_j) \forall a = q, b = p \\ 1 \text{ for } a = b \text{ and none above} \\ 0 \text{ for a not equal b and none above} \end{cases} \qquad (11.17)$$

An interesting factor of study is the impact of the mutation variance or step size. Consider that any individual is located at some point in the search space as shown in figure 1.2(a). Now a very small variance would mean that the individual would most probably be placed much near to its present location (Kita 2001). In such a manner we may predict that it would mostly be able to catch the local minima in its vicinity and walk towards it as a result of the mutation process. The success of mutation operation is very high. This success denotes the chances of the mutation operation leading the individual to a better fitness value as compared to its original fitness. However the convergence would be very slow as the individual is restricted to a very small move in the fitness landscape (Arnold and Beyer 2001; Babu and Murthy 1994; Back 1996; Beyer 1990).

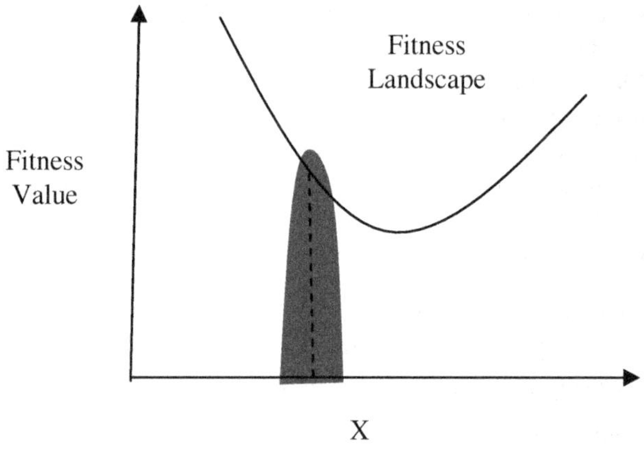

Fig. 11.2(a) Small Step Size

Now consider the same fitness landscape with a larger variance. This is shown in figure 1.2(c). In this problem the individual is now allowed to move in a reasonably large space. The individual has a much wider access. In this context the probability that the mutation results in a fitter individual is fairly low. This is due to the fact that the individual may lie at some other side of the minima and hence lose its optimality. Large movements in the fitness landscape are always risky in nature due to the non-predictability of the fitness landscape. However the individual in this case may have a very fast convergence where it might even land up far away from the minima.

The intermediate case of these two forms is a much simpler case that combines the discussed ideologies of both the cases discussed above. This is shown in figure 1.2(b).

Evolutionary Strategy may hence be applied using the discussed operators and the discussed concepts in similar lines to the evolutionary algorithms. This forms an adaptive and simple means to solve the problems of system evolution or optimization in specific. The widespread use of this technique in problem solving is much attributed to the high convergence and optimization that is achieved in their use. The ES is able to effectively solve many large fitness landscapes and these are able to figure out the optimal points in them.

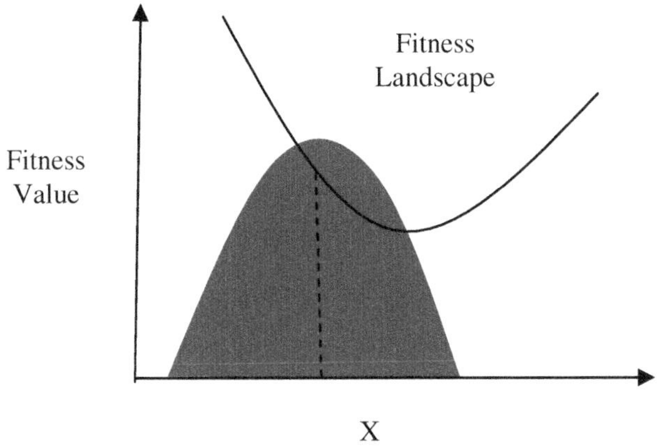

Fig. 11.2(b) Medium Step Size

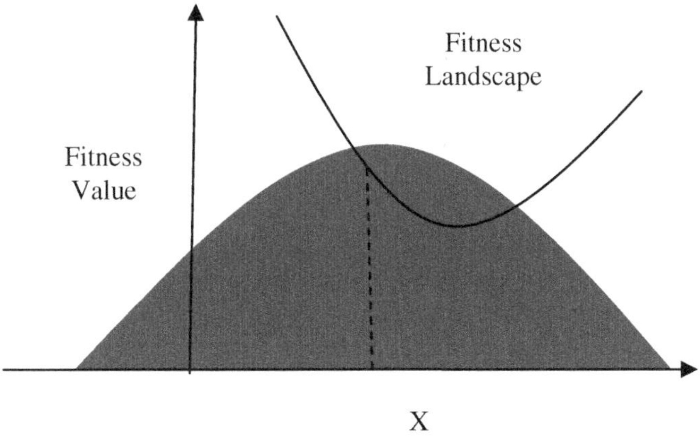

Fig. 11.2(c) Large Step Size

11.6 Chapter Summary

In this chapter evolutionary strategies were discussed and presented. This forms a major constituent of evolutionary algorithms, the other being evolutionary programming and genetic algorithm. The chapter introduced the adaptive nature of the evolutionary strategy and the mechanism by which the various individuals adjust their parameters as per the specific location in the fitness landscape. The entire algorithm executed as per the evolutionary norms where the individuals improve their fitness along with generations in an iterative manner. The chapter discussed the mechanism in which extra strategy parameters are embedded into the individual for the adaptive behavior. Then the recombination operator was discussed. Here the exchange of information between parents was carried out to make the children of the higher generation population. Then mutation operator was discussed. Here we discussed the mechanism of using the strategy parameters and adapting them as per requirements for better performance.

References

[1] Angeline, P.J.: Genetic programming and emergent intelligence. In: Complex Adaptive Systems archive Advances in genetic programming, pp. 75–97. MIT Press, Cambridge (1994)

[2] Angeline, P., Pollack, J.: Competitive environments evolve better solutions for complex tasks. In: Proc. of the Fifth Intl. Conf. Genetic Algorithms, pp. 264–270. Morgan Kaufmann, San Mateo (1993)

[3] Arnold, D.V., Beyer, H.-G.: Local performance of the (μ/μI,λ)-ES in a noisy environment. In: Martin, W., Spears, W. (eds.) Foundations of Genetic Algorithms, vol. 6, pp. 127–141. Morgan Kaufmann, San Francisco (2001)

[4] Babu, G.P., Murty, M.N.: Clustering With Evolution Strategies. Pattern Recognition 27(2), 321–329 (1994)

[5] Back, T.: Evolutionary Algorithms in Theory and Practice. Oxford University Press, Oxford (1996)

[6] Bäck, T., Hoffmeister, F., Schwefel, H.P.: A Survey of Evolution Strategies. In: Proc. of the Fourth Intl. Conf. on Genetic Algorithms, pp. 2–9 (1991)

[7] Back, T., Hammel, U., Schwefel, H.P.: Evolutionary Computation: Comments on the History and Current State. IEEE Trans on Evol. Comput. 1(1), 3–17 (1997)

[8] Back, T., Rudolph, G., Schwefel, H.P.: Evolutionary Programming and Evolution Strategies: Similarities and Differences. In: Proc. of the Second Annual Conf. on Evolutionary Programming, pp. 11–22 (1995)

[9] Beyer, H.-G.: Simulation of steady states in dissipative systems by darwin's paradigm of evolution. J. Non-Equilib. Thermodyn. 15, 45–58 (1990)

[10] Beyer, H.-G.: Some aspects of the 'evolution strategy' for solving tsp-like optimization problems. In: Männer, R., Manderick, B. (eds.) Parallel Problem Solving from Nature, vol. 2, pp. 361–370. Elsevier, Amsterdam (1992)

[11] Beyer, H.-G., Deb, K.: On self-adaptive features in real-parameter evolutionary algorithms. IEEE Transactions on Evolutionary Computation 5(3), 250–270 (2001)

[12] Eiben, A.E., Smith, J.E.: Introduction to Evolutionary Computing. Natural Computing Series. Springer, Heidelberg (2003)

[13] Eiben, A.E., Hinterding, R., Michalewicz, Z.: Parameter control in evolutionary algorithms. IEEE Transactions on Evolutionary Computation 3(2), 124–141 (1999)

[14] Fogel, D.B.: Evolving Artificial Intelligence. Ph.D. Thesis, University of California, San Diego, California (1992)

[15] Goldberg, D.E.: Genetic Algorithms in Search, Optimization, and Machine Learning. Addison-Wesley, Reading (1989)

[16] Hansen, N., Ostermeier, A.: Adapting arbitrary normal mutation distributions in evolution strategies: The covariance matrix adaptation. In: Proc. of the IEEE Intl. Conf. on Evol. Comput., Piscataway, New Jersey, pp. 312–317 (1996)

[17] Hansen, N., Ostermeier, A.: Convergence properties of evolution strategies with the derandomized covariance matrix adaptation: The (μ/μI,λ)-CMA-ES. In: Zimmermann, H.J. (ed.) 5th European Cong. on Intelligent Techniques and Soft Comput., EUFIT 1997, pp. 650–654. Verlag Mainz, Aachen (1997)

[18] Kita, H.: A Comparison Study of Self-Adaptation in Evolution Strategies and Real-Coded Genetic Algorithms. Evol. Comput. Archive 9(2), 223–241 (2001)

[19] Koza, J.R.: Genetic Programming: On the Programming of Computers by Means of Natural Selection. The MIT Press, Cambridge (1992)
[20] Michalewicz, Z.: Genetic Algorithms + Data Structures = Evolution Programs. Springer, Berlin (1992)
[21] Mitchell, M.: An Introduction to Genetic Algorithms. MIT Press, Cambridge (1996)
[22] Schwefel, H.P.: Evolution and Optimum Seeking. Wiley, New York (1995)
[23] Schwefel, H.P., Rudolph, G.: Contemporary Evolution Strategies. In: Morán, F., Merelo, J.J., Moreno, A., Chacon, P. (eds.) ECAL 1995. LNCS, vol. 929, pp. 893–907. Springer, Heidelberg (1995)
[24] Whitley, L.D.: A genetic algorithm tutorial. Statist. Comput. 4, 65–85 (1994)
[25] Yao, X., Liu, Y.: Fast evolution strategies. In: Angeline, P.J., Reynolds, R.G., McDonnell, J.R., Eberhart, R. (eds.) EP 1997. LNCS, vol. 1213, pp. 151–161. Springer, Heidelberg (1997)

List of Figures

Fig. 11.1. Evolutionary Strategy Individual Exploration
Fig. 11.2(a). Small Step Size
Fig. 11.2(b). Medium Step Size
Fig. 11.2(c). Large Step Size

Chapter 12
Other Evolutionary Concepts

Abstract. Evolutionary Algorithms in numerous forms present powerful problem solving tools. These tools have been modified and adapted to various kinds of problems as per the requirements. Genetic Algorithms, Evolutionary Strategies and Genetic Programming happen to be the basic classifications of these algorithms. In this chapter we further discuss some of the widely used models of Evolutionary Algorithms that are extensively used for evolving systems and their optimization. All these differ in their methodology of problem solving. We first present Differential Evolution that uses the differences between individuals for their optimization. The chapter would then present Artificial Immune Systems that are an analogy from the natural immunity system prevalent to fight diseases. Here we discuss the self and non-self methodology of classification. Then we present Co-evolution where the different individuals of the population pool help each other for the evolution. The other topics of discussion include Cultural Algorithm where the evolution is biased by a culture or belief space. We also discuss about Cellular Automata in this chapter.

Evolution is a novel concept that has immense potential to engineer systems that is otherwise done by humans by trying various kinds of designs, models and parameters. The ease with which the Evolutionary Algorithms (EA) are able to evolve systems is a major reason behind their ample of use in many real life applications. It is rather simple task to employ an evolutionary system to make the desired system. This has largely led to a great ease in system design and construction. Optimization specifically is a key problem. Many of the known problems are optimization in nature where the algorithm is supposed to find the correct values of some of the parameters on which the system performance depends. The EAs are very helpful for all these problems. They are able to give the best results in the smallest possible time. The iterative nature of these algorithms is especially very helpful in problem solving. In this manner the results of these algorithms is far better than any randomized algorithm that one may think of.

These algorithms find numerous applications in different problem domains. This has further resulted in many models that are built over the evolutionary approach. The EAs are hence very versatile in nature with a large variety of choices that one may select from. Each of these varieties is based upon some biological inspiration that guides its overall design. The various models find applications in different contexts.

A. Shukla et al.: Towards Hybrid and Adaptive Computing, SCI 307, pp. 249–273.
springerlink.com

We have discussed the major varieties of EAs in the previous few chapters. We started with the Simple Genetic Algorithm (SGA) in chapter 3. This algorithm used crossover and mutation in a simple genotype representation of the individual to generate population from one generation to the other. Various other genetic operators were used as well. Later we studied the Swarm Intelligence in chapter 9. Here the natural analogy of ant colonies, flocking birds, etc. was used to make systems where individuals imitate similar behavior to find the most optimal point in the fitness landscape. Chapter 10 presented another major variety of EA. This was Genetic Programming. Here we evolved programs that were executed to solve the problem and give the performance analysis. We also presented Grammatical Evolution in the same chapter. Chapter 11 presented the Evolutionary Strategies that were simple and adaptive systems for optimization and problem solving.

While all of these algorithms present powerful techniques for solving problems that come under the evolutionary approach, there are some more concepts that need to be discussed. All these are models that run on similar lines to the models presented and discussed so far. We discuss Differential Evolution, Artificial Immune System, Co-evolution, Cultural Algorithm and Cellular Automata in this chapter.

12.1 Differential Evolution

The first algorithm of study is the Differential Evolution (DE). This algorithm tries to adapt the mutation operator and fix its value to the optimal value as per the problem context. In all other terms this algorithm behaves in a manner similar to the GA (Brest *et al* 2006; Abbass 2002; Abbass *et al* 2001; Lin and Lampinen 2003; Qin and Suganthan 2005; Jakob and Rene 2004). Hence DE algorithm is more adaptive towards the problem and corresponding scenario as compared to the traditional GA. We have already discussed the motivation of adapting mutation operation in chapter 11 titled Evolutionary Strategies.

The representation of the individual in this algorithm is similar to that of the GA. The various parameters are arranged in a vector of numerals where each numeral represents some of the characteristics of the solution that the individual represents. The DE runs on the basis of difference in values of the attributes of the individual of the algorithm. It tries to use this information of the difference in attribute values for the adaptation of the mutation operator. Initially the diversity between the individuals is relatively large. Hence the difference between the values of some attributes is relatively high. As the algorithm continues, the diversity decreases and finally all individuals converge to some point. As the diversity reduces, the difference between attribute values also reduces. Hence the difference between the values of attributes represented in an individual is taken to be a source of adaptation of the mutation operator.

Much of the algorithm is similar to the GA. The general algorithmic structure of the DE more or less involves the terms used in GA. We discuss the various parts of this algorithm in the coming sections.

12.1.1 Attribute Difference D_{abcd}^j

The first major concept introduced in this algorithm is the difference in attribute values. This is calculated by random sampling (Storn and Price 1996, 1997). Let us say that we randomly pick up 4 individuals from the population pool. Let these individuals be represented by *a*, *b*, *c* and *d*. Let the difference in attributes between these individuals for any attribute *j* be given by the term D_{abcd}^j. This may be computed by using equation (12.1).

$$D_{abcd}^j = (X_a^j - X_b^j) + (X_c^j - X_d^j) \tag{12.1}$$

Here X_i^j denotes the attribute value of the jth attribute in the ith individual.

It can be clearly seen that this term selects 4 individuals. The difference in attribute values between both the pairs is computed and added. The first pair consists of *a* and *b*. The second pair consists of *c* and *d*. The difference of attribute values are noted and added for the jth attribute.

The relation between the diversity and attribute difference may again be easily verified. The initial generations witness high population diversity. The individuals are places very far off in the fitness space. Hence the difference in attributes may be very large. At the later generations, the individuals are closer and near the global minima. In such a case the difference in attributes are much smaller. To compute this term we took a random sampling of 4 individuals or 2 pairs of individuals. It may be seen that this may not return the true results as represented by the population diversity. Many times a highly diverse population pool may return a very small value of this parameter. The opposite may also be true. This is because of the choice of selection of the random individuals. Hence the diversity does not strictly decide the value of this parameter. This further adds some randomness to the algorithm which plays a role in its optimal working.

A simple means to understand this parameter is by using concepts of vectors. Consider the fitness space where the fitness value is plotted against the various parameters. For the ease of understanding, we plot the fitness as contours and take the individuals with 2 dimensions. This is shown in figure 12.1. Consider the same 4 individuals *a*, *b*, *c* and *d*. Now these 4 individuals represent points in the fitness landscape of the problem. These may be taken as 4 vectors. The difference between any two vectors in the vector space may be easily computed in the vector space using vector arithmetic. Further, the difference vectors may be easily added in the same vector space to give the final vector representing D_{abcd}. This vector represents the difference vector. It denotes the difference in attributes for various attributes or axis.

12.1.2 Mutation

The major factor of study in this algorithm is the mutation. Here we try to modify the attribute values of the individual in order to place it at a better position in the fitness landscape. This operator does the task of exploring the surroundings in the fitness landscape and tries to find a place of good fitness value. The changes

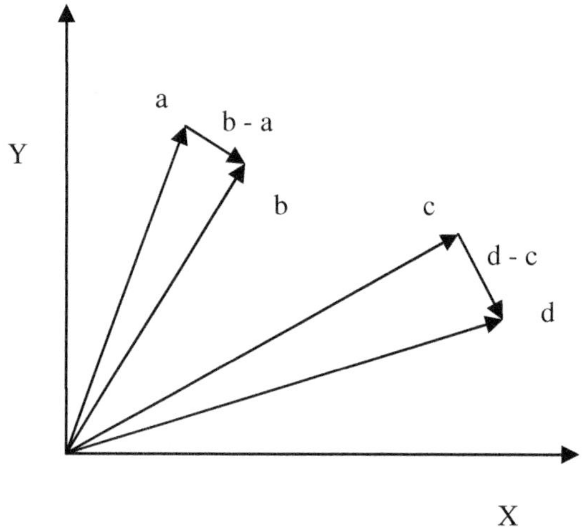

Fig. 12.1 Difference vectors in Differential Evolution

introduced in the individual may not always result in reaching a point with higher optimality. However the goal of the EA is to finally construct the most optimal individual or in other words to finally find the most optimal point into the fitness landscape of the problem.

Consider the individual i that has to be mutated. This may be represented by $X_i <X_i^1, X_i^2, X_i^3, \ldots., X_i^n>$ where n is the dimensionality of the problem or the total number of attributes. The mutation may be carried independently for all of the n attributes in the problem. The mutation of the attribute j of the individual may be done by the addition of some noise into its present value. This noise is proportional to the difference in attribute values. This may be given by equation (12.2).

$$X_i^{j'} = X_i^j(t) + F.D_{abcd}^j \tag{12.2}$$

Here t is the generation. a, b, c and d are random points. F is the magnification factor.

The factor F magnifies the difference vector. This factor is specified at the start and kept constant in the algorithm run. This decides the convergence of the algorithm. A very large value of this parameter magnifies the difference to a very large extent. As a result the algorithm is unable to converge to some point in the fitness landscape. This causes a very large wilderness in the mutated movements of the individual. Very small values of this parameter may normally result in premature convergence. In such a case the algorithm may converge to local minima. In some other cases it may not converge to minima at all. Hence this parameter plays a major role in the optimal algorithm working and needs to be fixed judiciously as per the problem and computational costs.

12.1.3 Crossover

The other operator is crossover that generates new individuals from the parents. This operator uses a constant known as the *crossover rate* (CR). This constant determines the total contribution of the crossover operator in the evolution process. The CR lies in the range of 0 to 1. A CR of 0 means that the crossover has no contribution in the process and the entire contribution in the generation of the individual comes from the mutation alone. Similarly a CR of 1 means that the entire contribution comes from crossover alone and nothing from mutation.

The evolution involves the making up of the complete individual from some parent. Consider the individual *i* that may be represented by $X_i <X_i^1, X_i^2, X_i^3, ..., X_i^n>$ where *n* is the dimensionality of the problem or the total number of attributes. Now *CR%* of the individuals are generated by the crossover operator. We hence randomly select *CR%* attributes in the individual representation of the individual. These are copied straight away from the parent into the next generation. The others undergo mutation.

Since the entire selection is based on random numbers, it is possible that all the attributes in the population pool undergo mutation and the total contribution of crossover is zero. This is undesirable as the newly generated child needs to carry some characteristics from the parent. We hence specify a certain index *k* randomly before generating the child. This index is directly copied from the parent to the child without any checking. In this manner the child is assured to have some characteristics of the parent.

The crossover operation directly copies the parent attribute to the child attribute. This may be given by equation (12.3).

$$X_i^{j'} = X_i^j(t) \tag{12.3}$$

Here *t* is the generation.

12.1.4 Selection

Another major factor in the use of this algorithm is the selection. The basic methodology is to select the fitter individuals to go to the higher generations and stop the weaker individuals from coming to the higher generations. This increases the fitness of the individuals in the population pool of the algorithm. Also the better characteristics are identified which may be further computed for optimization. In this problem the mutation is applied to every individual to generate a new set of population pool. The mutation may result in a better fitness individual or vice versa. We consider both the mutated and the un-mutated individual. The fitter individuals from both the individuals go to the next generation and the weaker ones are stopped. This means that a mutation operation is made operational only when it is successful and results in a better fitness. In case the mutation is not successful in increasing the fitness, its effect is cancelled. This again plays a key role in directing the algorithm towards the correct direction in the algorithm run. The individual is prohibited to run out in wrong directions that might lower its fitness value. This may be given by equation (12.4)

$$X_i(t+1) = \begin{cases} X_i(t) & F(X_i(t)) > F(X_i^{'}) \\ X_i^{'} & \text{otherwise} \end{cases} \quad (12.4)$$

12.1.5 General Algorithm

In the previous sections we closely noted the various steps of the algorithm. Most of the terminologies and operators used were similar to the GA. The implementation however varied as per the specific implementation of this algorithm. The overall structure of the DE algorithm may be given by figure 12.2.

12.2 Artificial Immune System

The Artificial Immune System (AIS) is another novel concept of evolutionary algorithms. This is an inspiration from the natural immunity system. In practical life we are exposed to various diseases. The immunity system within us is capable of fighting these diseases. Whenever any undesirable pathogen attacks the body, the natural immune system of the body identifies it. It then uses some mechanism to fight against it. In this manner the human body is saved from the evil effects of the pathogens.

The AIS is an attempt to imitate the same functionality of the AIS. Here we try to discriminate between the self and the non-self particles. The self are the desirable particles that belong to the desirable class. The non-self particles are the undesirable particles that belong to the foreign class. The problem is to identify the non-self particles from a set of given particles. In other words we need to classify the given set of particles as self and non-self. This requires a proper learning from the past trends that illustrates the characteristics of the self particles and that of the non-self particles. The classification may be carried out with this learning (Aickelin and Cayzer 2002; Kim and Bentley 1999a, 1999b, 2001, 2002; Gasper and Collard 1999; Hofmeyr and Forrest 2000; Forrest et al 1993; de Castro and Timmis 2002).

We have emphasized a lot over the adaptive nature of the systems in this text. The same applies to this problem as well. Nature puts us before a variety of disease against which we are expected to immunize ourselves. In a similar manner the AIS may come across numerous categories of self and non-self particles. The body is usually more immunized against the recently encountered diseases. In the similar manner the AIS behaves as an adaptive system. The recent characteristics usually have a much larger impact on the entire system than the other particles. As the learning goes on and on, the newer characteristics may replace the older ones making the algorithm stronger towards the recent trends as compared to the older ones.

A large application of this algorithm is in the intrusion detection in the networks. In this problem the activity may be constantly monitored and characterized. Based on the characteristics observed, we may need to decide whether the activities are normal or indicate some intrusion over the network. In case some intrusion is likely, the system may be equipped with actions to prevent damage by the intrusion.

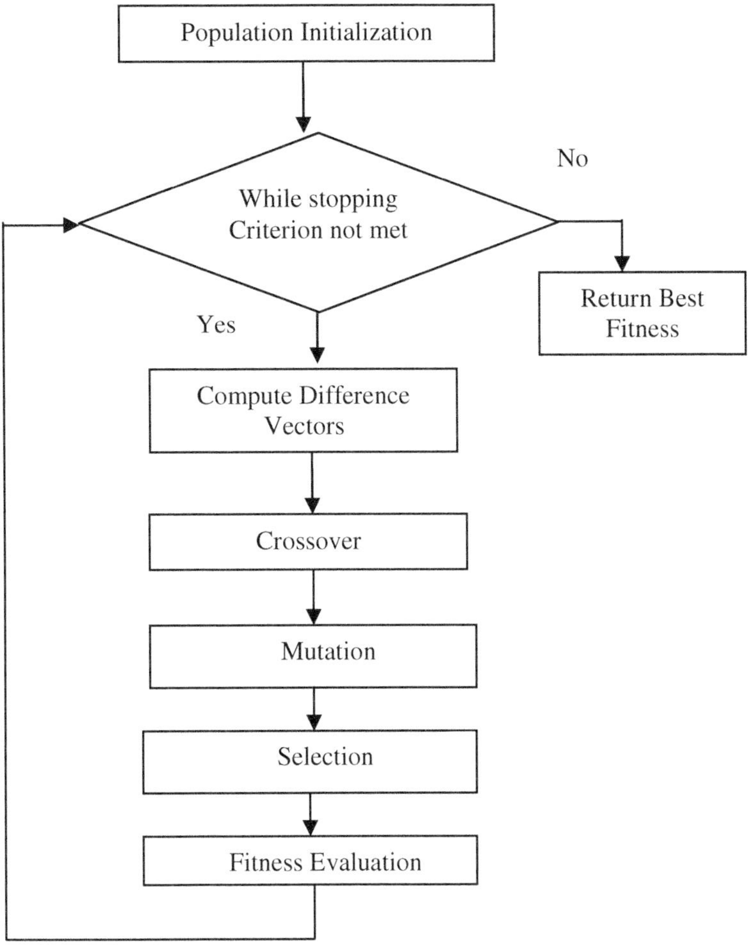

Fig. 12.2 Differential Evolution Operation

12.2.1 Representation

The task in this problem is the classification of the self and the non-self particles. A particle in this problem represents the entire feature as per the problem design. This may be taken from the phenotype representation of the problem as used in the GA. This representation may be easily converted into a set of attributes representing the genotype representation of the problem. Alternatively, a widely used mechanism of representing the particle is the bit vector representation. In this form the entire feature is coded into bits consisting of 0s and 1s. This string is used for the task and represents the complete particle of the system. The entire task of classification or the detection of the non-self is carried out using this representation. The problem given is coded as a sequence of these representations that needs to be classified or determined. The solution again makes use of the same representation. This consists of a set of detectors, each of which represents a particle in the same representation. In this manner the detection of the non-self molecules is carried out (Kim and Bentley 2001).

12.2.2 Detectors

The other concept used in this algorithm is that of the detectors. These are the particles that help in the detection of the non-self particles and separate them out from the set of self and non-self particles. There are a collection of detectors used in the AIS. Each detector represents a particle in the feature space. This particle does the task of detection by the measurement of the similarity between itself and the presented particle. This similarity is used as a measure of the non-self nature of the particle. The detectors usually follow a negative selection scheme. As per this they try to identify the non-self particles. These detectors hence try to place them in the regions where the non-self particles are located in the feature space. They equip themselves with similar characteristics as the non-self particles. There might be different kinds of non-self particles corresponding to the different kinds of characteristics that the immune system is expected to detect. This is done by the placement of different types of detectors in the entire feature space. Multiple detectors carry forward the detection of different types of non-self characteristics. In this manner the entire system is constructed and worked upon.

Consider a simple feature space as shown in figure 12.3. This shows the self and the non-self particles that are to be classified (Hoffmeyer and Forrest 2000). It is natural that we may place some detectors in the preferential regions that help in detecting the non-self particles and enable the detection to be effectively carried out.

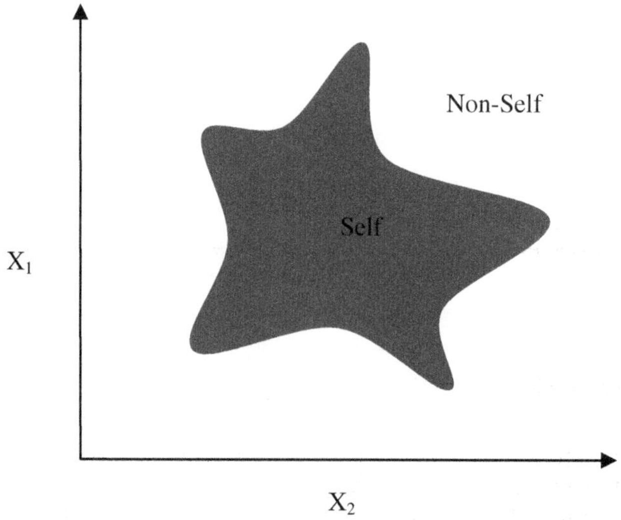

Fig. 12.3 Feature Space for Artificial Immune System Problem

12.2.3 Comparing Particles

As it is clear from the previous sections, one of the major task of the algorithm is to compare the presented particle with the detectors. If the comparison is above some predefined threshold, we may easily classify the particle as a non-self particle and vice versa. This comparison between the particles may be carried out using a variety of ways. The manner of comparison largely depends upon the fact that the algorithm represents the genotype or the problem representation in a numeral form or a bit string form (Forrest *et al* 1993).

Suppose that the problem is represented as a vector of numerals. Now the comparison involves finding the similarity between two numeral strings say $X <X_1, X_2, X_3, ..., X_n>$ and $Y<Y_1, Y_2, Y_3, ..., Y_n>$. Here n is the dimensionality of the feature space or the number of attributes represented in the particle. This may be easily taken as the Euclidean, Manhattan or any other norm. The usage of Euclidean norrn is common as is given in equation (12.5).

$$D(X,Y) = \sqrt{\sum_{i=1}^{n}(X_i - Y_i)^2} \qquad (12.5)$$

The similar techniques may be applied to the bit string representation as well to find the similarity between two particles. Here we may use the Hamming distance between the two particles to measure their similarity. Another commonly used technique used is the *r-contiguous bits*. In this technique two strings are said to match with each other if they have r bits in continuation common between them. Here r is a modifiable parameter that denotes the matching threshold and may be fixed as per the problem demands. Keeping r as n would make the detector very specific where it would only detect the particles exactly similar to it. Similarly keeping r very small would make it match to nearly everything. Hence this parameter may be decided judiciously. This matching is shown in figure 12.4.

12.2.4 Training

The system needs to be trained so as to adjust the various parameters and detectors of the system. This is done by a training procedure. As a general training methodology, the sequence containing the various instances of particles is fed into the training algorithm based on which the training algorithm is asked to adjust its state.

The initial system may be generated randomly. Here random detectors are produced that contain random entries in all their attributes. Now the sequence of the particles is supplied to the training algorithm. These are assumed to be belonging to the self part while training. The matching of each of the particle in the sequence is done by each of the detectors. In case any of the detectors matches with the string, it is eliminated. Since it is known that the particle belonged to the self part, the detector cannot match it. The detectors matching that string are erroneous and are hence eliminated from the system. In this manner the sequence of particles

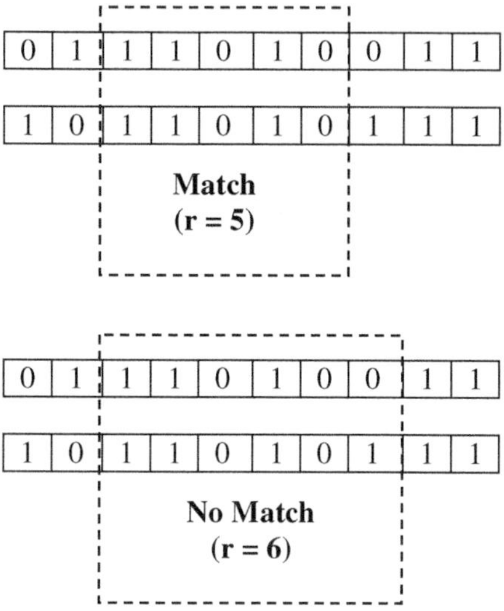

Fig. 12.4 Effect of increasing r on matching

used for training would witness elimination of the wrong detectors that matched with the particles of the self region.

Any detector that survives a toleration period of T is said to be mature. This is the detector that could not be matched by any of the particles and hence it lies at the region in the feature space where the self particles are not found. This region is hence likely to be belonging to the non-self particles and is a potential region for the placement of the detectors that match the non-self particle. This matured detector represents a potential region of non-self particles.

This mature detector continues to find its similarity with the other particles in the sequences. If it matches with any of the particles in the sequence, then an anomaly is said to have occurred. Since this is a new string that was not encountered in the toleration period of the detector, it is possible that it belongs to the non-self region. For this reasons, the detector is not eliminated. Rather it is stated that an anomaly has occurred.

Many times the anomaly may be wrong. This means that the string might actually be a self particle. This cannot be easily detected by the system. This is hence many times done by the human intervention. A human is asked to make the decision regarding the nature of the particle. This can be either self or non-self. Based on the human decision the system decision is made. If the human says that the particle is a non-self particle, the detection of the system is regarded as non-self. However if the human says that it is a self particle, the detector is eliminated. This puts an additional human intervention to the system but ensures an effective decision in case the human can easily make the decision within a small stipulated amount of time. This human intervention given as an external indication to the system for its behavior in the detection of the self and non-self particles is called as *costimulation*. This plays a great role especially in the intrusion detection systems where the security is of a very high value. Costimulation puts an additional constraint on the trained detector to be verified by an external signal before it can be effectively marked as a detector for the detection of the non-self particles (Hoffmeyer and Forrest 2000).

12.2.5 Memory Storage

The various detectors may be stored in permanent associative memory for a fast response of the system. When a detector has detected a non-self particle, it is largely possible that it encounters the same particle again. Hence these particles are stored into the permanent storage for easy access. These particles are especially stored for accurate and fast response to the non-self particles. These can be easily activated as they possess a very small toleration period.

The associative nature of the memory is also important. The amount of memory is always limited in size. Hence we can only store a small amount of information regarding the experience of non-self particles that were encountered previously. It is tried to store a set of diverse detectors for better recognition. The system prefers the newly encountered non-self particles rather than the older non-self particles. In this manner the system keeps forgetting about the past trends and on the contrary learns the new trends that it encounters in the process.

Whenever more than a single detector detects a non-self particle, there is a competition among these for storage in the memory. The detector that has a larger matching strength is given preference.

12.3 Co-evolution

Cooperative Evolution or Co-evolution is another novel evolutionary concept. In this type of evolution an individual's performance is not only rated by its own fitness, but also by its cooperation in the evolution of the other members of the population. Hence in this type of evolution the individuals of the population cooperate with each other and collectively evolve the most optimal solution to the problem. The cooperation not only leads to the overall optimality in the evolution of the solution, it rather also enables the general boost in the fitness value of the individuals in the population. The cooperative manner of working of the various individual results in a better evolution called as the cooperative evolution of the population (Stanley and Miikkulainen 2004; Pedrajas *et al* 2002; Wiegand *et al* 2002; Rosin and Belew 1996; Shi and Krohling 2002; Ficici and Pollack 2000; Potter 1997).

The cooperation plays a big role in the natural evolution process as well. We all develop as per our surroundings in a cooperative environment. We not only develop as a result of our own experiences, but also develop by the experiences of the other people whom we encounter and who cooperate with us. The continuous sharing of ideas, guidance and collective work forms the basis of the natural evolution. This cooperation where an individual helps the other individuals to develop is of ample importance in the natural evolution. We all take help in our weaknesses and help others in their weaknesses. In this manner a collaborative framework results in the development of all individuals in the population. This results in a better evolution as compared to the evolution where only the individuals are concerned about their own development.

The basic fundamental of co-evolution is cooperation. Hence we need to devise a mechanism where an individual would be rewarded for cooperating with the other individuals in the population by means of the fitness value. In all other terms this algorithm would be similar to the conventional Genetic Algorithm (GA). We study one of the examples of this approach in evolution of an Artificial Neural Network (ANN) in chapter 13.

12.3.1 Decomposition

One of the major issues in the implementation of this approach is the decomposition of the problem. It may not always be possible to evolve the entire individual that solves the problem. We instead try to fragment the individual in parts. Each part is evolved in isolation with regular interactions. This changes the entire evolutionary model that we have been studying where the individual represented a complete solution and improved with time. In the co-evolutionary model the individual represents a part solution of the problem. The entire problem may be solved by a collection of fragments which collectively makes the complete solution. The selection of the various parts or fragments to make the complete solution is an intelligent decision and lies largely on the specific contributions of the fragments in solving the problem.

Collaboration is a major factor when we deal with such kinds of problems with solution fragments evolving in isolation to make the complete solution. Now an individual needs to interact with the other individuals of the population. The entire population is sometimes called as the ecosystem. The individual needs constant interactions with the ecosystem to enable it know its fitness and modify its structure to maximize the fitness. The contribution of each individual in the ecosystem must be unique and effective. This can only be possible when the different individuals constantly interact with others within the framework of the ecosystem and assess their contributions against that of the others. However too much of collaboration may also make the problem very computationally expensive. This puts a computational limitation over the problem.

Since the different individuals of the population represent the part solution and the entire solution can only be built by consulting the various part solutions, it is natural that the entire model benefits from cooperation among the different members of the population pool. All the individuals guide in directing the developments of the better characteristics in the other individuals to enhance the overall efficiency and accuracy of the solution that comes up. This is facilitated by a judicious and cooperative credit assignment policy that assigns the fitness cost to the individuals. We study this in the next section.

12.3.2 Fitness Evaluation

The fitness function is used to compute the goodness of any solution in the natural ecosystem. This is unlike the normal fitness function used in GA where the individual is benefited by its own performance. The co-operative strategy of problem solving takes into account the performance of the individual as well as its cooperating nature. The individuals in the population pool that are cooperating and help others in attaining good overall performance of the solution to the problem are given higher fitness. This aids in the collaboration between the individuals and further encourages a collaborative or co-operative behavior among the individuals of the population pool.

In order to compute the fitness of any individual, we must consider its interaction with the other individuals in the population pool. It is computationally very expensive to enable the interaction of every individual with all other members of the population pool. This is because of the large number of individuals can interact in numerous combinations. It is computationally not feasible to facilitate all such interactions to compute the fitness at every generation of the co-evolutionary algorithm. We hence first build a sample population that incorporates various types of populations found in the population pool or the natural ecosystem. This becomes a small sampled population pool that may closely imitate the behavior of the entire pool. The various individuals of the population interact with this pool in order to determine their fitness values.

To compute the fitness of any individual i, we first place it into the sampled population pool. The entire solution is made by the individual i and the rest of the individuals in the sampled population. The interaction makes the complete solution to the problem whose performance may be measured. This gives an idea to us regarding the performance of the individual in making effective solutions. The key factor behind this is that the individual needs to be rewarded for cooperating. Hence if the individual has characteristics that are good and not generally found in the other individuals of the population, the fitness assigned may be high. This is because the individual is contributing key characteristics to the population pool without which the performance of the individual would have been sub-optimal. Its individual performance may not be very high, but it may be assigned a higher fitness for the key contribution of characteristics.

12.3.3 Replenishment

The other major factor that is usually not applied in the GA and makes the co-evolution different from GA is the replenishment of the old unproductive species from the population pool or the natural ecosystem. As the evolution goes on and on, it is natural that the species may become stagnated. They may possess the same characteristics that are well found in the population pool and the entre algorithm may have converged to some point. In such a case there is no appreciable improvement in the fitness value of the individual along with generations. This is largely due to the fact that the characters possessed by the individuals become tuned and do not give further improvement. In such a case the general tendency of this algorithm is to remove the unproductive individuals and replace them with better individuals that may give better characteristics to the problem. It may be possible that they lead to better solution by their contributions.

The replenishment is carried out for all the stagnated individuals when the algorithm does not show an appreciable performance boost for some generations. In such a case the algorithm is said to have been stagnated and further improvement is less likely by continuing the evolution process. In this scenario, we first task is to identify the individuals whose productivity has been lapsed and they are now not producing. These are the individuals whose fitness value has not improved for the past *k* generations. This is given by equation (12.6).

$$f(X_i, t) - f(X_i, t\text{-}k) < C \tag{12.6}$$

Here *k* is a parameter and may be fixed as per the problem and problem design. It denotes the total number of stagnated generations. *C* is a small improvement in fitness value that is negligible as per the problem definition. X_i is any individual. *t* is the current generation.

All such identified unproductive individuals are simply removed and replaced by new random individuals that are inserted into the population pool. It is likely that these random individuals contribute some good characteristics into the population pool that enables a better solution generation. As the generation goes on and on, these may also become unproductive with time. These may further be replenished in due course of time. The entire algorithm keeps executing in this manner. This replenishment strategy used by the co-evolution enables it to stop convergence at some local minima and makes the best use of computation to go towards the global minima.

12.3.4 Other Evolutionary Operators

The other evolutionary operators are the same as compared to the GA. They play mostly the same role in the evolution process of the co-evolutionary algorithm as played by the GA. In fact these operators are common to most of the evolutionary approaches discussed in the book.

These include scaling where the individuals are given some expectation values as per their fitness measures. The other operator is selection that selects the individuals that participate in the evolutionary process. The 2 commonly used genetic operators are mutation and crossover. The mutation adds new characteristics to the individual by deformation from its present value. The amount of deformation largely depends upon a constant known as the mutation rate. Crossover mixes two individuals to generate a new individual. In this manner it tries to combine the best characteristics of the individuals to generate even better individuals with better fitness value. The crossover plays a big role in convergence while the mutation tries to stop premature convergence.

In this manner the entire algorithm runs for every iteration. These iterations are termed as generations due to their analogy with the natural process where the evolution happens along with the generations of the individual. Every lower generation population individuals lead to the development of the higher generation population individuals. The co-evolution algorithm uses the same principles for its evolution along with generations.

12.4 Cultural Algorithms

Cultural Algorithms (CA) are another class of evolutionary algorithms extensively used for various kinds of problems. This algorithm uses the concept of culture to direct the evolution process in the correct direction. The culture has an effect on the direction and manner of evolution of the individuals in the population. It prohibits the evolution in the wrong direction by prohibiting the individuals to generate in the poor fitness regions. It tries to generate feasible solutions that have high fitness values. The culture plays a vital role in this coordination of the entire evolution process. The culture changes along with time as newer and newer individuals are generated and compute their fitness and feasibility. The culture here tries to model the problem in a manner to direct the evolutionary process (Reynolds and Chung 1996; Ho and Tay 2004; de Jong 2006; Fuks 1997).

The natural analogy of this algorithm is with the culture that we live and develop in. The culture always has an effect on the development and evolution of the people. It drives the development of their characteristics in such a way that benefits from the learning and experience of the others. As the individuals develop along with time, they have an effect on the culture they live in as well. The culture hence keeps getting modified along with time. This further has an effect on the evolution of the people within that culture. In this manner the growth, development and evolution of the human race is largely directed and benefited by the underlying culture. The motivation is to use similar concepts to develop oneself and maximize the evolution using the underlying culture and its recordings.

12.4.1 Belief Space

The cultural algorithms hence model two spaces. The first space is the population space where the individuals are generated and evolved. We may use a conventional genetic algorithm with genetic space for this purpose. This part of the algorithm runs like a conventional GA with the operators for selection, reproduction and fitness evaluation.

The other space is known as the belief space. This space models the culture of the algorithm that drives the population. This is a generalization of the entire feature or the genetic space. It tries to most generally determine the feasible, infeasible, good and bad regions in the population space. This is used for directing the evolution process of the individuals in the genetic population that evolve along with time and generations.

A constant coordination and exchange of information is required between the two spaces that enable an optimal evolution of the individuals and an optimal working of the entire algorithm (Coelho and Becerra 2003, 2004). This is done by the various functions that carry forward the algorithm specific task. The manner in which these functions are used may be specific to the problem and the kind of representation that is used. We discuss the major such functions and representations here. These may be summarized by using figure 12.5.

12.4.2 Belief Space Representation

The representation of the spaces and their operation may be done by numerous ways. It may be easily seen that any evolutionary algorithm may be used in the population space with its own operators and working methodology. The GA forms just one method of evolution in the population space. The similar is the case with the belief space where the different types of problems may use their own type of representation that aid in their optimal evolution. A commonly used representation is the *Version Space Representation* that we discuss here. The other representations may be taken and studied from the numerous invariants of the algorithm and their applications.

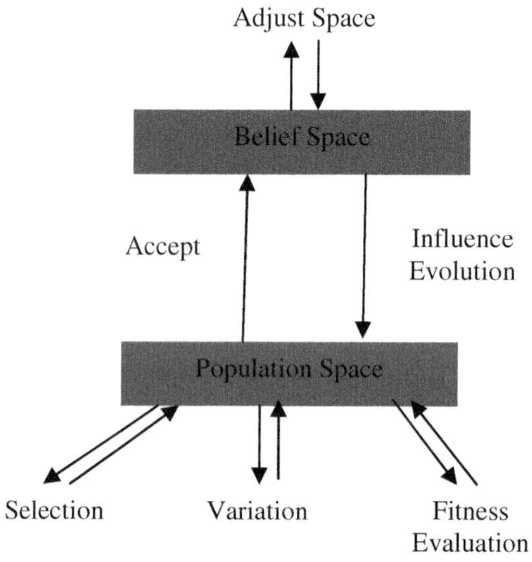

Fig. 12.5 Belief and Population space in Cultural Algorithm

We take a much generalized to specialized representation of the belief space. This may be represented in form of a tree where the upper root nodes represent a much generalized solution and the lower leaf nodes represent a much specific solution or individual (Reynolds 1994). The root is usually a dummy node representing a highly generalized solution of the problem. This can correspond to any possible solution in the genetic space. The leaf is the most specific representation that determines the exact and specific solution. Let the representation of the individual be in the form of string of bits or 0s and 1s. We assume that there are 2 bits that make the complete individual. The representation in the form of a tree may be

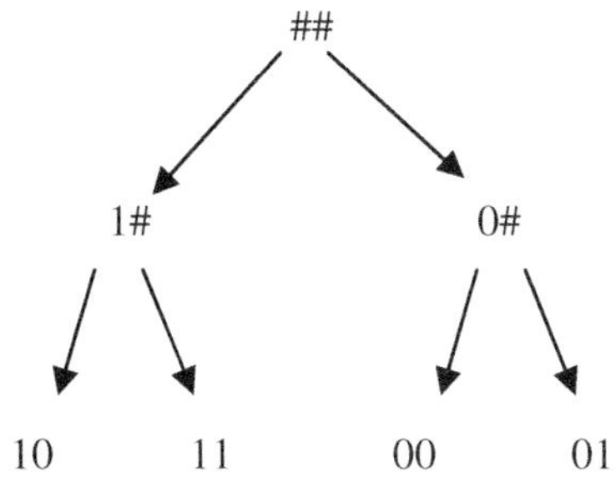

Fig. 12.6 Belief Space Representation

easily given by figure 12.6. Here the various nodes P_is may denote the state. The presence of a # means that it can be substituted by either 0 or 1.

The belief space may hence represent any solution or the class of solution that are generated in the evolutionary process. This space is used for the identification of good and bad regions in the evolutionary process that helps in the directing of the evolution.

12.4.3 Acceptance

The first major operator or function that carries forward the task of coordination of the two spaces is the acceptance. We may easily compute the fitness value of any individual in the population space using the fitness function that is pre-defined as per the problem requirements. This gives us an idea of the goodness of the solution and its feasibility. Based on this value we define another function that is known as the acceptance. This function determines whether the solution is good enough or not. This designates an acceptance value to the solution that affects the algorithm. The acceptance is computed for all the individuals in the population. The fitness function is a computation of the individual fitness. In the belief space this is done by the acceptance.

12.4.4 Influence

The carried out in the evolution process is to use the belief space for the guidance of the evolution process in the population space. This is done by the influence operator. This operator or function influences the evolution by altering the individuals of the population pool as per the recordings or the representations of the belief space. In this step the belief step is used as a directing force for the population space and its individuals.

12.4.5 Adjust

The other operator is adjust that updates the representation of the belief space. At every generation, new individuals are generated and their fitness and acceptance values are computed. This is a valuable data for the belief space that may use to adjust and modify its information about the problem using these recordings. This operation is carried out using the adjust operator. This makes the belief space constantly update itself and use this information for directing the evolution in the population space.

The manner in which the adjustment is carried out is specific to the problem of study. The beliefs in the belief space represent and store the information regarding the goodness of the various types of solutions. This storage is determined by the manner in which the belief space is represented that is specific to the problem of study.

The representation we discussed in the above section may be updated using simple updating principles. In this representation a belief is in the form of a tree segment or a sub-tree that denotes the acceptability of all the nodes within that sub-tree. This is stored in the belief space. This forms a condensed manner of storage of the beliefs where we can store a large amount of information using these generalized representations. The same can be used for easy checking and directing in the influence stage.

Assume that the node *00* represents an acceptable node as measured by the acceptance function. Now suppose we use a Breadth First Search to compute the acceptability of the related nodes. Assume that we find that *10* represent an acceptable node and *01* represents an unacceptable function. Thus we may easily compute that *#0* is the generalized representation of the acceptable nodes or solutions. This representation may further combine with the existing acceptable representation stored in the belief space to make more compact representations representing the acceptability.

12.5 Cellular Automata

The last concept that we study in this chapter is the cellular automata. This is quite a different approach as compared to the various evolutionary algorithms discussed in this chapter so far. We discuss the algorithm to present the completeness of the book in terms of the commonly used systems for problem solving (Maerivoet and deMoor 2005; Toffoli 1984; Gardner 1971; Worsch 1999; Wolfram 1983; Drossel and Schwabl 1994; Floreano and Mattiussi 2008).

The cellular automaton is analogous to the living cells from the natural world. The cells are a very simple structure that performs limited computation in the living beings. In fact the unicellular organisms are the simplest organisms in the natural world. They display only the basic functions of life and are able to live with these. However doing large volumes of task is possible by a collection of these cells or a regular processing of these cells. This enables the cells to perform computation needed for the most complex tasks that are solved by the living beings.

The cellular automata is an algorithm that is by itself a very simple system. This system is however able to solve many problems that require intense modeling and computation. These are hence valuable tools for problem solving. The cellular automata works on the concepts of states. The state denotes the present condition of the cell of the system. The states are always finite in nature that the system can use. The algorithm uses a set of rules called as the transition rules. These rules specify the conditions in which the system makes a transition from one state to the other. This collection of rules may be collectively stored in a table called as the transition table.

We discuss the various concepts associated with the cellular automata one by one. The entire working philosophy in cellular automata involves the use of transition rules to make a transition of the system from one state to the other until the stopping criterion is met. This is given in figure 12.7.

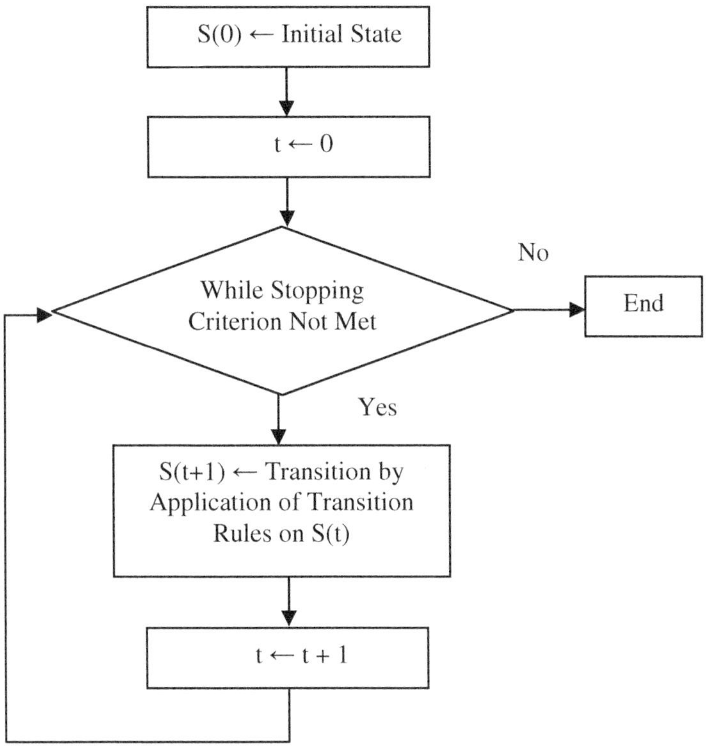

Fig. 12.7 Cellular Automata

12.5.1 States

The basic working of the algorithm involves states where the cell of the system is presently. The state may have practical relevance as per the problem being considered and its modeling for the use of cellular automata. The system always has only a finite number of states in its execution. This number may be small or large and depends upon the problem complexity. The smaller number of states are much easier to work upon and model than a system that has a vast number of states. The states of a system may hence be referred by a finite set of elements given by equation (12.7)

$$S = \{S_0, S_1, S_2, \ldots.S_n\} \tag{12.7}$$

Every cell of the system always has a predefined state at the start (S_0). This is known as the initial state of the system. The system makes transformation from one state to the other with every iteration based on the transition rules.

12.5.2 Neighborhood

The other related concept in cellular automata is the neighborhood. The state is associated with a part of the system or the system cell. There are many cells that are arranged in some manner that make up the complete picture of the system. It is natural that we would not be able to work over this complete picture or frame rules for this bigger picture. While framing rules we only consider a small part of this complete system. This is defined by the neighborhood of the system. Only the cell along with its neighborhood is considered for the application of every transition rule. This is repeated for all the cells along with their neighborhood in a unit iteration of the algorithm. The neighborhood of any cell S_i is denoted by $N(S_i)$. Here $N(.)$ is called as the neighborhood function. The number of elements in the neighborhood of any cell is termed as the size of the neighborhood.

The neighborhood may be conveniently assumed as a window surrounding the cell which is considered while framing or applying the transition rules. This window is wrapped across the entire system for every iteration. The size of the window or neighborhood is of importance. A larger size gives more flexibility in designing the rules where a large part of the system may be considered. The number of rules are however quite large and the system becomes very complex. Most of the problems on the contrary may be solved by a small window size. Here the number of rules as well as their complexity is small. These simple rules are applied over the window multiple times in multiple iterations till the stopping criterion is met, or till the time no change is encountered by the iteration. This solves the problem.

Consider the problem where some transformations have to be made on a binary picture. Here each pixel may be either black or white. Each pixel may hence be stated to possess two states of black and white. The initial state corresponds to the initial form of the picture. Let this be given by figure 12.8.

12.5.3 Transition Rules

The transition of the state of the cell from one state to the other is given by a set of rules known as the transition rules. Each rule specifies the initial state before the rule is applied and the final state after the application of the rule. The initial state is a condition that is necessary for the execution of the rule. The rule is fired only when the condition is true, that is the system is in the same state which serves as the initial state of the system. Upon the firing of the transition rule, the cell changes its state from the current state to the state specified by the rule. A transition table is a collection of all transition rules that are noted as a table.

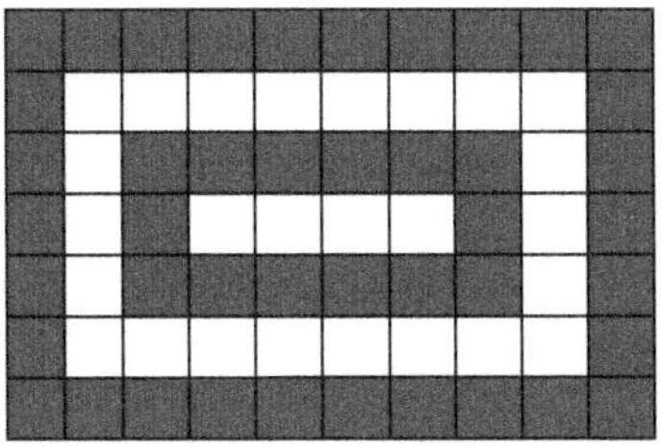

Fig. 12.8 Initial Picture given to Cellular Automata

Transition Function is a function that incarnates all the transition rules and gives the final state of a cell. It takes as its input the current state and gives as its output the next state after the rules have been applied. The transition function is used for every cell to compute its next state for a unit iteration of the algorithm. This is given by equation (12.8).

$$S_i(t+1) = \phi(Sj(t): j \varepsilon N(S_j)) \tag{12.8}$$

Here t is the current iteration number, S_i is the next state at iteration $t+1$, S_j is current state of all the neighbors at iteration t. $\phi(.)$ is the transition function.

The number of transition rules may sometimes get very large. We may hence use wildcard characters for denoting the system rules. These are very helpful for

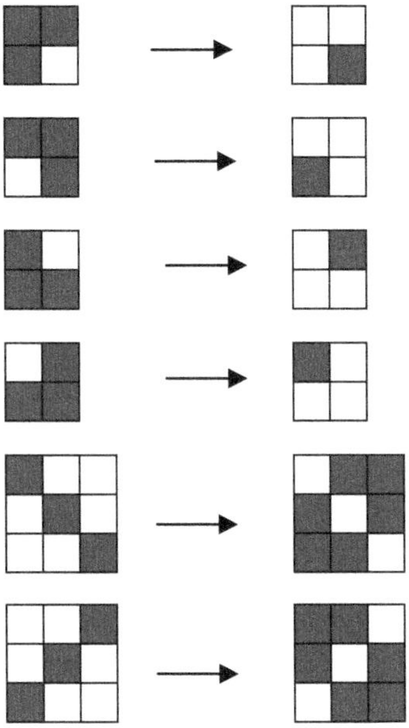

Fig. 12.9 Transition Rules

writing the rules of any system. They enable the writing of the rules as per the transition logic without a lot of manual labor when the system has small logic.

Consider the same problem of picture. Some transition rules for the problem may be given by figure 12.9.

12.5.4 Space Time Diagrams

The entire algorithm operates in iterations. At each iteration some rules are fired that change the state of the cells. In this manner the system continuously changes from one state to the other along with iterations.

We have to specify a criterion based on which the system may stop iterating. This criterion is fixed depending upon the problem and the system demands. We

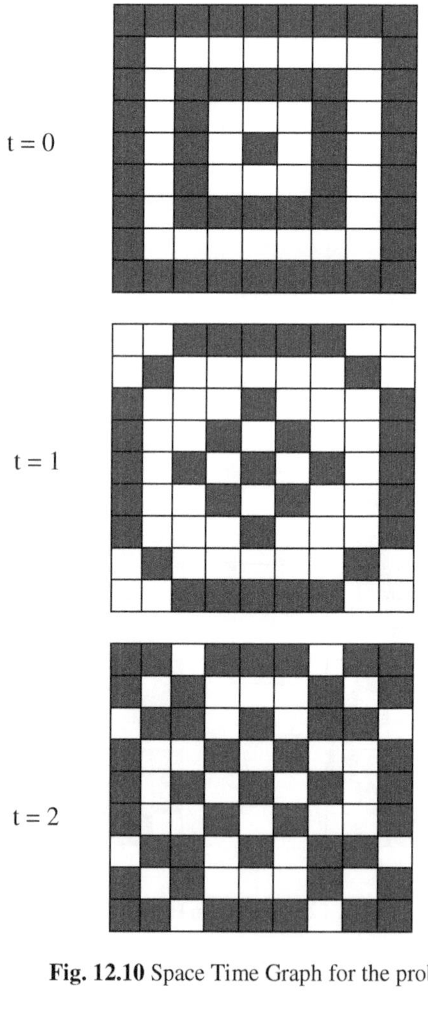

Fig. 12.10 Space Time Graph for the problem

may instead specify a maximum number of iterations as the stopping criterion. In other cases the stopping criterion may be when the iterations do not change any of the cells state or there is a loop in the change of states and the system keeps oscillating between these states.

The condition of the system at the various iterations or times may be plotted continuously as a graph with time as one of the axis. These are called as the space time diagrams. These denote how the condition of the system changes along with time. These may be used to visualize the system along with time. The space time graphs give a magnificent view of how the algorithm proceeds to solve the problem along with time or iterations.

Consider the rules and the initial state represented in the earlier problem. The various states of the system at different time iterations are given in figure 12.10.

12.6 Chapter Summary

This chapter presented a variety of algorithms on the lines of evolutionary algorithm. The first algorithm discussed was differential evolution. This algorithm made use of difference between the individual vectors in the feature space to adapt the mutation process. As a result the evolution yielded good convergence towards global minima. The next algorithm discussed was Artificial Immune System. These systems made use of the self and non-self discrimination fundamental with a matching technique. These systems easily help in intrusion detection and related applications. The other algorithm was cooperative evolution or co-evolution. Here the various individuals of a population aided in the improvement of each other in a cooperative manner. We then discussed the cultural algorithms. Here a new space called as the belief space was modeled and was used for directing the evolution of the population space. At last we studied about cellular automata. Here different transition rules were used for making the system change from one state to the other along with time.

References

[1] Abbass, H.: The Self-Adaptive Pareto Differential Evolution. In: Proc. Con. on Evol. Comput., pp. 831–836 (2002)

[2] Abbass, H.A., Sarker, R., Newton, C.: PDE: A Pareto-frontier Differential Evolution Approach for Multi-objective Optimization Problems. In: Proc. of the 2001 Cong. on Evol. Comput. (2001)

[3] Aickelin, U., Cayzer, S.: The Danger Theory and Its Application to Artificial Immune Systems, pp. 141–148 (2002)

[4] Brest, J., Greiner, S., Boskovic, B., Mernik, M., Zumer, V.: Self-Adapting Control Parameters in Differential Evolution: A Comparative Study on Numerical Benchmark Problems. IEEE Trans Evol. Comput. 10(6), 646–657 (2006)

[5] Coello, C.A., Becerra, R.L.: Evolutionary Multiobjective Optimization using a Cultural Algorithm. In: IEEE Swarm Intell. Sympos., Piscataway, NJ, pp. 6–13 (2003)

[6] Coello, C.A., Becerra, R.L.: Efficient Evolutionary Optimization through the use of a Cultural Algorithm. Engg. Optim. 36(2), 219–236 (2004)

[7] de Castro, L.N., Timmis, J.: An artificial immune network for multimodal function optimization. In: Proc. of the 2002 Cong. on Evol. Comput., vol. 1, pp. 699–704 (2002)
[8] de Jong, K.A.: Evolutionary computation: a unified approach. MIT Press, Cambridge (2006)
[9] Drossel, B., Schwabl, F.: Formation of space-time structure in a forest-fire model. Physica A: Statistical and Theoretical Physics 204(1-4), 212–229 (1994)
[10] Ficici, S.G., Pollack, J.B.: A Game-Theoretic Approach to the Simple Coevolutionary Algorithm. In: Deb, K., Rudolph, G., Lutton, E., Merelo, J.J., Schoenauer, M., Schwefel, H.-P., Yao, X. (eds.) PPSN 2000. LNCS, vol. 1917, pp. 467–476. Springer, Heidelberg (2000)
[11] Floreano, D., Mattiussi, C.: Bio-Inspired Artificial Intelligence: Theories, Methods, and Technologies (2008)
[12] Forrest, S., Javornik, B., Smith, R.E., Perelson, A.S.: Using genetic algorithms to explore pattern recognition in the immune system. Evol. Comput. archive 1(3), 191–211 (1993)
[13] Fuks, H.: Solution of the density classification problem with two cellular automata rules. Physical Review E 55(3), R2081–R2084 (1997)
[14] Gardner, M.: On cellular automata, self-reproduction, the Garden of Eden and the game life. Scientific American 224(2), 112–117 (1971)
[15] Gasper, A., Collard, P.: From GAs to artificial immune systems: improving adaptation in time dependent optimization. In: Proc. of the 1999 Cong. on Evol. Comput., vol. 3, pp. 1859–1866 (1999a)
[16] Ho, N.B., Tay, J.C.: GENACE: An Efficient Cultural Algorithm for Solving the Flexible Job-Shop Problem. In: Proc. Cong. on Evol. Comput., pp. 1759–1766 (2004)
[17] Hofmeyr, S.A., Forrest, S.: Architecture for an Artificial Immune System. Evol. Comput. archive 8(4), 443–473 (2000)
[18] Jakob, V., Rene, T.: A comparative study of differential evolution, particle swarm optimization, and evolutionary algorithms on numerical benchmark problems. In: Proc. of the 2004 Cong. on Evol. Comput. (2004)
[19] Kim, J., Bentley, P.J.: Towards an Artificial Immune System for Network Intrusion Detection: An Investigation of Dynamic Clonal Selection. In: Proc. of the 1999 Cong. on Evol. Comput., vol. 2, pp. 1244–1252 (2002)
[20] Kim, J., Bentley, P.: Negative Selection and Niching by an Artificial Immune System for Network Intrusion Detection. In: Late Breaking Papers at the 1999 Genetic and Evol. Comput. Conf., pp. 149–158 (1999b)
[21] Kim, J., Bentley, P.J.: An Evaluation of Negative Selection in an Artificial Immune System for Network Intrusion Detection. In: Proc. of the Genetic and Evol. Comput. Conf., GECCO, vol. 1, pp. 1330–1337 (2001)
[22] Liu, J., Lampinen, J.: A Fuzzy Adaptive Differential Evolution Algorithm. Soft Comput. 9, 448–462 (2003)
[23] Maerivoet, S., de Moor, B.: Cellular automata models of road traffic. Physics Reports 419(1), 1–64 (2005)
[24] Pedrajas, N.G., Martinez, C.H., Perez, J.M.: Multi-objective cooperative coevolution of artificial neural networks (multi-objective cooperative networks). Neural Netw. archive 15(10), 1259–1278 (2002)
[25] Potter, M.A.: The design and analysis of a computational model of cooperative coevolution. PhD thesis, George Mason University (1997)
[26] Qin, A.K., Suganthan, P.N.: Self-adaptive differential evolution algorithm for numerical optimization. In: 2005 IEEE Cong. on Evol. Comput., vol. 2, pp. 1785–1791 (2005)

[27] Reynolds, R.G.: An Introduction to Cultural Algorithms. In: Proc. of the 3rd Annual Conf. on Evol. Programming, pp. 131–139. World Scientific Publishing, Singapore (1994)

[28] Reynolds, R.G., Chung, C.J.: A Self-adaptive Approach to Representation Shifts in Cultural Algorithms. In: Proc. of IEEE Intl. Conf. on Evol. Comput., Nagoya, Japan, pp. 94–99 (1996)

[29] Rosin, C., Belew, R.: New Methods for Competitive Coevolution. Evol. Comput. 5, 1–29 (1996)

[30] Shi, Y., Krohling, R.A.: Co-evolutionary particle swarm optimization to solve min-max problems. In: Proc. of the Evol. Comput. Conf., pp. 1682–1687 (2002)

[31] Stanley, K.O., Miikkulainen, R.: Competitive Coevolution through Evolutionary Complexification. J. Artif. Intell. Res. 21, 63–100 (2004)

[32] Storn, R., Price, K.: Minimizing the real functions of the ICEC 1996 contest by Differential Evolution. In: IEEE Conf. on Evol. Comput., pp. 842–844 (1996)

[33] Storn, R., Price, K.: Differential Evolution – A Simple and Efficient Heuristic for Global Optimization over Continuous Spaces. J. Global Optim. 11, 341–359 (1997)

[34] Toffoli, T.: Cellular automata as an alternative to (rather than an approximation of) differential equations in modeling physics. Physica D: Nonlinear Phenomena 10(1-2), 117–127 (1984)

[35] Wiegand, R.P., de Jong, K.A., Liles, W.C.: Analyzing cooperative coevolution with evolutionary game theory. In: Proc. of the Evol. Comput. Conf., vol. 2, pp. 1600–1605 (2002)

[36] Wolfram, S.: Statistical mechanics of cellular automata. Reviews of Modern Physics 55(3), 601–644 (1983)

[37] Worsch, T.: Simulation of cellular automata. Future Generation Computer Systems 16(2-3), 157–170 (1999)

List of Figures

Fig. 12.1. Difference vectors in Differential Evolution
Fig. 12.2. Differential Evolution Operation
Fig. 12.3. Feature Space for Artificial Immune System Problem
Fig. 12.4.Effect of increasing r on matching
Fig. 12.5. Belief and Population space in Cultural Algorithm
Fig. 12.6. Belief Space Representation
Fig. 12.7. Cellular Automata
Fig. 12.8. Initial Picture given to Cellular Automata
Fig. 12.9. Transition Rules
Fig. 12.10.Space Time Graph for the problem

Part IV
Hybrid Computing

Chapter 13
Evolutionary Neural and Fuzzy Systems

Abstract. Artificial Neural Networks are valuable tools for problem of machine learning and problem solving. A major problem in the use of neural networks is that the architecture needs to be fixed. Further the training algorithm is needed to fix the various parameters. The training algorithm may many times give a sub-optimal performance by getting trapped in local minima. In this chapter we make use of the optimization powers of the evolutionary algorithms and use them for the construction of the neural network. We would first present the application of evolutionary algorithms in setting the weights and biases of the neural networks. We later make use of evolutionary algorithms for fixing the architecture as well along with the weights and biases. Here we would describe a connectionist approach where the optimal connections evolve. We would also describe an incremental evolution technique in the same problem. We then make use of Grammatical Evolution for evolving the neural networks. At the end we give a similar treatment to the Fuzzy Systems as well.

Artificial Neural Networks (ANNs) are an inspiration from the biological neuron and try to imitate its functioning in artificial intelligent systems. These networks carry the task of machine learning with the help of some training data. The system learns the inputs and sometimes associated with the outputs. These learning are in the form of rules or logic that the system extracts on its own by looking at the training data. The ANNs are hence widely used agents for carrying the task of learning and generalizing the learning to new data at the time of testing.

There are many models of ANNs which are used in various contexts and scenarios. The prominent ones include Multi-Layer Perceptron with Back Propagation Algorithm, Radial Basis Function Networks, Self Organizing Maps, Learning Vector Quantization, etc. Radial Basis Function Networks use neurons at various points in the input space. The Euclidian norm is used for computing the activation of each neuron and accordingly the outputs are produced by a weighted addition of all neuron outputs. The Self Organizing Maps carry out unsupervised learning. In these networks the inputs are clustered and represented in a smaller dimensional space of the Self Organizing Feature Map. Similarly the Learning Vector Quantization carries out a supervised learning for the classification of the inputs. We studied the various ANN concepts as well as the various models in chapter 2.

Another class of systems is the Fuzzy Inference Systems (FIS). These systems carry the task of problem solving or mapping up of the inputs to outputs with the help of rules or logic that is supplied by the human. These systems use rules for

A. Shukla et al.: Towards Hybrid and Adaptive Computing, SCI 307, pp. 277–306.
springerlink.com

finding the output to every given input using the Fuzzy Operators. These systems also find extensive applications in many real life applications, especially when a fuzzy behavior in the manner of working of the system is prominent. We studied the mechanism of solving problems using these systems in chapter 4.

13.1 Issues in Use of Neural and Fuzzy Systems

The beauty in the use of evolution in the neural and the fuzzy systems is only because the conventional systems fail to give optimal performance. One of the major tasks is hence to study the limitations and then to build the systems that can remove the same. We do these under two separate heads. The first assumes the architecture is already known and the task is only to fix the parameter values. The second studies the limitations the fixed architecture.

13.1.1 Fixed Architecture

Each model of ANN had a training algorithm that carried out the training of the ANN. Here the various weights and biases of the ANNs were set. The weights and biases are the memory of the ANN that enables the ANN to correctly compute the output of any applied input. These weights and biases hence need to be judiciously computed so as to maximize the performance of the ANN. Consider the error space of the ANN. Here the performance or error of the system is plotted against the ANN weights and biases. Many of the training algorithms try to adjust the weights and biases so as to give a better performance at the next run. This usually means a slight change of some weight in the direction of the minima. But this strategy may lead the algorithm to get struck at some local minima. Hence this may not be very good strategy to work with. We need robust training algorithms that can find out the most optimal point in the error space or the global minima and escape from any local minima. This is made possible using the optimization power of the Evolutionary Algorithms.

Similar observation may usually be seen in case of the Fuzzy Systems as well. The various membership values, weights of the rules, etc. need to be fixed to the best value for the optimal performance. This is normally a problem as the conventional mechanism of fuzzy inference design has no manner of determining the optimal values. The continuous cycles of trying some parameter values, noting the errors and again trying to set other values may be very time consuming and suboptimal. This task may again be done by using the evolutionary algorithms.

The evolutionary algorithm is used as a general term here that may be implemented by any of the discussed approaches in chapter 3 or chapters 9-12. The performance of the various evolutionary algorithms may be quite different from one another. This is due to the data behavior, complexity and shape of the fitness landscape or in this case the error space, etc.

13.1.2 Variable Architecture

Another major problem associated with the ANNs is their fixed architecture. The architecture of the ANN needs to be clearly defined before the start of the training algorithm. This is kept fixed in the entire procedure of training and testing. The architecture is defined or set by the human or the system designer. The person sees the performance of the system for some combination of values defining the neural architecture. Then these values are altered and re-run until the desired performance is met. As the task is entirely human driven the total number of combinations tried are fairly low. Also apart from the final performance, there is relatively less information that we carry forward from the previous runs of the training algorithm.

The case with the Fuzzy Systems is quite similar. Here also the complete architecture of the Fuzzy Systems needs to be defined at the start itself. This is fixed in the process of working with the Fuzzy Systems. This architecture is human designed and hence may not be optimal in its performance.

These limitations in the neural and fuzzy systems give rise to a new class of algorithms. In case of neural networks these are termed as the Evolutionary Neural Networks (ENN) (Angeline et al 1994; Cho and Shimohara 1998; Happel and Murre 1994; Yao 1993, 1997) and in case of the Fuzzy Systems these are termed as the Evolutionary Fuzzy Inference Systems (EFIS) (Isibucchi *et al* 1995; Rastegar *et al* 2005; Shi *et al* 1995). We discuss these under the paradigm of evolutionary systems and then we study specific cases of these systems. Numerous good models for the same problem can also be found in (Bullinaria 2007; Fieldsend 2005; Funabiki *et al* 1998; Han et al 2004; Jung and Reggia 2006; Kala *et al* 2009c, 2010; Pedrajas *et al* 2003; Rivera *et al* 2007; Ye *et al* 2005)

13.2 Evolutionary Systems

In this section we present a general class of systems called the evolutionary systems. The above discussed limitations exist in most of the general systems that we engineer. Evolution is seen as a solution to these limitations. The first major factor here is the use of optimization powers of the evolutionary algorithms for the task of optimization of the performance of these systems. Consider the neural network as an example. The various weights and biases represent the parameters that can be codified. These may easily serve as the parameters of the evolutionary algorithms which may fix their values for the best performance. In this manner the evolutionary systems perform some kind of learning similar to one that was being done by the training algorithms of the neural networks. This learning optimizes the neural network for performance similar to the training algorithms.

We have further seen that many times the complete architecture of the system needs to be optimized. In chapters 9 to 12 we studied numerous ways in which the evolution could be carried out. The novelty now lies in the manner in which we map the problem into a form specific to the evolutionary algorithm of use. This is the phenotype to genotype mapping. Once this mapping has been done, it becomes reasonably simple mechanism for the evolutionary algorithm to carry forward its conventional cycle of optimization. In case of Genetic Programming we may have to frame the system in tree based architecture. Grammatical Evolution would demand the Grammatical Rules or the language specifications, etc.

One point that we mentioned in each of the discussed evolutionary algorithms was customization. Evolution is a general concept constituting individual, population, reproduction and other terms and operators. Many systems need customized operators that are capable of better evolution of the problems. These are the customized problem specific operators. Many other times we require a complete customization of the entire algorithm. This has a customized representation of the individual, customized evolutionary operators and customized evolution. The hint regarding the choice of operators and their customization is provided by the system itself. We usually know by our understanding of the system the manner in which the lower generation or pool of generation may give a higher generation or the optimal strategy in which the evolution may be carried out. Customized evolution is usually better than the non-customized evolution.

We next study the various specific cases of evolutionary neural and fuzzy systems. Much of the discussion would take specific neural, fuzzy or evolutionary systems as example. It may however be noted that any kind of evolution may be applied to any neural or fuzzy system. The important factor is how to model the problem as per the evolutionary approach of choice.

13.3 Evolving Fixed Architecture Neural Network

In this section we would study the mechanism by which we may tune the weights and biases of a fixed architecture neural network. This plays a role similar to the training algorithm. The specific model that we consider in this section is of the Mulri-Layer Perceptron, though the discussions may be generalized to other neural models as well. Further the Simple Genetic Algorithm would be used in all our discussions in this section (Kala *et al* 2009a, 2009b).

As discussed, the conventional training algorithms of the neural networks are prone to get struck at local minima. This is because of the gradient approaches, or similar approaches of tuning the parameters that they employ. The numerous training algorithms of the ANNs are mainly single agent systems. Here a single copy of the ANN is maintained and various changes are made in it to achieve good optimization. This further adds to the problem of getting struck at the local minima which cannot be eliminated. Further we may only be able to explore a very limited part of the entire complex error space. The most optimal point may be located far from the space where the current search operation and optimizations are taking place. This limits the overall performance of the network. Even the local minima might be fair enough for many problems. The performance at the local minima might be fair enough as per the limited requirements of the system. Some other problems may present a fairly simple error space that may be easily explored. This happens with the presence of single minima in the error space that is simple to hunt.

Most of the other problems may not have such an advantage. Here the problems are very complex and a high degree of performance is required. In such a case a multi-agent approach may be motivating enough to mine the entire complex fitness landscape in search of the global minima. We have seen how this is performed with the help of Genetic Algorithms.

The Genetic Algorithms use a variety of individuals, each of which presents an ANN. Each of the ANN or individual represents a point in the error space with some degree of performance or fitness value. These algorithms use the relative fitness or performance to generate or move the various individuals in this error space to finally enable convergence at the optimal value. We study the various aspects of these algorithms one by one.

13.3.1 Individual Representation

The foremost task in this algorithm is to represent the genetic individual. Here we see how to represent the ANN from its phenotype representation into the genotype representation to be used by the Genetic Algorithm. The representation usually adopted is in the form of a vector of real numbers. These hold the various weights and biases of the various connections and neurons. The architecture of the system is known in prior and fixed. Hence the total number of weights and biases may be easily calculated. This decides the size of the individual of the evolutionary process. The mapping of the various weights and biases may be done in any order from the genotype to the phenotype representation.

Consider the case that the ANN to be trained had a single hidden layer. Let this layer consist of L hidden neurons. Further let the number of input neurons be I and output neurons be O. Now we know that there would be a total of I x L number of weights from the input to the hidden layer. Similarly there would be a total of L x O weights from the hidden to the output layer. These I x L weights form the first part of the individual and the L x O weights form the next part of the individual. The next things to be represented in the individual are the biases for the neurons of the hidden layer as well as the output layer. There would be L number of hidden layer biases that are stored in the individual representation as the third part in form of real numbers. The biases of the output neurons are stored in the fourth or the last part of the individual. These are also real numbers. This makes the complete individual to be used for the optimization purposes consisting of all the weights and biases. The general architecture is given in figure 13.1. It may be noted that we have specifically taken the example of a single hidden layer as most of the commonly used neural networks in real life applications have just a single layer with limited number of neurons.

13.3.2 Genetic Operators

Once the problem has been formulated in the form of a genotype representation, the other task to carry out is its evolution along with generations. This is done with the help of various genetic operators. The initial population may be randomly generated. This is used for the generation of the higher and higher generation of populations.

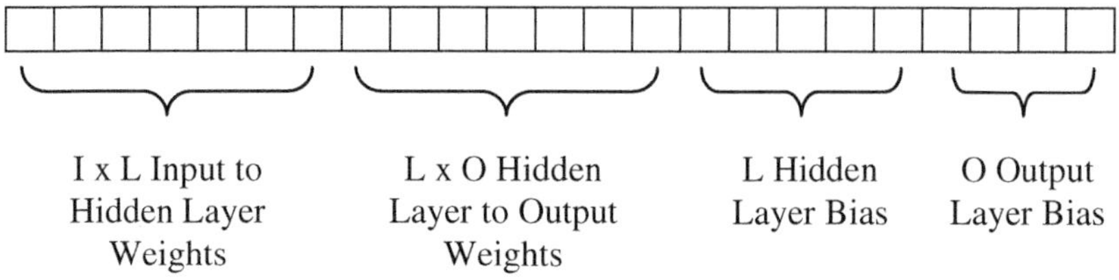

Fig. 13.1 Individual Representation

The evolution of ANN is now a conventional problem and all the standard genetic operators are applied. The common choices in the operator selection includes a rank based scaling where individuals are weighted by their ranks rather than the errors. A uniform stochastic selection technique is commonly used for selection. The other operators include uniform crossover, Gaussian mutation and a small elite count. The task of these operators is to carry forward the evolution as a conventional optimization problem where the various weights and biases are optimized.

13.3.3 Local Search

The discussed method where the ANN was mapped into a genotype representation sounds a fine solution to optimize the ANN with the help of optimization power of the GA. This however assumes some degree of simplicity of the fitness landscape. The actual fitness landscape in case of a highly complex problem that ANN represents may be very complex. There might be a large number of hills and valleys all around with much less trends that can be guessed. The landscape would further be too extensive due to the large size of the individual denoting the weights and biases. Any real life application has limited number of neurons which are not very large. But the total number of weights usually becomes very large because of their multiplicative nature. Each of the weight and bias forms a new axis in the search space or the error space. This drastically adds the problem complexity for the GA search for the minima.

Now imagine a highly complex fitness landscape (or error space as per the ANN methodology). The various GA individuals are the various search agents deputed at various parts of the vast fitness landscape. The GA at every stage tries to guess the structure of the fitness landscape by the relative fitness of the various individuals or search agents. It then tries to pull or push the various individuals in this fitness landscape so as to form a higher level of generation from a lower level. In this manner the algorithm goes to convergence. But in a complex landscape the fitness of an individual may not be a representative of the general fitness of that region. Changing any one of the parameters might largely affect the entire fitness value.

These problems in the evolution of ANN play a very big role in the use of GA. The entire evolutionary process is hence not optimal. It may get trapped in any local minima. Many times the system might behave random in nature. Further as we shall see in the later section, the evaluation of the fitness function is quite a time consuming operation in case of the ANN. This makes the entire evolution process very slow and time consuming. This further adds to the problems in the evolution of the ANNs.

As a result the attempt is to use a local search strategy for every individual to explore the most optimal points in the near surroundings from its present location in the fitness landscape. Hence at the time of fitness assignment, every individual represents a point in the fitness landscape that is better than the general points in the surroundings. This makes the evolution much better as the evolution has a general idea of the fitness of a region. Further the problem is reduced in complexity now. The task is only to find the better region in the fitness landscape. The task of finding the exact minima would be done by the local search strategy.

The training algorithm of the ANN is the local search strategy adopted for finding the minima. In case of the MLP this is the Back Propagation Algorithm (BPA). The training algorithm specified to the GA serves as the training algorithm for this training. It fixes the values of the various weights and biases, or in other words moves the genetic individual representing the ANN to a better position in the vicinity. The learning rate and momentum may normally be kept to lower values, since now the training algorithm is only supposed to do a local search.

13.3.4 Fitness Evaluations

The last task in the evolution of the ANN or its training by GA is to formulate the fitness function. The fitness function assigns a score to every genetic individual representing the ANN. The fitness is used by the GA to assess the usefulness of the various individuals as per their current positions in the fitness landscape.

The Genetic individual is in form of a genotype representation where it holds a set of real numbers. The first task to be performed is to convert it into its phenotype representation where it represents a neural network. The ANN may undergo some cycles or epochs of training by the training algorithm if a local search strategy is used. The training data of the algorithm is used for this purpose. The ANN hence obtained is tested against the training data itself. The performance given by the ANN is the training data is a measure of its fitness value. This is returned by the fitness function.

The complete methodology of the GA and ANN used for the evolution of a fixed architecture ANN or the training of the ANN is given in figure 13.2.

13.4 Connectionist Approach in Neural Network Evolution

The training of the fixed architecture ANN is only a part of the total job that fixes the values of the various weights and biases. We must be able to evolve the complete ANN along with its architecture. This is a larger part of the problem since the complete evolution is a highly difficult task considering the large dimensionality of the search space, in case we are able to develop a method to map the problem into genotype representation by some means. In this section we again consider the MLP model of the ANN and GA as the evolutionary approach (Kala *et al* 2009a, 2009b).

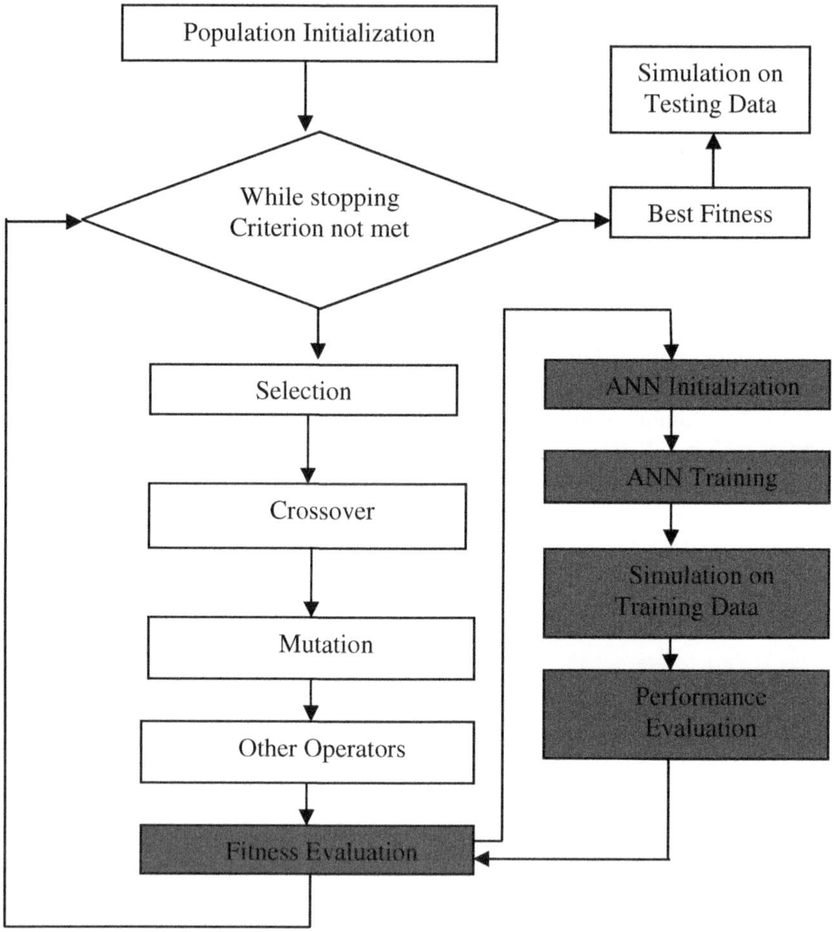

Fig. 13.2 Fixed Architecture Evolutionary Neural Network

13.4.1 Problems with Large Number of Weights

One of the major problems in ANNs is to decide the number of hidden neurons. This is a major factor that decides the learning and generalizing capability of the ANN. A network having too many neurons may be able to learn, but the results may not be generalized to the unknown inputs. This makes the performance of the system fairly low for the testing inputs. If the neurons are too less, the learning may not take place at all, though the small amount of learnt rules may be general enough to be extended to the testing data set.

Consider the ANN curve fitting approach to solve problems. Here the ANN tries to imitate a curve. The various decisions or the system outputs are based on this curve that the ANN figures out. The weights of the ANN decide the shape of the curve. Each weight has some implication to the ANN shape in some or the other way. Very small number of neurons or associated weights is able to imitate reasonably simple curves. These usually have very small number of turns in the input space. As the number of neurons and the associated number of weights increase, the curve starts getting more complex. This has more number of turns. This gives it a very sensitive nature, where there is a large change in the output value for even a small change in the input value. This is the reason why large number of neurons gives a good performance in the training input and a poor performance in the testing input. The curve is made sensitive enough to fit all the training inputs, no matter how many sharp turns are needed in the input landscape for the fitting. The testing inputs may naturally not follow these trends.

The other problem associated with the large number of weights is the time and data requirements. A large number of weights mean a larger computation. This increases time in both the training as well as the testing phases. Further the larger number of weights may require more training samples for the tuning. This again adds to the training time.

13.4.2 Limited Connection Neural Network

We need to have as less neurons as possible. The problem with the ANNs is that the addition of even single neurons adds a lot of weights. The problem or the data in hand might not be trained in a lower number of neurons. Addition of a single neuron might add a lot of weights and the associated problems. This inhibits the optimal performance of the ANN.

These problems are largely due to the fact that we assume a fully connected structure of the ANN. Here every neuron needs to be connected to the other neurons. We hence try to eliminate this assumption by making a limited connection neural network. Here the connection may or may not exist between the two neurons of the ANNs. This adds a lot of flexibility as well as complexity control in the entire design perspective of the neural network. We would see next how we use the evolutionary powers of the ANN to evolve such a network.

Another aspect of this neural network is that the connections are expensive in terms of the performance. Larger connections are undesirable and must be avoided. Hence performance is not the only criterion for the evaluation of these networks, the networks with smaller number of connections may be preferable at some performance cost. Since the performance is only measured by the training data, the larger networks are likely to give better performances. The performances however would not be good for the testing inputs from the testing data sets.

Another major factor now is the number of neurons. In such an approach there might be neurons not connected to any input and may hence be regarded to be completely absent from the network, giving a picture of a much smaller and compact network. The number of neurons in this network hence does not carry a much larger significance, since the connections are the driving factor of the network rather than the neurons. The number of neurons is a much passive factor.

13.4.3 Individual Representation

The first step we study in this evolution of the neural network is the individual representation. Here we look at the manner in which a neural network may be cast into a genotype representation for the evolution to take place. In this system, the individual is supposed to store the various connections along with the weights and biases. This may be trivially done using any ordering between the various connections, weights and biases.

Again let us consider a neural network with a single hidden layer and I input neurons, L hidden neurons which are the maximum neurons a hidden layer may have and O output neurons. Let us take the entire genotype representation to be stored in the form of a vector of real values. In all there can be a maximum of $I \times L$ connections between the input to hidden layers and $L \times O$ number of connections between the hidden and the output layer. In this approach the first part of the genotype representation consists of this information. Here the first $I \times L$ numbers denote the presence or absence of a connection between the neurons. This gene can only take 0 or 1 as the allowed values. A 0 means absence of the connection and a 1 means presence of the connection. The same is true for the next $L \times O$ numbers in the vector.

The second part of the gene stores the weights. This consists of $I \times L$ and $L \times O$ numbers that denote the weights between the input and hidden layer as well as the hidden to output layer. A weight is present even if a connection is physically not present. This is simply ignored in the working of the algorithm. This storage fixes the size of the individual to a constant value.

The third and the last part of the individual consist of the information about the biases. This consists of L numbers in sequence that denote the biases of the hidden layer and O number in sequences that denote the biases of the output layer.

In this manner the entire network may be represented using this approach. The general structure of the individual is shown in figure 13.3.

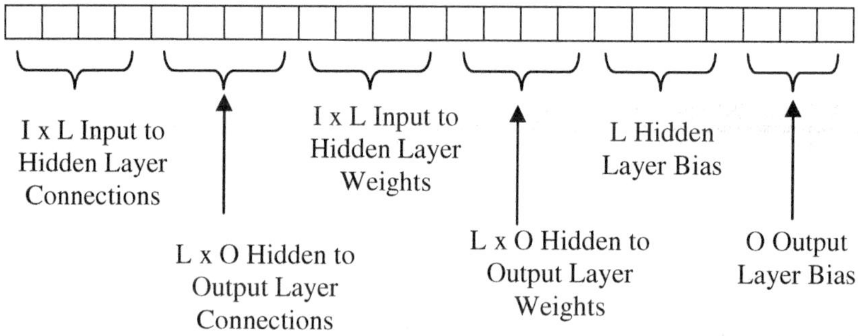

Fig. 13.3 Individual Representation

13.4.4 Genetic Operators

The task of the GA is to optimize the various connections, weights and biases for optimal performance. The conventional GA operators that we studied in section 13.3.2 may be used for this purpose.

An important factor here is the connections. The numerals in the genetic individual representing the connections may be either 0 or 1. This limitation applies in the various genetic operators that we carry out mainly mutation and crossover. The mutation and crossover may be preferably done in this part of the individual as if it was a bit representation and not a real number representation. Alternatively some other mechanism may be determined to limit the numbers in this part to either 0 or 1.

13.4.5 Fitness Evaluation

The last part of discussion for these systems is the fitness evaluation. Here we have to first make an ANN as per the individual representation. The specifications of the neural network including the weights, connections and biases are taken from the individual representation. Consider the above representation with a single hidden layer; I input neurons, maximum of L hidden neurons and O output neurons. We first make a fully connected neural network. The weights are then assigned. In every weight we first check for its existence. If it is existence with an entry of 1 against its position in the individual, the weight is assigned. In case it is un-existing, it may simply be assigned a weight of 0. In this way the complete neural network is ready. The tuning of the weights of the existing connections may be done using a training algorithm or the BPA. This serves as a local search strategy. Then the performance over the training data set may be taken.

The other factor of these networks is that we need to prefer the smaller networks. This is because they are likely to give better performance in the testing data even though they might not perform well over the testing data. This is done by assigning a penalty for each connection. The net performance of the ANN is the total accuracy as per the database with some penalty for every connection. This is given by equation (13.1)

$$F(I) = P(I) - \alpha\, C(I) \tag{13.1}$$

Here I is the individual, $F(I)$ is the fitness, $P(I)$ is the performance and C(I) is the total number of connections. α is the penalty constant.

It is natural from the equation that the connections are acceptable if the performance boost that they give to the network exceeds the additional penalty with the more number of connections. This prohibits very large networks. This is because after a considerable number of connections, the total penalty would exceed the increase in performance. Consider any ANN with an optimal architecture as per the design methodology and principles. It is natural that the addition of some connections might not bring a very vast change in the network performance over the training data. These however may encounter a much larger penalty caused by the additional connections they try to introduce. In the lower phases the addition in performance would be larger than the associated penalty.

Deciding the correct connection penalty α is an important task. This depends upon the problem, data and the largest size of the ANN possible. Further this depends upon the learning accuracy and generalization needed in the system. The ultimate goal is to achieve a high recognition score in the testing data set that makes the final performance of the system.

13.5 Incremental Evolution of Neural Network

The next topic of our interest is the incremental evolution of the neural network. The basic aim in the evolution of the ANN is to be able to find out the complete architecture as well as the associated weights and biases automatically using some evolutionary technique. We hence need to build or engineer the ANN completely for the most optimal performance.

Many things we see around are developed incrementally. Life started with very simple living creatures. Along with time more complex forms generated. These were formed over the learning and adaptation of the simple life forms. It is hence not sudden that the humans got made in the passage of time which represents a very sophisticated life form with large intelligence.

Incremental evolution of the neural networks follows the similar norms. Here also we first try to evolve or form the simple ANNs. As generations pass, we allow the generation of more complex networks for solving the same problem. The allowable complexity of the problem keeps increasing along with the generations. In this manner the ANN keeps taking more complex forms.

The smaller network structures are indeed better than the larger ones. In this evolution we follow the same penalty based evolution where the larger networks are penalized in their fitness values. We know that the entire evolutionary process follows the Darwin's concepts of survival of the fittest. Using the same concept we may easily conclude that the smaller networks would get promoted from one generation to the other and the larger ones would be penalized. This means the survival of optimal architecture network generation after generation.

We study the various concepts and evolutionary technique of this network in the next few sub-sections.

13.5.1 Individual Representation

The first major factor in the use of these networks is the individual representation. Here the task is mapping up of a neural network in a genotype representation. This consists of the entire specification of the ANN in terms of the number of hidden layers, number of neurons in each of the layer, weights between various pairs of neurons, and biases. As the number of neurons is variable, the individual size may also vary. This depends upon the architecture that the individual represents. The maximum number of hidden layers and neurons per hidden layer may sometimes be fixed for implementation reasons.

Consider that the ANN is limited to a single hidden layer. This is a fair assumption since many real life applications are easily able to solve the problem with the use of a single hidden layer. The number of neurons in the hidden as well as the output layer is fixed. Let these be *I* and *O* respectively. We now have a variable number of weights and biases that we need to store. Let any individual possess *L* number of hidden neurons. The representation of this individual is the same that was considered in section 13.4.3 with the absence of the connection information. This is a variable size individual. The number *L* may also be stored in the individual for implementation reasons. At any instant of time or generation we need to ensure that this number of neurons must be less than the maximum allowable number $L_{max}(t)$ as per the current generation *t*. This is given by equation (13.2)

$$L \leq L_{max}(t) \tag{13.2}$$

13.5.2 Increase in Complexity

We have limited the total number of neurons possible in the hidden layer to $L_{max}(t)$. This goes with our basic motivation of incremental evolution where the ANN keeps increasing in complexity along with time and generations. Increasing $L_{max}(t)$ would allow the generation of larger ANNs. We hence increase this factor along with time. This factor grows as the generations increase. A simple method to increase this factor would be as given in equation (13.3).

$$L_{max}(t+1) = L_{max}(t) + \Delta L \tag{13.3}$$

Here *t* is the generation and ΔL is a unit increase in the maximum allowable neurons per unit generation. In case the maximum value of the neurons in the hidden layer is fixed as a result of problem or computational assessment, this value may be given by equation (13.4).

$$\Delta L = \frac{L_{max}(G) - L_{max}(0)}{G} t \tag{13.4}$$

Here $L_{max}(G)$ is the maximum possible number of neurons in the ANN, assuming the ANN initially starts with no neurons i.e. $L(0)=0$. *G* is the maximum number of generations for which the algorithm runs.

This is an interesting application where the allowable complexity increases along with generations. Similar to the manner in which the complexity increases with generations, we can also change the other evolutionary parameters. Let us take the number of individuals as an example. We may sometimes need that the number of individuals increases along with time. This may be to enable a better exploration of the search operation. We know that the individuals with smaller network architecture may be having fine fitness values. Hence more individuals may be used for more exploration when some individuals are already deputed at the lower levels or lower number of neurons. The number of individuals may be changed in accordance with the equations (13.3) and (13.4). This gives us equations (13.5) and (13.6).

$$N(t+1) = N(t) + \Delta N \tag{13.5}$$

$$\Delta N = \frac{N(G) - N(0)}{G} t \tag{13.6}$$

Here $N(t)$ represents the number of individuals at generation t.

13.5.3 Crossover

The other operation of concern is crossover. This operator tries to mix two separate ANNs and generate a new ANN. Here we place a restriction that only the ANNs of same architecture would be considered for the crossover. In this scenario the evolution process must select the ANNs such that both of them have same number of neurons in their hidden layer. This requires a change in the selection mechanism that supplied the individual for the crossover, giving a preference to the fitter individuals. The crossover in between these ANNs is simple. Since the architecture is same, the crossover is carried between the corresponding weights and biases by taking their weighted mean. In this a fitter individual has more contribution that a weaker individual in the making of the child ANN. The total number of individuals contributed by this operator to the next generation depends upon the crossover rate. We assume that a total of $c\%$ individuals of the total number of individuals in the higher level are contributed using this operator.

13.5.4 Mutation

The other major operator of use in this technique is mutation. This tries to modify the various attributes or parameters of the ANN for better performance. The mutation is responsible for the exploratory nature of the ANN. Due to this the ANN explores the surrounding areas in the error space in search for the optima. The magnitude of exploration depends upon the mutation constant which defines the maximum amount of change that any parameter may have. Mutation may also be carried out using a Gaussian mutation technique. Using this technique, the chances of small change in values is very likely and large change is very unlikely. This is controlled by the scale and the spread of the Gaussian curve denoting the probability distribution of the changes in parameter magnitude. Mutation is applied independently to the selected ANNs or individuals in a conventional manner. This operator only changes the weights and biases of the ANN and not the architecture. We assume that a total of $m\%$ individuals of the total number of individuals in the higher level are contributed using this operator.

13.5.5 Grow

The incremental evolution may have a series of user defined operators that enable a better evolution. All these are specific to the problem and designed by the designer understanding the evolution demands. A good operator must drive the convergence to the correct direction as well as guide the evolutionary process for generating better individuals. One such operator that is presented in the incremental evolution of the ANN is grow.

This operator simply adds a new neuron in the hidden layer of the ANN. In this manner it makes the lower order ANNs promote to the higher order ANNs. It is evident that after spending a lot of time in the lower order, an ANN must have done a lot of optimization. Further optimizations may not result in a good performance. Hence we may consider adding a new neuron to enable it check its capability at a higher level. The added neuron adds new weights and biases. We assume any random values to these while generation of this new individual from the lower value. We assume a total of *r%* individuals of the total number of individuals in the higher level are contributed using this operator.

It may be noted that each of the discussed operator results in destruction of the original individual that had some performance and fitness value. This may mean a destruction of a good individual. However as we shall see later, the top few individuals are passed straight away to the next generation by elite operator. This saves them from genetic operation destruction. Further any individual has chances of getting selected for all of the genetic operators. In this manner the fitter individuals may have multiple children from them produced by same or multiple genetic operators.

13.5.6 Elite

The next operator we design for the evolution process is elite. It is very likely in the genetic evolution process that the original good individuals get destroyed. This is because of the disruptive effects of the operators. A good individual may get selected for an operation. Some changes are made in its structure or other parameters hoping to enhance its fitness value. However this may lead to an even lower fitness value. The use of GA is to ultimately return one individual with the best fitness value or the best performance. Hence we may not afford to lose best individuals in this manner. Elite transfers best few individuals from a lower generation to a higher generation directly. This is especially important in an incremental approach adopted in this evolution. The higher generations may result in more complex networks. The simpler ones may however be more optimal. It is hence necessary to transfer the better networks from one generation to other to enable it survive till the end. We assume a total of *e%* individuals of the total number of individuals in the higher level are contributed using this operator.

13.5.7 Add

The other genetic operator we study is add. This operator generates a new random individual of the largest possible size and adds it to the population pool of the next generation. We know that the maximum number of neurons at any generation t can be $L_{max}(t)$. This operator generates a new individual with the required number of input and output neurons and $L_{max}(t)$ number of hidden neurons. This network is given random number of weights and the biases. These may be tuned in the later phases or generations of the evolutionary process.

There is a minimum number of neurons that an ANN must have for good performance. At the initial generation this operator carries out this task of prompting the larger networks since the lower ones may not be large enough for solving the problem. At the higher generations, a large number of individuals may be occupied at smaller sizes networks as they give good performance. It is necessary to try out larger networks as well. This is because the possibility of a good boost in performance by addition of a single neuron may not be ruled out. This requires few genetic individuals. This operation is carried out by this operator.

We assume a total of $a\%$ individuals of the total number of individuals in the higher level are contributed using this operator.

13.5.8 Fitness Evaluation

The fitness evaluation technique of this algorithm is same as that discussed in section 13.4.5. Here as per the individual specifications a new network is created. This ANN is trained for a few cycles by Back Propagation Algorithm. This forms the implementation of the local search strategy in these algorithms. Later we see the performance of the network on the training data. The larger networks with more neurons are penalized by some performance penalty. The total performance of the network is measured by using equation (13.7).

$$F(I) = P(I) - \alpha\, L(I) \tag{13.7}$$

Here I is the individual, $F(I)$ is the fitness, $P(I)$ is the performance and $L(I)$ is the total number of neurons in the hidden layer. α is the penalty constant.

13.6 Grammatical Evolution of Neural Network

The next model we consider in this class of evolutionary algorithms is the grammatical evolution of the neural network. In this section we would evolve a Multi-Layer Perceptron solely by the evolutionary technique of Grammatical Evolution (GE) (O' Neil 2001, 2003; Ryan 1998).

We studied about the GE in chapter 10. This technique uses a Grammar for the conversion of a phenotype representation into its genotype representation. The individual is represented as a set of numerals. These numerals represent the instruction set that helps aids in the generation of the problem specific phenotype solution based on the grammar. The generation of the solution starts with the initial or the starting symbol that is defined by the grammar. Grammar contains the terminating and non-terminating symbols. At every step we read the next instruction from the sequence of numerals in the genotype representation of the individual. The first non-terminating symbol is selected from the solution in its current state. We find the number of ways in which the non-terminating symbol may be replaced by a look up at the grammar that is defined and fixed. The read numeral decides the replacement. This goes on and on until there are no non-terminating symbols in the solution or until there are no numerals left in the genotype representation and the genotype representation is not allowed to wrap around more. The leftover non-terminating symbols may be replaced by default terminating symbols.

The evolution of the ANN by GE involves the specification of a grammar that would enable the conversion from a genotype representation into the phenotype representation. In this manner the problem would get converted into a standard problem of optimization with the help of Genetic Algorithm. Further this would make a linear representation of the ANN. The linear representations are usually better and easier to handle due to the ease in the application of the various genetic operators.

In this approach we would first define the Grammar and then discuss the evolution of the ANN by this mechanism.

13.6.1 Grammar

The Grammar defines the base of the entire procedure of GE. The grammar is referred for the conversion of the individual from the genotype to the phenotype representation which is an existent neural network to be used for performance evaluations. A good grammar is the basis of the GE. The grammar is a natural depiction of the generation rules of the solution of the problem. This depends upon the problem being considered. Further we have studied in chapter 10 how the Backus Naur Form (BNF) is used for the specification of the rules that form the most important part of the grammar. The grammar to be used in case of the evolution of the MLP with the help of GE is discussed next. Here again we assume that the MLP to be evolved has a single hidden layer. We further assume for simplicity that the system has only one output. In other words the output layer is assumed to be of a single neuron. The systems with multiple outputs may be solved by a generalized version of this system. The activation functions are assumed to be constant and same for all the neurons. For implementation purposes this may usually be taken as *sigmoid* function. The network is supposed to compute the number of neurons in this layer along with the corresponding weights and biases. The number of neurons in the input and the output layer are known as per the problem definition being solved with the ANN.

The Grammar may be given by equation (13.8).

$$G = \{T, N, P, S\} \tag{13.8}$$

Here G is the grammar, T is a set of terminating symbols, N is the set of non-terminating symbols, P is a set of production rules and S is the start symbol.

The terminating symbols are given by equation (13.9)

$$T = \{0, 1, 2, 3, 4, 5, 6, 7, 8, 9, I_1, I_2, I_3, \ldots, I_n\} \tag{13.9}$$

Here 0-9 are digits. I_1, I_2, I_3, … I_n are the inputs to the system. n is the total number of inputs of this system.

The non-terminating symbols are given by equation (13.10)

N = {ANN, neuron, weight, sum , bias, input_list, number, digit_list, digit}(13.10)

The start symbol is the ANN. This is formally given by equation (13.11)

$$S = ANN \tag{13.11}$$

The set of production rules are framed as per the concept and developmental methodology of the ANNs. These are given by equation (13.12).

<ANN> ::= <neuron> (13.12)
| <neuron><ANN>

<neuron> ::= <weight> *x* F(<sum> + <bias>)

<sum> ::= <weight> *x* <input_list>
| <sum> + <sum>

<input_list> ::= $I_1 \mid I_2 \mid I_3 \mid \ldots \mid I_n$

<weight> ::= <number>

<number> ::= <digit_list>.<digit_list>
| -<digit_list>.<digit_list>

<digit_list> ::= <digit >
| <digit ><digit_list>

<digit> ::= 0 | 1 | 2 | 3 | 4 | 5 | 6 | 7 | 8 | 9

Here *F(x)* is the activation function that is constant for all neurons in the ANN.

13.6.2 Genetic Operators

The individual in this evolution process is simply a sequence of integers. Since the generation of the actual ANN takes place on the basis of the underling Grammar, it may not be easy to comprehend what the integers mean in the neural context. But these integers represent the grammatical instructions and are hence taken similar to the conventional parameters that the conventional GA individual holds in the genes. The optimal genetic representation is the one that generates the optimal ANN when decoded using the grammar. Similar would be the presence of local and global optima. The initial population is generated by random integers filling the entire individual. The maximum size of the individual or this integer sequence is always fixed.

After this representation of the individual as a set of integers, all conventional genetic operators may be used. The same default style of the application of mutation, crossover, elite and other parameter would lead to optimal solutions using this approach. In this manner a linear representation of the ANN helps in easy implementation and optimization.

13.6.3 Fitness Evaluation

The last part of discussion as per our evolutionary implementation methodology is the fitness evaluation of the genetic individual that guides the entire evolutionary process. This is done in a number of steps. The first step is the decoding of the genetic individual using the grammatical rules. This generates the ANN. This may then be tested for fitness evaluation using the methodology we have been discussing in the various models. The basic objective of the fitness evaluation is same in these systems as compared to the previous systems.

13.7 Evolution of Fixed Architecture Fuzzy System

We now leave the neural network domain and enter into the domain of Fuzzy Inference Systems (FIS). The fuzzy systems use the concepts of fuzzy arithmetic and fuzzy sets for the problem solving. The fuzzy sets are the sets where every member of the set has a membership value. This value denotes the degree to which the element is a member of the set. Hence there is some degree to which it belongs to the set and some degree to which it does not belong to the set. This is unlike the crisp set theory where the element may be completely a member, or may not be a member at all. FIS are driven by a set of fuzzy rules. This makes it easy to model and solve problems by specifying the underlying rules which must be known in prior.

The fuzzy systems have a large amount of parameters that need to be optimized. This optimization is facilitated when we have some data whose inputs and outputs are known. The FIS may easily be trained in a supervised mode with the use of this training data using GA. The role played by the training algorithm here as well is similar to that played in the previous methods with ANNs. The training algorithm tunes the various parameters such that the resulting system behaves better especially in the data that it has been trained with. If the architecture of the FIS is generalized enough, these parameters would yield a good performance in the other cases as well.

This problem assumes that the architecture of the FIS is fixed and has already been determined. Hence the rules are precisely known and have been formulated and given to the system. Further for each of the inputs and outputs, the number of membership functions (MFs) is known. Here we further assume that the MF is of a fixed type that is mentioned in advance for each of the MF. Usually this may be taken as Gaussian in nature. Every MF corresponding to both inputs and outputs has some parameters. The number and role of parameters depends upon the function of use. The Gaussian MF may have two parameters denoting its location in the entire input as well as the spread. Similarly the other curves may have different number of parameters along with their meanings and expected value ranges. Since we have assumed a fixed architecture FIS, the number and type of MFs is fixed and cannot change in the evolutionary process.

The evolutionary algorithm is supposed to find out an optimal set of parameters for all the MF parameters, such that the system has a high overall performance.

This is similar to tuning the weights and biases of a fixed architecture ANN. We assume that the architecture has been carefully formulated such that the system would behave well not only for the training data sets, but also for all the other data cases that the system may face in future.

We studied the manner in which the FIS are designed, developed and tested in chapter 4. There is no natural manner by which these algorithms may be trained. This forms a major limitation in the optimal performance of the FIS in these problems. The system needs to be again and again designed, simulated, tested and altered by looking at the errors. This puts up a lot of manual work until the system gives a satisfactory performance over the known cases. This is especially poor if some training data is available. The evolutionary fuzzy inference system hence gives a big tool for the design and development of fuzzy systems that give a high performance as these are automatically trained.

The evolution of the fixed architecture fuzzy inference system forms a part of the major problem where we evolve the complete FIS. The architecture of the FIS may be sub-optimal as it is designed manually under the human limitations. Hence the automation of this process, where an evolutionary techniques evolves the fuzzy system with time and generations may be optimal in nature.

In this section we present a mechanism to fix the MF parameters of the FIS by using Genetic Algorithm. The major concepts in the generation are discussed one by one. The solution discussed is a conventional manner of the application of the GA.

13.7.1 Individual Representation

The first major step is the proper representation of the individual. Here we need to map a fixed architecture FIS from its phenotype representation to a vector of numerals representation or the genotype representation. Since the architecture is fixed, the number of MFs, their types, etc. is all fixed. The only thing to be optimized and represented in the genetic individual is the parameters of the MFs.

Let us suppose that the inputs to the system are $I_1, I_2, I_3 \ldots I_n$. Let us further assume that any input p is made up of a number of MFs. Let these MFs be MI^1_p, MI^2_p, MF^3_p ..... MI^{ip}_p. Further let the system have outputs $O_1, O_2, O_3 \ldots O_m$. Let us represent the MFs corresponding to any output q by MO^1_q, MO^2_q, MO^3_q ..... MO^{oq}_q. We know that each MF would be having a set of parameters. The number of parameters depends upon the choice of MF used for representation. In order to make the GA individual we first mention the parameters of the input MFs and then the output MFs. We iterate the inputs (or outputs) in order. At each input (or output) we iterate through the MFs in order. Each MF occupies some genes in the genotype representation. The parameter is given a gene. Since the total number of inputs, outputs and corresponding MFs are constant, the total length of the genetic individual is fixed.

13.7.2 Genetic Operators

The next major task in the optimization of the FIS by GA is to use the genetic operators for evolution. These operators apply to every population to generate a higher level population. The higher populations are fitter and solve the problem better than the lower populations. In this algorithm we have already casted the individual into a numeric vector. All conventional GA operators may hence be used for the optimization. The common choices include rank based scaling, uniform stochastic selection crossover, Gaussian mutation and elite.

13.7.3 Fitness Evaluation

The other important task in the use of GA for FIS optimization is to frame a fitness function. We know that the architecture of the FIS is fixed. The various parameters of the FIS MFs may be read from the individual whose fitness is to be

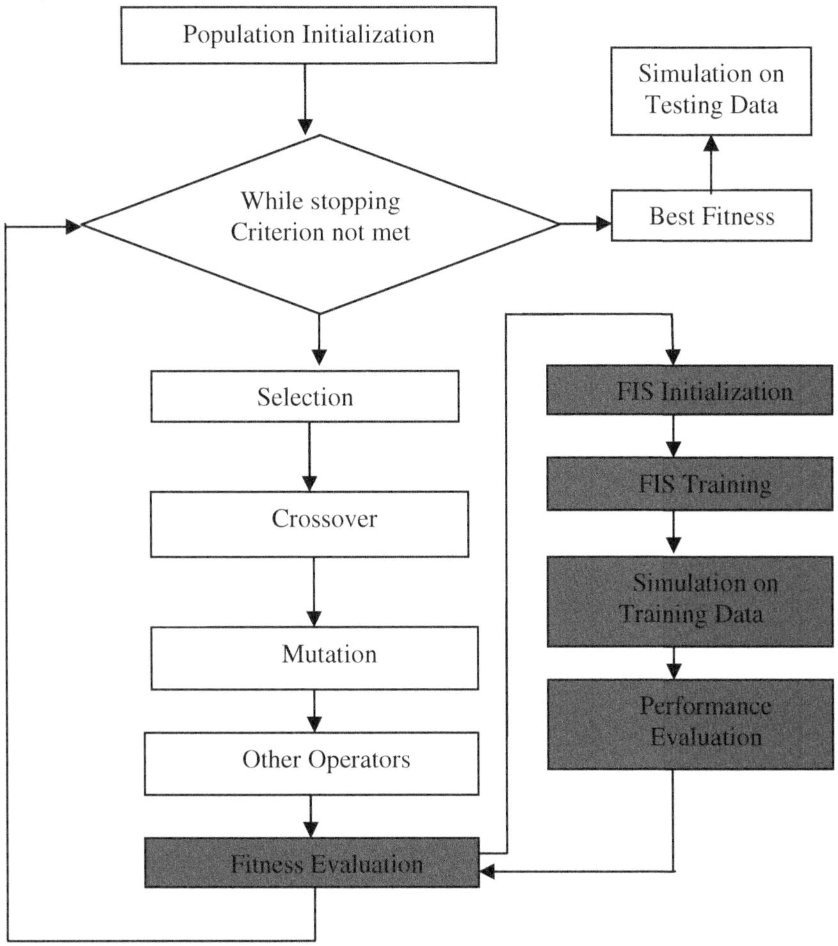

Fig. 13.4 Fixed Architecture Evolutionary Fuzzy Inference System

found out. In this manner we may construct a FIS. The next task is to pass the training data to this FIS and note the errors or the system performance. This is the measure of the fitness of the individual. The ultimate task of the GA is to optimize this performance or fitness.

The complete strategy is given in figure 13.4.

13.8 Evolving Variable Architecture Fuzzy System

The major limitation in the discussed approach is that the architecture needs to be fixed. This is a major limitation as far as the practical implementation is concerned. This is because of the fact that constituting optimal system architecture is itself a major problem in the use of FIS. The humans are prone to construct sub-optimal architectures. It further requires a lot of skill and expertise to analyze the system behavior and to set the system architecture accordingly. We hence need to extend the evolutionary algorithms to be able to completely evolve the FIS without having human to specify the architecture.

The evolution of the FIS largely follows the similar concepts that we discussed in the evolution of ANN. There must be a means by which we can cast a variable size FIS into the genotype which is the representation of the genetic individual. This would then make use of standard or custom built genetic operators to facilitate the evolutionary process. At every generation the optimization improves and the resulting FIS are better than the ones in the previous generations. The best FIS is returned after the evolution is regarded over or after the stopping criterion is met.

In this section we take the problem of the evolution of a variable architecture FIS. We would be making use of GA for carrying this optimization. The major problem here is individual representation. We need a means to represent the FIS as a vector of numerals that would be optimized using the optimization capabilities of the GA.

13.8.1 Individual Representation

The complete specification of the individual in this case consists of the specification of the number of MFs for every input and output, specification of the individual MFs and the rules. It must be kept in mind that the inputs and outputs are always fixed and known as per the problem definition. We only need to optimize the number, type and parameters of FIS along with the rules. Each of these is discussed in the subsequent sections.

13.8.1.1 Representation of Number of Membership Functions

The first part of the genetic individual stores the total number of MFs for the various inputs and outputs. Let us suppose that the system has a total of n number of inputs. Let these inputs be I_1, I_2, I_3 … I_n. Further let the system has a total of m number of outputs. Let these outputs be O_1, O_2, O_3 … O_m. We assume that the maximum number of MFs for any input or output can be α. Let the number of MFs for each of the inputs be $i_1, i_2, i_3, \ldots i_n$. Further let the number of MFs for each of the outputs be $o_1, o_2, o_3, \ldots o_m$. The first part of the individual storing the number of individuals stores $< i_1, i_2, i_3, \ldots i_n, o_1, o_2, o_3, \ldots o_m >$. Here each numeral is an integer that lies between 0 and α. It may be easily seen that these are $n+m$ genes where n is the number of inputs and m is the number of outputs.

13.8.1.2 Representation of Type and Parameters of Membership Functions

We have represented the number of MFs for every input and output. The next part of the individual contains the specification of the MFs. The specification of every MF holds the type of the MF, its location in the input axis and the other parameters that are specific to the MF of use. The various types of MFs that the system may use are coded as integers. Every MF type is assigned an integral equivalent. These are used for lookup between the genotype and the phenotype representations. The location along with the other parameters of the MF is usually a set of numbers that may occupy separate genes in the individual. In this manner any MF may be easily represented in the genotype representation.

We again assume here that β genes may conveniently represent all the parameters of any MF. This may usually have a value of 2 to 4 as per the list of allowable MFs for the problem. So 1 gene is needed for representing the type of MF and a maximum of β are needed for specifying the parameters. These $1+\beta$ genes are reserved for every MF. A MF may not require all the bits. Say a MF has only 2 parameters and 1 for storing the type of MF. Still $1+\beta$ genes are reserved for it. The first few may be actually used and the rest may be assumed to be filled with garbage values, in case they are not required.

We assumed that there can be a maximum of α MFs for any input or output. We always store the information for α MFs. This is even though the actual number of MFs used by the FIS may be less. The unused MFs are simply ignored at the time of fitness evaluation. There were a total of $n+m$ inputs and outputs. This means that every individual needs to store a total of $\alpha \, x \, (n+m)$ MFs in it. Here each MF has $\beta+1$ genes. This means that the individual needs a total length of $\alpha \, x \, (n+m) \, x \, (\beta+1)$ genes for the MF specification.

The various MFs may need an internal sorting after every genetic operation. This is because the rules are always formed with some ordering of the MFs. The MF termed *low* is always before the MF *high*. This means that the various MFs need to be arranged or sorted as per their locations in the input axis. This sorting makes it easy for the GA to carry out optimization as it limits the total genetic space.

13.8.1.3 Representation of Rules

The last part that the FIS optimizes and needs to store is rules. A rule consists of antecedents and consequents. The antecedents are a collection of rules along with their MF with which that input participates in the rule. Similarly, the consequents are a collection of outputs along with the MF with which a particular output participates in the rule. Any of the input or output may optionally have a negation by a *NOT* operator. A general rule may hence be given by equation (13.13).

$$\text{If } I_1 \text{ is } MI_1 \wedge I_2 \text{ is } MI_2 \wedge \ldots\ldots \wedge \text{Input}_n \text{ is } MI_n \text{ then } O_1 \text{ is } MO_1 \wedge O_2 \text{ is } MO_2 \wedge \ldots\ldots \wedge O_m \text{ is } MO_m \quad (13.13)$$

Here $I_1, I_2, I_3, \ldots, I_n$ are the inputs and $O_1, O_2, O_3, \ldots O_m$ are the outputs. $MI_1, MI_2, MI_3, \ldots, MI_n$ are the MFs corresponding to the inputs $I_1, I_2, I_3, \ldots, I_n$. Similarly $MO_1, MO_2, MO_3, \ldots, MO_n$ are the MFs corresponding to the inputs $O_1, O_2, O_3, \ldots, O_n$. Here it is not necessary that all the inputs and outputs participate. It is possible that some of the input and outputs are absent from the equation (13.13). Similarly any of the input or output may optionally have a *NOT* clause preceding it.

This forms the basis of the representation of the rule into the genotype representation. We assume that the system may have a maximum of γ rules. All these are coded into the genes and appended into the genetic individuals. All these rules are of the same length of $n+m$.

For the representation of a rule in the individual we look at all the inputs in the antecedents and outputs in the consequents. The MF associated with the input or output would be one in the list of the MFs we specified in section 3.8.1.2. The number of MFs in possession with the input or output was specified in section 3.8.1.1. We assumed that this could take any value subjected to a maximum of α.

We state the MF out of the available MFs with the input or output in the genotype of the genetic individual. The MFs are given integral equivalent values corresponding to the order in which they occur in the list of MFs stored in section 3.8.1.2. Hence every input and output has some MF associated whose integral equivalent or order is sought. If any of the input or output does not participate in the rule, it is given a value of 0. If any of the input or output has a *NOT* operator that precedes it, its integral equivalent is negated.

Assume for any input the antecedent is "I_p *is* MF^i_p". The integral equivalent of the MF would be taken for the input I_p which is i. If suppose the antecedent was "I_p *is NOT* MF^i_p", we would have taken the integral equivalent as $-i$. If this input is completely absent in the rule, we would have taken an equivalent of 0.

In this manner we collect the integral equivalents of all the inputs and outputs. These are mentioned in order for all the inputs and then the outputs as genes of the individual. There are a total of n inputs and m outputs. This makes a total of $n+m$ genes for the individual for every rule. Further the maximum number of rules may be γ. This means that the total space reserved in terms of number of genes in the genetic individual would be $\gamma \times (n+m)$. Here each gene stores a number. The total number of rules also need to be stored in the individual. This would again make the rest of the rules stored in the individual out of the γ rules useless. They still need to be stored for casting the variable size FIS into a constant size genetic individual.

The complete individual to be used for the evolutionary process contains the entire information appended one after the other mentioned in the preceding sub-sections.

13.8.2 Genetic Operators

The genetic operators to be used in this evolutionary scheme are the same as any conventional GA. Since the individual has already been casted into a fixed length individual storing all integrals, the conventional operators work to produce good optimization. The operators must additionally sort the MFs within any input or output which is a constraint put in the individual for better performance and better convergence.

It may be seen that the individual is maintained of equal length by all parts. At the time of crossover the MF of any of the inputs or outputs lie over each other between the two parents. Hence the crossover always happens between the same MF. Similar is the case with rules. Always a rule crossover takes place with another rule as they lie over each other in the individual. It is not possible that the rules of one parent gets exchanged with the MF of other parent, or say the MF parameter of some output of one parent gets exchanged with another output of the other parent. This plays a major role in good convergence and better characteristic inheritance by the child from the parents.

13.8.3 Fitness Function

The method of implementation of the fitness function is similar in these systems to any of the approaches of evolutionary neural networks. The first step we need to do is to read the individual and form the FIS out of it. The number of inputs and outputs are already known. The total number of MF against each of the input and output is read. Then the various MFs are formed by reading the values from the genetic individual. This makes the complete FIS MF structure. The next step is to read and construct the MFs rules. The information about the various rules is again available with the individual. In this manner the FIS is ready.

The next step to be performance is the evaluation of the generated FIS. This is done by passing the training data set to this FIS as inputs and attaining the system generated outputs. These are matched with the targets to compute the error. This forms the measure of the performance or the fitness of the individual.

13.9 Evolving a Learning Rule

Finding the most appropriate weights of the ANN that leads to an optimal performance of the entire network is a major and most important problem. The use of conventional training algorithms is able to use some techniques so that at every epoch the weights are modified so as to give better performances. After a series of epochs, the network gets trained to get an optimal performance. The other mechanism is the use of GA for training or setting up of these weights. The GA assumes this problem as an optimization problem where the network weights need to be optimized so as to maximize the performance of the neural net over the training data set (Shukla *et al* 2010).

Both these approaches have limitations that we have discussed numerous times throughout the chapter while forming various evolutionary systems. The GA is likely to miss the global optima due to a very complex nature of the fitness landscape. This is largely due to the large number of weights and biases that are a consequence of even a very small number of neurons. The conventional GAs do not give a reasonably good performance in cases of very complex shapes of the fitness landscapes with multiple optima, which is a natural phenomenon in case of training of neural network.

The model specific training algorithms are formulated with some presumptions. The mechanism of weight change can never be perfectly formulated by using statistical techniques due to the exhibition of complex phenomenon by a number of parameters. Further the issues of fitness function complexity, global and local optima, etc. are existent. Every training algorithm makes some or the other assumption in order to make the best approximate regarding the mechanism governing the change of weight. This mechanism is then repeated with all pairs of inputs (and outputs in case of supervised learning). The entire process is repeated for a number of iterations or epochs. The assumptions existent in the formulation of the weight update rule hence get magnified in the process. It is not necessary that the training data set obeys all the assumptions. Even slight alteration from the assumption might result in magnification of the error. This further leads to an undesired result that is sub-optimal in nature.

We must hence try to use the evolutionary powers to evolve this weight update mechanism. The weights in a network may exist between any pair of neurons. The neurons between which the weights exist may further possess different activation functions. The different weights may possess different jobs in the overall problem solving by the network. Hence the ideal updating strategy would be different for each of the weights.

In this section we try to use the optimization power of the EA for formulation of this learning strategy. It is natural that since the EAs could be used for many complex evolution and optimization tasks like evolution of ANN and evolution of FIS, they must be able to evolve or optimize the most optimal learning strategy of every weight in the system. This would greatly enable the weight to learn the training data in an optimized manner. Hence we wish to optimize the learning of the training data set. This task involves evolution of a learning rule as well as setting the various parameters as per the learning rule. Each weight optimizes the strategy that best carries out the learning in it.

13.9.1 Learning Rule Modeling

Learning rule depicts the strategy that any weight uses for modification in its value, so as the entire network gives the best performance on the training data. Every weight in the system may have its own learning rule that is optimized as per its location, task, parameters and other features in the network. We consider the back Propagation Algorithm (BPA) for each of discussion. We know that the delta rule or the gradient descend approach governs the mechanism of weight updating in a conventional Multi Layer Perceptron (MLP). Here we calculate the change to be carried out in the value of any weight by means of a set of indicators that serve as the inputs to the equation computing the change in weight for any pair of input and output. Consider that the MLP is of a single hidden layer. This means that there are in all three layers in the network consisting of the input layer, the hidden layer and the output layer. The change of weight Δw_{ij} in the BPA for any weight w_{ij} from node i to node j is governed by the values of the activation a_i of neuron i, the activation of output o_j of neuron j, the training signal on output t_j at neuron j, and the current weight on the weight w_{ij} from node i to node j. This is given in equation (13.14). The new weight w'_{ij} is given by equation (13.15)

$$\Delta w_{ij} = f(a_i, o_j, t_j, w_{ij}) \tag{13.14}$$

$$w'_{ij} = w_{ij} + \Delta w_{ij} \tag{13.15}$$

Ideally this function may be simple, complex or of any nature that cannot be predicted. We assume that a linear learning rule is capable of giving a highly optimal system performance. Hence we may assume the change in weight shown in equation (13.14) to be a function of the input variables and all their pairwise products. The learning rule may hence be assumed to be as given in equation (14.16)

$$\Delta w_{ij} = k_0\,(k_1 w_{ij} + k_2 a_i + k_3 o_j + k_4 t_j + k_5 w_{ij} a_i + k_6 w_{ij} o_j + k_7 w_{ij}\, t_j + k_8 a_i o_j + k_9 a_i t_j + k_{10} o_j t_j) \tag{13.16}$$

Here the various k_m $(1 < m < 10)$ are constants. The learning rule performance depends upon these values. k_0 is called as the scale parameter. This parameter decides how much the weights can change on any cycle.

It is natural that the most optimal learning rule may not follow equation (13.16). It may be more complex in nature. But here we have to make some assumptions. The optimization of the parameters of the assumed learning rule may give a fair enough performance. The most optimal learning rule may not be achievable by any other method. Further the equation (13.16) is capable of imitating a large variety of input surfaces in the input space of the learning rule. In its optimized format it is capable of very closely following the ideal learning rule.

Further we could have taken more terms in equation (13.16). This would have made the learning rule more generalized that could have better imitated the most ideal behavior. But the additional terms might have an overhead on the evolutionary optimization by making the fitness landscape more dimensional and complex.

13.9.2 Evolutionary Optimizations

Once a general framework of the learning rule has been made, the other task is to optimize it. The optimization of the learning rule is done with the help of GA. The optimization problems of the GA would enable find out the best values of the parameters k_m that optimize the weight training. This is done for every weight in the system. It is important to note here that the optimization is only on the learning rule of the weights. This technique in the discussed format assumes that the network architecture is known in prior and the same has already been initialized. Further the network needs to be trained after the learning rule has been optimized for each of the weight. The learning of the network uses the weight specific learning rule for every weight.

The individual representation in the problem of optimization of the learning rule may be done by directly placing the various optimization parameters one after the other in a double vector format. Alternatively the bit string representation may be used as well.

The conventional evolutionary operators of the GA may be used. This consists of the selection, crossover, mutation, elite, etc. The optimization is carried out for a few cycles or generations. After the stopping criterion has been met, the optimized learning rule for every neuron may be easily formulated using the parameters that the algorithm returns.

13.10 Chapter Summary

The chapter presented various models of fusion of evolutionary algorithms and neural or fuzzy systems. The resulting systems had potential for better performance as compared to the individual systems. The evolutionary neural networks use evolutionary power of the evolutionary algorithms to evolve the optimal neural network. In this category we first discussed the evolution of fixed architecture neural network. The entire neural network was coded into the genetic individual which evolved using conventional evolutionary operators. Neural training algorithm was used as a local search strategy for the evolution. The other model was the connectionist approach of the neural network. This evolution tried to optimize not only the weights and the biases, but also the manner in which the neurons are connected to each other. Penalty was given to the networks with larger connections. The next model discussed was the incremental evolution of the neural network. This model had a variable number of maximum neurons in the hidden layer. These neurons kept increasing along with the generations of the evolutionary algorithm. Next the chapter presented Grammatical Evolution of the neural network. The problem involved the specification of the Grammar for the evolution process. Based on this Grammar the individuals could be converted from the genotype to the phenotype representation. A similar discussion was presented for the fuzzy systems as well. Here we first discussed the evolution of the fixed architecture fuzzy system. In this system the various parameters of the membership functions were coded into the evolutionary individual which optimized with time. The other model was the evolution of the fixed architecture fuzzy system. In this system we mainly discussed the manner in which the various constituents of the fuzzy inference system could be encoded into an evolutionary individual. This included the number, type and parameters of the membership functions as well as the rules. Towards the end the evolution of the learning rule was discussed that tried to evolve the optimal learning strategy of the neural network.

References

[1] Angeline, P.J., Saunders, G.M., Pollack, J.B.: An Evolutionary Algorithm that Constructs Recurrent Neural Networks. IEEE Trans. Neural Netw. 5(1) (1994)

[2] Bullinaria, J.A.: Using evolution to improve neural network learning: pitfalls and solutions. Neural Comput. & Applic. 16, 209–226 (2007)

[3] Cho, S.B., Shimohara, K.: Evolutionary Learning of Modular Neural Networks with Genetic Programming. Applied Intelligence 9, 191–200 (1998)

[4] Fieldsend, J.E.: Pareto Evolutionary Neural Networks. IEEE Trans. Neural Netw. 16(2), 338–354 (2005)

[5] Funabiki, N., Kitamichi, J., Nishikawa, S.: An Evolutionary Neural Network Approach for Module Orientation Problems. IEEE Trans. Syst. Man Cybern. – Part B Cybern. 28(6), 849–855 (1998)

[6] Han, S.J., Kim, H.J., Cho, S.B.: Evolutionary Learning Program's Behavior in Neural Networks. In: Pal, N.R., Kasabov, N., Mudi, R.K., Pal, S., Parui, S.K. (eds.) ICONIP 2004. LNCS, vol. 3316, pp. 236–241. Springer, Heidelberg (2004)

[7] Happel, B.L.M., Murre, J.M.J.: The Design and Evolution of Modular Neural Network Architectures. Neural Netw. 7, 985–1004 (1994)

[8] Ishibuchi, H., Kwon, K., Tanaka, H.: A learning algorithm of fuzzy neural networks with triangular fuzzy weights Source Fuzzy Sets and Systems. Special issue on fuzzy neural control 71(3), 277–293 (1995)

[9] Jung, J.Y., Reggia, J.A.: Evolutionary Design of Neural Network Architectures Using a Descriptive Encoding Language. IEEE Trans. Evol. Comput. 10(6), 676–688 (2006)

[10] Kala, R., Shukla, A., Tiwari, R.: A Novel Approach to Classificatory Problem using Grammatical Evolution based Hybrid Algorithm. International Journal on Futuristic Computer Applications (2010)

[11] Kala, R., Shukla, A., Tiwari, R.: Comparative analysis of intelligent hybrid systems for detection of PIMA indian diabetes. In: Proceedings of the IEEE 2009 World Congress on Nature & Biologically Inspired Computing, NABIC 2009, Coimbatote, India, pp. 947–952 (2009a)

[12] Kala, R., Shukla, A., Tiwari, R.: Hybrid Intelligent Systems for Medical Diagnosis. In: Shukla, A., Tiwari, R. (eds.) Intelligent Medical technologies and Biomedical Engineering: Tools and Applications, IGI Global (2009b)

[13] Kala, R., Shukla, A., Tiwari, R.: Handling Large Medical Data Sets for Disease Detection. In: Shukla, A., Tiwari, R. (eds.) Biomedical Engineering and Information Systems: Technologies, Tools and Applications, IGI Global (2009c)

[14] O'Neill, M., Ryan, C.: Grammatical evolution. IEEE Trans. Evol. Comput. 5, 349–358 (2001)

[15] O'Neill, M., Ryan, C.: Grammatical Evolution. Kluwer, Boston (2003)

[16] Pedrajas, N.G., Martínez, C.H., Pérez, J.M.: COVNET: A Cooperative Coevolutionary Model for Evolving Artificial Neural Networks. IEEE Trans. Neural Netw. 14(3), 575–596 (2003)

[17] Rastegar, F., Araabi, B.N., Lucast, C.: An evolutionary fuzzy modeling approach for ANFIS architecture. Evol. Comput. 3, 2182–2189 (2005)

[18] Rivera, A.J., Rojas, I., Ortega, J., del Jesus, M.J.: Anew hybrid methodology for cooperative-coevolutionary optimization of radial basis function networks. Soft. Comput. 11, 655–668 (2007)

[19] Ryan, C., Collins, J.J., O'Neill, M.: Grammatical Evolution: Evolving Programs for an Arbitrary Language. In: Banzhaf, W., Poli, R., Schoenauer, M., Fogarty, T.C. (eds.) EuroGP 1998. LNCS, vol. 1391, pp. 83–95. Springer, Heidelberg (1998)

[20] Shi, Y., Eberhart, R., Chen, Y.: Implementation of Evolutionary Fuzzy Systems. IEEE Trans Fuzzy Systems 7(2), 109–119 (1999)
[21] Shukla, A., Tiwari, R., Kala, R.: Real Life Applications of Soft Computing. CRC Press, Boca Raton (2010)
[22] Yao, X.: Evolutionary artificial Neural Networks. Int. J. of Neural Systems 4(3), 203–222 (1993)
[23] Yao, X.: A New Evolutionary System for Evolving Artificial Neural Networks. IEEE Trans. Neural Netw. 8(3), 575–596 (1997)
[24] Ye, B., Zhu, C., Guo, C., Cao, Y.: Generating Extended Fuzzy Basis Function Networks Using Hybrid Algorithm. In: Wang, L., Jin, Y. (eds.) FSKD 2005. LNCS (LNAI), vol. 3613, pp. 79–88. Springer, Heidelberg (2005)

List of Figures

Fig. 13.1. Individual Representation
Fig. 13.2. Fixed Architecture Evolutionary Neural Network
Fig. 13.3. Individual Representation
Fig. 13.4. Fixed Architecture Evolutionary Fuzzy Inference System

Chapter 14
Modular Neural Networks

Abstract. Modular Neural Networks are use of a number of Neural Networks for problem solving. Here the various neural networks behave as modules to solve a part of the problem. The entire task of division of problem into the various modules as well as the integration of the responses of the modules to generate the final output of the system is done by an integrator. In this chapter we first look at the various Modular Neural Network models. Here we would mainly study two major models. The first model would cluster the entire input space with each module responsible for some part of it. The other model would make different neural networks work over the same problem. Here we would be using a response integration technique for figuring out the final output of the system. The other part of the chapter would present Evolutionary Modular Neural Networks. We would first present a simple genetic approach and then a co-evolutionary approach for this evolution of the entire Modular Neural Network.

The Artificial Neural Networks (ANNs) are a good means of problem solving. These networks are able to learn from the historical data, tune their parameters, and ultimately give correct outputs to the given inputs. The ease of use in problem solving with these networks has largely enabled them to be extensively used for a variety of problems of both classification as well as functional prediction. A variety of models of ANNs are available that all solve different kinds of problems and using different methodologies.

The ANNs give good results for most of the problems. But this largely depends upon the complexity of the problem being solved with the help of ANNs. If the problem is simple enough, the ANN would be easily able to perform the necessary task. In case of the classificatory problem, this may be taken as the case when the various classes have a fairly low intra class separation and a fairly high intra class separation. In terms of a functional prediction problem, this may be taken as a simple input space with less number of turns or other complex shapes in the input landscape. The simple problems may be solved easily for any number of data items in the training data set. They would likely result in a good performance as per the requirements of the user.

But this may not be true in many real life applications. The small and simple networks might only give some degree of performance. Further performance boost might be needed in the problem. In many other problems, the mapping of the inputs to outputs is reasonably complex by nature. These cause a problem for the conventional neural networks. The networks fail to give the desirable performance.

A. Shukla et al.: Towards Hybrid and Adaptive Computing, SCI 307, pp. 307–335.
springerlink.com

A good solution is always to work over the parameters and to re-design the inputs and outputs. But this may not be possible always. Say the problem was of speaker recognition. We may not be able to formulate better attributes for the system because of the black box approach where these attributes are just given to the system for recognition. In such a context the hold only lies with the ANN to make the best attempt to give the maximum efficiency.

Many times the data as well gets too large. When a high amount of training data is given to the ANN for training, it requires a lot of computation. This increased computation may result in lowering of the epochs so as to complete the entire training within finite time. This results in sub-optimal network being returned after training. Many times the data may be so large in size that the network can only be trained for a few epochs. It is natural that the network would be unable to formulate rules or learning in a few epochs. This is a great limitation in large training data sets.

The large data sets may further bring in new rules that they represent. In such a case the problems associated with the ANNs increase to an even larger degree. Now the ANNs are supposed to learn these rules as well that the new data represents for overall optimal learning by the system. This further increases the complexity of the ANN and the problem. It now becomes very computationally expensive for the ANN to figure out all the possible rules for the mapping of the inputs to the outputs.

The Modular Neural Networks (MNNs) are a key to solve these problems. These networks advocate the use of more than one ANNs for problem solving. As a result of the collective performance of the various networks, we are able to remove the limitations that are dominant in the use of the ANNs.

14.1 Modular Neural Networks

A single ANN is many times not able to give a good performance as per the problem and requirements. We hence may need more than one ANN to solve the same problem. This is done by the use of Modular Neural Network (MNN). MNN makes use of the different ANNs as modules that solve the complete problem. Each of the ANN represents a module and is responsible for solving some part of the problem. In this manner a collective effort of the various ANNs is taken. The task left is the division of the problem into the various modules. Here each ANN or module is given some input corresponding to its role. The various modules solve their part of the problem independently. These compute their results. The computed results come to an integrator. The role of the integrator is to combine the various individual responses from the various ANNs and to generate a common response that is the output of the system. Hence the complete system may be said to have two parts that work in a master slave mode. The first consists of an integrator that divides the problem and gives it to the various modules. Later this collects the responses of the individual modules and computes the final output of the system. The other part is the modules, each of which is an ANN. In this manner the MNNs divide the problem, solve them and integrate them. This working methodology of the MNN is shown in figure 14.1.

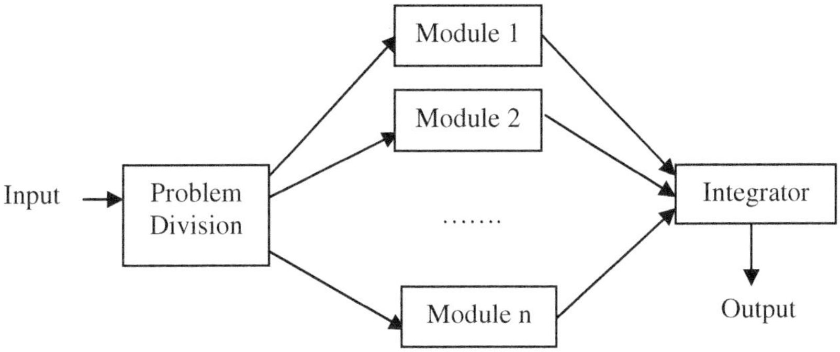

Fig. 14.1 General Architecture of Modular Neural Network

We would study the basic working of these three steps of the MNN in the next three sub-sections. This would form the basis of our discussion over the specialized models of MNN in the later sections.

14.1.1 Problem Division

The first major task is to divide the problem into parts. We mentioned that the complexity of the problem is a major problem in the use of the ANN and numerous modules or ANNs are thus required for solving this problem of complexity. The entire set of inputs with all the possible values makes the entire task very complex. The first method commonly used in these systems is to delegate different modules for different kinds of inputs. Using this approach each ANN or module has a reasonably small complexity that it handles.

The other mechanism is to make all the various modules solve the same problem. In this mechanism there is naturally no division of problem that takes place. Rather the division stage is supposed to simply pass the inputs to the various modules. At the later stages we solve the same problem by a number of ways and then try to make the best use of the outputs generated by the various modules.

14.1.2 Modules

Each of the modules in this approach is an independent ANN. This ANN has its own training procedure and the training is performed independent of all the other modules. At the time of testing also these networks are given a completely independent set of inputs as figured out by the integrator. They use their own procedures for computing the outputs of the inputs that they are provided with. Any model of the ANN may be used as per the problem requirements and the MNN design. The possibility in the use of various models for various tasks is known as Mixture of Experts.

14.1.3 Integration

The last job to be carried out by the integrator is integration. In this mechanism we look at the responses of the various modules. On the basis of the outputs that the various modules or ANNs generate, the decision regarding the final system output is made. Integration is the mechanism to combine the various outputs into a single output of the system. The integration again depends upon the design of the MNN and the manner in which the division of the problem has taken place. If the division assigned every module some range of inputs which it works upon, only the invoked module would give an output. This is the output of the system. On the contrary if the division was carried out making the various modules work over the same problem, we would need some mechanism to compute the final output by these outputs. This is done by the integrator using a variety of methods as we shall study later in the chapter.

14.2 Ensemble

The first model that we study under the head of the MNNs is ensemble. Ensembles make use of multiple ANNs to solve the same problem. Each of the ANN is given the complete part of the problem input. Each ANN solves the problem using its own approach. All these ANNs or modules return a solution to the integrator which then computes the final output (Dietterich 2000; Hansen and Salamon 1990; Jacobs 1991; Krogh and Vedelsby 1995; Melin *et al* 2003, 2004; Melin and Castilo 2005; Castilo and Melin 2002; Waterhouse *et al* 1996).

The ensemble approach relies on redundant solution of the problem by a variety of ANNs or modules. It may be easily seen that all the modules are redundant and solve the same problem. Each module or ANN makes use of its own mechanism of problem solving. Now we know that the problem may either be classification or functional approximation. We handle both these cases separately.

14.2.1 Ensembles for Classification

The classification problems are the ones in which we need to compute the correct class to which an input belongs with the help of input attributes. The problem is the construction of the correct decision boundaries that separate the various classes. In case the attributes are such that the inter class separation is very high and intra class separation is very low, any mechanism would be able to solve the problem. However in real life data, the various classes may not be so neatly separated. Further there would be a vast amount of data spread all around the input space that needs to be classified into. We now may not be able to build boundaries that can clearly separate the classes. The decision boundaries further must be able to assume flexible shapes as well as have a great generalizing ability where the testing data follows the trends of the training data. A sample training data over a two dimensional input space, along with the ideal decision boundary is shown in figure 14.2.

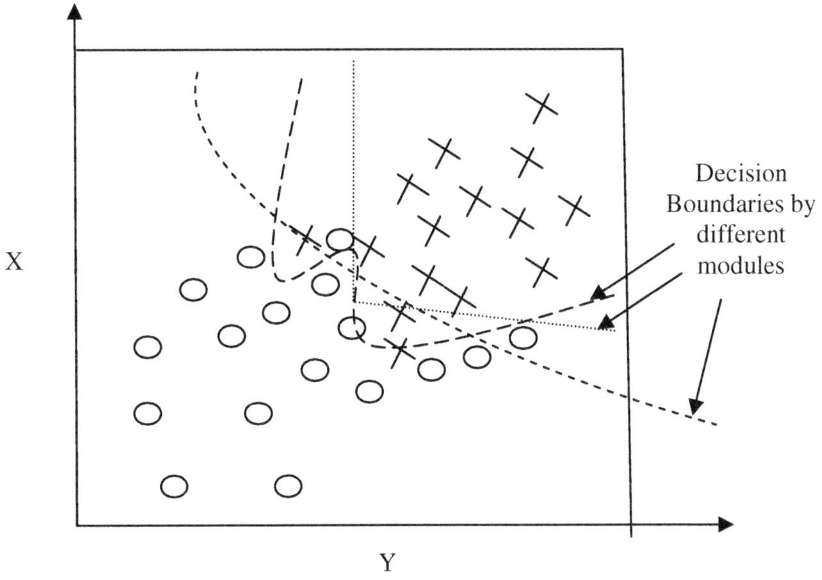

Fig. 14.2 Use of Ensembles for Classification

It is natural that due to computational and modeling reasons no single ANN or model may be able to predict the correct decision boundary. Each model of ANN or the same model of ANN with different state may only try to predict the decision boundary to some extent. There would be regions in the input space where this decision boundary coincides with the ideal decision boundary. There would be other regions in the input space where this decision boundary is far from the ideal decision boundary. While the data well away from the decision boundary may be fairly easily classified, the problem comes with the data that lies on or close to the decision boundary. Here a little up and down of the decision boundary changes the fate of the input. This is where the different models and the different states of the same model differ in carrying out the classification.

We have already seen the mechanisms used by the various ANNs in modeling of this decision boundary. The Multi-Layer Perceptron (MLP) trained with Back Propagation Algorithm (BPA) uses a curve fitting approach for each class where it tries to draw a curve that has a maxima where the class is found and a minima where the class is not found. The decision boundary marks the transition stage of the maxima to minima. Ideally this is a binary function with a sharp turn from the maxima to minima at the decision boundary, but the MLP with BPA is only capable of having a smooth transition unless the number of neurons is exceptionally high.

A similar approach is used by Radial Basis Function Networks (RBFN) which delegate neurons at various stages of the input space for the mapping of the inputs to outputs. In case of the classificatory problems, a neuron itself or a small pool of neurons tries to completely cover a space such that all the instances in the input space where this class lies are easily accommodated and the rest of the space is rejected in the imitated function that is similar to the one used in MLP with BPA.

The other are the specific classificatory models. The Learning Vector Quantization delegates neurons in the input space. Each of which represents a sub class. The idea is to have a number of big or small subclasses to cover the entire region where a class is located. The decision boundaries here are the points in the input space that lie at an equal distance between the sub-classes of two different classes. Similar is the working of the Self Organizing Maps with the major difference that they carry out the learning in an un-supervised mode.

Hence the different models have different manners of construction of the decision boundaries which plays a major role in the classification of the input into one of the classes. A good flexibility and generalizing ability of this decision boundary is of key importance.

The similar is the nature of two different states of the same model. The difference in states means the differences in the network parameters after training. This may be the weights and biases in case of the MLP with BPA, location of the hidden neurons in case of Radial Basis Function Networks, Learning Vector Quantization, Self-Organizing Maps, etc. These differences naturally cause the decision boundaries to change. The change in the decision boundaries means that it changes in terms of location and shape which affects the overall performance of the ANN.

We know that every ANN has an optimal structure in which it gives the best performance in the training data set. This optimal structure further has advantages in terms of running and training time as well. Every training algorithm along with the human expert designing the system is expected to generate this optimal structure of the ANN. Yet we know that a single structure may not be able to optimally solve the entire problem. A particular ANN state may be good for a number of inputs at some part of the input space. It may however completely neglect the other part of the input space. Some other network may have a very poor performance at many regions of the input space. It may however solve the problem reasonably well for some part of the input space that may not be easy for other ANNs with different states.

Hence we realize the inability of a single state of a single model of the ANN. This motivates the use of redundant ANNs for solving the same problem. We know that the different ANNs may give different answers. Answers to some inputs would be correctly given by a set of ANNs and not by others. The answers to some other inputs might be correctly given by another class of ANNs. The major task is to use all the answers generated by the various ANNs to compute the correct output of the system. In this strategy we would make use of the best strategy of the various ANNs. Each of these ANN is a model of the MNN. Hence the various modules have a correct picture of some part of the input space and a wrong view of some other part of the space. The integrator tries to make the best view of each of these modules to get the best view of the entire input space. We study the methodology and the various ways in which the integrator does this task in the coming sub-section.

14.2.2 Ensembles for Functional Approximation

The other major variety of problem that we face is the functional approximation. Here the system is supposed to approximate a function. The output is usually in forms of continuous real values. The functional approximation problems may be easily solved by any curve fitting mechanism that is usually adopted by the MLP with BPA and RBFNs as studied in section 14.2.1. In this section we would concentrate on the use of ensembles in these problems. Figure 14.3 shows the curve fitting mechanism with a single input attribute.

Consider the input space for any functional approximation problem. Here the input space contains the system output as a surface that spans across the various axes. The problem of the ANN is to imitate a function that implements this surface. The input given to the system at the time of training is in the form of points in the input space whose output values are known. The ANN is supposed to fit these points in a curve that best matches the desired output surface. This curve fitting approach may be done by the various ANN models. The end objective is that the formulated surface must be generalized enough to fit the testing data fairly well. A problem that may usually be predominant in these systems is the presence of noise. Because of noise the exact fitting may be undesirable and may produce wrong results. The generality is of a higher importance. This means that the network must be small enough so that the assumed output surface matches to the best possible means to the actual surface and fits the testing data to the best possible degree.

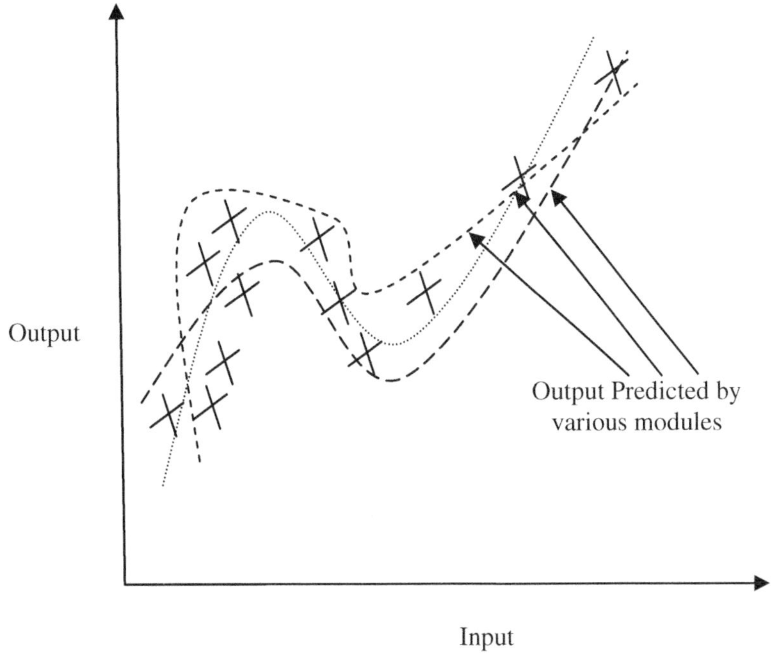

Fig. 14.3 Ensembles for functional prediction problems

As discussed for classification, the various models have different mechanisms of curve fitting and imitating the actual curve be it MLP with BPA, Radial Basis Function Networks, Recurrent Neural Networks, etc. All these make the best attempts to imitate the curve to the best possible means, making the resulting curve simple and general enough for the testing data to give a high performance. Again the curves of the different ANNs may closely follow the desired curve at some regions in the input space and may differ largely at some other regions of the input space.

The same scenario exists with the use of different states of the same model. A single architecture may not be able to imitate the entire curve. It may be further possible that it possesses too few neurons and is only able to learn some generalized rules, or it possesses too large neurons for poor generalizing capability. A single architecture may again not be able to solve the problem well for the entire input space. This is even more emphasized by the fact that the actual output being tried to imitate for the various inputs depends upon the problem and the actual physical system which may have a characteristic behavior at various regions in the input space.

Here again the redundant solution of the problem by the various systems is a source of large information about the manner in which the various systems interpret and fit the various training data to imitate some output surface. The problem is given to all the various ANNs which are the modules of the MNN. The solution provide by each of the module is used by the integrator which tries to combine these solutions using the best possible strategy so as to get the final output surface as close as possible to the desired one.

14.2.3 Integrator

We have already stated the basic working methodology of the ensembles. The input is given by the system to each of the modules. Each module is similar in regard to the inputs and the outputs. Each module represents an independent ANN of its own which is trained separately using the same training data. The training may hence be carried out in parallel among the various ANNs or modules. The various ANNs must have some difference in them. This difference is in terms of the model of the ANN, ANN architecture, or the present state. The difference enables the different ANNs to be logically different and contributing to solve the problem. The difference in model and architecture is simple to develop. The difference in states may be developed by different training parameters at the time of training of the network. In this manner all the modules or ANN solve the same problem and make their distinct contributions. All the outputs are communicated to a central integrator that carries the next task of computing the final output of the system based on the system module outputs. The overall methodology of the system is given by figure 14.4.

There are many ways in which the integration may be carried out in this approach (Auda *et al* 1995). This further depends upon the problem being classificatory or functional prediction in nature. The integration manner would determine the design and behavior of the ANN as well, as the ANN output is supposed to aid

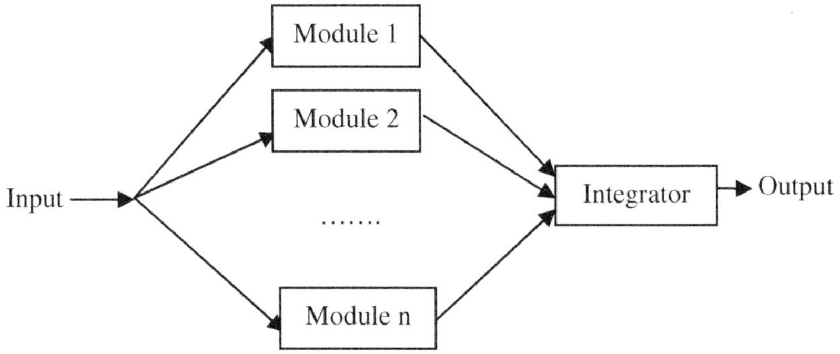

Fig. 14.4 General Architecture of Ensembles

the integration task. We discuss three major integration techniques here. These are polling, probabilistic polling and gating. The first two would be used for classificatory problems and the third would be used for the functional prediction problem.

14.2.3.1 Polling

The first method of integration that we consider is polling. This method of integration is followed in the classification problems. In this method each of the modules gives as its output the class to which the input may belong as per its knowledge and methodology. The integrator gets the various classes that are potentially the output of the system by the various modules. The task of the integrator is to decide the system output as a means of voting between the various modules. Each module casts one vote in favor of the class which is its output. The votes for the various classes are collected and the class getting the largest vote is regarded as the final system output by the integrator. In case of a tie between the modules, any one of the classes may be randomly chosen out of the classes involved in a tie. Alternatively any one module may be given a preference as per its design, training or other reasons and its output may be regarded as the final system output by the integrator (Shukla *et al* 2009). This is shown in figure 14.5.

All the modules that we train are optimal to a reasonably good extent and hence give a good performance for a large degree of inputs. It is hence natural that these modules give the correct output for most part of the input space and fail to give the correct output for some regions. The voting mechanism selects the class with the maximum vote count. Hence the regions in the input space where a module gives poor performance is overcome by the other modules and the system mostly gets the correct output.

We mentioned that the various modules largely differ in the decision of inputs lying on or close to the decision boundaries. This is the region where the different models put some region on different sides of the decision boundary by the manner in which they model the decision boundary. Hence majority of the classes that are a part of the system to which the input may belong to do not participate at all in the voting mechanism. In other words these do not get even a single vote. Only the

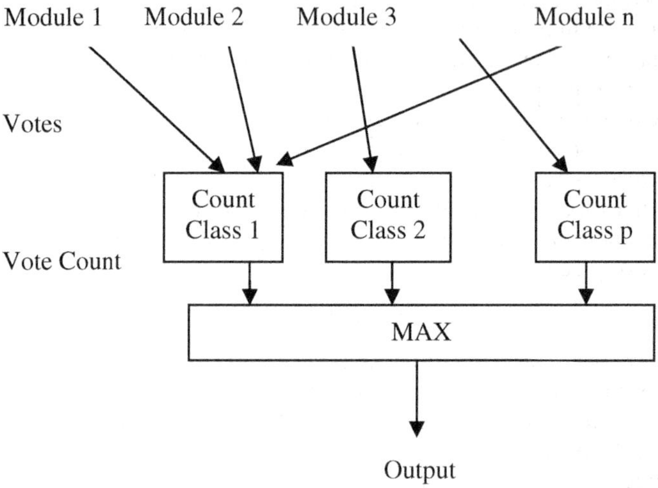

Fig. 14.5 Polling Integration

classes having their presence nearby the applied inputs may be participative and the various modules may cast their vote in favor or against them. If an input is quite far from a decision boundary, all votes may be casted in favor of a single class that is the actual output of the system.

These systems hence present promising solutions to the entire problem of classification with the help of neural networks. The underlying system is a collection of neural networks that is able to give a better performance as a result of the simultaneous and redundant use of multiple networks for the same problem. The problem of classifying the inputs lying on or near the decision boundary is still difficult for any system. Only better worked out attributes may be able to work in this problem. The best we can do from a machine learning perspective is to reduce the internal inefficiencies of the various techniques using hybrid models.

Now since the key of perfectly classifying the various inputs ultimately lies in the correct selection of attributes, it would be wise to realize and understand the alterations in the current attributes or additions in the attributes that must be carried out for facilitating the classification. We assume that training data set to be spread enough to cover all possible inputs that the system may face in real time use or at any stage at the time of testing. It is not necessary that all classes must be having problems in classification where they intermingle with the other classes or maintain small inter-class separation. Further only some inputs in the data base must be having this problem. If we add an attribute or cleverly alter an attribute that has a high inter class to intra class separation ratio, it is possible that classification is enhanced. However it is much important to note that the wrongly classified inputs must be benefitted from this rather than all the inputs. Inputs that can correctly be classified by the available attributes do not need any additional help of more or better attributes. However we need attributes that can separate the inputs of two classes that are being misclassified. An understanding of the misclassified class may help many times. In most of the other situations we fail to figure out good attributes that ultimately leaves to good machine learning techniques to extract the maximum out of the present system.

14.2.3.2 Probabilistic Polling

The other integration technique that we study in this chapter is the probabilistic polling mechanism. The polling that we studied made every module decide the winning class by its own means that was nominated for the final output by the system. Careful work by the various modules resulted in a careful figuring out of the winning classes. It was assumed that most modules make the correct decision and hence the class winning out by the voting mechanism is the actual class. This method of voting however has a problem in implementation. The various modules can give a single vote in favor of winning class as per its decision.

In most intellectual decisions this may not happen. Consider the various modules as the experts that solve a problem of classification of the given input. The above discussed approach simply organizes a voting amongst these experts. In actual decision making however these experts would like to discuss, pass on their beliefs, share them and based on all of these the final decision may be made. This gives a lot of flexibility in decision making by the modules or experts in which they can pass better information and collaborate for the decision making, rather than just passing the final verdict as per their understanding.

In this mechanism, every module gives as many outputs as there are classes in the system. Each output measures the probability of the occurrence of any class as the final output class. The probability lies in between 0 to 1. An output of 1 by any module for any class means that as per the recordings of the module that particular class is surely the class to which the input perfectly maps to. An output of 0 by any module for any class means that as per the recordings of the module that particular class is surely not the class to which the input perfectly maps to. Every module hence computes the probability for each of the classes in the system. This probability vector comprising of all the probabilities is passed to the integrator for computing the final output (Kala *et al* 2010).

Ideally the probabilities must be a perfect 0 for all classes which belong to a particular input and 1 for all classes that belong to a particular input. In implementation however the probability would not be this crisp number. Rather the probability would be very close to 0 for classes that lie far away from the input and very close to 1 for classes that lie near the input. If multiple classes are equidistant from some input, they may have equal probabilities of occurrence which depends upon the state and model of use and the manner in which it models the decision boundaries.

Consider the problem is hand is that the input needs to be classified to any one of the m classes. In this approach every module outputs m numerals each of which lies in between 0 and 1. Let the output of any module i be $<o^i_1, o^i_2, o^i_3, \dots o^i_m>$. Here the number o^i_j represents the probability of the input belonging to the class j as per module i computation and knowledge.

In case of the use of Multi-Layer Perceptron with Back Propagation Algorithm we may simply train the system such that the ANN has as many outputs as the classes in the system. In the outputs of the training vector we specify 1 for the class to which the input belongs and 0 for the other classes. The testing directly makes the ANN output the probability vector that may be given to the integrator.

The integrator does the task of integration of these probabilities. This is simply a polling mechanism where the votes are probabilistic in nature. The integrator first adds up all the probabilistic votes for each of the classes. This is simple addition over the probabilities given by the various modules in favor of any class. The final probabilistic sum for any class *i* is given by equation (14.1).

$$P_i = \sum_{j=1}^{n} o_j^i \tag{14.1}$$

Here *n* is the total number of modules.

The next task is to compute the class that gets the maximum sum in this process. The class corresponding to the maximum sum is returned as the final system output. This is given by equation (14.2).

$$\text{Output} = k: P_k > P_i \ \forall\ k \neq i \tag{14.2}$$

This system is summarized with the help of figure 14.6.

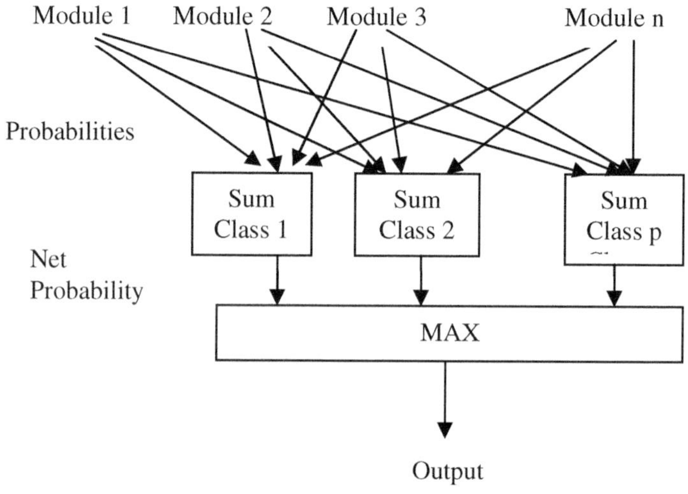

Fig. 14.6 Probabilistic Count Integration

It may again be seen that most of the classes would have very little count in the final probability vector and only competing classes would have appreciable values based on which the final decision be taken by the integrator.

14.2.3.3 Gating

The last method that we study is gating. This holds mainly for the functional prediction problems; however the concept may be extended to the classification problems as well. In gating we try to combine the various outputs of the modules and generate the final output of the system by a simple function. The easiest implementation of this is to make the integrating function take a weighted mean of the various outputs of the modules. This gives the final system output of the MNN. This concept is shown in figure 14.7.

The outputs of the various modules may be taken as the outputs of the experts that use their own methodologies for solving the problem. Each of this output represents the actual output of the system as computed by the module. Since the various modules are unable to solve the complete problem on their own, it is natural that this output generated by a single module may not be correct. It may deviate from the actual system output by some magnitude. The magnitude largely depends upon the ANN model, architecture and state. This further depends upon the fact how well the ANN solves the problem in the specific region of the input space. We have used a variety of modules to solve the same problem. All the generated solutions deviate from the actual output by some magnitude. Some might under-estimate the output and the others might over-estimate it. In this manner we may expect that a weighted addition of these may give us a good overall output. The various shortcomings of one may be balanced by the other.

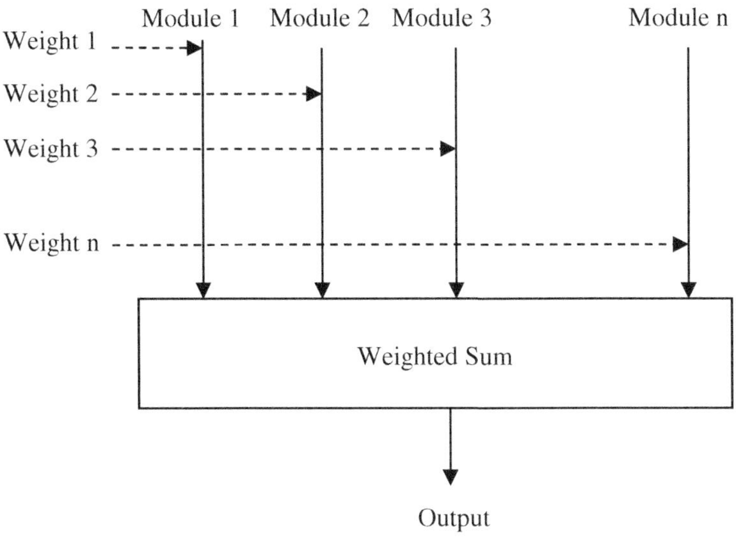

Fig. 14.7 Gaiting Integration

The various modules may not give equal performance. Some might be very close to the ideal surface to be imitated, others might have large deviations. Hence the contributions of the various modules are not taken to be the same. Rather we assign a weight to each of the modules. The weight may be determined by the general ability of the network in solving the problem.

Consider that the MNN had a total of n modules. Let the ith modules have an output o_i. Let the weight associated with this module be g_i. The output of the system using this strategy is given by using equation (4.3).

$$O = \sum_{i=1}^{n} g_i o_i \tag{4.3}$$

14.3 Modularity in Computation

The other model we study in this chapter is where we try to introduce modularity in computation. In this approach the different modules do not do a redundant work of solving the same problem using their own methodologies that we saw in the ensemble approach. Rather each module carries out a task independently that contributes towards the overall solution of the problem. In this manner we can decompose a big problem into a set of smaller problems. A big ANN might not be able to solve a bigger problem. However it may be able to solve a smaller problem. The various solutions may then be integrated using some integration mechanism. Many of the real life applications that we encounter all around have problems of very high complexity. These problems need some kind of modularity being introduced such that the problem may be solved (Fogelman-Soulie 1993; Fu *et al* 2001; Gruau 1995; Jenkins and Yuhas 1993; Kala *et al* 2009a, 2009b, 2009c, 2010; Ronco and Gawthrop 1995; Schmidt and Bandar 1997; Schyns 1991).

Expecting a very large ANN to solve the entire problem may be a solution, since it is capable of modeling and solving problems with a reasonably high degree of complexity. However a large ANN has numerous connections that require a lot of data and computation. Besides there are usually a lot of chances of having the network poorly trained in a large sized network. Handling smaller networks is much easier than the handling of the larger networks. Hence the solution of having a very large network may not be easy. Here we rather insist on the modularity. In other words the various ANNs possessing smaller sizes may be able to work independently of each other and solve some independent module or part of the problem. It is as per the design of the MNN to decide which ANN is supposed to solve what part or module of the problem. Since the individual ANNs are small in size, these may be easily designed, trained and tested as per the task these are expected to carry out.

14.3.1 Divide and Conquer

The general methodology of solving problem in this manner is known as divide and conquer. This approach consists of two stages. The first stage is the divide phase. In this phase we divide a problem into a number of smaller problem of a similar nature. Smaller means that the problem must be smaller in terms of input ranges, size, or complexity. In other words solving the smaller problem must be generally easier for the system than solving the original problem. By similar we mean that all the various smaller problems must be of the same general form. If we make the smaller problems different in nature, a specialized technique may be needed to be engineered for each and every sub problem. The division in a similar set is easy to design. Any module may be taken as a general case and the design may be extended to all the modules with small changes.

Many times the divided problem or the smaller problem may still be very large. In such a case the problem needs to be divided to an even larger extent by similar approach. This way we may divide the problems into smaller and smaller sizes indefinitely till the solution can be easily framed by the known simple methods. The problem in its smallest parts is easily solvable by a known mechanism.

The other part of this technique is the conquer phase. In this stage we combine the results of the various sub problems and use them to form the solution of the main problem. Since the solutions to the smaller problems are known, it must normally be easy to develop the solution of the bigger problem.

In case the problem went over reputed division from one level to the other by constant division from a larger size to a smaller size, the same procedure would take place for the conquer stage as well. Smallest problems may be trivially solved by the available mechanisms. We need to combine these solutions to build a solution for a larger problem. These solutions of a larger problem may still be combined for a result of a still larger problem. In this manner we may be ultimately able to solve a large problem.

The use of the divide and conquer approach is that it enables solution to problems that are really complex in nature and behavior. These we may not be able to solve using any of the available methods. The complex problem solving with ANNs is one such problem where a single ANN fails. Hence we need to somehow divide the problem into multiple ANNs. We discuss them next.

14.3.2 Hierarchy of Experts

The general architecture of MNNs which try to modularize the entire task of computing is with the use of multiple ANNs arranged in a hierarchical manner. Each ANN is an independent module that takes specific inputs and generates specific outputs. These ANNs or modules are hence called as experts that specialize in their part of the problem that contributes towards the overall problem solving by the system. The outputs generated by the various modules may be used for decision regarding the further processing or may be given as inputs to the other modules. In this manner a hierarchal processing of information takes place from one hierarchy to the other hierarchy (Jordan and Jacobs 1994; Melin and Castilo 2005; Melin *et al* 2003). This is shown in figure 14.8.

The modular processing of information in this manner is largely the key to solve any complex problem. The behavior of each of the modules and their processing depends upon the manner in which the problem has been designed and the manner in which the division of the problem is carried out. This task is done by a human expert possessing some knowledge about the problem as well as the manner in which the solution may be formed. The use of multiple ANNs as modules in the overall architecture of the MNN hence requires planning and design perspective to problem solving. Mostly the systems developed would have fewer hierarchies as larger levels of hierarchies may be very difficult to design as well as build. We discuss a specific example of this approach next.

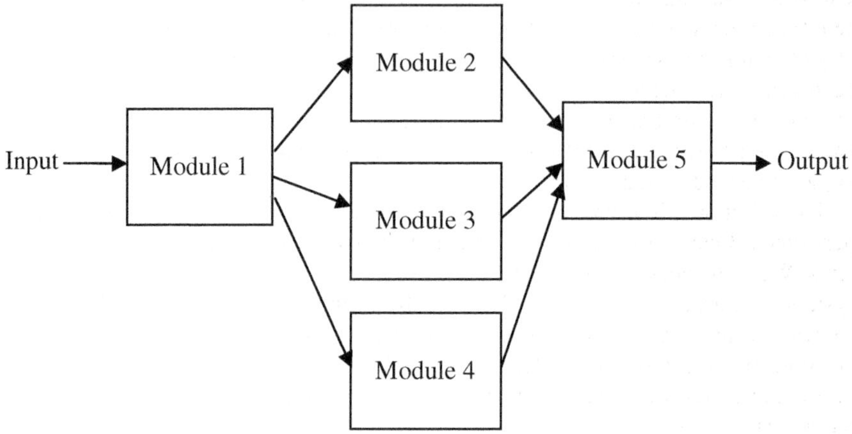

Fig. 14.8 Hierarchy of Experts

14.4 Modularity in Input Space

In this approach we would divide the entire input space into modules. Each module would be given a region in the entire input space. It is expected to solve only the inputs that come within that region. In this manner every module or ANN has a dedicated part of the input space that it is trained and tested with. The division of the problem simply looks at the input and decides which module or ANN is supposed to solve the problem. This module is invoked for the particular input whereas all the other modules are inactive. The entire input vector is passed over to this module as its input. This module solves the problem using its own means. The final result is communicated to the integrator. The integrator gets a response from a single module. This may hence be simply returned as the final output of the system. The input space division may take place by the use of Self Organizing Maps, clustering algorithms, or any other neural network in a supervised or unsupervised mode. The complete architecture of these systems is shown in figure 14.9.

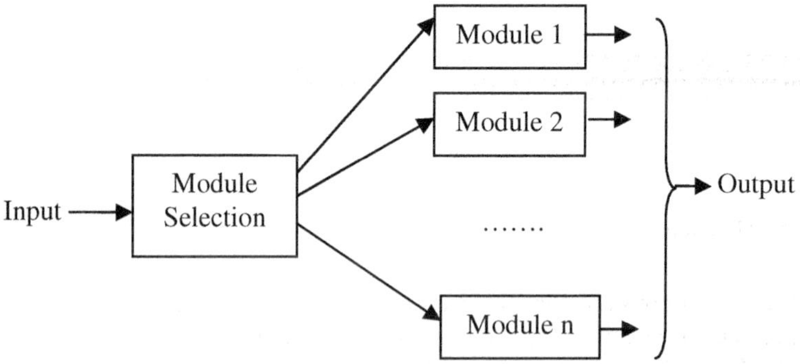

Fig. 14.9 Modular Network for Modularity in Input Space

14.4.1 Division of Input Space

The first problem in these systems is the division of the input space. The input space represents the various input attributes of the systems in its various axes.

The division of this space may be easily carried out by any clustering technique. The clustering is done on the basis of the training data as it is taken to be a sample of all the inputs that the system may be exposed to in real life. The clustering creates cluster centers that represent the central input of the inputs belonging to that cluster. Based on these cluster centers, the cluster boundaries may be drawn. These boundaries fragment the entire input space into regions or clusters. The number of clusters is equal to the number of modules to be used in the approach. Each cluster has its own ANN that is used for figuring out the outputs from the applied inputs. This is the final output of the system.

The clustering may be done using any mechanism. This includes the Fuzzy C-Means Clustering, k-Means clustering, etc. Let us assume that the cluster centers given by the clustering algorithms are C_1, C_2, C_3,..., C_n. Here every centre represents a point in the high dimensional input space. An input is said to be belonging to the cluster whose center lies closest to that input. Hence any input *I* is mapped to the cluster *j*, where *I* may be given by equation (14.4).

$$\text{Cluster} = j : \| I - C_j \| < \| I - C_i \| \ \forall \ j \neq i \qquad (14.4)$$

$\|x\|$ is the Euclidian norm. Any other norm may also be used.

Figure 14.10 shows the clustering for an input space with two dimensions.

Another method of implementation is with the use of Self-Organizing Maps. These are natural clustering agents that carry clustering in an unsupervised neural learning mechanism. These can hence handle a large amount of data reasonably well. They carry out the mapping of the input space into a lower dimensional feature map. The neuron or neuron groups may be mapped to the modules of the MNN in this approach. Many other unsupervised learning models follow same ideology and may hence be used for clustering or the selection of the module from the input by division of the input space.

The other mechanism to carry out the entire process of division may be done with the help of supervised learning. Any of the supervised learning neural network architectures may be used for this purpose. The supervised learning means that the human expert would have to make rough mappings of the inputs to the modules. The easiest manner may be to make the human expert classify the various training inputs into the various modules based on his own criterion or understanding of the problem. We may later expect that the network learns this mechanism by which the mappings are made by the human expert. These may be generalized to the other inputs as well. Now whenever any input is applied, the system uses the trained network to compute the module to which this input needs to be mapped to, or the module that needs to be invoked.

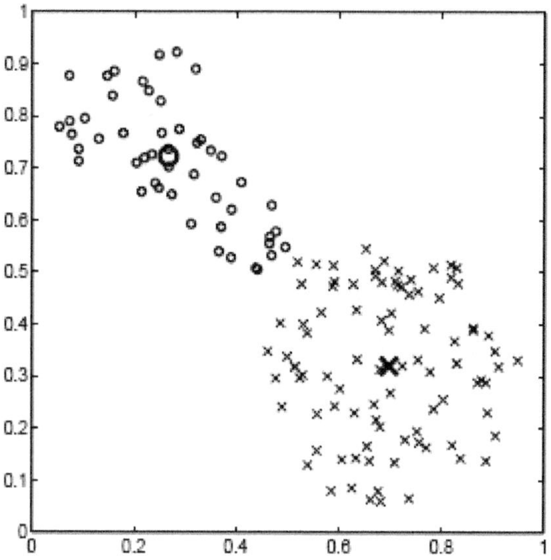

Fig. 14.10 Formation of 2 clusters in the input space

14.4.2 Training and Testing

The next task is to work over the modules. Each of the modules of this network is an ANN. This needs to be trained and tested as per the working methodologies of the ANN. We discuss these steps here.

The training of each of the module takes place by the inputs that belong to it. Hence the entire training data is first passed over the division of input space mechanism to get mapped to some or the other module. Each ANN now has its own training database that consists of the type of inputs that it is supposed to deal with. This network is trained only with these inputs.

At the time of testing, an input is applied to the MNN. The mapping criterion is again used for the decision regarding the module. The specific ANN or module as decided by this criterion is given the input for processing or testing. The outputs generated by this ANN are the final outputs of the system.

As we did for the ensembles, we discuss how division of the input space facilitates the classification and functional prediction problems. We additionally discuss the role of the modules in the system generalizing capability.

14.4.3 Classification

The first major type of problem that we consider is classification. Here we need to construct the decision boundaries around the various classes in the input space. The problem would normally come when the number of classes is too high or the decision boundaries are too complex. The large number of classes may be very difficult for a single system to separate especially when the inter classes distances are fairly low. Each class may intermingle in a complex manner with the other class lying nearby in the input space. The system is supposed to imitate all these decision boundaries among the classes. This makes the entire task very big. Similar is the problem when the decision boundary has a complex shape. A single system is expected to record this behavior in the entire input space. This makes the task too big for the system which may fail to perform to imitate all the complexity the decision boundary represents.

When we cluster the input space, we primarily reduce this complexity. A cluster is given a separate ANN for the problem solving. The clusters divide the training and the testing data. As a result every cluster would have a limited number of classes. Now the task is easier. The system just needs to know the classifying mechanism between a limited numbers of classes that happen to lie in the region designated to the ANN. There would be classes that lie around the clustering space. These would have two or more ANNs that carry their detection. The original problem is shown in figure 4.11(a). The simplified versions of the problem given to different neural networks are shown in figure 4.11(b), 4.11(c) and 4.11(d)

Every ANN further has a limited part of the decision boundary that it needs to imitate. This would be much less complex than the entire decision boundary showing complex shapes difficult to imitate.

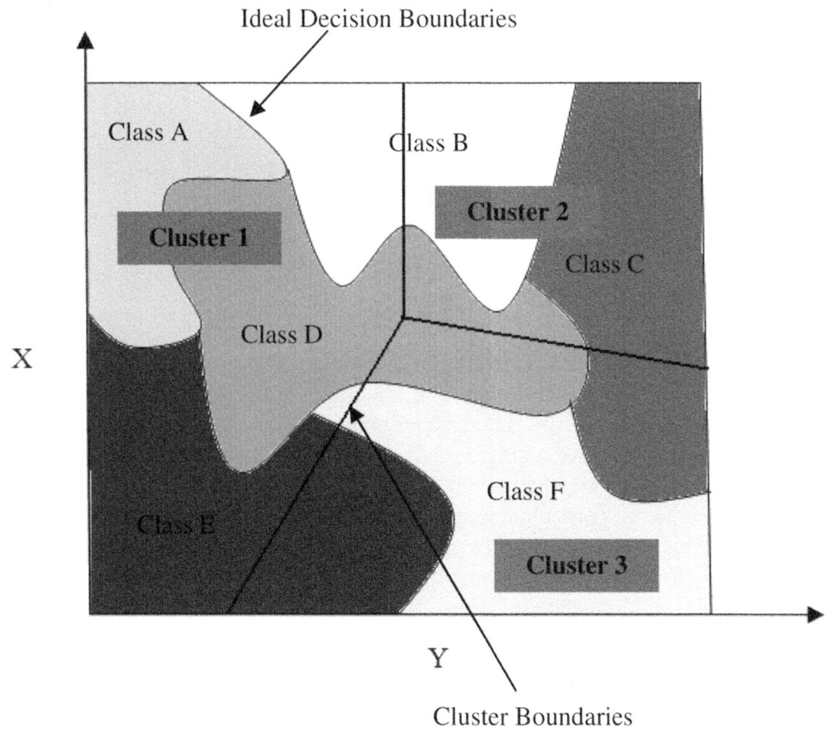

Fig. 14.11(a) Original Problem of classification

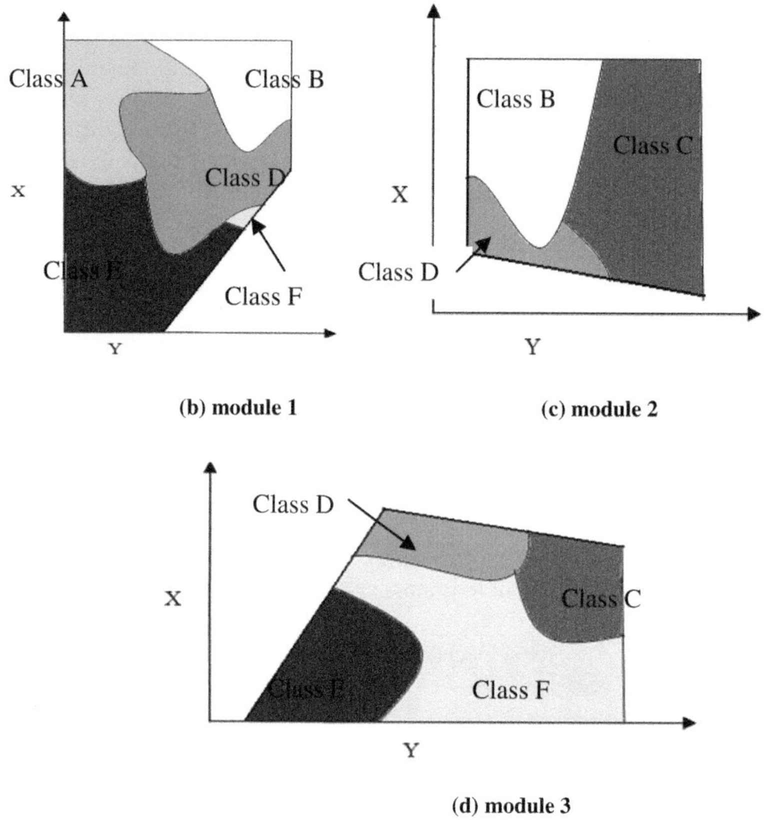

Fig. 14.11(b-d) Simpler problem corresponding given to different modules

14.4.4 Functional Prediction

In the functional prediction problems we have a curve fitting approach. Here we try to form a function whose surface matches the desired output surface. This is done by various means by the various models of the ANN. The complete surface may be too complex for a single ANN. The complexity may be in the form of the number of turns or the changes in output for the changes in any input. The sensitive curves have sharp changes in the output value for small changes in input which is a highly complex surface to be imitated by the ANN. A single ANN may not be able to imitate a high complexity function. Very large ANNs might be very difficult to train and handle. These may further require extremely large training time as well.

The division of the input space is of a high value in such cases. Rather than giving the complete output surface to the ANN, we give only a part of it. This would carry much less complexity and sensitive nature as compared to the original problem. This may be conceived as if we have sliced a part of the output surface, expanded it and given for training and testing to an independent ANN. The complexities would naturally be reduced and training would be easy. This may further take much less time for training than what would have been taken by the entire system comprising of a single ANN.

14.4.5 *Problem Generality*

We have seen that the division of the input space is of a large value as it enables effective learning in cases where it might not have been conventionally possible. This however does not mean that it does not have any adverse effect and multiple and indefinite division would be useful for the overall performance of the system. Recall our discussion over the generalizing capability of the systems that we discussed in chapter 1. A system needs to have a good and global view of the entire surface so as to give the best results. This is the reason why the number of neurons needs to be limited for a Multi-Layer Perceptron and the spread needs to be high for a Radial Basis Function Network. Giving only a localized view of the system or making the problem more local would result in adverse effects of noise, poor generalizing capability and too sensitive prediction for problems with otherwise smooth functions.

If a problem is naturally complex and cannot be solved using a single ANN, it would be wise to break it into smaller problems with an ANN for each of these. However over division may result in the problem having very poor overall generality and hence a poor performance in the testing data set. Hence the decision regarding the number of modules must be done judiciously.

14.5 Evolutionary Modular Neural Network

We discussed in chapter 13 that specifying the architecture of the neural network is a major problem as it is done by the humans who may only be able to try out a limited combinations. This largely leads to sub-optimality. The other problem we considered was that the training algorithms usually get struck at local a minimum which limits the performance of the network. In chapter 13 we presented evolutionary neural networks which solved this problem to a reasonably large extent. The same problem exists with the ANNs as well because the individual modules of the MNNs are ANNs that need to be well trained and must possess an optimal architecture.

Hence in this section we focus upon the evolutionary modular neural network. These are modular neural networks that evolve along with time and generations to attain the optimal architecture. This marks a powerful hybrid algorithm. It uses the power of optimization from the evolutionary algorithm, the modularity from the modular neural networks and the problem solving ability of the artificial neural networks. These systems hence are powerful tools. At the same time it must be realized that the evolution of MNNs is more difficult and complex than the ANNs in terms of the optimal form. This is because of the very complex search space. This is however very simple than the evolution of a very complex neural network that has a large number of connections and neurons (Auda and Kamel 1997; Happel and Murre 1994; Fieldsend 2005; Funabiki 1998; Happel and Murre 1994).

The evolution of the MNN may be done in means and ways similar to the ones discussed in chapter 13. We need a correct selection of the evolutionary as well as the MNN model. This further needs a proper means of representation of the individual and sometimes the formulation of the problem specific operators. The rest of the operation may be trivially carried out as per the evolutionary mechanism with the MNN performance evaluation in the fitness function.

In the subsequent sections we discuss the work of Cho and Shimohara (1998) who used Genetic Programming for the evolution of the Modular Neural Networks. In this approach a tree-like structure is used for the individual representation. Here every genetic individual is a MNN.

The MNN that the authors have proposed processes information by activations between the nodes. These activations may be inhibitory or exhibitory. The MNN contains four types of nodes these are R-nodes, V-nodes, A-nodes and E-nodes. The R-nodes are the ones that have fixed valued connections to the other R-nodes that lie in the same module. These however have variable or modifiable nodes between the members of the other modules. These nodes in the complete MNN architecture are responsible for carrying information or excitation from other modules or send information or excitation to the other modules. V-nodes are the ones that have inhibitory connections going out to the other nodes. The various V-nodes compete for this excitation. A V-node receives the excitation from a particular R-node. In this manner these nodes pair with each other in the MNN. The A-nodes are the ones that are excited by one R-node and are inhibited by all the V-nodes in the module. Activation of A-node is positive function of the amount of competition in a module. E-node activation is a measure of the level of competition going in a module. The details of the manner in which these nodes carry out the task of classification of any input to compute the correct output can be found in Cho and Shimohara (1998). Here we would be mainly interested in the task of evolution of the proper architecture of this type of MNN with the discussed characteristics.

Hebb Learning rule is used for the optimization of the various inter-modular weights in the MNN. This rule uses the past values of the weights and finds their new values based on the activations produced by the various nodes. This may be repeated for a few iterations till the weight is converged to some value.

The evolution is carried out till the network meets the minimum performance set. The genetic operators are used for the generation of a higher generation population from a lower generation population. We discuss the various aspects of this algorithm in the coming sections.

14.5.1 Individual Representation

The first task is the mechanism by which we represent a genetic individual. Here the optimization is supposed to evolve the complete architecture of the MNN. Two points of optimization that are represented in the genetic individuals are the manner in which the various modules are connected to each other as well as the total number of nodes in every module.

Every module used in the evolution is designated some numeric code. The structure of the manner in which the various modules are connected to each other is given in form of a tree like structure. Every node in this tree structure represents a module of the MNN. The module number is contained by the tree node. The various arcs stand for the inter-module connections. Suppose a parent has arcs to its children in the tree like genetic individual. This is the representation of a physical connection between the corresponding modules in the MNN. The total number of nodes in every module may be separately mentioned in the genetic individual. Consider a general genetic individual showing the inter-modular connections as given in figure 14.12(a). The equivalent MNN is shown in figure 14.12(b). It is possible that some modules would not be the part of the working MNN that the individual represents. These may not be linked to the inputs directly or indirectly by other connections. These may be simply ignored.

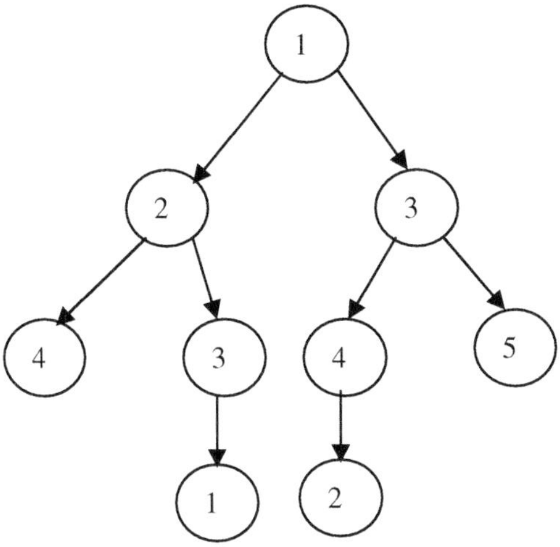

Fig. 14.12(a) The Genetic Individual

14.5.2 Genetic Operators

The genetic operators help in the generation of a higher generation of population pool from a lower generation. The conventional Genetic Programming parameters may be used since the problem has been effectively converted into a genetic programming optimization problem. This consists of a selection scheme which may be conveniently taken as uniform stochastic selection with rank based scaling. Crossover is used to interchange sub-trees between two trees. Mutation is used for altering the trees by some amount depending upon the mutation rate. Additionally insert is used to insert a sub-tree and delete is used to delete a sub-tree from an individual.

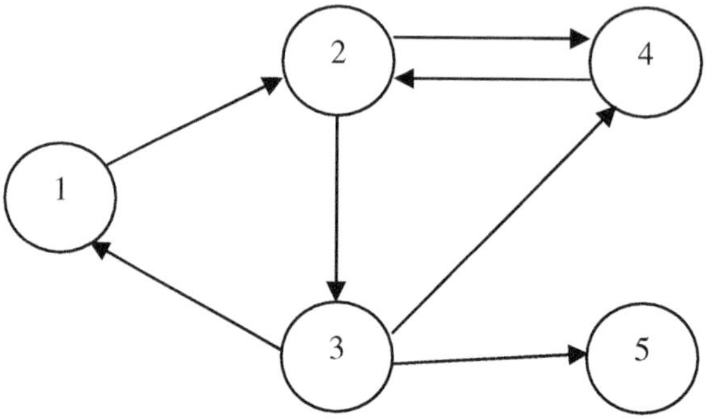

Fig. 14.12(b) The Genetic Individual

14.5.3 Fitness Evaluations

The fitness evaluation of the algorithm consists of reading the individual from its genotype representation and converting it into a MNN. This MNN is then trained using the Hebbs learning law to figure out the inter-modular connection weights. The final network is trained for the performance evaluation. This performance decides the overall fitness of the genetic individual that needs to be optimized by the Genetic Programming.

14.6 Co-evolution of Modular Neural Network

In this section we make use of the optimization power of the cooperative convolution in evolving the MNN. Cooperative convolution believes the adoption of a cooperative strategy in the evolution process. In this strategy an individual is not only judged on the basis of its performance. It is also judged on how well it cooperates and helps in the evolution of the other individuals of the population. This evolutionary technique believes in the convolution of the entire population where all the individuals help each other to attain higher performance.

In this section we see how this methodology of cooperative convolution would help in the evolution of the MNN. Here we make the various modules of the ANN cooperate with each other for the generation of the best modules. The MNN simply selects the best modules that solve the problem.

Here we would summarize the works of Pedrajaz *et al* (2003) who used Co-evolution for the evolution of the MNN and called the algorithm as COVANET. Here the authors proposed the evolution to be at two levels. The first level aimed in the construction of good modules as per the requirements. The second level was at the MNN level. This aimed at the correct selection of the modules for an overall high performance in solving the problem. The first level or the module level used Evolutionary Programming as the evolution technique and a variable size neural network is evolved in the evolutionary process. At the second level an invariant of

Genetic Algorithm is used as the evolutionary technique. We discuss both the levels individually.

14.6.1 Module Level

The first level of the implementation of the algorithm is to generate high performance modules for the solution of the problem. Here the evolutionary process tries to optimize a variable size neural network. There are a fixed number of subpopulations that are maintained for the evolution to take place at this level. This plays a major role in diversity preservation and retaining of diverse modules that may make different modules of the entire MNN.

In this mechanism, the individual may be coded using any suitable technique such that all the parameters appear into the genotype representation. The total number of neurons is variable which may further make the individual size as variable. The initial population contains a rich mix of ANNs with varied size in terms of the hidden layer neurons and a varied number of connections in between the neurons.

The next major task is the specification of the genetic operators that generate higher generation population from the lower generation. This technique does not use the conventional crossover operator. Instead replication operator is used. At any time the best *P%* of the individuals are passed straight from the lower generation to the higher generation. The rest *(1-P)%* of the individuals are generated by mutation of the *P%* of the individuals.

The mutation rate depends upon the relative fitness of the individual as compared to the other individuals in the population pool. Mutation is done in two ways. These are parametric mutation and structural mutation. The parametric mutation uses a stimulated annealing mechanism for local search in the search space. Using this, the parameters are modified in the individual. Structural Mutation changes the structure of the individual. It has operators like adding a hidden layer node, deleting a hidden layer node, adding a connection and deleting a connection.

The fitness evaluation of the module must ensure an incentive for cooperation as per the philosophy of the MNN. The fitness evaluation of this level hence takes into account three factors. These are *substitution*, *difference* and *best*. We discussed that the MNN is a collection of these modules. In an MNN the individuals are taken from the various subpopulations. Since the total number of subpopulations is fixed, the total number of modules in the MNN is also fixed.

In *substitution* we first select the best *k* MNNs. These are selected from the optimizations at the other or the network level which we discuss in the next section. Each MNN has an individual from each subpopulation. To measure this factor we substitute the individual from the corresponding subpopulation in the selected MNNs with the individual whose fitness needs to be found. The performance of the MNNs with substituted individual is measured. This is matched against the original performance of the selected MNNs. Here we measure how well the module performs against the best module of that subpopulation and this encourages competition among the various modules of a subpopulation.

The *difference* fitness evaluation factor removes the particular module from all the MNNs formed at the second or network level. Here we measure the difference in performance that we get by deletion of this module. This factor tries to encourage different modules developing different characteristics. A module that possesses similar characteristics than the available modules is not of use to the system. These are hence assigned poor fitness values.

The best evaluation technique selects the best *k* MNNs from the individuals of the network level optimizations where the module is found. It returns the average performance of these MNNs. This factor judges the performance of this module in the best MNNs. This prevents the module from being penalized in case it belonged to a MNN with an overall poor performance. We hence only select the best *k* MNNs.

The large networks in terms of both the number of connections as well as the number of neurons are given some penalty. This is done by degrading their fitness by some value directly proportional to the number of neurons as well as the number of connections.

14.6.2 Network Level

The other level in which the algorithm works for optimization is at the network level. In this level the algorithm tries to select the best modules optimized from the module level and forms a network or MNN out of it. The various individuals in this level come from the various subpopulations that were maintained at the module level. The individual at this level is hence a collection of integers. These integers denote the modules that make up the complete MNN. This level uses an invariant of the Genetic Algorithm called as the steady state genetic algorithm. This algorithm is more aggressive in selection criterion. This further has a slower evolution than the module level. This algorithm differs from the standard genetic algorithm in terms of implementation of the crossover operator and fitness assignment, the details of which are explained in Pedrajaz *et al* (2003).

A two point crossover has been used for the evolution. Mutation consists of the exchange of modules between the individuals of this level that represent MNNs. These must belong to the same sub-population. The fitness assignment at this level is directly the performance of the MNN that the individual represents. This may be computed by reading the individual, fetching the modules, making the MNN and judging its performance.

The two discussed levels operate hand in hand. A good evolution of both of them is necessary for the overall optimization by this algorithm. The inter-relation between the two levels clearly shows that both the levels help each other to evolve and attain optimal values.

14.7 Chapter Summary

The chapter explored the concept of modular neural networks and discussed the various models of the same. We first took a motivation for solving complex problems using modularity. The task of problem division, solution by individual modules and integration by an integrator were discussed. The ensembles make use of redundant solution to problem. Here a number of modules or neural networks are employed to work on the same data and give results. The results are combined by an integrator. The chapter presented how redundant information in form of solution by various modules could be used for better problem solving. Both classification and functional approximation problems were dealt individually. A number of integration mechanisms were presented. This included polling, probabilistic polling and gating. The other part of the chapter presented the concept of introduction of modularity in computation. Here different modules performed different parts of the same problem. The concept of divide and conquer was used. The model presented was hierarchy of experts where different experts or modules are arranged in some hierarchy for problem solving. Then a model was discussed that divided the entire input space into clusters and used a separate module for each cluster. The different modules hence worked over different parts of the input space. The problem got simplified as a result. The cases of both classification and functional prediction problems were presented. The other concept discussed was evolution of modular neural network. This made use of the evolutionary ability of the evolutionary algorithms to evolve a complete modular neural network. The first model discussed used genetic programming for the same task. Every module had a variety of nodes and connections that were used for problem solving. The genetic programming individual was a representation of the complete modular design. Numerous operators were used for carrying out evolution. The last topic of discussion was the cooperative evolution of the modular neural network. In this model the various modules acted as the individuals of the evolutionary process. These all cooperated with each other for better evolution. Co-evolution was used as the evolutionary algorithm. In this approach the individuals possessing rare skills not possessed by the other modules were suitably rewarded by the fitness function.

References

[1] Auda, G., Kamel, M.: CMNN: Cooperative Modular Neural Networks for pattern recognition. Pattern Recognition Letters 18, 1391–1398 (1997)

[2] Auda, G., Kamel, M., Raafat, H.: Voting schemes for cooperative neural network classifiers. In: Proc. of the 1995 Int. Conf. on Neural Netw., Perth, Australia, pp. 1240–1243 (1995)

[3] Castillo, O., Melin, P.: Hybrid Intelligent Systems for Time Series Prediction using Neural Networks, Fuzzy Logic and Fractal Theory. IEEE Trans. on Neural Netw. 13(6), 1395–1408 (2002)

[4] Cho, S.B., Shimohara, K.: Evolutionary Learning of Modular Neural Networks with Genetic Programming. Applied Intelligence 9, 191–200 (1998)

[5] Dietterich, T.: Ensemble methods in machine learning. In: Kittler, J., Roli, F. (eds.) MCS 2000. LNCS, vol. 1857, pp. 1–15. Springer, Heidelberg (2000)

[6] Fieldsend, J.E.: Pareto Evolutionary Neural Networks. IEEE Trans. Neural Netw. 16(2), 338–354 (2005)

[7] Fogelman-Soulie, F.: Multi-modular neural network-hybrid architectures: a review. In: Proc. of 1993 Int. Jt. Conf. on Neural Netw. (1993)

[8] Fu, H.C., Lee, Y.P., Chiang, C.C., Pao, H.T.: Divide-and-Conquer Learning and Modular Perceptron Networks. IEEE Trans. on Neural Netw. 12(2), 250–263 (2001)

[9] Funabiki, N., Kitamichi, J., Nishikawa, S.: An Evolutionary Neural Network Approach for Module Orientation Problems. IEEE Trans. Syst. Man Cybern. – Part B Cybern. 28(6), 849–855 (1998)

[10] Gruau, F.: Automatic definition of modular neural networks. Adaptive Behavior 3(2), 151–183 (1995)

[11] Hansen, L.K., Salamon, P.: Neural network ensembles. IEEE Trans Pattern Analysis Machine Intell 12(10), 993–1001 (1990)

[12] Happel, B.L.M., Murre, J.M.J.: The Design and Evolution of Modular Neural Network Architectures. Neural Netw. 7, 985–1004 (1994)

[13] Happel, B., Murre, J.: Design and evolution of modular neural network architectures. Neural Netw. 7, 985–1004 (1994)

[14] Jacobs, R.A., Jordan, M.I., Nowlan, S.J., Hinton, G.E.: Adaptive mixtures of local experts. Neural Comp. 3, 79–87 (1991)

[15] Jenkins, R., Yuhas, B.: A simplified neural network solution through problem decomposition: The case of the truck backer-upper. IEEE Trans. on Neural Netw. 4(4), 718–722 (1993)

[16] Jordan, M.I., Jacobs, R.A.: Hierarchical Mixtures of Experts and the EM Algorithm. Neural Com. 6(2), 181–214 (1994)

[17] Kala, R., Shukla, A., Tiwari, R.: Handling Large Medical Data Sets for Disease Detection. In: Shukla, A., Tiwari, R. (eds.) Biomedical Engineering and Information Systems: Technologies, Tools and Applications, IGI Global (2009a)

[18] Kala, R., Shukla, A., Tiwari, R.: Hybrid Intelligent Systems for Medical Diagnosis. In: Shukla, A., Tiwari, R. (eds.) Intelligent Medical technologies and Biomedical Engineering: Tools and Applications, IGI Global (2009b)

[19] Kala, R., Shukla, A., Tiwari, R.: Comparative analysis of intelligent hybrid systems for detection of PIMA indian diabetes. In: Proc. of the IEEE 2009 World Cong. on Nature & Biologically Inspired Comput., NABIC 2009, Coimbatote, India, pp. 947–952 (2009c)

[20] Kala, R., Vazirani, H., Shukla, A., Tiwari, R.: Fusion of Speech and Face by Enhanced Modular Neural Network. In: Proc. of the Springer Int. Conf. on Information Systems, Technology and Management, ICISTM 2010, Bankok, Thailand, March 2010, pp. 363–372 (2010)

[21] Krogh, A., Vedelsby, J.: Neural network ensembles, cross validation, and active learning. In: Gesauro, G., Touretzky, D.S., Leen, T.K. (eds.) Advances in Neural Information Processing Systems, vol. 7. MIT Press, Cambridge (1995)

[22] Melin, P., Acosta, M.L., Felix, C.: Pattern Recognition Using Fuzzy Logic and Neural Networks. In: Proc. Int. Conf. Artif. Intel., Las Vegas, USA, pp. 221–227 (2003)

[23] Melin, P., Castilo, O.: Hybrid Intelligent Systems for Pattern Recognition Using Soft Computing. Springer, Heidelberg (2005)

[24] Melin, P., Gonzalez, F., Martinez, G.: Pattern Recognition Using Modular Neural Networks and Genetic Algorithms. In: Proc. of Int. Conf. Artif. Intel., Las Vegas, USA, pp. 77–83 (2004)

[25] Pedrajas, N.G., Martínez, C.H., Pérez, J.M.: COVNET: A Cooperative Coevolutionary Model for Evolving Artificial Neural Networks. IEEE Trans. Neural Netw. 14(3), 575–596 (2003)

[26] Ronco, E., Gawthrop, P.: Modular neural networks: A state of the art. Center Syst. Contr, Univ. Glasgow, Glasgow, U.K., Technical report: CSC-95 026 (1995)

[27] Schmidt, A., Bandar, Z.: A Modular Neural Network Architecture with Additional Generalization Capabilities for Large Input Vectors. In: Proc. of the Int. Conf. on Artif. Neural Netw. and Genetic Algorithms, Norwich, England (1997)
[28] Schyns, P.G.: A modular neural network model of concept acquisition. Cognitive Science 15, 461–508 (1991)
[29] Shukla, A., Tiwari, R., Meena, H.K., Kala, R.: Speaker Identification using Wavelet Analysis and Modular Neural Networks. J. of Acoustic Society of India 36(1), 14–19 (2009)
[30] Waterhouse, S., Mackay, D., Robinson, T.: Bayesian methods for mixture of experts. In: Touretzky, D.S., et al. (eds.) Advances in Neural Information Processing Systems, vol. 8. MIT Press, Cambridge (1996)

List of Figures

Fig. 14.1. General Architecture of Modular Neural Network
Fig. 14.2. Use of Ensembles for Classification
Fig. 14.3. Ensembles for functional prediction problems
Fig. 14.4.General Architecture of Ensembles
Fig. 14.5. Polling Integration
Fig. 14.6. Probabilistic Count Integration
Fig. 14.7. Gaiting Integration
Fig. 14.8. Hierarchy of Experts
Fig. 14.9. Modular Network for Modularity in Input Space
Fig. 14.10. Formation of 2 clusters in the input space
Fig. 14.11(a). Original Problem of classification
Fig. 14.11(b) Simpler problem corresponding given to different modules: module 1
Fig. 14.11(c) Simpler problem corresponding given to different modules: module 2
Fig. 14.11(d) Simpler problem corresponding given to different modules: module 3
Fig. 14.12(a). The Genetic Individual
Fig. 14.12(b). The Genetic Individual

Chapter 15
Hybridizing Neural and Fuzzy Systems

Abstract. The neural networks are excellent means of learning where training algorithms may be used for the tuning of the various parameters of the neural network. The fuzzy systems are extensively used for their fuzzy approach to problem modeling and solving. In this chapter we would present how the problem modeling capabilities of the fuzzy systems combines with the learning ability of the neural networks to create the Adaptive Neuro Fuzzy Inference Systems. We later see how these systems may be evolved using an evolutionary approach to make evolutionary neuro fuzzy systems. The other part of the chapter would focus upon the mechanism of fuzzy neural networks. These are neural networks that take fuzzy inputs and generate fuzzy outputs. Here we would transform the various neural computations into fuzzy arithmetic for problem solving. The neural networks are many times regarded as black boxes. We hence need specialized mechanisms to extract out rules from these networks for understanding and implementation. This would be discussed as the last part of the chapter.

The fuzzy systems are good problem solving agents. These systems use the concepts of fuzzy sets to compute the output of any applied input. The basic purpose of these systems is to use rule based logic for problem solving. The rules are used for the purpose of specifying the system behavior to any input. The rules are evaluated based on fuzzy algebra. For the same reasons the inputs first need to be fuzzified and the fuzzy outputs need to be later defuzzified to produce the final system output. The rules are the most important constituents of the fuzzy systems as the entire behavior of the system depends upon how well these are framed as well as how realistic these are.

The Artificial Neural Networks (ANNs) are good means of machine learning. These systems map the inputs to the outputs with the help of layer wise processing. Every layer works over the information it receives and transfers the outputs to the next layer for the further processing. In this manner the information is reputedly processed by multiple layers and multiple neurons before being returned as the output of the system. The knowledge of these ANNs is in the form of weights that connect the various neurons.

The ANNs have a specialized training algorithm that enables it to tune the weights. This tuning is done by reputedly passing the training inputs and sometimes also the outputs through the network and correcting the errors. After a few cycles of training with the entire training data, the system is able to give a good

A. Shukla et al.: Towards Hybrid and Adaptive Computing, SCI 307, pp. 337–359.
springerlink.com

performance and compute the correct outputs to the inputs given. If the generality of the network is high, the outputs are almost correct for the unknown input as well. In this manner the training helps the ANNs to attain optimal state which plays a major role behind its widespread use in applications.

The first major point of discussion in this chapter would be the Adaptive Neuro Fuzzy Inference Systems (ANFIS). The fuzzy systems are good in problem solving by their fuzzy approach. These however cannot learn from the available data. In ANFIS we construct a fuzzy system over a multi layered neural architecture. This enables us to use the neural training algorithm for the learning and evolution of the fuzzy systems.

ANFIS may have sub-optimality due to various discussed reasons like poor training algorithm that is likely to get struck at some local minima, ineffective shape of the underlying fuzzy system, etc. We hence discuss the evolutionary ANFIS systems that implements evolutionary concepts in the ANFIS architecture. Here we use the optimization power of the evolutionary algorithms for the proper as well as optimal evolution of the ANFIS.

This is not the only manner in which the ANNs and fuzzy systems may combine. The ANN process information in a layered manner to generate the outputs from the inputs. These have crisp inputs, crisp processing and crisp outputs. The Fuzzy Neural Networks are special types of neural networks that bring this entire concept into fuzzy domain. These neural networks have fuzzy inputs and fuzzy outputs. Further there are fuzzy neurons that carry out processing of the inputs using the fuzzy arithmetic. These form another major class of the application of fuzzy and neural systems that brings the fuzzy capabilities into neural networks.

The ANNs are usually regarded as black box approaches. This is because of the fact that we apply the inputs and obtain the corresponding outputs. To a very large extent it is unknown what do the various weights and neurons stand for and what logic do they represent. This makes the entire task of problem solving using ANNs completely uncontrollable. This motivates the extraction of proper rules from these networks that represents the system in the form of some set of rules or logic. This further bridges the gap between a neural approach that entirely relies on learning and the fuzzy approach that largely relies on the rules.

15.1 Adaptive Neuro Fuzzy Inference Systems

The first system that we take in this chapter is the Adaptive Neuro Fuzzy Inference System (ANFIS). We learnt in chapter 4 about the Fuzzy Systems. One of the major problems in these systems was that they were usable to learn from the training data. Hence we need to manually adjust the system parameters on seeing the kind of error and the manner in which it may potentially be corrected. This made the entire task of working with fuzzy systems problematic due to the large human intervention in figuring out the rules, membership functions, etc. as per the requirements of the training data. This hence needs to be automated using some mechanism.

We saw the application of the Genetic Algorithm is a potential application to make this system automated using the optimization capabilities of the Genetic

Algorithms. This was discussed under the head of evolutionary Fuzzy Inference Systems in chapter 13. The evolutionary approach may however be slow in many cases, especially when the training data is very large. It is a valuable tool especially for the cases where we need to escape from local minima and reach the global minima.

The ANNs have a good training procedure where the weights and biases of the various neurons of the neural networks are automatically tuned by the training algorithm. This tuning of the weights takes place in multiple cycles or epochs. The motivation behind ANFIS is to use this learning procedure that is already built and well-studied for the neural networks. This requires that we must be able to represent the Fuzzy Inference System (FIS) in the form of a Multi-Layer Perceptron (MLP) or ANN. This would enable us to tune the various weights of the ANN which would represent the parameters of the FIS (Dubois and Prade 1980; Jang 1991, 1993; Jang and Sun 1995; Abraham 2001; Juang and Lin 1998; Kala *et al* 2010a, 2010b; Rutkowski 2004).

The general architecture of the ANFIS is given in figure 15.1. This shows the ANFIS with multiple layers that perform various operations for the processing of the input to produce the output.

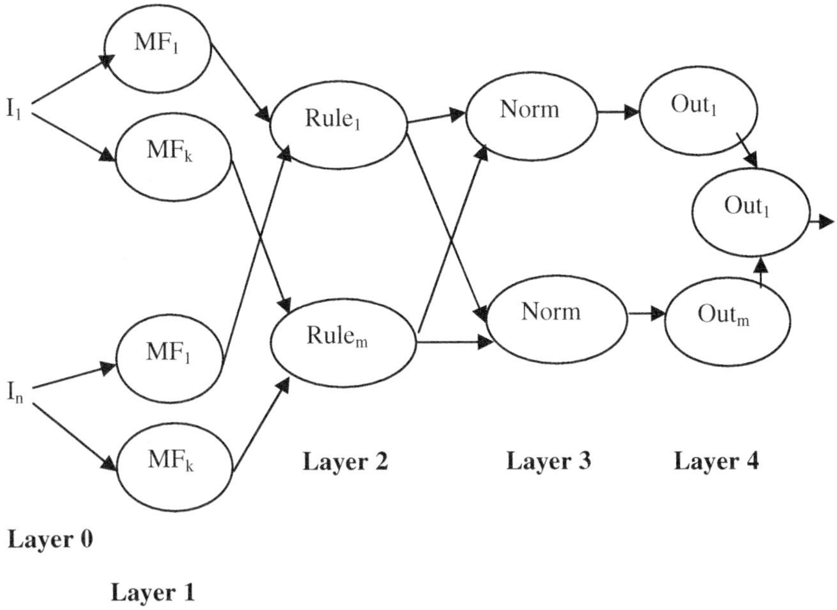

Fig. 15.1 ANFIS Architecture

Looking at figure 15.1, we can easily state that the complete system may be viewed in two ways. The first view is that of the FIS. In this view there is a FIS. It performs all the operations in the same manner as we studied in chapter 4. The input is applied to the system. It is then fuzzified. This now becomes the fuzzy input. The next task is the application of the rules. Each rule acts over the fuzzified inputs to produce the fuzzified outputs. The various fuzzy outputs are later on aggregated by the aggregation operator to produce a single fuzzy output. This is later on used as an input of the defuzzification operator that returns the final output of the system. In ANFIS also the same procedure takes place as it must be clear with the notations used in figure 15.1.

The other manner that we can look at the figure 15.1 is by a neural network approach. This figure represents a MLP where the various neurons are interconnected to each other in the form of connections. There are various neurons that are used for the information processing. Each neuron performs a weighted addition and has an activation function that further processes the information. The various ANFIS parameters are conveniently modeled as weights and biases of this ANN. This enables the use of the ANN training algorithm for the training of the system. This carries out the tuning of the various weights. In this manner the FIS parameters get automatically tuned by using the neural training procedure. The Back Propagation Algorithm (BPA) that was extensively used for carrying out the training of the MLP can be conveniently used for the training of the ANFIS as well for the parameter optimization.

As a result of this training, the ANFIS is able to give a better performance over the training data. The role of the ANN after the parameter tuning is largely over and the resulting system may be viewed as a Fuzzy Inference System. The resulting FIS is able to correctly map the training inputs to the corresponding training outputs. In case the architecture of the system is such that the generality is high, the system would give the correct outputs to the testing inputs as well.

We study some of the major concepts behind these systems in detail.

15.1.1 Layered Architecture

The first major task in the use of ANFIS was to represent the FIS in a layered architecture to enable the use of ANN training algorithm. We hence discuss how the various layers of the ANN carry out the same information processing that is carried out by the FIS. We look at the entire architecture from a Fuzzy point of view to study as well as verify the same operations are carried in the same manner.

15.1.1.1 Layer 0

The first layer is the input layer. This layer is a dummy layer that does not carry out any computation. The role of this layer is to simply provide the inputs to the other layer or layer 1. We discussed the same concepts in the conventional ANNs where the input neurons were only supposed to provide the inputs to the subsequent layers. The inputs of this layer are distributed to all the neurons of the first layer by the interconnections as seen in figure 15.1.

15.1.1.2 Layer 1

The next layer is the membership function layer. This layer does the task of fuzzification of the crisp inputs applied in layer 0. Each of the applied inputs is broken down into the membership values of the various membership functions. The number of membership function for each of the applied inputs depends upon the design of the FIS. Each of the neuron of this layer represents a membership function and is connected to the input whose membership function it represents. We studied in chapter 4 that fuzzification was the first step to be carried out in the problem solving approach with FIS. Say an input had membership functions as *low*, *medium* and *high*. In such a case we would connect the input into three neurons with each neuron having its own membership function.

The membership function serves as the activation function of the neuron. The input is directly given to the membership function. In other words the interconnecting connection from the input to the membership function neuron has a fixed weight of unity. The neuron further has a single input whose membership value is to be computed.

Let the applied input of the ith input neuron be I_i. Let us further assume that this input needs to be divided into a variety of membership values. Let the jth membership value be denoted by O^1_{ij} and the corresponding membership function be given by $\mu_{ij}(x)$. We also know that from a neural perspective O^1_{ij} is the output of the jth neuron connected to the ith input in layer 1. Also $\mu_{ij}(x)$ is the activation function of this neuron. The relation between O^1_{ij} and the applied input I_i may be easily given by equation (15.1).

$$O^1_{ij} = \mu_{ij}(I_i) \tag{15.1}$$

This output would form the input to the next layer that would be transferred using the interconnecting weights as seen in figure 15.1.

15.1.1.3 Layer 2

The next layer is the *AND* Layer. Here each rule of the system is presented as a neuron. This layer hence has as many neurons as there are rules in the FIS. The inputs to any rules are the membership functions that participate in a rule. We know that every input participates in a rule with a single membership function. Say the rule is "*If input$_1$ is low and …*". This means that the membership function corresponding to *low* in input$_1$ would be participating in the given rule. The membership functions participating in a rule are the one connected to it in the neural architecture. Every neuron in the *AND* layer has connections from a membership function neuron of the previous layer if that membership function participates in the rule. If it does not participate, no connection is made.

Here we have assumed that the rules consist of the various membership functions joined with the help of the *AND* operator. This is a valid assumption since most of the design of the FIS makes use of the *AND* separated expressions as the antecedents. Since we have only considered the antecedents of the rules, this layer may be called as the *AND* layer. This layer carries exactly the same operation in the FIS architecture. It reduces the series of *AND* separated membership functions in the antecedents of the rule to a single value by using the fuzzy *AND* arithmetic operator. The fuzzy *AND* operator may be conveniently taken as the product operator.

Let the output of the kth rule or the kth neuron be o_k^2. Here we are supposed to take the product of the various inputs that are given to this neuron. This is shown in equation (15.2).

$$o_k^2 = \prod o_{ij}^1 \tag{15.2}$$

From the neural approach, these neurons may be taken to be special neurons that compute the product of the inputs in place of their sum and have an identity activation function that returns the same output as is the input given to it. Hence in all only the product of inputs is performed.

15.1.1.4 Layer 3

The next layer in the ANN architecture of the ANFIS is the normalization layer. The output of the *AND* layer is the value of the antecedents of any rule. This signifies the total firing power of the rule. A large value means that the rule is more dominant in deciding the final output of the system. A less value means that the input is loosely related to the particular rule and this rule has less dominant role in deciding the final output.

This layer does the task of normalizing these values, so that the final firing powers lie within the range of 0 to 1 and all the firing powers of the various rules collectively sum to unity. This normalization is done by the division of every firing power with the sum of all firing powers from the *AND* layer. Let the output of this layer be o_i^3. This may be easily given by using equation (15.3).

$$o_i^3 = \frac{o_i^2}{\sum_j o_j^2} \tag{15.3}$$

15.1.1.5 Layer 4

The next layer is used for the calculation of the output of every rule. Hence there are as many neurons in this layer as there are rules in the system. Each neuron represents a rule of the system. This layer is adaptive. This means that we can tune this layer as per the training data. This adds a lot of flexibility or adaptability to the system where we can modify the weights and biases of this layer for the tuning of the network.

The output of this neuron is taken as a general weighted sum of the inputs that were originally given to the system. This is additionally multiplied by the corresponding weights that the neuron gets from the previous layer. This forms the output of these neurons. The inputs given to the neurons are hence all the system inputs as well as the normalized firing power from the previous layer. The complete architecture of these neurons may hence be given by figure 15.2.

Let the output of this neuron be o^4_i for the ith rule. This may be easily given by equation (15.4).

$$o_i^4 = o_i^3 \left(\sum_j p_j I_j + b_j\right) \tag{15.4}$$

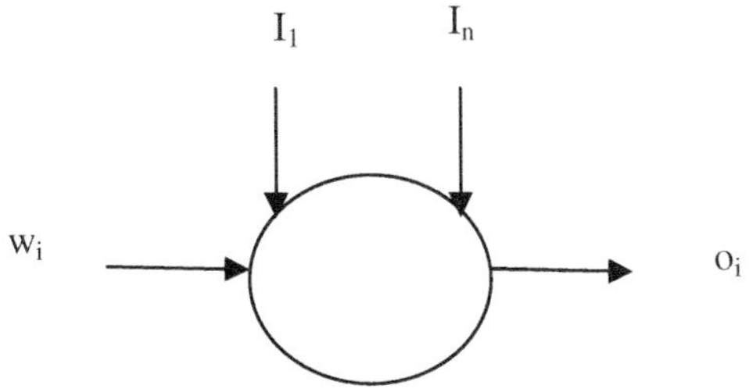

Fig. 15.2 Layer 4 neuron

In this manner the computation by every rule is done by the ANFIS. We can easily see that this is not the computation that we studied in chapter 4 under FIS. The model here deviates from the model presented in chapter 4. This model of the FIS is a linear model that has the individual outputs of the rules as linear weighted sums of the inputs. An extra term of bias is added to control the outputs. This model is a much simplified model that is easier to train and test using the neural architecture. This model enables easy learning and testing of the ANFIS. This specific step is the equivalent implication operator where the implication is performed using the multiplication or the product operator over the antecedents and the consequents of the rule. Using this model the general fuzzy rule may be conveniently be given by the equation (15.5).

$$\text{if } I_1 \text{ is } MF_1 \wedge I2 \text{ is } MF_2\ldots\text{.. then } O_1 \text{ is } p_1.I_1+p_2I_2+\ldots \quad (15.5)$$

15.1.1.6 Layer 5

The last layer is the output layer. This layer has as many neurons as there are outputs in the system. The task of this layer is to give the final output of the system. We have the outputs of the various rules from the previous layer. Each output is an indicator of the kind of output the system must have. The task of this layer is to compute this final output.

Again we may see that a conventional FIS studied in chapter 4 gave complete membership functions as their outputs. These needed to be aggregated and defuzzified. The ANFIS however restricts us from these bulky operations as the FIS needs to be trained from the database available. Besides there are always modeling constraints. The linear model that we have assumed is much simplified version which makes the computation of output reasonably simple. Here we just need to take a weighted sum of the outputs of the various rules. This forms the final output of the system.

Consider the output of any neuron *i* of this layer to be given by o^5_i. This may be computed using equation (15.6).

$$o_i^5 = \sum_j o_j^4 \quad (15.6)$$

In this manner we have simply represented an FIS in an ANN architecture that can carry out the same operation as carried out by FIS. This further enables learning using the training algorithms that apply to ANN. This is discussed next.

15.1.2 Training

The underlying motivation behind the entire ANFIS architecture was that it must be able to take up neural learning so as to optimize its architecture. Now since we have successfully implemented the FIS over the neural architecture, we must have an insight into how this structure can be trained by the available training data. This training may be easily done by the application of the Back Propagation Algorithm that is a commonly used algorithm for the training of the ANN. We additionally have a hybrid training algorithm that enhances the entire training procedure. We discuss these concepts in the coming sub-sections.

15.1.2.1 Back Propagation Algorithm

The Back Propagation Algorithm (BPA) is the first algorithm we discuss that carries out the training of the ANFIS. This algorithm has two stages of forward and backward propagation. In the stage of forward propagation the input is applied to the system and the corresponding output is computed. This is done using the conventional manner of problem solving by the particular ANN or FIS in this case. This happens in a layered manner with the previous layers feeding the inputs to the forward layers until the final output is computed by the last layer. The other stage is the backward propagation. In this layer we compute the error by comparing the actual and the desired output. This error is propagated backward from one layer to the previous layer. All the layers adjust their modifiable parameters in such a manner so as to give a better result the next time the same input is applied.

This complete step of forward and backward propagation is applied for all the inputs that the system has. This makes one training cycle or one epoch of the algorithm. The training is repeated with many of such cycles until the error is within the toleration limits or some other stopping criterion is met. The number of cycles is known as epochs. The algorithm is shown in figure 15.3.

15.1.2.2 Hybrid Training

The next algorithm for training is the hybrid training algorithm. In this algorithm we train the ANFIS by using a hybrid algorithm that trains both in the forward as well as the backward pass of the training algorithm. We essentially divide the training parameters of the algorithms into two types. These are premise parameters and the consequent parameters. The premise parameters are the ones that lie towards the left hand of the implication operator of any rule. The consequent parameters are the ones that lie on the right hand side of the rule. The premise parameters are nonlinear in nature whereas the consequent parameters are linear in nature.

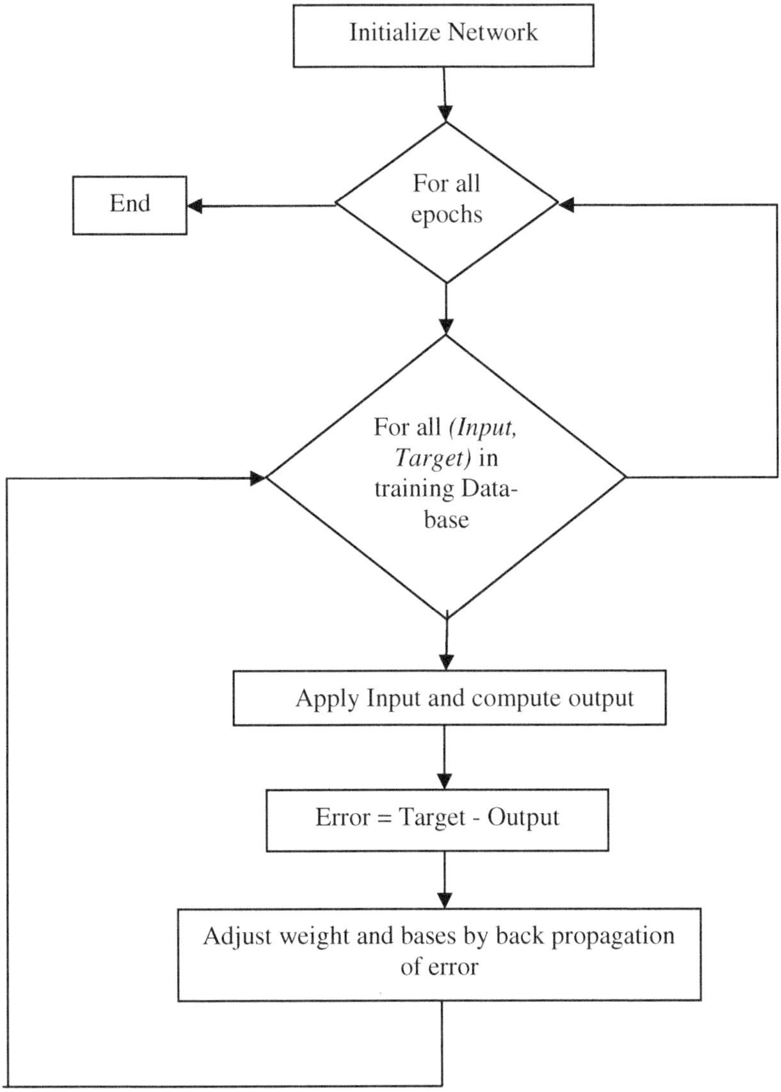

Fig. 15.3 Back Propagation Algorithm

In the forward stage the least-squares method is used to compute the error. This error is used for the modification of the consequent parameters. The premise parameters are kept constant in this stage of the algorithm. The other stage of the algorithm is the backward phase. In this stage the consequent parameters are kept as constant and the premise parameters are modified. This takes place in a conventional manner as the backward stage of the BPA.

The algorithm, just like BPA, follows a steepest descent approach is used for the modification. Here the various parameters are modified such that the error is much less in the next stage. If the total error is large, the modification would be large so as to make the system give almost correct output the next time the same input is applied. For smaller errors, the changes are small. These approaches normally use the slope or the gradient to predict the location of the minima, keeping all other parameters except it constant. The necessary modifications are hence carried out to reach this minimum which is the case when the correct output is given.

15.2 Evolutionary Adaptive Neuro Fuzzy Inference Systems

We studied the added powers that evolution brings in the neural as well as the fuzzy systems. These advantages of the evolution must be naturally present into the neuro fuzzy systems as well. In this section we try to add this advantage of evolution towards the ANFIS and we devise a mechanism that would enable the ANFIS to evolve itself along with time. We model a variable architecture ANFIS. This system would be able to compute the correct architecture as well as the correct parameter values of the ANFIS. In these systems we use the problem modeling of the ANFIS and the optimization power of the evolutionary algorithms to generate a final system for optimal performance.

The ANFIS used in this approach has a similar architecture to the one discussed in section 15.1. This consists of 5 layers. Each of these layers performs the desired tasks of the ANFIS and contributes towards the problem solving. The input is applied to layer 0 and the output is collected from layer 5. In this entire architecture, there are a number of parameters that decide the performance of the systems. Some parameters decide the structure of the ANFIS. These include the total number of membership functions per rule. The other parameters are non-structural that does the tuning of the ANFIS. These include the various weights and biases that we discussed in the previous section.

The manual working in ANFIS is especially difficult. The manual expert is supposed to set the ANFIS architecture. The major problem is to specify the number of membership functions for every input and sometimes the number of rules. This greatly impacts the total number of modifiable parameters that the ANFIS is supposed to optimize during the training phase. An increase in the number of membership functions exponentially increases the parameters available for modification which again greatly increases the training search space and the complexity. This makes the algorithm very time consuming and slow. The algorithm many times is not able to train itself optimally within decent time limits. The entire procedure may not be easy to be carried out again and again with different parameters. This necessitates the need of automated systems to evolve the architecture as well as the parameters of the ANFIS that give optimal performance as per the problem in hand (Fagarasan 1996; Ouyang and Lee 2000; Pedrycz and Reformat 2003; Pena-Reyes and Sipper 2001; Russo 2000).

Here we present the model proposed by Rastegar *et al* (2005). In this paper the authors had used a cooperative co-evolutionary approach for problem solving with ANFIS. The ANFIS system has numerous parameters that must be fixed. Besides we need to compute the most optimal structure. The model consisted of two evolutionary algorithms that worked in a cooperative manner for problem solving. The first approach is Genetic Programming (GP) that optimizes or evolves the architecture of the system. The second approach is Genetic Algorithm (GA) which further tunes the parameter values to enable the individual attain the optimal form. Both these algorithms interacted for the exchange of information.

15.2.1 Interaction between the Evolutionary Approaches

The interaction between the GP and the GA is a major factor in this model. The GP individuals represent ANFIS. The GP individual has parameters that need further optimization. This optimization is carried out by the use of GA. In this manner the GP uses the GA algorithm as slave optimization process. The optimized parameters are returned by the GA that may be used by the GP individual being optimized. This individual has a better fitness value as compared to the original individual before the GA optimizations.

The continuous evaluations of the GA would be too much computationally expensive. This is especially considering the fact that the GP would have a large number of function calls and for every call the GA optimization needs to be called. The computationally expensive nature of the GA would itself be a problem for the algorithm to be time effective. Hence the time needs to be shortened for the entire optimization to be completed timely.

For this time constraints in the model, the authors proposed the use of a database. The database stores the results and structures of all the GA optimizations. Every GA call ends with the storage of the results into the database. Now whenever the GP needs to optimize a structure, it checks whether the same structure has been optimized sometimes in the past. In case this optimization has happened in the past, the results are simply returned by the database for the use in the GP individual. In case this structure is not found in the database, the call to the GA is made.

15.2.2 Genetic Programming

The model proposes a combined usage of Genetic Programming (GP) and Genetic Algorithm (GA) for the task of optimization. The first task is hence to build a GP solution to the problem that optimizes the architecture of the ANFIS. The GP solves the problem of optimization by encoding the individual in tree based architecture. The GP in this problem is supposed to model an ANFIS structure.

The primary task of the GP is to fix the number of membership functions for every input along with all the membership function parameters. In chapter 4 we saw that for any input this task is to divide the entire input axis into a number of fuzzy regions. Each region denotes start and end value. This function primarily has some membership degree for the inputs that lie within this range. Further there is some region or some value of the input where this function has the highest membership value of 1. The number of membership functions may be visualized as the number of divisions we perform for this input axis. The various parameters of each of this membership function are an indicator of their effective width in the input axis. We further extend this concept by plotting a multi-dimensional graph with each axis denoting an input. The division of this axis may be carried out in the same manner. Naturally the space would be divided perpendicular to the input axis being divided such that the other axis do not have their values altered and these remain same for all the inputs across the axis being divided. This space is known as the feature space. The GP works over this space and carries out the division of the various axis of this space. This is similar to the process of dividing the individual input axis. The various membership functions and their fuzzy division across the input axis is shown in figure 15.4.

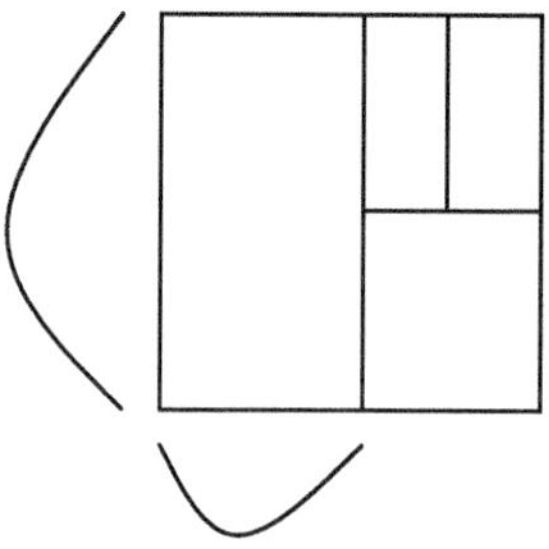

Fig. 15.4 Division of input space by fuzzy membership functions

15.2.2.1 Individual Representation

The tree based individual representation in this model consists of the leaf nodes and the non-leaf nodes. The non-leaf nodes contain either one of the inputs of the system. These are denoted by F_1, F_2, F_3, ... F_n. The presence of F_i represents the input i and denotes the ith axis in the feature space. This means that the input i is to be divided into two halves or that the feature space is to be divided across the ith axis. Using this concept we can divide the feature space across any input indefinitely that generates a large variety of discrete divisions denoting the membership functions. The undue division may be stopped by specifying the largest length possible of the genetic tree. The leaf nodes always contain the terminal instruction L. These nodes denote that the feature space cannot be further subdivided.

Consider a two dimensional feature space. This means that the system has two inputs. Let any individual of the GP be as denoted by figure 15.5(a). The corresponding discrete divisions of the feature space are given by figure 15.5(b). These may be obtained from the tree base structure by following the divisions node by node and implementing them onto the feature space.

Any membership function is marked by its center and width. Each space specification hence has its own center list which denotes the centers of the membership function. This list may be denoted by $<c_1, c_2, c_3, \ldots c_n>$. Similarly the width list may be given by $<\delta_1, \delta_2, \delta_3, \ldots\delta_n>$. When we divide a feature space into two halves, there is a natural change in the width and centers of the two membership functions that come up. The width of both the membership function is half of the original width across the axis of division. The centers get shifted by one fourth of the original width. The center of the left child membership function is shifted by

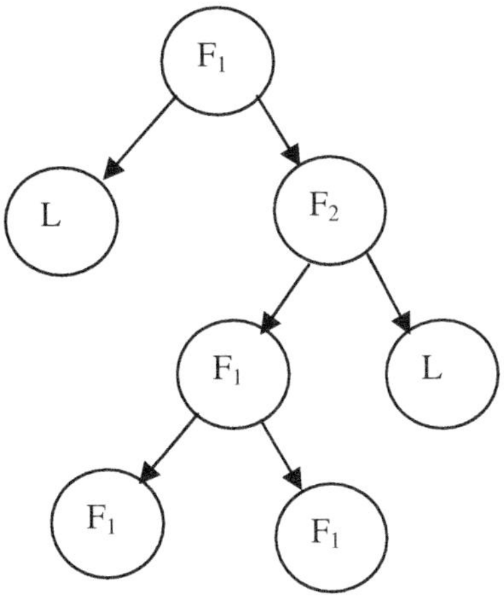

Fig. 15.5(a) Sample Genetic Programming Individual

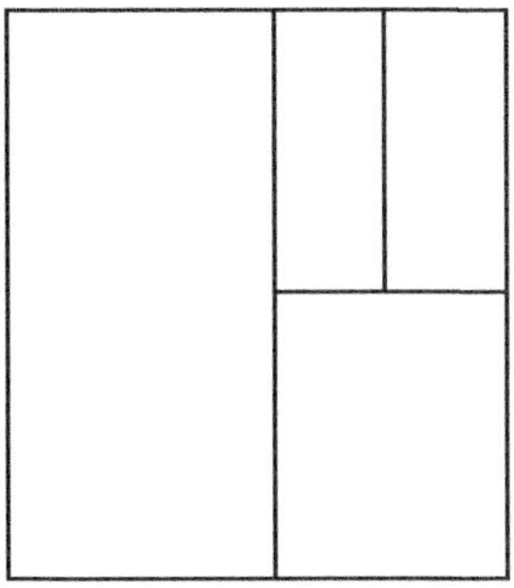

Fig. 15.5(b) Equivalent Workspace Division

one-fourth the original width (or one half of the final width) in the leftward direction across the axis being divided. Similarly the center of the right membership function is shifted rightward by one-fourth the original width (or one half of the final width) across the axis being divided.

In this manner any individual of the GP may be conveniently represented as a collection of membership functions. But these membership functions need to be fuzzy. We have divided the entire input axis into discrete divisions. The next and final step that we need to carry out is to make the boundaries fuzzy so as to completely represent these as a set of membership functions of ANFIS. All the membership function is assumed to be Gaussian. For the implementation of the Gaussian membership function we need to specify the center and spread. The center can be directly obtained from the discrete equivalent of the GP individual. The spread is directly proportional to the width of the discrete equivalent and may be given by equation (15.7).

$$\sigma_i = \lambda.\delta_i \tag{15.7}$$

Here λ is a constant and σ_i is the spread of the membership function.

15.2.2.2 Genetic Operators

The other part in the implementation of the GP is the genetic operators that carry out the entire task of evolution and generation of newer individuals of higher generations. The tournament selection scheme is used in the algorithm. This mainly consists of mutation and crossover. The implemented approach by the authors modifies the parameters stored in the various nodes of the GP individual in mutation. This carries out a change in the entire phenotype representation that consists of the set of fuzzy membership functions. This is known as the point mutation operator. The crossover is a conventional crossover technique that selects the sub-trees between individuals and exchanges them to generate the children from the parents.

15.2.2.3 Fitness Evaluation

The last point of discussion in the GP is the fitness evaluations. Here we assign a fitness count to the individuals of the GP. The performance of the resulting GP is an indicator of the fitness of the individual. This however might result in too large ANFIS structures with too many rules. We hence need to ensure that the number of rules is least in the final evolved system. This adds a lot to the generalizing power of the system. This is done by giving the system a penalty proportional to the total number of rules. This is similar to the penalty given to the evolutionary ANNs for extra number of neurons and connections that we studied in chapter 13.

15.2.3 Genetic Algorithm

The other part of the whole algorithm that carries out optimization is Genetic Algorithm (GA). The GA further optimizes the centers and the widths of the various membership functions. All the widths and spreads as represented in the GP individual are coded and stored in the form of a real valued GA individual. These are then allowed to be optimized by the evolutionary powers of the GA. Here we assume that everything else is fixed and cannot be changed in the complete evolutionary process. This makes it a conventional problem of the GA where some parameters are to be optimized. The authors implemented a tournament selection with a one point crossover and uniform mutation. The evaluation is simply the performance of the resultant ANFIS that the individual represents.

15.3 Fuzzy Neural Networks

The next topic of discussion is the fuzzy neural networks. In these systems also we combine or hybridize the fuzzy as well as the neural approaches, but from a different perspective. Much of the information we see around is fuzzy in nature. This means that its exact value may not be easily to be specified. We may only be able to denote the value to some degree of confidence. The neural networks are good agents of information processing. These networks take inputs and process them to produce the output. The learning enables these networks to adjust their state so as to give good performance to the available data. These networks hence must be able to take and process fuzzy inputs. This forms the motivation behind the use of fuzzy neural networks that add a fuzzy dimension to the conventional neural networks.

The fuzzy neural networks are hence able to take fuzzy inputs. These fuzzy inputs are processed as per the fuzzy arithmetic. This task of information processing is done by the help of fuzzy neurons. The various layers work layer after layer. Each layer carries out some computation and gives its output to the forward layer for further processing. The final output layer gives the system output. This output is also fuzzy in nature.

The term fuzzy nature of the inputs, outputs or any of the intermediate outputs means that there is a numerical output and some degree of membership or confidence of this output. This may be conveniently represented in the form p/x. Here x is the numerical input, output or intermediate output and p is the degree of membership or confidence. Using the notation of the membership functions this may further be represented by $\mu(x)/x$. Here $\mu(x)$ is the membership function.

The fuzzy nature of the inputs has a special relevance in the real life applications. Suppose that an experiment is performed. For this we make use of a number of sensors to measure a number of parameters needed for making the inferences or computing some result. Let us assume that a sensor failed to record the needed data. This means that the system can now not be used for computing the outputs as some data is missing. Suppose these recordings were to be used as inputs of the neural network. This means that we would neither be able to use this data for training and nor for testing. This would make the complete experiment worthless. In many applications this can mean a lot of loss of data. We would have hence liked to have some general idea of the output or the system behavior based on the available recording. In many real life application databases this problem persists where the complete set of attributes is not available. Rather only some attributes are available. In these problems it may sometimes be wise to have some idea regarding the output or to somehow use this data for the system training. The other common problem that may happen is that a number of sensors may be used for measuring the same attribute and all of them record different readings. In such a case we have to give a single input to the system, but instead we have a number of them available that may not match. Many times the differences in the recordings might be too large because of one or more instruments being faulty. In such a case again it is not possible to decide what the actual input is. Even if a single sensor is used or the various readings match, we may never be assured that the recorded reading is the actual value of the attribute. This is because of the measuring limitations in the various instruments used for the experimentation.

Hence the various inputs have a fuzzy nature. The fuzzy neural networks cater to the needs of these problems by enabling us to specify the inputs in a fuzzy manner. This produces a fuzzy output that is sufficient to enable us have some idea of the output. The degree of fuzziness in the output depends upon the fuzziness of the inputs. If a completely crisp input is given to the system, the corresponding output may be a crisp number. If the fuzziness of the inputs is very high, the outputs will also be of a very high fuzzy value. For this reason the inputs must be always with as less imprecision or fuzziness as possible. This is because the continued processing by the various layers adds impreciseness or fuzziness in the input. If the inputs has a little large fuzziness, the outputs might be so fuzzy in nature that it might be completely impossible to make any valuable inferences from them. It would appear as if the output may lie anywhere within a very large or sometimes the entire possible range dedicated to the output (Buckley and Hayashi 1993; Buckley *et al* 1993; Gupta and Qi 1991; Gupta 1992; Kosko 1991; Shukla *et al* 2010).

15.3.1 Fuzzy Arithmetic

The conventional ANNs carry out the processing of the inputs by using some arithmetic operators. This processing is done by the use of various neurons. The same concept applies to the fuzzy neural networks as well. Here also the task of processing is done by the fuzzy neurons. The major difference comes from the fact that these neurons make use of fuzzy arithmetic for the processing. We hence must be able model the fuzzy arithmetic for the tasks for which the conventional arithmetic is used. A neural network computation involves the use of addition, multiplication and operation on function. The multiplication is performed when the inputs are multiplied by the weights. The addition is performed when these multiplied values are added. The operation of function is when the resulting sum goes through the activation function. We hence see the fuzzy equivalents of these three operations in fuzzy arithmetic in the next sub-sections. We may further require the use of error function, derivatives, etc. which are not dealt in this book for simplicity.

15.3.1.1 Addition

The addition is a binary operator that takes two inputs and returns their sum in conventional arithmetic. In fuzzy arithmetic these two inputs are fuzzy in nature. These possess a numeric value and the membership value. The numeric value of the addition of two fuzzy numbers is simply the addition of the numerical values of both the inputs. The membership value of the addition is the minimal of the two fuzzy membership values. This minimal membership value is found for all the combinations of the two input membership functions. The maximum of this is denoted as the membership value of the output. This may be given by equation (15.8)

$$\mu_{A+B}(z) = \max\{\mu_A(x) \Lambda \mu_B(y) / z = x + y\} \tag{15.8}$$

Here $\mu_A(x)$ is the first input membership function; $\mu_B(x)$ is the second input membership function; $\mu_{A+B}(z)$ is the resulting membership function of the sum. ^ is the minimum operator.

15.3.1.2 Product

The next operator of interest is the product binary operator. This operator takes two fuzzy inputs and generates a fuzzy output. The numerical value of the output is the same as the product of the two input numerical values. The membership value is the also computed in the manner similar to addition where we take the maximal membership value across any combination of the inputs. This is given in equation (15.9).

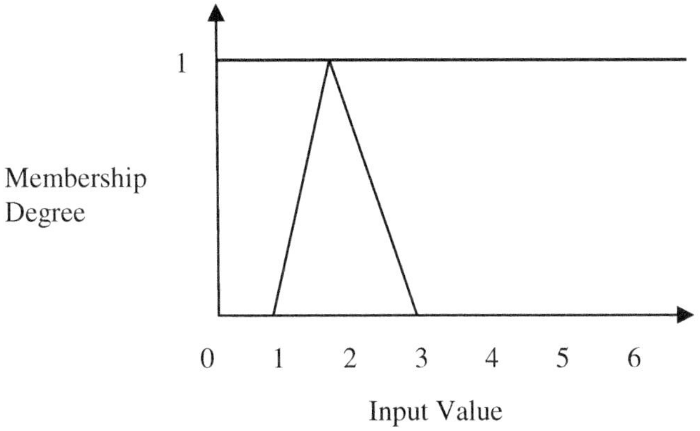

Fig. 15.6(a) Input 1

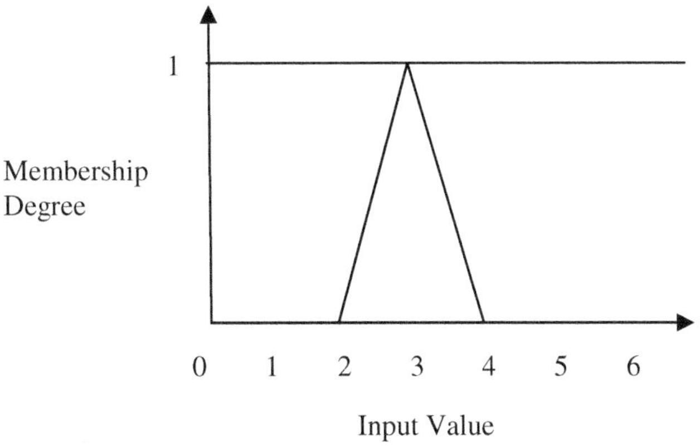

Fig. 15.6(b) Input 2

$$\mu_{AB}(z) = \max\{\mu_A(x) \Lambda \mu_B(y) / z = xy\} \quad (15.9)$$

Here $\mu_A(x)$ is the first input membership function; $\mu_B(x)$ is the second input membership function; $\mu_{AB}(z)$ is the resulting membership function of the product. ^ is the minimum operator.

Let the two inputs be given by figure 15.6(a) and 15.6(b). The resulting sum computed by these inputs is given by figure 15.6(c). The product is given by figure 15.6(d).

15.3.1.3 Function

The last operator of interest is the application of any function over the fuzzy inputs. This takes a single fuzzy input and generates a single fuzzy output. The numeric value of this is simply the numeric value of the application of the function

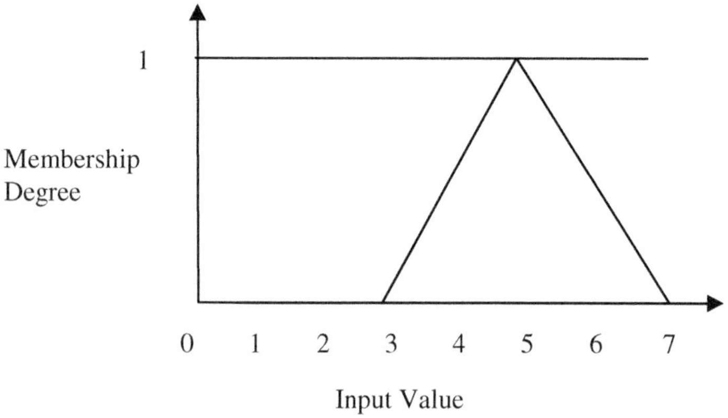

Fig. 15.6(c) Sum of Input 1 and Input 2

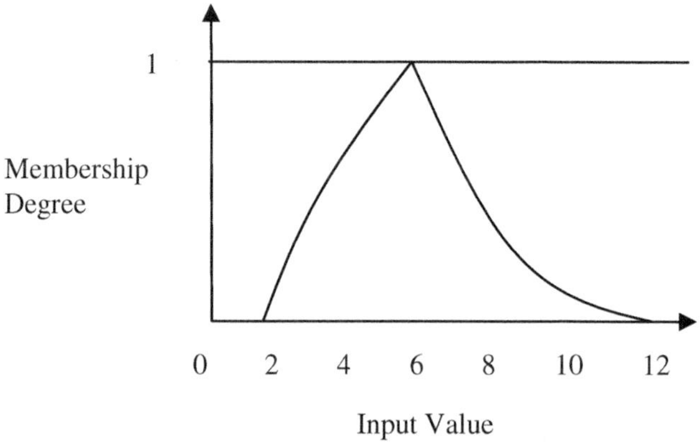

Fig. 15.6(d) Product of Input 1 and Input 2

to the numeric input. The membership value in this operation is the same as that of the input. This is given in equation (15.10) and shown in figure 15.7.

$$\mu_{f(net)}(z) = \max\{\mu_{net}(x) / z = f(x)\} \tag{15.10}$$

15.3.2 Training

The training of the fuzzy neural networks may be done using the Back Propagation Algorithm that was used for the conventional neural networks. In the fuzzy counterparts as well the training for a single input of the training data set has two parts of forward propagation and backward propagation. In the forward propagation we apply the inputs, compute the outputs and compute the errors. The error is communicated to the backward layers for adjustment of the weights and biases in the backward propagation. This is repeated for all the inputs for a number of training cycles or epochs. Here all the inputs, outputs, intermediate outputs and errors are fuzzy. In place of processing the entire input, we may wish to compute a part of it that is known as the α cut.

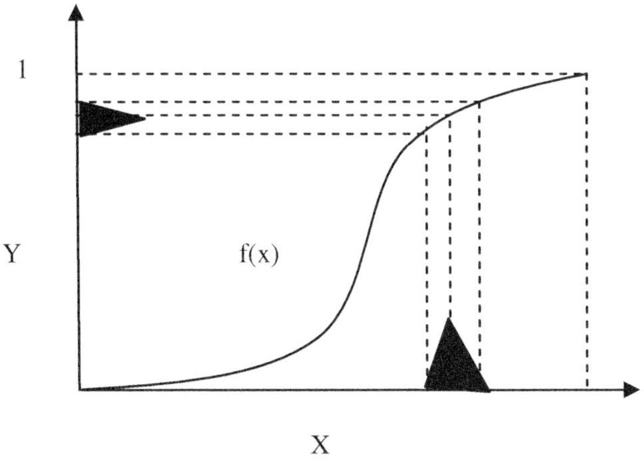

Fig. 15.7 Application of function

The α cut is another important aspect that comes into the use of fuzzy neural networks. In fuzzy algebra we normally prefer to perform the operation over the region in the input that has high membership value. This section is known as the level sets. Thc h level set of a fuzzy number consists of all elements whose membership value is greater than h. This forms a closed interval in the input. Consider X is any fuzzy set. The h level set of X may be denoted by $[[X]^L_h, [X]^U_h]$. Here $[X]^L_h$ denotes the lower level and $[X]^U_h$ denotes the upper level of the h level set. Using this concept we may further be able to implement the fuzzy arithmetic over the conventional neural network.

15.4 Rule Extraction from Neural Networks

The last topic of discussion in this chapter is the rule extraction of the neural networks. Throughout the text of this chapter we have realized the individual importance of the neural as well as the fuzzy systems. The system we discuss in this section marks another exciting application where we try to interpret the neural network with the help of fuzzy rules. In this section we would try to convert a conventional neural network into fuzzy rules (Baesens *et al* 2003; Andrews *et al* 1995; Towell and Shavlik 1993; Setiono and Liu 1995; Garcez *et al* 2001; Shukla *et al* 2010).

The neural networks are effective problem solving tools that can learn from the historical data and generalize these learning to the new or the testing data. The easy applicability of these systems makes them widely used tools for problem solving. These systems are hence usually used as black boxes where we give the inputs and collect the outputs without knowing the underlying knowledge. This black box approach may many times have associated problems as well. We may get the outputs without any understanding of the how these were computed. Further if an input represents new trends that may not be present in the historical data, the system would give poor outputs. This is even when some underlying logic was known.

We have discussed the disadvantages and advantages of rule based and the neural approaches numerous times in the book. We discussed that the fuzzy systems could be modeled in the neural architecture as Adaptive Neuro Fuzzy Inference Systems. In this section we discuss a reverse application. Here we try to make a fuzzy system based on rules out of a given neural network that has already been trained and tested as per the problem requirements. This would enable us understand the neural network and use the inferences for the understanding of the manner in which the various inputs map to the outputs.

A good fuzzy system must have as less rules and membership functions as possible. We discussed the use of the lower number of rules and membership functions in affecting the problem generality in chapter 4. Hence the extraction of relevant rules out of a trained neural network must interpret the neural network with as less rules as possible. It may again be seen that having too many rules would make the system look too much complex which may not be easy to interpret and understand. In the worst case the system might look like a large number of input output pairs have been formulated in the form of rules which would make the system more of a lookup table. It would naturally be impossible to make out anything from such an extraction. As discussed in the concept of problem generalization, if we do not extract a large number of rules, the errors may be high. This means that the extracted fuzzy system might not give very accurate outputs to the applied inputs. There would be some disparity between the outputs of the fuzzy system and that of the neural network.

Hence a good extraction of fuzzy rule is in such a manner that the extracted rules give as accurate results as possible in the least possible numbers. This is the same as discussed in neural and fuzzy systems. A good neural network gives the closest possible outputs in the least number of neurons and hidden layers. Similarly the best possible fuzzy system gives the closest possible outputs in the least number of rules and membership functions.

There are various algorithms that may be used for the task of extraction of the rules from the neural networks. The various algorithms are generally classified into two separate heads. These are decomposition algorithms and pedagogical algorithms. In the decomposition technique we try to decompose the various inputs into various parts. After this decomposition we try to separately analyze how the various classes of inputs are getting mapped to the outputs. We analyze which combination of inputs activates which hidden neurons and how these are further processed to produce what type of outputs. These recordings are framed in the form of rules. A number of such rules may be extracted depending upon the decomposition. This type includes the algorithms KBANN/M of N and RULEX.

The other category is the Pedagogical algorithms. These algorithms have a pedagogical approach for the extraction of the fuzzy rules. The neural network is taken as a black box and the various rules are tried for performance. These rules are framed on the algorithmic logic. This category includes the algorithms of Craven and Shavlik (1994). Genetic Algorithms, Heuristics, etc. may be applied to both these approaches for their optimization.

15.5 Chapter Summary

In this chapter we presented a variety of methods in which the hybridization of the fuzzy and the neural systems could be carried out. The advantages and disadvantages of both these systems was carried out that enabled understanding possible mechanisms of fusion of these individual systems. The first model to be discussed in this section was Adaptive Neuro Fuzzy Inference System. Here the motivation is to use the training algorithm of the neural networks into the fuzzy architecture for its training. We hence map the complete fuzzy model into a layered neural network. The complete system works as a fuzzy system built over the neural architecture. The training algorithms of the system include Back Propagation Algorithm and Hybrid Algorithm. The other topic discussed was evolution of adaptive neuro fuzzy inference systems. These systems added the evolutionary capability of the evolutionary algorithms into the adaptive neuro fuzzy inference systems. A mixture of genetic programming and genetic algorithm was used for the purpose. The other model discussed into the chapter was of fuzzy neural networks. These networks took fuzzy inputs, provided fuzzy outputs, and each neuron did fuzzy processing of information. The various arithmetic operators were reformulated into fuzzy equivalents which enabled the conventional neurons to take fuzzy inputs and produce fuzzy outputs. The last topic of discussion was rule extraction of neural networks. This model presented the mechanism in which rules could be extracted from a trained neural network for better understanding of the neural network working and its knowledge content.

References

[1] Abraham, A.: Neuro-fuzzy systems: state-of-the-art modeling techniques. In: Mira, J., Prieto, A. (eds.) IWANN 2001. LNCS, vol. 2084, pp. 269–276. Springer, Heidelberg (2001)

[2] Andrews, R., Diederich, J., Tickle, A.B.: A survey and critique of techniques for extracting rules from trained artificial neural networks. Knowl.-Based Syst. 8, 373–389 (1995)

[3] Baesens, B., Setiono, R., Mues, C., Vanthienen, J.: Using neural network rule extraction and decision tables for credit-risk evaluation. Manage. Sci. 49(3), 312–329 (2003)

[4] Buckley, J.J., Hayashi, Y.: Numerical relationship between neural networks, continuous functions and fuzzy systems. Fuzzy Sets Syst. 60(1), 1–8 (1993)

[5] Buckley, J.J., Hayashi, Y., Czogala, E.: On the equivalence of neural nets and fuzzy expert systems. Fuzzy Sets Syst. 53(2), 129–134 (1993)

[6] Craven, M.W., Shavlik, J.W.: Using sampling and queries to extract rules from trained neural networks. In: Proc. of the Eleventh Int. Conf. Machine Learning, San Francisco, CA, USA (1994)

[7] Dubois, D., Prade, H.: Fuzzy Sets and Systems: Theory and Applications. Academic Press, New York (1980)

[8] Fagarasan, F.: A genetic-based method applied in fuzzy modeling. In: Proc. of IEEE Int. Conf. on Evol. Comput., vol. 7, pp. 253–257 (1996)

[9] Garcez, A.S.A., Broda, K., Gabbay, D.M.: Symbolic knowledge extraction from trained neural networks: A sound approach. Artificial Intelligence 125, 155–207 (2001)

[10] Gupta, M.M.: Fuzzy logic and neural networks. In: Proc. 2nd Int. Conf. on Fuzzy logic and Neural Networks, Iizuka, Japan, pp. 157–160 (1992)

[11] Gupta, M.M., Qi, J.: On fuzzy neuron models. In: Proc. of Int. Jt. Conf. on Neural Netw., Seattle, pp. 431–436 (1991)

[12] Jang, J.S.R.: Fuzzy modeling using generalized neural networks and Kalman filter algorithm. In: Proc. Ninth Nat. Conf. Artificial Intell., pp. 762–767. AAAI, Menlo Park (1991)

[13] Jang, J.S.R.: ANFIS: Adaptive-Network-Based Fuzzy Inference System. IEEE Trans. on Systems, Man, and Cybern. 23(3), 665–685 (1993)

[14] Jang, J.S.R., Sun, C.T.: Neuro-fuzzy modeling and control. Proc. of the IEEE, Special Issue on Fuzzy Logic in Engineering Applications 83(3), 378–406 (1995)

[15] Juang, C., Lin, C.: An on-line self-constructing neural fuzzy inference network and its applications. IEEE Trans. Fuzzy Syst. 6, 12–32 (1998)

[16] Kala, R., Shukla, A., Tiwari, R.: A Novel Approach to Classificatory problem using Neuro-Fuzzy Architecture. Int. J. of Systems, Control and Communications (2010a)

[17] Kala, R., Shukla, A., Tiwari, R.: A Novel Approach to Classificatory Problem using Grammatical Evolution based Hybrid Algorithm. Int. J. on Futuristic Computer Applications (2010b)

[18] Kosko, B.: Neural Networks and Fuzzy Systems: A Dynamical Systems Approach to Machine Intelligence. Prentice-Hall, Englewood Cliffs (1991)

[19] Ouyang, C.S., Lee, S.J.: A hybrid algorithm for structure identification of neuro-fuzzy modeling. In: Proc. of IEEE Int. Conf. on Syst., Man., Cybern., vol. 5, pp. 3611–3616 (2000)

[20] Pedrycz, W., Reformat, M.: Evolutionary fuzzy modeling. IEEE Trans. Fuzzy Syst. 11, 652–665 (2003)

[21] Pena-Reyes, C.A., Sipper, M.: Fuzzy CoCo: A coevolutionary coevolutionary approach to fuzzy modeling. IEEE Trans. Fuzzy Syst. 9(5), 727–737 (2001)
[22] Rastegar, F., Araabi, B.N., Lucast, C.: An evolutionary fuzzy modeling approach for ANFIS architecture. Evol. Comput. 3, 2182–2189 (2005)
[23] Russo, M.: Genetic fuzzy learning. IEEE Trans. Evol. Comput. 4, 259–273 (2000)
[24] Rutkowski, L.: Flexible Neuro-Fuzzy Systems: Structures, Learning and Performance Evaluation. Kluwer, Dordrecht (2004)
[25] Setiono, R., Liu, H.: Understanding neural networks via rule extraction. In: Proc. of the 14th Intl. Jt. Conf. on Artificial Intell., pp. 480–485 (1995)
[26] Shukla, A., Tiwari, R., Kala, R.: Real Life Applications of Soft Computing. CRC Press, Boca Raton (2010)
[27] Towell, G.G., Shavlik, J.W.: Extracting refined rules from knowledge-based neural networks. Mach. Learn. 13, 71–101 (1993)

List of Figures

Fig. 15.1. ANFIS Architecture
Fig. 15.2. Layer 4 neuron
Fig. 15.3. Back Propagation Algorithm
Fig. 15.4.Division of input space by fuzzy membership functions
Fig. 15.5(a). Sample Genetic Programming Individual
Fig. 15.5(b). Equivalent Workspace Division
Fig. 15.6(a). Input 1
Fig. 15.6(b). Input 2
Fig. 15.6(c). Sum of Input 1 and Input 2
Fig. 15.6(d). Product of Input 1 and Input 2
Fig. 15.7. Application of function

Chapter 16
Parallel and Hierarchical Evolutionary Algorithms

Abstract. Evolutionary Algorithms are good optimizing agents that use the evolutionary concepts to evolve systems. The complex fitness landscapes are a problem that restricts the performance of these algorithms. Besides specifying the correct parameter is always very important. In this chapter we study the means by which we can execute a number of evolutionary algorithms in a variety of models to give an overall optimal performance to the entire network. We first discuss the Island Model Algorithm where the individuals are evolved in multiple islands with regular interaction. We then study a related algorithm called the Hierarchical Fair Competition algorithm. Here we would separate the various individuals into classes as per their fitness values. Each of these would have a separate evolution. We would then discuss the Nested Evolutionary Strategies where we fix the parameters of an Evolutionary Strategy by some technique. Towards the end we discuss the Hybrid Genetic Algorithm Particle Swarm Optimization. This algorithm generates individuals by twin usage of Genetic Algorithm and Particle Swarm Optimization.

Evolutionary Algorithms (EA) are widely used algorithms for the task of optimization and system evolution. These systems occupy a predominant role in each of the hybrid system that we discuss in this unit. These algorithms are able to tune any parameter of any system to enable a good overall performance. Besides, there is a widespread use of these algorithms in completely designing the architecture of a variable size system. Evolution hence becomes an important and indispensable part of any system. Any intelligent system that we build requires a lot of optimization that is effectively done by the EAs (Bäck 1996, Bäck and Hoffmeister 1991, Davis 1987; Deb and Goldberg 1989; Holland 1975; Mitchell 1999; Fogel 1995; Koza 1992; Michalewicz 1992; and Schwefel 1995).

These algorithms are of varied types and varieties. These are essentially classified into three types i.e. Genetic Algorithms, Genetic Programming and Evolutionary Strategies. Each class may be further sub-divided to make a variety of models. Each of the models uses the same concept of iterative optimization and generation of a higher generation solution from the lower generation solution. The solutions keep getting optimized as the generations or iterations increase. After the stopping criterion is met the algorithm optimization process is terminated.

Each of the evolutionary models has its own mechanism of modeling of the problem and further optimizing it. The genetic algorithm uses the individual representation in form of a vector of bits or numerals and specialized operators. The

A. Shukla et al.: Towards Hybrid and Adaptive Computing, SCI 307, pp. 361–381.
springerlink.com

genetic programming has the tree type program based individual representation that may be of variable sizes. The Evolutionary Strategy is an adaptive system with simple reproduction operators. The other systems include the Ant Colony Optimization that were inspired from the working of ant colonies, the Particle Swarm Optimization that were an inspiration from the flocking birds, Grammatical Evolution that used a linear grammar based program, co-evolution that believed in the ideas of cooperative evolution where every individual helped the entire population to evolve, etc. We discussed a variety of these approaches in unit 3 that was completely dedicated to evolution.

16.1 Fitness Landscape Complexity

The variety of EAs gives a good performance when the complexity is reasonably low. The ultimate goal of the EAs is to find the optimal point in the fitness landscape which comprises of the fitness value for all combination of values of modifiable parameters represented as axis. However, all real life applications may not have low complexity. The applications of EAs in system evolution that we discussed in this unit are examples of very complex problems which have a very complex fitness landscape (Banker 1985; Forrest and Mitchell 1993; Whitley 1989). The complexity of this landscape may be because of two major reasons. These are high dimensionality and sensitive fitness landscape. Both of them may usually be correlated with one leading to the other.

16.1.1 High Dimensionality

Suppose that the problem to be solved has a large amount of modifiable parameters that are represented as axis in the fitness landscape. In such a case the entire landscape becomes highly multi-dimensional in nature. As the number of dimensions increase, there is an increase in the total combination of values between the various parameters. Consider that we are asked to do an exhaustive search in this fitness landscape. Naturally in case the parameters have continuous values, the total number of points would be infinite. We however suppose that every parameter has a small value δ which is reasonably small and denotes the smallest level of precision needed. In such a case we would have a finite set of points as points closer than δ distance may not be evaluated. Now addition of any parameter is an increase of axis in this fitness landscape which multiplies the number of evaluations in this exhaustive search.

The EAs make the best value out of a small level of evaluations. The performance of the EAs would largely depend upon what proportion of the total number of points in the exhaustive search they are allowed to evaluate. Suppose the total number of evaluations possible is too high due to the large dimensionality. The EAs are allowed to have only some number of evaluations considering the computational limitations. The high dimensionality would make the proportional number of evaluations fairly low. We may be able to try out only a small part of all the evaluations possible. In other words we may be able to explore a reasonably low part of the entire fitness landscape. We would have to make out decisions only on the basis of these evaluations. This creates a problem in the use of conventional EAs.

16.1.2 Sensitive Fitness Landscape

The other problem that we term under high complexity is when the fitness surface has a very sensitive shape. We know that it is much simpler to search a fitness landscape that has single optima, than a fitness landscape that has multiple optima. In any algorithm the assessment of the present individual locations and their fitness values is done assuming that the place around the individuals is a single optima and the attempt is to move all of them towards this optima. In case there are multiple local optima, the location and the fitness value conveyed by the individuals may naturally be deceiving that would make the EA make the wrong generation of the higher generation individuals. This may be further clarified by the fact that a population of reasonably low number of individuals may be able to figure out a fitness surface which comprises of a single optima. One may expect the individuals to constantly march and ultimately reach the optima. In case there are too many local optima, the search becomes more random in nature. Now we may require a large number of individuals.

Too many local optima may be easily realized as a system having large changes in functional values with only a small change in a single attribute value, keeping the other attributes constant. A fitness landscape having lesser optima would mostly exhibit smooth changes when viewed from any dimension. This would enable easy location in the evolutionary search process. This is shown by the figures 16.1(a) and (b) which shows a function with low number of optima and high number of optima respectively, when the fitness landscape cut across any dimension.

16.1.3 Parameter Setting

The complexity is not the only problem associated with the EAs. Every EA uses a set of parameters for its functioning. In many algorithms this is the mutation rate, crossover rate, number of individuals, etc (Chakraborty and Dastidar 1993; Chakroborty *et al* 1996; Goldberg 1989; Goldberg and Deb 1991; Lobo *et al* 2007). These parameters play an important role in the optimal performance of the EAs. The problem lies that these need to be optimally fixed which may not always be possible. The human designer of the system may fix any value that may result in a sub-optimal performance or optimization. Further the various parameters may not work well by a constant value. We may need a high or low value at different times in the solution of the problem. We know that the crossover rate decides the convergence and the mutation rate decides the exploration in an evolutionary approach. Different needs may arise at different contexts of time in the search process. This necessitates the need of variable parameters or adaptive parameters that can adjust their values as per the situation the search operation is in.

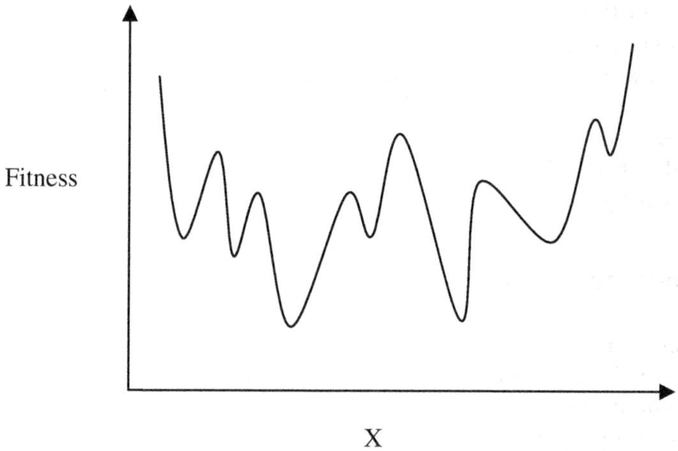

Fig. 16.1(a) Fitness Landscape with too many local minima

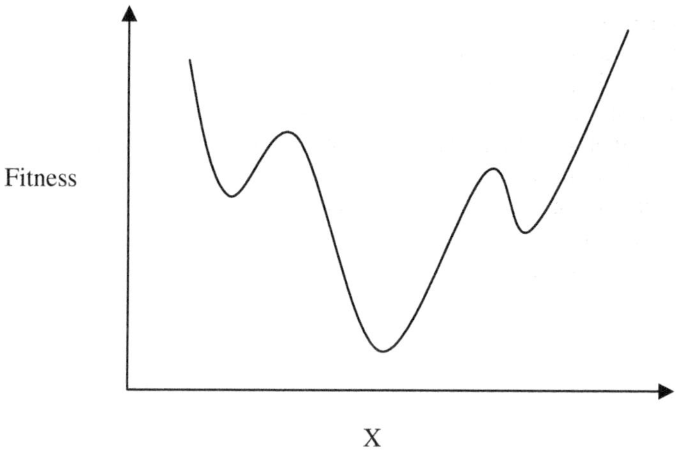

Fig. 16.1(b) Fitness Landscape with manageable number of local minima

16.1.4 Diversity

Diversity is another major factor of concern. As discussed the use of the EA in a conventional manner may give poor performance in case of multiple optima. Another way to view the same problem is by a measure of population diversity. Diversity is a measure of average separation between the individuals. As the algorithm starts the separation is high which means high population diversity. As the algorithm continues to execute the population diversity decreases. Finally all the individuals converge to some point making the diversity to a minimal value (Laumanns *et al* 2001a, 2001b; Oei *et al* 1991).

Consider that the fitness landscape has a couple of optima located at two distant locations. As the conventional algorithm proceeds, the individuals would get pulled towards one of these optima, depending upon the location of the best individual. This would wipe off all the individuals from the other optima. This may however lead to the convergence at the wrong minima. In reality, the global minima might be the other one. The correct strategy would have been to let the algorithm simultaneously converge individuals to both these optima. As the algorithm continues both of them would attract the nearby individuals for convergence. After the stopping criterion has been met, the fitter individual may be returned and that optima may be taken to be the global optima.

The conventional algorithms kill the diversity that is responsible for the poor performance in many fitness landscapes. Taking of preventive measures against this is hence important. This is known as diversity preservation. Using this mechanism we prohibit the algorithm from undue killing of the population diversity. The major factor behind the diversity loss is crossover operator. This mixes any two individuals and generates new individuals or children at some intermediate point in the fitness landscape. It may be possible that two very distinct individuals, both possessing good fitness values (lying at two different optima) get crossed over to generate the individual at some intermediate point. This individual may naturally not be near some optima or possess good fitness value. In order to maintain diversity we hence apply a restriction to the crossover operation. Two individuals I_1 and I_2 may only participate in crossover if they are separated less than some threshold amount in the fitness landscape. This means that the crossover is only allowed if equation (16.1) holds true.

$$| I1 - I2 | \leq \eta \tag{16.1}$$

Here η is the distance threshold and $| x |$ is the norm.

16.2 Island Model Algorithm

The first method that we discuss in this chapter is the Island Model Algorithm. We apply this model for the Genetic Algorithm (GA) (Gordon *et al* 1992). In this approach we run a number of GAs in parallel. Each GA runs on its own set of population and evolution process. The entire population pool in this algorithm may be broken down into smaller sets, each set having some individuals. Each set of population in this scheme is called as an island. The various islands evolve in isolation from each other for some time. This enables the various individuals to get optimized as per their own evolution scheme. Later on the information is exchanged in between the various islands. This enables the various islands to benefit from the optimizations of the other island. We look at the various concepts of the algorithm in the coming sub-sections.

16.2.1 Islands

The island is a collection of individuals. Many islands together make the population of the discussed GA. An island may hence be reargued as a population pool.

The various islands have their independent evolution by their own copy of GA. These are not externally disturbed by the other islands until the information exchange phase is called. Since these represent a part of the entire population, the number of individuals in an island may be lower than those in a conventional GA. Inside the island a conventional GA is executed. This consists of the standard scaling, selection, crossover, mutation, elite and other operators. The different GAs in the various islands may be invoked with the same genetic parameter or with different genetic parameters. This depends upon the design of the algorithm. These parameters denote the strategy of evolution. The different islands carry out the task of optimization independently. Hence these usually result in the development of different characteristics which means that in a multi-optima landscape, each of them may converge towards some different optima.

16.2.2 Information Exchange

The isolated evolution may result in enhancing different characteristics among the different islands or population pools. This is good in terms of diversity reservation and controlling associated problems of complex landscapes. However the ultimate goal of the algorithm is to compute the global optima. This requires a lot of exchange of information in between the various population pools. The aim is to transfer the best characteristics from one population pool to the other population pool. These good characteristics would interact and multiply as per the evolutionary process. This would enable the other population to further enhance its fitness by the newly added characteristics.

The exchange of information in this algorithm takes place by the transfer of fittest individuals from one island to the other island. In some implementations the individual may not be completely transferred, rather a copy of it may be given to the other population pool. In many others the individual is completely transferred which might result in slight lowering of the fitness of the particular population pool in case the imported individual has lesser fitness value. The original pool may hence still possess the individual. As the fittest individual goes to the other population pool it carries with it the best characteristics that were developed so far by the individuals of the island. This individual interacts with the other individuals in the crossover operations. This gives a chance for these best characteristics to get mixed in the population of the higher generations. This enables the island to further optimize itself using the best characteristics originally developed and those conveyed by the imported individuals.

One of the commonly used models of individual exchange is the ring model. Here the various islands are arranged in a ring–like architecture. These evolve in isolation with each other. At the time of information exchange, each island passes its fittest individual to the next island in the ring. In this manner a complete exchange of individuals takes place. This constitutes one exchange epoch of the algorithm. Numerous such epochs together make the entire global information exchange where every island is benefitted by the optimized characteristics of the other island. This is shown by figure 16.2.

16.2.3 Algorithm Outline

The entire algorithm operates in two stages. The first stage deals with the regular evolution of the various population pools. The other stage is when the various islands exchange information. The total span of time when the islands operate in isolation with each other and optimize themselves is known as the isolation period. The entire algorithm consists of a series of execution of these stages one after the other. Single execution of these stages is called as one epoch of the algorithm. The general outline of the algorithm may hence be given by figure 16.3.

The isolation period is an important factor that needs to be fixed judiciously. Too less of isolation period would result in too frequent and too much exchange of information amongst the islands. This would make the algorithm behave more like a conventional GA. Having too large isolation time is also harmful. This results in all the islands developing their own characteristics. This makes the algorithm more random in nature where the various islands only operate in isolation with no mutual information exchange. The non-exchange of information may lead to many islands operate with bad characteristics, in case these are not present in the initial pool or develop with time. Their optimality would naturally be very poor. This means a few islands would operate without making any contribution. This would waste a large amount of individuals that could have better been parts of some fitter island.

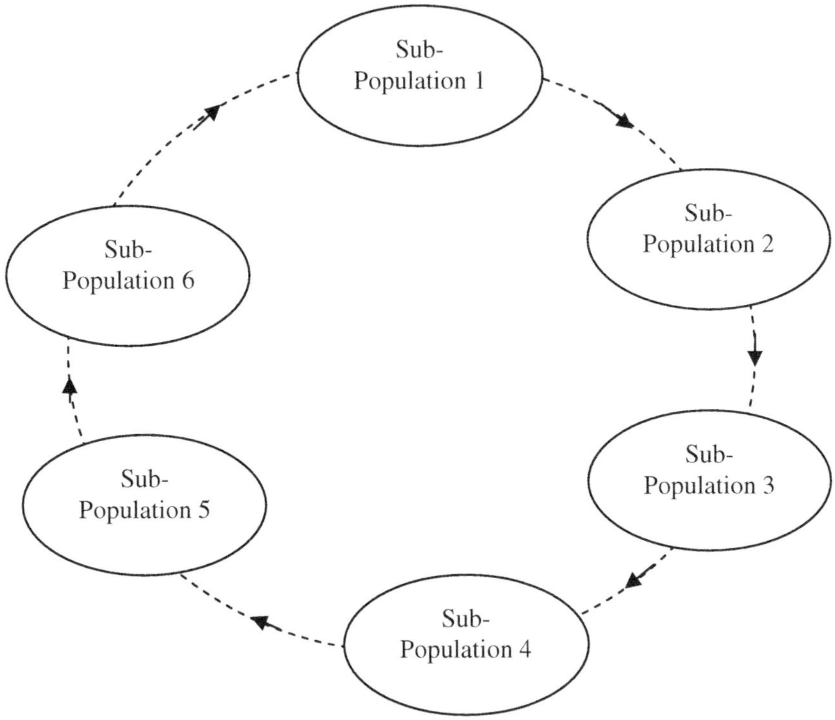

Fig. 16.2 The ring of sub-populations

The concept of island can be easily seen from the manner in which the human life evolves in nature. The various communities develop themselves. Their individuals mate within themselves. The interaction is chiefly between the individuals of the same community. A community may be seen as a geographical area where the individuals are under constant interaction. This results in generation of good individuals. These are the ones who expertise in some or the other skill. The various communities develop different skills between the people or individuals. But the various communities do not only operate in isolation. There is some amount of interaction between the communities at a global level. This interaction is much less in nature and is facilitated by the individuals who are experts of their domain. This gives a chance for a community to know about new characteristics and develop the same. In this manner the entire global population develops in islands or population pools.

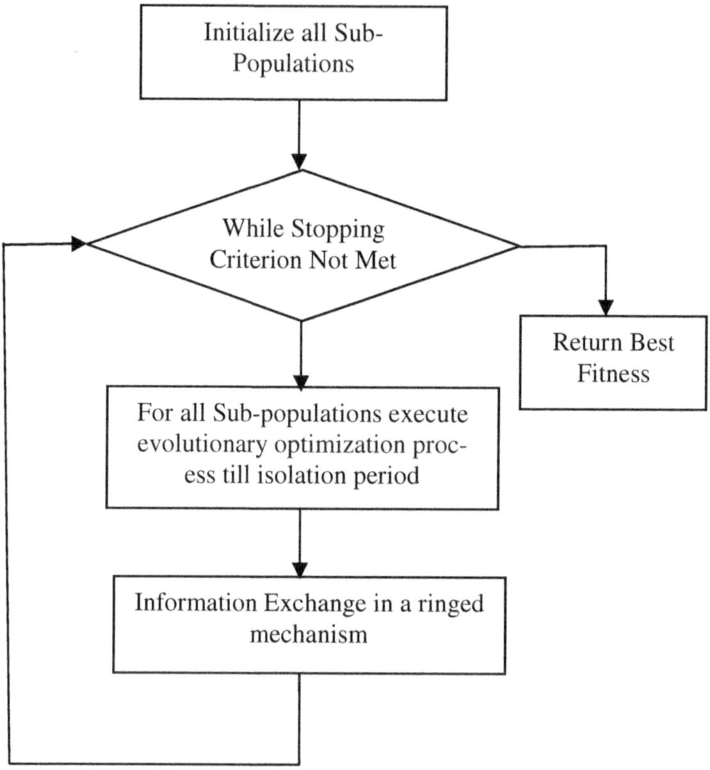

Fig. 16.3 Island Model Algorithm

16.3 Hierarchical Fair Competition

The other model that we discuss in this chapter is the Hierarchical Fair Competition (HFC) (Hu and Goodman 2002; Hu *et al* 2002). The basic motive of this model is the same as discussed in island model. We carry out the parallel implementation of the EA by using a number of population pools. This model however makes the division between the various individuals on the basis of their fitness values. Each pool hence has a range of values. All the individuals in this pool lie within this range. At any time of the evolution process, an individual may attain a better fitness value. If this value les above the allowed threshold of this pool, the individual is immediately promoted to the higher level pool. In this manner the entire algorithm operates to carry forward the optimization. The basic concept behind this algorithm is that every individual must compete against the individuals of the same level of fitness. Here competition is a reference to the evolution process where the individuals compete under the survival of the fittest criterion. Hence the entire fitness scale is partitioned into levels and each level is allocated to some population pool. The individuals of this are evolved separately. We look at the various aspects of this algorithm in the next sub-sections.

16.3.1 Sub-populations

The different individuals in this scheme evolve in different islands or sub-populations. The division of the individuals is done on the basis of the fitness of the individuals. This is done in order to enable each individual compete with the individuals with like fitness values. Each sub-population has two thresholds. The first is the admission threshold. This states the minimal fitness value that an individual must possess in order to be admitted in the population pool or the sub-population. The other is the export threshold. This states the maximal fitness value that an individual may have in this population pool. If an individual has a fitness

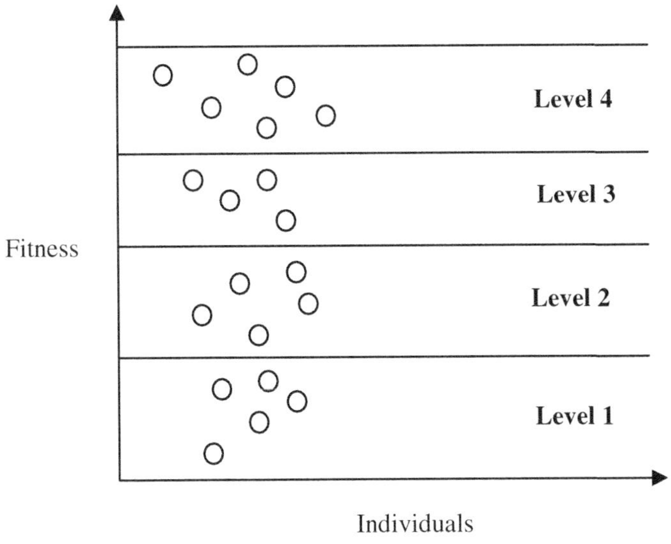

Fig. 16.4 Division of fitness range

value above this threshold, it is transferred from this pool to another pool whose criterion the individual best fulfills. In this manner the entire fitness is divided into regions with each region denoting a sub-population.

The division of the fitness range by the various population pools is shown in figure 16.4.

Each of the sub-population has its own evolution process that may depend upon the choice of designer. Each evolves independent of the other sub-populations till the individuals need to be imported and exported. The evolutionary parameter may be different for the different sub-populations. A lower order sub-population with poor fitness value mainly does the task of exploration. It tries to explore better area in the fitness landscape in search of new characteristics with optimal performance. The higher level population with better fitness values mainly tries to carry out convergence to good and optimal values. In this manner both the exploration and convergence is carried out by the algorithm. This results to a very large extent in an optimal performance.

The number of individuals is an important parameter that decides the performance of any EA. In this approach it is known that the number of individuals with poor fitness values may be reasonably higher than the number of individuals with high fitness values. Hence it is sometimes wise to fix the number of individuals in the various levels in increasing order. This means that the higher levels corresponding to higher fitness values have only a few individuals. The lower individuals with poor fitness values have much higher number of individuals.

16.3.2 Transfer of Characteristics

The other major point of concern for the algorithm is the manner in which the characteristics are transferred from one population pool to the other. The exchange of information among the various population pools takes place in the form of migration of individuals from one level to another in case they qualify from the export threshold of the present level and the admission threshold of the other level to which they are to be migrated. An individual on qualifying the export threshold selects the highest level it qualifies as per the admission threshold.

Whenever an individual needs to be exported, it is deleted from the current level sub-population and placed in a buffer of the other sub-population to which it is to be exported. This buffer may be sorted in decreasing order to indicate the preference of admission to the pool. An individual is only admitted to the actual population pool from the buffer when the GA of that population pool demands. The individual may replace an existing individual from the population pool, in case it possesses a better fitness value. The exporting of individuals to the other population pool results in lowering the total number of individuals of that pool. For this a sub-population seeks its admission buffer to find individuals there. It then admits the necessary number of individuals in its pool. In case sufficient number of individuals is not available, the empty locations of individuals in this population pool might be randomly filled. It is important to note here that an individual in this method can only be jumped from a lower level to a higher level. The individuals are hence not demoted from higher to lower levels.

The transfer of individuals to the fitness levels and their admission buffers are shown in figure 16.5.

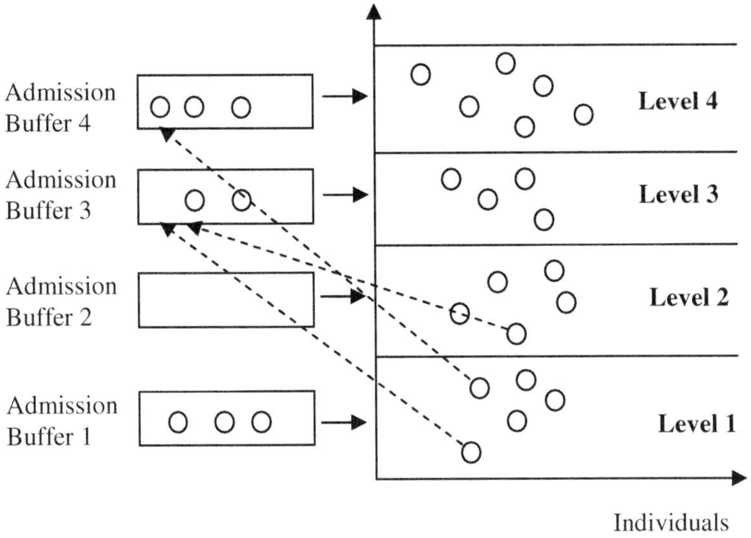

Fig. 16.5 Admission Buffers of various levels

16.3.3 Algorithm Outline

The algorithm in all other respects is similar to the island model discussed in section 16.2. The algorithm carries the optimization of the various population pools using their own modeling of the EA. If an individual happens to exceed the fitness thresh hold, it is promoted by the algorithm to a higher level. The level depends upon the highest sub-population that the individual qualifies as per their admission threshold. In this manner the algorithm keeps performing the task of optimization as the generations increase. Finally the best individual in the fittest population pool is returned as the optimized individual.

This algorithm also has a very strong underlying biological analogy. We all develop from our childhood to a stage where we become mature and sophisticated selves. At any stage we compete for survival. This makes the complete journey of development in stages. We clear any stage before succeeding to the next stage. The competition at any stage is only restricted to the individuals at that particular stage. This is easily observed as school children compete with school children and college students compete with college students. Further at many times jumping stages might be allowed. If an individual qualifies a higher stage, he may be directly admitted there irrespective of his or her age. This development framework enables the optimal development of any individual.

16.4 Nested Evolutionary Strategies

The next model of study is the nested evolutionary strategies. We studied about the Evolutionary Strategies (ES) in chapter 11. This evolutionary model used simple operators and usually worked with small number of individuals to generate a higher generation of population from a lower generation. The algorithm resulted in a decent exploration and convergence that enabled the search and convergence to global minima in the exploratory process. This made evolutionary strategies widely used tools for the optimization tasks. We further saw that this evolutionary technique of solving problems was adaptive in nature where the various parameters were judged as per the present context of the algorithm in order to make every generation in the best possible manner from the previous lower level generation. This adaptability further enabled good performance to these algorithms.

Nested Evolutionary Strategy adds a layer of computation over the conventional implementation of the ES (António 2006; Herdy 1991; Arnold and Macleod 2006; Arnold and Castellarin 2009). This mechanism also goes on the lines of parallel implementation of the EA where a number of executions of the EA are carried out in parallel. Here we execute a number of ES in parallel. All of these are executed in isolation with each other. The total number of generations that these ES execute is known as the isolation period.

The nested ES use the various sub-populations of individuals as an individual of a master ES. This master ES carries out the task of directing the evolution of the slave ES. Hence there is a master and slave relation between the two ESs. This is discussed separately in the next sub-sections which is followed by the algorithm outline.

16.4.1 Inner Evolutionary Strategies

The inner ES is a conventional ES that consists of some individuals that together carry out the task of optimization. The best manner to understand this is by the notation of the ES. The ES was generally denoted by using the notation $(\mu ,^{+} \lambda)^{\gamma}$. This means that the population consists of μ individuals. At every generation we first carry out crossover followed by mutation to generate a total of μ individuals. Then a total population pool is made up out of the μ parents and λ children that totals to $\mu + \lambda$ individuals. We then select total of μ individuals for the next generation based on fitness value. The other individuals are killed. In case only, is used in place of the '+' operator, this means that the population pool would only consist of the children for the selection and not the parents. The best μ children would go to the next generation. For this method $\lambda \geq \mu$. This entire task of generating children from parents and selecting the best for the next generation is carried out till γ generations.

16.4.2 Outer Evolutionary Strategies

The outer ES is an ES formulated over the inner ES. Each inner ES is an individual of the outer ES. The outer ES carries forward the task of directing and coordinating the inner ES for better optimizations. This may be carried out in a variety of mechanisms that we would study in the next sub-sections.

The outer ES is in itself an ES and may hence be denoted by the same mechanisms. An individual of this ES is itself a sub-population of individuals of a conventional ES. We denote this population by μ'. Each of the individual of this μ' individuals are in itself a sub-population consisting of μ individuals as discussed in section 16.4.1. The task is to use the evolutionary operators to generate λ' individuals that are potential individuals of the higher generation. The algorithm again makes a pool of individuals consisting of λ' children and μ' parents. The best of the μ' individuals are selected out of these. In case the , operator is used the pool only consists of the children out of which the best μ' individuals go to the next generation. This constitutes one generation of the ES. The outer ES may be repeated for a total of γ' generations.

An important factor here is how we assign the fitness value of individuals of this ES. Each individual represents an entire ES of its own. For assigning the fitness values of the individuals of the master ES, we first execute the slave ES to optimize the various individuals. We then compute the average fitness of the individuals of the slave ES. This average is regarded as the fitness value of the master ES.

The notation of the nested ES is build around the notation of the conventional ES. Let us assume that there are a total of μ' sub-populations. Let us further assume that each of these sub-populations has a total of μ parents. The various sub-populations have their conventional runs with μ parents and λ children for γ generations. This can be easily denoted by $(\mu \,{,}^{+} \lambda)^{\gamma}$. Here γ is the isolation period. Let us assume that these μ' sub-populations generate λ' sub-populations as candidates for the next generation. Let the selection be done using the '+' operator. Let the algorithm is carried for a total of γ' iterations. This can be denoted by the expression $[\mu' \,{,}^{+} \lambda'(\mu \,{,}^{+} \lambda)^{\gamma}]^{\gamma'}$.

16.4.3 Algorithm Outline

The general algorithm consists of the implementation of the outer ES. Fitness evaluation is a part of the implementation of the outer ES. At the time of fitness evaluation the outer ES makes a call to the inner ES. The inner ES is initialized and executed as per the representation of the outer ES. After the isolation period is lapsed the fitness value is averaged and noted. This continues for all the generations of the outer ES. This is shown in figure 16.6.

The relation between the inner and the outer ES largely depends upon the manner of use. In other words the manner in which the outer ES directs the inner ES is important as it decides the working of the slave ES and contributes to the total optimization process. We study three specific applications of these in the coming sub-sections. These are for parameter optimization, mixed integer optimization and min-max optimization (Rudolph 1997).

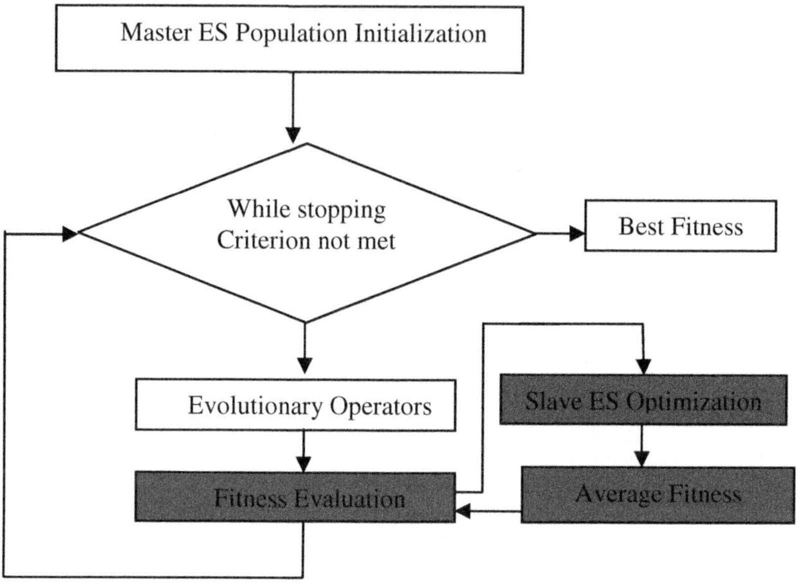

Fig. 16.6 Nested Evolutionary Strategies

16.4.3.1 Parameter Optimization

The fixing of the correct parameters is always a problem in the evolutionary approaches. This many times results in sub-optimal performance. We used EA for fixing the parameters of neural and fuzzy systems in numerable number of ways as discussed in chapters 13, 14 and 15. This added a lot of optimization to the conventional systems. We can hence even add the power of optimization of EA to fix the parameters of the other EA. In this way we hope the system, which is an EA to work and evolve with the correct parameter values.

This approach made use of nested ES. The underlying operational unit was the slave ES. The step size and number of generations is one of the major parameters in ES. This needs to be judiciously fixed. We may hence apply the master ES to fix these values for the slave ES. The generations were called as isolation period in the above discussion.

The master ES is hence expected to find out the correct step size as well as the number of generations for the slave ES. The master ES individual hence contains these in its individual representation. The rest of the algorithm proceeds exactly as discussed above. The specific individuals of the slave ES may be randomly generated for the fixed number of generation and step size.

The algorithm for fixing the step sizes is given in the work of Herdy (1992).

16.4.3.2 Mixed Integer Optimization

Mixed Integer problems are the optimization problems where some of the modifiable attributes are discrete in nature and the rest are continuous in nature.

The discrete attributes in the individual representation of any EA always cause a problem as they make the fitness landscape lose its shape and make it irregular. It may hence be wise to separate the discrete and the continuous attributes for carrying out the optimization. We may use the slave for the optimization of the continuous attributes and the master for the optimization of the discrete attributes. In this manner the slave EA has a fine fitness landscape that may be reasonably easy to optimize for every combination of the discrete attribute values.

In the implementation with the nested ES we adopt the same technique of optimization of the continuous attributes by slave and the discrete attributes by the master. The master ES hence has an individual representation that specifies all the discrete values. The population pool of the slave ES is made with only the continuous attributes. The salve assumes the discrete parameters to be fixed to the values supplied by the master.

The master ES sets some values of the discrete attributes for every individual. The slave ES takes these values fixed and optimizes the other continuous values. It may be assumed that the value returned by the slave ES is the most optimal value for any combination of the continuous attributes. Hence the master ES knows the goodness or the fitness of the discrete value combination it represents. In this manner the master ES carries out the optimization of the discrete attributes while the optimization of the continuous attributes is done by the slave ES.

16.4.3.3 Min-Max Optimization

The min-max problem is of the form given in equation (16.2).

$$\min_{x}\{\max_{y}\{f(x, y)\}\} \tag{16.2}$$

Here the problem is finding the minimal value of a function that tries to maximize its value. Here also we use two EAs in a nested architecture. The master ES tries to minimize the function given in equation (16.3).

$$\min_{x}\{g(x)\} \tag{16.3}$$

The evaluation of this function *g(x)* requires some value of the attribute *y*. The slave ES carries out the task of finding the optimal value of *y* for which the inner function is maximized as per the requirements of the algorithm given in equation (16.4).

$$g(x) = \max_{y}\{f(x, y)\} \tag{16.4}$$

This algorithm operates in a manner similar to the one discussed in section 16.4.3.2 for the mixed integer optimization problem. The outer ES is responsible finds out the optimal values of *x* and the inner ES finds out the optimal values of *y*. For any value of *x* set by the outer ES, the inner ES tries to supplement it with the most optimal value of *y*. We may assume the value of y fixed by the inner ES is the optimal value for the specific value of *x*. In this manner the entire task of optimization is carried out.

16.5 Hybrid Genetic Algorithm Particle Swarm Optimization

The next algorithm for study that beautifully mixes the optimization powers of two evolutionary algorithms is Hybrid Genetic Algorithm Particle Swarm Optimization (HGAPSO) (Juang 2004; Robinson *et al* 2002; Shia *et al* 2005). Both these algorithms are optimizing agents that enable the search for the global optima in the given fitness landscape. Hence the important aspect in the hybridization of both these algorithms is that both these must assist each other for the proper optimization. We hence first study these algorithms separately and then present the algorithm outline of the entire algorithm framed by the fusion of both these algorithms.

16.5.1 Genetic Algorithm

Genetic Algorithm (GA) uses a set of individual that make a population for carrying out the optimization. The algorithm operates in iterations or generations. The fitness of the individuals generally improves with iterations. The algorithm uses a set of operators for the generation of a higher generation population from the lower generation. The process starts with the selection of the individual for the genetic operators. The two widely used operators are crossover and mutation. Crossover uses two parent individuals for generating child individual. Mutation results in altering some genes of an individual to get the next generation individual. Elite operator is used to transfer the best individuals from one generation directly to the other generation without modifications.

One of the major problems associated with the GA is of premature convergence. We know that the crossover plays a role in the convergence of the individuals of the algorithm and mutation promotes exploration by prohibiting convergence. Elite operator further results in convergence as some individuals are directly transferred which may be predominant in crossover as well. Increasing the mutation rate and decreasing the crossover rate may many times be seen as a solution to this problem, but this may add a lot of randomness to the algorithm. Making mutation high may result in the individuals being displaced by a very large margin in the fitness landscape. This may result in completely random nature to the entire algorithm. The major problem in crossover is caused due to the crossover operator that mixes two individuals or parents to form children that are placed at position intermediate to the parent individuals. Many times this may result in a lot of movements by the individuals towards the fitter individuals in the fitness landscape. This results in the premature convergence.

The GA is on the other hand very effective in locating the global optima. It is less likely that this algorithm gets struck at local optima. It can search out for the global optima in a reasonably complex fitness landscape as well. This is attributed to the multi individual deployment as well as the nature of the genetic operators that judiciously place the individuals at key locations. This results in uncovering the global optima.

16.5.2 Particle Swarm Optimization

The other algorithm of use in the hybrid strategy is Particle Swarm Optimization (PSO). This algorithm also uses a set of individuals for the optimization of the problem. The individuals in this case represent the moving particles that explore the fitness landscape for the search of the global minima. Each particle has a position in the fitness landscape and a velocity which represents the speed with which the particle walks in every dimension of the fitness landscape. The particle velocity and position are updated at every iteration. The update is based upon the best position recorded by the particle so far, globally known best position and the past velocity of motion. The maximum velocity is restricted to some value so that the particle does not jump the global optima.

The PSO has a good exploratory nature where the individuals can easily explore their surroundings to uncover better places. In this manner the particles try to continuously seek better fitness positions until they all converge to some point in the fitness landscape. This exploratory nature of the PSO results in relevant improvement in the particles at each iteration. The attempt of the particle to move towards the best location recorded so far results to go towards the local optima. At the same time the attempt to move towards the globally best position recorded so far attracts the particles for exploration for the global optima.

16.5.3 Algorithm Outline

The HGAPSO intends to combine the optimizing power of both the individual algorithms i.e. GA and PSO. These individual algorithms fuse together to enable better performance of the hybrid algorithm that has nature of both the constituent optimization algorithms.

In this algorithm we make a common pool of individuals for both GA and PSO. The entire algorithm has an iterative nature that operates in generations. The task is chiefly to generate a higher generation from a lower generation. We take the first half of the individuals as the elite individuals. These are the only individuals that participate in the evolutionary process. The other individuals are deleted. The selected individuals are modified as per the PSO. This modification results in the avoidance of the premature convergence that is a problem with the conventional GA. This further adds the exploratory abilities to the algorithm to move towards the global optima.

These individuals constitute half of the next generation population. The other half individuals come from the GA. The GA acts on the individuals after being modified by the PSO. The GA uses its crossover and mutation operators for the generation of the other half of the population for the next generation. In this manner the algorithms proceeds in every generation, until the stopping criterion has been met.

The working of the algorithm is shown in figure 16.7.

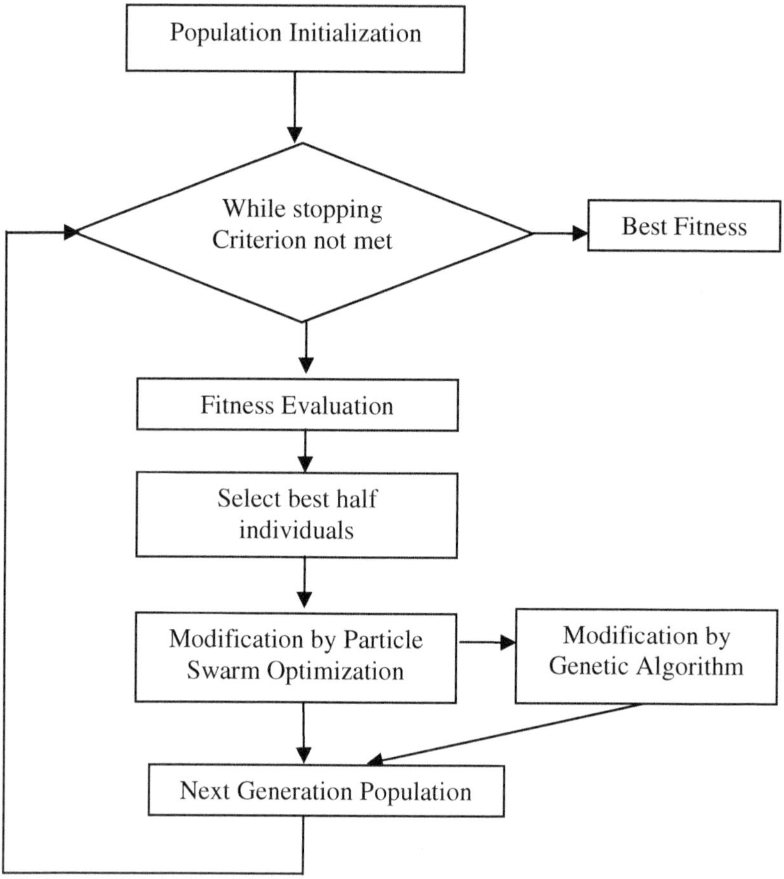

Fig. 16.7 Hybrid Genetic Algorithm Particle Swarm Optimization

16.6 Chapter Summary

This chapter focused upon the use of more than one evolutionary algorithm for carrying out better optimization especially in complex landscapes. The complex landscapes may have numerous local minima with a rapid change in fitness value on a small change in parameter value. Various issues related to complex fitness landscape were presented. The first model discussed was the island model algorithm. Here the entire population comprised of a ring of sub-population. Each sub-population evolved independently till an isolation period, after which an exchange of information was carried out. This process is repeated till the stopping criterion is met. The next model was the hierarchical fair competition. Here the entire fitness range was divided into ranges. Each range had an admission buffer to accept new individuals. The individuals could jump to any level of fitness, provided they pass the admission threshold. The other model of discussion was nested evolutionary strategy. The two nested strategies operated in a master slave mode. For every fitness evaluation of the master evolutionary strategy, a number of executions of the slave were carried out. The last topic of discussion was hybrid genetic algorithm particle swarm optimization. In this model both genetic algorithm and particle swarm optimization were used for seeking the global minima. The best half population was selected which is subjected to PSO optimizations. This constitutes first half of the next generation population. The other half is generated by Genetic Algorithm over the same individuals.

References

[1] António, C.A.C.: A hierarchical genetic algorithm with age structure for multimodal optimal design of hybrid composites. J. Structural and Multidisciplinary Optimization 31, 280–294 (2006)

[2] Arnold, D.V., Castellarin, A.S.: A novel approach to adaptive isolation in evolution strategies. In: Proc. of the 11th Annual Conf. on Genetic and Evol. Comput., pp. 491–498 (2009)

[3] Arnold, D.V., Macleod, A.: Hierarchically organised evolution strategies on the parabolic ridge. In: Proc. of the 2006 Genetic and Evol. Comput. Conf., pp. 437–444 (2006)

[4] Bäck, T.: Evolutionary Algorithms in Theory and Practice: Evolution Strategies, Evolutionary Programming, Genetic Algorithms. Oxford (1996)

[5] Bäck, T., Hoffmeister, F.: Extended selection mechanisms in genetic algorithms. In: Belew, R.K., Booker, L.B. (eds.) Proc. of the Fourth Intl' Conf. Genetic Algorithms. Morgan Kaufmann, San Francisco (1991)

[6] Baker, J.E.: Adaptive selection methods for genetic algorithms. In: Grefenstette, J.J. (ed.) Proc. of the First Intl. Conf. on Genetic Algorithms and their Appl., Erlbaum, Mahwah (1985)

[7] Chakraborty, U.K., Dastidar, D.G.: Using reliability analysis to estimate the number of generations to convergence in genetic algorithm. Inf. Process. Lett. 46, 199–209 (1993)

[8] Chakraborty, U.K., Deb, K., Chakraborty, M.: Analysis of selection algorithms: A Markov chain approach. Evol. Comput. 4(2), 133–167 (1996)

[9] Davis, L.: Handbook of Genetic Algorithms. Van Nostrand Reinhold, New York (1987)

[10] Deb, K., Goldberg, D.E.: An investigation of niche and species formation in genetic function optimization. In: Schaffer, J.D. (ed.) Proc. of the Third Intl. Conf. on Genetic Algorithms. Morgan Kaufmann, San Francisco (1989)

[11] Fogel, D.B.: Evolutionary Computation: Toward a New Philosophy of Machine Intelligence. IEEE Press, Los Alamitos (1995)

[12] Forrest, S., Mitchell, M.: What makes a problem hard for a genetic algorithm? Some anomalous results and their explanation. Mach. Learn. 13, 285–319 (1993)

[13] Goldberg, D.E.: Genetic Algorithms in Search, Optimization, and Machine Learning. Addison-Wesley, Reading (1989)

[14] Goldberg, D.E.: Sizing populations for serial and parallel genetic algorithms. In: Schaffer, J.D. (ed.) Proc. of the Third Intl. Conf. on Genetic Algorithms. Morgan Kaufmann, San Francisco (1989)

[15] Goldberg, D.E., Deb, K.: A comparative analysis of selection schemes used in genetic algorithms. In: Rawlins, G. (ed.) Foundations of Genetic Algorithms. Morgan Kaufmann, San Francisco (1991)
[16] Gordon, V.S., Whitley, D., Böhn, A.: Dataflow parallelism in genetic algorithms. In: Männer, R., Manderick, B. (eds.) Parallel problem solving from nature, pp. 533–542. Elsevier, Amsterdam (1992)
[17] Herdy, M.: Reproductive isolation as strategy parameter in hierarchically organized evolution strategies. In: Manderick (ed.) Parallel Problem Solving from Nature - PPSN II, pp. 207–217. Elsevier, Amsterdam (1991)
[18] Herdy, M.: Reproductive isolation as strategy parameter in hierachically organized evolution strategies. In: Parallel Problem Solving from Nature, Proc. 2nd Int. Conf. on Parallel Problem Solving, pp. 207–217. Elsevier, Amsterdam (1992)
[19] Holland, J.H.: Adaptation in Natural and Artificial Systems. University of Michigan Press, Ann Arbor (1975)
[20] Hu, J.J., Goodman, E.D.: The hierarchical fair competition (HFC) model for parallel evolutionary algorithms, Evol. Computt., 2002 (2002)
[21] Hu, J.J., Goodman, E.D., Seo, K., Pei, M.: Adaptive Hierarchical Fair Competition (AHFC) Model For Parallel Evolutionary Algorithms. In: Proc. of the Genetic and Evol. Comput. Conf., pp. 772–779 (2002)
[22] Juang, C.F.: A Hybrid of Genetic Algorithm and Particle Swarm Optimization for Recurrent Network Design. IEEE Trans. on Syst., Man, and Cybern.—Part B: Cybern. 34(2), 997–1006 (2004)
[23] Koza, J.R.: Genetic Programming: On the Programming of Computers by Means of Natural Selection. MIT Press, Cambridge (1992)
[24] Laumanns, M., Thiele, L., Deb, K., Zitzler, E.: On the convergence and diversity preservation properties of multi-objective evolutionary algorithms. TIK-Report No. 108, Computer Engineering and Networks Laboratory (TIK),Zurich, Switzerland, Swiss Federal Institute of Technology (ETH), Zurich (2001a)
[25] Laumanns, M., Rudolph, G., Schwefel, H.P.: Mutation control and convergence in evolutionary multi-objective optimization. In: Proc. of the Seventh Int. Mendel Conf. on Soft. Comput. (2001b)
[26] Lobo, F.G., Lima, C.F., Michalewicz, Z.: Parameter Setting in Evolutionary Algorithms. Springer, Heidelberg (2007)
[27] Michalewicz, Z.: Genetic Algorithms + Data Structures = Evolution Programs. Springer, Berlin (1992)
[28] Mitchell, M.: An Introduction to Genetic Algorithms. MIT Press, Cambridge (1999)
[29] Oei, C.K., Goldberg, D.E., Chang, S.J.: Tournament Selection, Niching and the Preservation of Diversity. Illinois Genetic Algorithms Laboratory, Illinois, IL, IlliGAL Report no. 91011 (1991)
[30] Robinson, J., Sinton, S., Rahmat-Samii, Y.: Particle swarm, genetic algorithm, and their hybrids: optimization of a profiled corrugated horn antenna. In: Proc. IEEE Int. Symp. Antennas and Propagation Soc., vol. 1, pp. 314–317 (2002)
[31] Rudolph, G.: Evolutionary Strategies. In: Handbook of Evolutionary Computation. IOP Publishing Ltd./Oxford University Press (1997)
[32] Schwefel, H.P.: Evolution and Optimum Seeking. Wiley, Chichester (1995)
[33] Shia, X.H., Lianga, Y.C., Leeb, H.P., Lub, C., Wanga, L.M.: An improved GA and a novel PSO-GA-based hybrid algorithm. Information Processing Letters 93(5), 255–261 (2005)
[34] Whitley, L.D.: The Genitor algorithm and selection pressure: Why rank–based allocation of reproductive trials is best. In: Schaffer, J.D. (ed.) Proc. of the Third Intl. Conf. on Genetic Algorithms. Morgan Kaufmann, San Francisco (1989)

List of Figures

Fig. 16.1(a). Fitness Landscape with too many local minima
Fig. 16.1(b). Fitness Landscape with manageable number of local minima
Fig. 16.2. The ring of sub-populations
Fig. 16.3. Island Model Algorithm
Fig. 16.4. Division of fitness range
Fig. 16.5. Admission Buffers of various levels
Fig. 16.6. Nested Evolutionary Strategies
Fig. 16.7. Hybrid Genetic Algorithm Particle Swarm Optimization

Part V
Hybrid Intelligent Systems

Chapter 17
Fusion Methods in Biometrics

Abstract. The use of speech as a biometric gives limited accuracy in the problems of speaker recognition and verification. The need of better recognition scores has resulted in the fusion of speech with other biometric modalities. This chapter discusses the fusion of speech with face which gives a high recognition score at the same time making the system convenient to be used for the user. We discuss three distinct ways to carry out this fusion. The first method is by directly mixing the attributes. This method has problems of excessive dimensionality of the resultant system. Hence many attributes from both modalities need to be deleted. The other method we discuss is the application of modular neural networks with division of attributes. In this technique the various attributes are divided between the various modules. The results are combined by an integrator. The last method is the use of clustering based division of input space by the system.

Speech is a widely used biometric identity that is widely used for the recognition and the verification problems. The speech signal can be easily pre-processed and analyzed for the extraction of the various features. These features, in most of the general scenarios work well to carry out recognition or verification. The various features extracted usually have a fine inter class to intra class separation ratio. This makes it easy to construct the decision boundaries to carry out classification. The speech based systems may be conveniently used by the user who just needs to utter a few sentences for the purpose of recognition or verification. The other computations are done by the system itself to give the correct output (Campbell 1997; Farrell *et al* 1994; Martin and Przybocki 2000; Naik 1990; Reynolds 2002).

The speech based systems in most of the conditions receive fine accuracies. However the conditions may not always be ideal. Many times the speech signal may be interfered by noise, this causes a problem for the recognition as the actual signal may be quite different from the noisy one (Mammone 1996; Rose *et al* 1994). In many other cases the available data requires a high scalability or an increased level of security. These may be higher than the offerings of the speech based systems. This necessitates in the use of better approaches to bring up the accuracies. We discussed various issues and models in the use of speech in chapter 5 and 6.

The other biometric modalities may not always be a solution as all of them have the same flaws to variable extents. All of them may have the problem of noise and reduced performance due to high volumes of data. This puts a threshold on the accuracies that they can offer. The other biometric modalities may include face, lips, ear, palm print, etc.

A. Shukla et al.: Towards Hybrid and Adaptive Computing, SCI 307, pp. 385–399.
springerlink.com

Face is another interesting biometric modality. The face of the human has various features that if captured may be easy to be used for the recognition. The face based biometric systems use a variety of methods for carrying out the recognition (Er *et al* 1999; Lu *et al* 2003; Phillips *et al* 2005; Shukla *et al* 2009a, 2009b, 2009c; Turk and Pentland 1991). These systems either use dimensionality reduction tools like Principle Component Analysis (PCA) (Jollife 2002; Smith 2002) or Regularized Linear Discriminant Analysis (R-LDA) (Huang and Hwang 2005; Lu *et al* 2005). These algorithms may additionally use morphological feature extraction tools that extract features like the distance between the eyes, distance between eyes and nose, etc (Cards 1991; Margaros *et al* 1986). The resultant vectors are then trained and tested using any classificatory machine learning tool like Radial Basis Function Networks, Learning Vector Quantization, etc. We studied the various methods and algorithms related to this in chapter 8.

17.1 Fusion of Face and Speech

The limited accuracies of both face as well as speech necessitate their fusion to make a bi-modal system with face and speech as the two biometric modalities. With this we aim to have added advantages of both the systems. We expect that the resultant system would give higher recognition scores as compared to both the individual systems. The disadvantages or loopholes in the speech of a person may get rectified by the face and vice versa. In this manner we hybridize two biometric systems to make a more robust biometric system (Jain 2005; Fierrez-Aguilar et al 2003; Snelick 2005; Fox *et al* 2003; Ross and Jain 2003; Jain *et al* 1999).

The fusion of face and speech is of special important. The biometric systems are needed to automate the task of recognition thereby saving time and adding to security. These systems cannot themselves have a very complex form that the users are unwilling to use. Also these cannot be too complex that they take a lot of time for processing. This may be annoying to the users.

A user of a fused system of face and speech is supposed to say some utterances and pass on. During this time the video camera may easily record and process the face. At the same time the audio microphone may record and process the speech signal. If the system was a recognition system, the system would be able to identify the person. In case the system was for verification, the system would be able to find out whether the person is genuine or not. Based on this decision he may be allowed to go or be restricted. In the worst case the person may be required to repeat the process once more, if the recognition or verification fails at the first instant.

The ease of use is hence important for the user. Hybridizing completely different biometric system may not be liked by the user. It may appear to him as if he is being made to pass against a series of security checks. This would further increase time. Repeating the entire process once would be painful.

The fusion of these biometric systems is prevalent in the manner we identify people in our daily lives. We may be able to identify people when we observe them from a distance. We may even be able to identify them on hearing them over the phone. But by both these means we have some uncertainty regarding our decision. Many times we may not be able to make decision regarding the identity with fine clarity. But when we hear the person speak while seeing his face, the certainty is reasonably high. The identification of a person largely takes place by observing his face, gestures, speech, etc. Hence the multimodal identification is much better than the unimodal identification in everyday life. The artificial intelligent systems however have a limit to the computation that they may be able to perform. This restricts the number of modalities that we may include in the recognition in the artificially engineered systems.

In this chapter we present three models to carry out the fusion of these two modalities of face and speech. In the first model we pool together the attributes of both modalities and then select some of these to be used for recognition. In the second model we divide the attributes in the different modules of the modular neural network. Each of these modules does the recognition and the final decision is made by the integrator. In the third model we cluster the input space and each cluster is given its own classifier for the recognition.

17.2 Pre-classification Fusion

The first model that we consider makes a pool of attributes of the face and speech. This is known as the pre-classification method of fusion as we first pool the attributes or carry out the fusion and then carry the task of classification. The selected pool of attributes from this pool is given as the inputs to the classifier that does the final task of deciding the class to which the input belongs. In the context of recognition of face and speech, we extract the attributes from both face as well as speech. The attributes from both these modalities are combined to form a common list of attributes. We select a few attributes out of these attributes. These form the input to the classifier. The classifier based on these inputs gives as its output the identified person. In this section we discuss the experimentation and results reported in the work of Shukla and Tiwari (2007, 2008). The information about the attributes used can also be found in (Kumar *et al* 2009; Rabiner 1993; Shukla *et al* 2009b).

17.2.1 Fusion Process

The inputs of the fused system are the combined inputs from both the speech and face recognition system. This pooling of attributes adds up attributes in the system. We know that the problem is classificatory in nature. Hence the problem deals with the determination of suitable decision boundaries in the feature space. The addition of the attributes to the system, which we carry out in this technique, means an addition of the number of axis in the feature space. The added axis in the classificatory problem always results in easier formulation of the decision boundaries. Two inputs may seem to be very close to each other in the feature space. But if we add a new axis, it might completely separate them. It is possible that these inputs occupy the opposite extremes in this axis making it very simple and easy to construct the decision boundaries.

An important aspect here is that the addition of axis could even have been possible by increasing the number of attributes in the speech alone. This would have saved from complicating the system by an addition of a completely new modality with the additional hardware and processing costs. However, the addition of attributes or axis must ensure that good attributes are added in the system that they themselves have high contribution in the recognition. These attributes must ensure a low intra-class separation and a high inter-class separation. The addition of more speech attributes would not make much of a difference as these attributes may not be good as per the stated criterion. These attributes may hence not contribute to the separation of the classes in the feature space.

The addition of features from face adds a completely new paradigm to the entire recognition task. These attributes are measured for a completely new modality. Hence these do not convey any kind of redundant information regarding the classification which may be derivable by the system. Also the information is of high quality that results in a better separation between the classes. These attributes are themselves able to effectively classify the classes well. The limitations in speech are fulfilled by face and the limitations of face are fulfilled by speech. Further the facial photograph of the person is independently recorded and processed. Hence even if the noise is high that results in poor speech signal recording, the person may be identifiable by the system. The noise plays a much lesser role in these systems.

The pooling of attributes however has a bad impact on the system as well. Speech and face biometric usually use a large number of attributes when operating in the single modality mode. The pooling of the attributes hence creates a vast amount of attributes in the resulting system. This results in too large a data for the system to handle. This would make the system computationally very slow for both the training as well as testing. The training especially might be needed to stop at sub-optimal optimization of finish in a few epochs only to be finished in a finite amount of time. Many times the network may completely fail to train. Many systems might require a large amount of training data to be effectively trained. For all these associated reasons, a large dimensionality is undesirable. Hence we do not select all the pooled attributes. We rather select only a few of the best attributes form both face as well as speech. Only these attributes are used for the task of recognition. The other attributes are simply eliminated.

The general concept of this fusion is given in figure 17. 1.

17.2.2 Inputs

The specific inputs given to the system constitute the selected inputs from both speech as well as face. The entire attribute set of features of speech consisted of a total of 11 features. These are time duration, number of zero crossing, max cepstral, average PSD, pitch amplitude, pitch frequency, peak PSD and 4 formants (F1-F4).

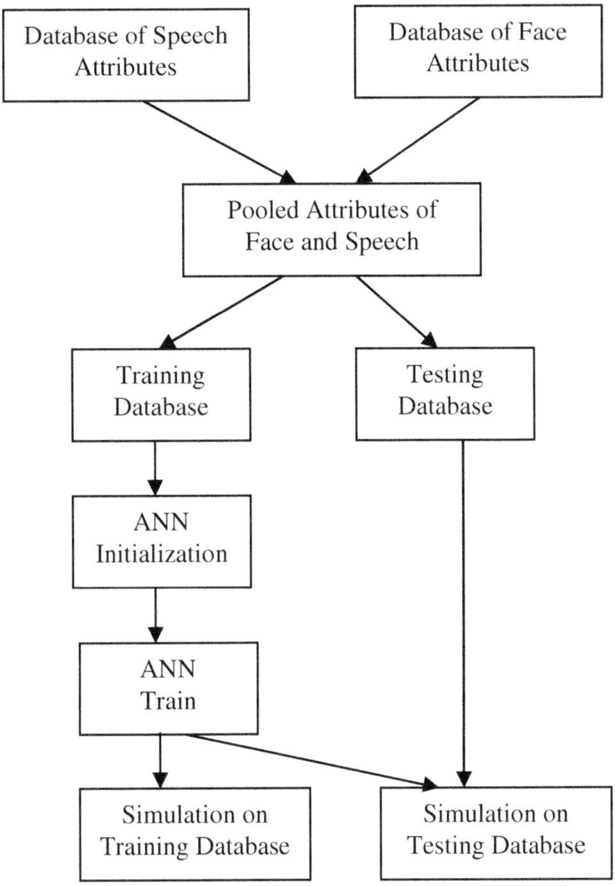

Fig. 17.1 Pre-Classification Fusion of Face and Speech

Similarly the feature set of face consisted of 13 features. These were These are length of the first eye, width of the first eye, position of the center of the first eye in x and y axis, length of the second eye, width of the second eye, position of the center of the second eye in x and y axis, length of the mouth, width of the mouth, position of the center of the mouth in x and y axis, distance between eye and eye, distance between eye and mouth.

The total pool of the attributes hence consisted of 24 attributes. The method of extraction of these features was taken in chapter 6 for speech and chapter 8 for face. The selection of the attributes from this pool was done randomly in nature.

17.2.3 Classification

The classification was done with the help of Artificial Neural Network. The Multi Layer Perceptron model with Back Propagation Algorithm for training was used.

The method of application of this algorithm is similar to the method discussed in chapter 6 for the recognition of the speech. This algorithm has two phases of training and testing. The complete data set is also divided into training and testing. In the training stage the training data is used to train the network using Back Propagation Algorithm. This is a supervised learning of the network. The testing is performed on the trained network with the help of the testing data set. In testing data set the testing input is given to the system and it is asked to compute the output. This output is matched with the actual output to compute the efficiency of the system.

17.2.4 Experimentation

The discussed model was implemented and tested in the work done by Shukla and Tiwari (2008a, 2008b). They extracted the features from both face and speech. These features were pooled in a common database. The relevant features were selected and the others were eliminated. The database was divided into training and testing and used for the same tasks as discussed in section 17.2.3. The author reported an accuracy of 99% using this approach. This clearly marks a boost in the performance level when compared to the performance with the use of speech alone presented in chapter 6.

17.3 Post-classification Fusion

The other type of fusion is the post classification fusion. Here we first classify the data by a set of classification techniques. Later the task of fusion is carried out where the various decisions of the classifiers are fused with each other. This type of system uses the concept of Modular Neural Networks (MNN). We saw while working with the ensemble techniques in the MNN that a set of systems were used as classifiers. An integrator was used that integrated the responses of the individual MNNs and made the final decision regarding the output of the system. Hence in this system we maintain a set of modules, each of which is a classifier in its own. We also implement an integrator to carry out the fusion task.

In this section we discuss the model proposed by Kala *et al* (2010).

17.3.1 Problem Division

The first task as per the MNN approach is to divide the problem into a set of modules. We know that each of the MNN is a classifier by itself that does the classification as per the inputs given to it. The stage of problem division deals with figuring out the correct inputs to each of these modules so that it is able to carry out the classification.

One of the major problems with the pre-classification fusion was the large number of attributes that result in the problems related to high dimensionality. This made us delete some attributes from the attribute list. This results in a permanent loss of information from the system. The deleted attributes may be carrying some vital piece of information that may have been useful for carrying out the classification. Many times an attribute may be able to separate out two inputs that all the other attributes might fail to separate. The deletion of attributes may hence be of a great loss to the system.

We hence need an intelligent method such that we do not delete the data as well as we are able to get saved from the problem of high dimensionality. We do this by introducing modularity at the attribute level of the problem. We distribute the attribute among the various modules of the MNN. In this manner each attribute gets into some MNN or the other. This saves the attribute from being deleted from the system. Also any module is given a manageable number of attributes. Due to this the MNNs do not have the associated problem of high dimensionality. In this manner we are able to save attribute as well as control the dimensionality problem.

The division of attributes is an important task that must be wisely done. In the MNN approach, each module must be able to effectively carry out the classification giving decent accuracy even if it was operating alone. This means that the attribute set of any module must be good for recognition by itself. Further giving too many attributes may result in a lot of computational cost and other problems of dimensionality. This necessitates a check on the total number of attributes per module as well. We need to further ensure that every attribute gets into some or the other module. It is possible that an attribute belongs to more than one module, but deletion of attributes must be avoided.

In the pre-classification fusion we stated that the addition of the attributes was the addition of axes in the feature space that resulted in better classification as it was easier to make the decision boundaries. This however had a poor effect due to the large dimensionality. In the post-classification fusion we work with multiple feature spaces. Each of the feature space has its own classifier and the mechanism of making decision boundaries. The various classes may look impossible to divide in some feature space which may be a reasonably simple task as per another feature space. The different classes may have different locations and properties in the different feature spaces. The major task is to use the observations in the different feature spaces to get a consolidated idea of the class to which any input belongs. This task is done by the integrator. This is especially important if some class has vague existence in some parts in all the feature spaces. Here we need to make a judicious decision based on the individual observations in the various feature spaces. The various feature spaces plotted in a 2-dimenional graphs for a synthetic data is shown in figure 17.2.

This technique naturally looks poorer than the pre-classification fusion. It is better to have a single feature space as the addition of axes would give a better view regarding the class separation than maintaining different feature spaces that may not be easy to be separated. However dimensionality is the major point of concern in the use of this type of fusion.

17.3.2 Module Inputs

Based on these principles we need to distribute the attributes in this problem of fusion of speech and face. The list of attributes for both face and speech is the same as discussed in section 17.2.2. These attributes are first pooled and then distributed into modules based on the discussed principles. The model uses 4 modules in the MNN. Based on the high success of the face and speech individually as a biometric, the division first separates the face and speech features. The first two modules are dedicated to speech and the other two modules are dedicated to face. The next task is the division of these attributes between the modules.

The division of the speech features between the first two modules is done equally. The attributes given to the first module are time duration, max cepstral, pitch amplitude, peak PSD, F2 and F4. The rest of the attributes are given to the second module. This consists of number of zero crossing, average PSD, pitch frequency, F1 and F3.

The last two modules are for the face features. Here also an equal distribution is done. The third module contains the attributes length of the first eye, width of the first eye, location of the center of the first eye in x and y axis, length of the mouth, and distance between eye and eye. The rest of the attributes are given to the fourth module. This includes length of the second eye, width of the second eye, location of the center of the second eye in x and y axis, width of the mouth, location of the center of the mouth in x and y axis, distance between eye and mouth. This division of the attributes between the various modules is shown in figure 17.3.

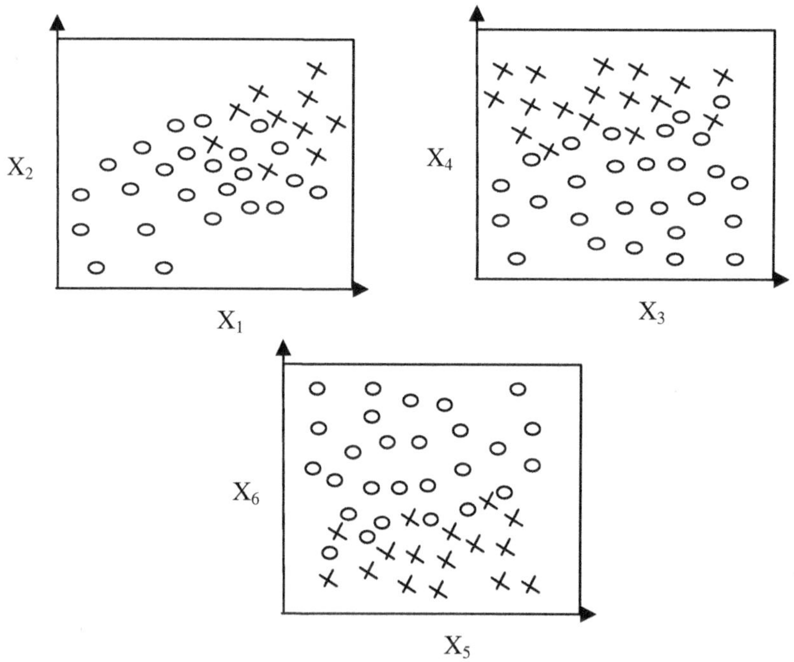

Fig. 17.2 Different Views of the Input Space

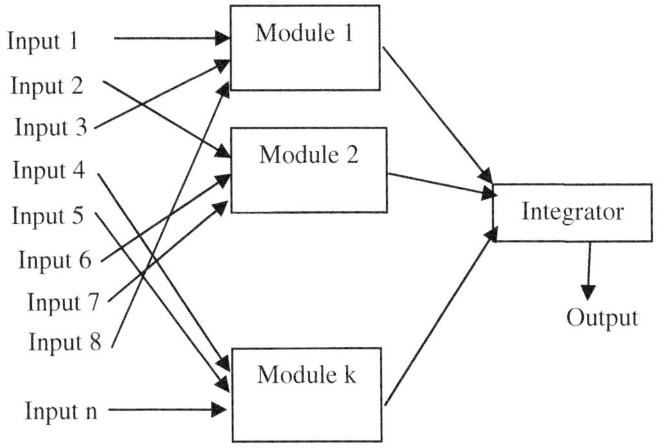

Fig. 17.3 Division of attributes

17.3.3 Modules

The modules of the MNN are classifiers that are given some input for the task of classification. Here the modules are an implementation of the Multi-Layer Perceptron trained with Back Propagation Algorithm. Each of the modules is given its designated attributes in the training and testing phase. The outputs of the module are conveyed to the integrator that further carries forward the task of classification and deciding the class to which the inputs belong.

The outputs of system are probability based as discussed in chapter 6. In this approach any module gives as many outputs as there are classes in the system. Each output denotes the probability of the input belonging to that class. The entire probability vector is given by the modules to the integrator for further processing.

Suppose that the system had a total of n classes. Every input would hence give n outputs. Let these outputs be $<o_1, o_2, o_3, \ldots, o_n>$. Any output o_i denotes the input belonging to class i. In case the input belongs to some class, its corresponding output would be very high and very close to 1.

The passage of the entire probability vector gives a lot of flexibility to the module to convey the results. This was discussed under the ensemble approaches in chapter 14.

17.3.4 Integrator

Once the various modules have done their computation and given their outputs in the form of probability vector, the next task is to decide the final output of the system. For this the integrator is made. The integrator computes the sum of all probabilities for every class. It then selects the class having the largest probability count. This is declared as the winning class. The input is returned to be classified to this class. This is shown in figure 17.3.

17.3.5 Experimentation

In this section we discuss the experimentation reported in the work of Kala *et al* (2010). The discussed algorithm was implemented in the manner discussed. The system consisted of four modules. The complete database was split into training and testing data sets. Each of the data set inputs were divided into the inputs and outputs of the various modules. Each of the modules was trained separately by its own training inputs and outputs. The final system consisting of the trained modules and integrator was passed through the testing data. The outputs were matched against the known targets to compute the final accuracy of the system. The system gave an accuracy of 97.5%. It may be easily seen that this was better than the individual accuracies of both speech and face which were computed to be 90.0% and 92.5% respectively.

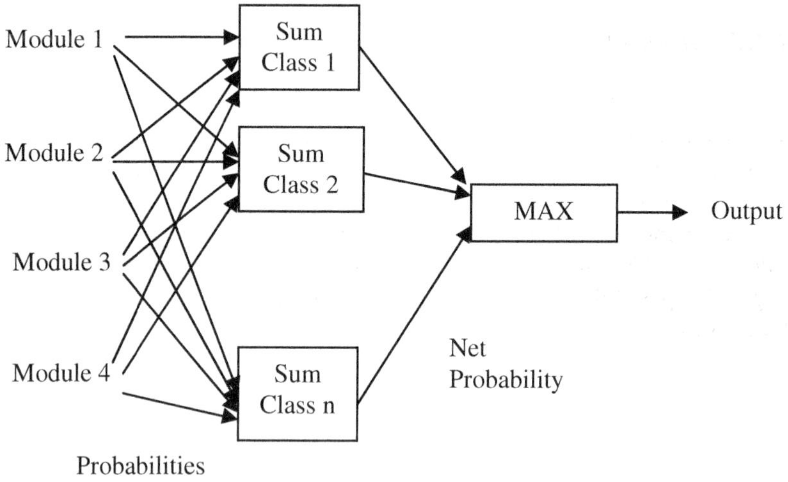

Fig. 17.4 Probabalistic Poling Integration

17.4 Division of Input Space

The next model that we build in this fusion method is based on the computational modularity in the Modular Neural network (MNN). In this example as well we develop a MNN that can carry out the task of recognition of the person by using both face as well as speech features. In this model we carry out the task of division of the problem, development of independent modules for problem solving and integration mechanism as per the fundamentals of the MNN. Since the problem is division based, only one of the modules in the MNN would be made active. This would be used for solving the problem. The other modules would be inactive and not used for the problem solving. The integrator hence gets only one output from

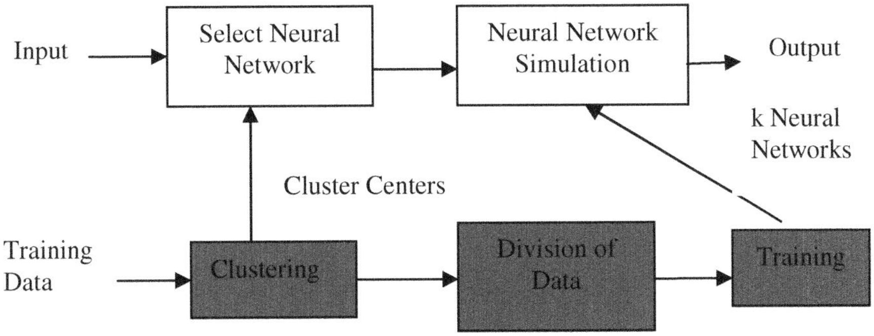

Fig. 17.5 Algorithm for fusion with division of input space

one module. This is regarded as the final output of the system. The integrator hence has a reasonably easy job to be done as it directly returns the output of the active module.

The general model used for problem solving in this chapter is derived from the work of Kala *et al* (2009). Here the authors had used the model for solving the problem of face recognition. The algorithm is shown in figure 17.5.

17.4.1 Problem Division

The first task is to distribute the problem into the modules. Here the task of division is to select the proper module that would carry the task of recognition out of all the available modules. This is done be selecting the cluster in the feature space that is closest to the applied input. Every cluster has a module associated to it that does the computation of the modules that belong to it.

The problem is hence to cluster the feature space by some mechanism. For this the Fuzzy C Means clustering is used. This clustering mechanism takes as inputs the entire training data that are represented as points in the feature space. The number of clusters to which the inputs need to be clustered into needs to be specified in advance. This number corresponds to the total number of modules that the system needs to have in the MNN being designed. Recall our discussion in chapter 14 that a higher number of modules may result in the problem losing its generality. At the same time lower number of modules may suffer from the associated problems of high problem complexity, high dimensionality, large training time, etc. The number of clusters or modules may be fixed accordingly.

The Fuzzy C-Means clustering algorithm forms fuzzy clusters. This means that all the data belong to all the clusters. There is no strict association of any data to any one cluster only. This further means that we may not be able to make strict boundaries that separate the various clusters. These boundaries have a fuzzy nature. In this clustering every data belongs to every cluster by some degree. This degree is known as the degree of membership between the data and cluster. Using this technique we may easily compute the degree of membership between any pair of data and cluster. The membership degree is very close to 1 if the data lies very close to the cluster center. On the other hand the degree of membership is almost 0 if the data is very far off from the cluster center. The Euclidean norm may be used to compute the closeness of any data to cluster center. The cluster center is simply the mean of the various data items weighted by their membership values.

The clustering algorithm is iterative in nature. Here at the staring the algorithm assumes random cluster centers and computes the associations. As the algorithm proceeds the cluster centers keep attaining their ideal values. After sufficient iterations there is no much change in the cluster centers. This is the time when the algorithm is said to be converged.

The final output of the clustering algorithm is the data centers as well as the membership values of the various data to the various centers. The algorithm we develop would naturally need crisp boundaries instead of fuzzy boundaries. This is because of the fact that specifically a single module needs to be selected out of the available modules. We are hence only interested in the cluster centers.

Whenever any new input is given to the system, the job is to decide the module that would be activated. For this we compute the distance of this input to all the cluster centers in the feature space. The cluster at the closest distance is selected and the corresponding module is invoked.

17.4.2 Modules

The modules are the problem solving units in the MNN. In the proposed model the modules consisted of the use of Multi-Layer Perceptron (MLP) trained with Back Propagation Algorithm (BPA). Each module of the MNN was trained using the training data set that lied in its own cluster in the feature space. The module only gave outputs of the classes that were present in the cluster. In this manner each module was suppose to separate a lesser number of classes that made the entire task of recognition much simpler. Further the reduced amount of training data with lesser problem complexity made the training much faster.

For the task of identification only a small set of inputs out of the complete set of inputs were taken. The inputs taken were the same that were taken in pre-classification fusion discussed in section 17.2. The outputs of each of the modules of the system returned the probabilities of the existence of the classes as discussed in section 17.3. The maximum of these probabilities was taken and the corresponding class was returned as the final output of the module.

Since only one of the modules is active, the integrator gets a single output for every input. This output is returned as the final output of the system.

17.4.3 Experimentation

The testing of the system was done using the same data as used in section 17.2. The data was divided into training and testing data sets. The training data was clustered using the Fuzzy C-Means Algorithm. The cluster centers were saved for the further processing.

The training data was divided into various clusters by finding the cluster center with the least Euclidian distance. Then the training data corresponding to each of the cluster was used for the construction and training of the neural network. This made a series of neural networks with one neural network corresponding to each of the modules.

Later the testing data was passed for the selection of the module. The corresponding module was given the entire input vector. This was used for the computation of the corresponding output. The computed output is the class to which the input classifies to. The system after training and testing gave an accuracy of 95% in the testing data set.

17.5 Chapter Summary

Use of a single biometric modality may not always yield good results. This therefore necessitates the use of more than one biometric modality for better biometric recognition and verification. This chapter presented the need and means to carry forward this fusion of the biometric modalities. The fusion was essentially classified under two headings of pre-classification fusion and post classification fusion. In the pre-classification fusion, pooling of the attributes from the various biometric modalities was done. A selected list of attributes is selected for the fusion to make the entire problem less dimensional in nature that has its own problems. The other discussion took place on the post classification fusion. Here the classification is carried out first and then the fusion is done. The first model discussed here was the use of modular neural network with probabilistic integration. The various attributes were distributed between different modules. Each module returned the probability score of the occurrence of each of the class. This was summed to get the net probability count. The class with largest probability was declared as the winner. The last model discussed divided the entire input space into clusters. Each cluster was a separate module that carried the task of classification.

References

[1] Campbell, J.P.: Speaker Recognition: A Tutorial. Proc. IEEE 85(9), 1437–1462 (1997)

[2] Card, S.K., Mackinlay, J.D., Robertson, G.G.: A morphological analysis of the design space of input devices. ACM Trans. Inform. Syst. 9(2), 99–122 (1991)

[3] Er, M.J., Wu, S., Lu, J., Toh, H.L.: Face Recognition with Radial Basis Function(RBF) Neural Networks. IEEE Trans. on Neural Nets 13(3), 697–710 (1999)

[4] Farrell, K.R., Mammone, R.J., Assaleh, K.T.: Speaker Recognition using Neural Networks and Conventional Classifiers. IEEE Trans. Speech Audio. Process. 2(1), 194–205 (1994)

[5] Fierrez-Aguilar, J., Ortega-Garcia, J., Garcia-Romero, D., Gonzalez-Rodriguez, J.: A Comparative Evaluation of Fusion Strategies for Multimodal Biometric Verification. In: Kittler, J., Nixon, M.S. (eds.) AVBPA 2003. LNCS, vol. 2688. Springer, Heidelberg (2003)

[6] Fox, N., Cohn, J.F., Gross, R., Reilly, R.B.: Person Identification Using Automatic Integration of Speech, Lip, and Face Experts. In: Proc. ACM 2003 Multimedia Biometrics Methods and Applications Workshop, WBMA, pp. 25–32 (2003)

[7] Huang, S.Y., Hwang, C.R. (2005) Kernel fisher's discriminant analysis in gaussian reproducing kernel hilbert space. Technical report, Institute of Statistical Science, Academia Sinica, Taiwan (2005)

[8] Jain, A.K.: Biometric Recognition:How do I know who you are? In: Kalviainen, H., Parkkinen, J., Kaarna, A. (eds.) SCIA 2005. LNCS, vol. 3540, pp. 1–5. Springer, Heidelberg (2005)

[9] Jain, A.K., Hong, L., Kulkarni, Y.: A Multimodal Biometric System using Fingerprint, Face and Speech. In: Proc. of Second Int. Conf. on AVBPA, March 1999, pp. 182–187 (1999)

[10] Jollife, I.T.: Principle Component Analysis. Springer, USA (2002)

[11] Kala, R., Shukla, A., Tiwari, R.: Fuzzy Neuro Systems for Machine Learning for Large Data Sets. In: Proc. of the IEEE Intl. Advance Comput. Conf., pp. 541-545, Patiala, India (2009)

[12] Kala, R., Vazirani, H., Shukla, A., Tiwari, R.: Fusion of Speech and Face by Enhanced Modular Neural Network. In: Proc. of the Intl. Conf. on Info. Syst., Technol. and Manag., ICISTM 2010, pp. 363–372, Bankok, Thailand (2010)

[13] Kumar, R., Ranjan, R., Singh, S.K., Kala, R., Shukla, A., Tiwari, R.: Multilingual Speaker Recognition Using Neural Network. In: Proc. of the Frontiers of Res. on Speech and Music, FRSM 2009, Gwalior, India, pp. 1–8 (2009)

[14] Lu, J., Plataniotis, K.N., Venetsanopoulos, A.N.: Face Recognition Using LDA-Based Algorithms. IEEE Trans. Neural Networks 14(1), 195–200 (2003)

[15] Lu, J., Plataniotis, K.N., Venetsanopoulos, A.N.: Regularization studies of linear discriminant analysis in small sample size scenarios with application to face recognition. Pattern Recognition Lett. 26(2), 181–191 (2005)

[16] Mammone, R.J., Zhang, X., Ramachandran, R.P.: Robust Speaker Recognition. IEEE Signal Process. Mag. 13, 58–71 (1996)

[17] Martin, A., Przybocki, M.: The NIST 1999 Speaker Recognition Evaluation - An Overview. Digital Signal Process 10(1), 1–18 (2000)

[18] Naik, J.M.: Speaker Verification: A Review. IEEE Comm. Mag. 28(1), 42–48 (1990)

[19] Phillips, P.J., et al.: Overview of the Face Recognition Grand Challenge. In: Proc. of the 2005 IEEE Conf. on Compu. Vision and Pattern Recog., CVPR 2005, vol. 1, pp. 947–954 (2005)

[20] Rabiner, L., Juang, B.H.: Fundamentals of Speech Recognition. Prentice Hall International, Englewood Cliffs (1993)

[21] Reynolds, D.A.: An Overview of Automatic Speaker Recognition Technology. In: Proc. of the IEEE Intl. Conf. Acoustics, Speech, and Signal Process (2002)

[22] Rose, R.C., Hofstetter, E.M., Reynolds, D.A.: Integrated Models of Signal and Background with Applications to Speaker Authentication in Noise. IEEE Trans. Speech Audio Process. 2(2), 245–257 (1994)

[23] Ross, A., Jain, A.: Information fusion in biometrics. Pattern Recognition Lett. archive, 2115–2125 (2003)

[24] Shukla, A., Tiwari, R.: Fusion of face and speech features with ANN for speaker authentication. IETE Tech. Rev. 4(5), 359–368 (2007)

[25] Shukla, A., Tiwari, R.: A Novel Approach Of Speaker authentication By Fusion of Speech and Image Feature using ANN. Intl. J. Info. Comm. Technol. 1(2), 159–170 (2008)

[26] Shukla, A., et al.: Intelligent Biometric System using PCA and R-LDA. In: Proc. IEEE Intl. Global Cong. on Intel. Syst., Xiamen, China (2009a)

[27] Shukla, A., et al.: Face Recognition using morphological method. In: Proc. of the IEEE Intel. Advance Comput. Conf., Patiala, India (2009b)
[28] Shukla, A., et al.: Face Recognition based on PCA, R-LDA and Supervised Neural networks. Intl. J. of Engg. Res. Ind. Appl. 2(3), 243–255 (2009c)
[29] Li, S.: A Tutorial on Principle Component Analysis (2002), http://www.cs.otago.ac.nz/cosc453/student_tutorials/principal_components.pdf
[30] Snelick, R.: Large-Scale Evaluation of Multimodal Biometric Authentication Using State-of-the-Art Systems. IEEE Trans. Pattern Analysis and Machine Intell. 27(3), 450–455 (2005)
[31] Turk, M.A., Pentland, A.P.: Face Recognition Using Eigenfaces. In: IEEE Conf. on Comput. Vision and Pattern Recog., pp. 586–591 (1991)

List of Figures

Fig. 17.1. Pre-Classification Fusion of Face and Speech
Fig. 17.2. Different Views of the Input Space
Fig. 17.3. Division of attributes
Fig. 17.4.Probabalistic Poling Integration
Fig. 17.5. Algorithm for fusion with division of input space

Chapter 18
Multimodal Biometric Systems

Abstract. The uni-modal biometric systems making use of a single biometric modality have a limited performance that restricts their applicability in real life scenarios. The multimodal biometric systems make use of two or more modalities that together achieve much higher performances. In this chapter we discuss the means to fuse three modalities to make a more robust system. We first discuss the fusion of speech, lip, and face. This system uses Hidden Markov Models for the classification and an integration technique called as late integration for decision making from the three modalities. We then discuss the fusion of face, speech and fingerprint. Here each of the individual biometric modalities would make use of modular neural network which would then be combined using a fuzzy integration technique. The last model we discuss would carry the fusion of fingerprint, face and hand geometry. This system uses a variety of fusion techniques including a sum rule, linear discriminant function and decision trees.

Biometric recognition and verification is a major challenge for the researchers to get higher accuracies in more robust conditions with more scalable inputs (Campbell 1997; Jain 2005; Martin and Przybocki 2000; Naik 1990; Er *et al* 1999). The natural counterpart of these systems is, on the contrary, reasonably well which gives very high performance scores. We all are easily able to identify people by their faces or voice with a high degree of confidence. Hence biometrics especially is far apart the imitation of the natural counterpart which may largely be attributed the wasteful practices in each step of recognition (Bunke and Kandel 2002).

Humans see the person and the various parts of the brain are activated to carry various tasks. Each unit contributes for the analysis of the entire image of the person. In this manner the decision is made. On the contrary the artificial intelligent systems use other practices. The preprocessing to remove noise may itself cause some damage or loss of information which is further carried by the segmentation. This kills many of the good features that facilitate recognition. The largest loss is at the feature extraction unit. The humans do not recognize faces by measuring the distances between eyes and other such features, but the artificial systems do that. This wastes a lot of good features. One of the major reasons of the limited accuracy is the loss of information which is again attributed to the limited computational power that the artificial systems may process (Konar 1999).

Also the humans get trained to locate and process good features. The module of feature enhancement in humans is naturally developed as the human grows from

A. Shukla et al.: Towards Hybrid and Adaptive Computing, SCI 307, pp. 401–418.
springerlink.com

infant to child. It is able to catch the eyes, lips and other interesting features in the face automatically. The artificial systems on the contrary need to be externally assisted to do so. This is by the use of image processing algorithms which again have their limitations, assumptions and wasteful practices. All this limits the recognition of the intelligent systems by a large amount.

18.1 Biometric Modalities

Numerous biometric modalities have been identified that are extensively used for the biometric systems. This includes speech, face, signature, etc. All these biometric modalities are fused in some or the other manner to make the multi-modal systems. These modalities may be classified into two heads of physiological biometric modalities and behavioral biometric modalities. The physiological modalities remain constant with time whereas the behavioral modalities change along with time. We studied about this classification in chapter 1.

The biometric systems are the systems that use these modalities for recognition or verification. The simplest systems are the ones that make use of a single biometric modality. These are the single modal systems. The systems making use of more than two modalities are multi-modal systems. These may be in turn classified as bi-modal systems that make use of two biometric modalities, tri-modal systems that make use of three biometric modalities and so on. Each of these has its own properties, advantages and disadvantages. We have a look at these systems in the next sub-sections.

18.1.1 Bi-Modal Systems

The use of the intelligent systems for biometric recognition demands high accuracies. The inability of a single biometric modality had led us to fuse two modalities making bi-modal biometric systems. We built numerous models out of the fused biometric systems in chapter 17. In all these systems speech was fused with face (Kala *et al* 2009, 2010; Shukla and Tiwari 2007, 2008). This resulted in a big rise in the accuracies. Many instances of speech may be noise prone and hence unusable for recognition. These were recognizable by the high quality of face and vice versa. This enabled the systems to be more robust against the real life data. The fusion is not limited to speech and face alone. Fusion may be carried out between fingerprint and palm, face and iris, face and ear, etc.

Biometric recognition using bi-modal biometric systems made use of selection of the features and their extraction as the first step. This enabled the construction of the single modality biometric systems. The next task was the fusion of the modalities. For this we made use of pre-classification fusion as well as post-classification fusion. The pre-classification fusion involved the pooling of the features or attributes, out of which the best few were selected. We saw that this technique resulted in a large dimensionality which further resulted in the associated problems of dimensionality. In post classification fusion each of the individual modalities returned some probability or decision. All these probabilities or decisions were later integrated using the integrator.

18.1.2 Tri-Modal Systems

The high success of the bi-modal systems is a motivation to fuse an even higher number of modalities to get even higher accuracies. This is done in the multi-modal biometric systems. The multi-modal systems carry out fusion of two or more biometric modalities for effective decision making. In this book however we have separately discussed the bi-modal systems in chapter 17 and tri-modal systems in the present chapter as the addition of a single modality changes the system complexity and the associated issues by a large amount. The tri-modal biometric systems would have too many attributes that come from each of the three participating modalities. Hence the pre-classification fusion cannot be done in these systems. Only selecting few good attributes might result in throwing away of a reasonably large information which may carry important features. Further the selected features may not be good enough to themselves carry out the complete classification. Selecting all the attributes would make the complete problem highly dimensional which any classifier may not be able to solve (Fierrez-Aguilar *et al* 2003; Snelick 2005; Jain *et al* 1999; Melin *et al* 2003).

In the tri-modal biometric systems we fuse three biometric identities. The choice of the biometric identities that need to be fused is entirely made based on system design. The system should be such that it is easy to use by the user with a high degree of convenience. Making the process very slow where a lot of recording time is needed would do away with much of the benefits in the use of biometrics.

The tri-modal biometric systems use post-classification information fusion where independent modules are used for each of the biometric modalities and their findings are later on fused using an integrator. Each of the modality that the humans possess is independent of each other. The fingerprint of a person has almost no correlation with his speech, retina, etc. All the biometric modalities hence go independent of each other. Only by looking at a person we may not be easily able to guess the voice that he possesses or the type of finger print that he may have. This emphasizes on the fact that modularity is a natural phenomena in the biometrics and hence using a modular technique where each modality is independently analyzed by a module may be useful.

18.1.3 Higher Modality Systems

Using four and more biometric identities is usually not done. This is chiefly because the resulting system would become highly complex to use. It would be very inconvenient for the user to carry out all the recordings of the various modalities.

Further in case the system is of biometric verification and the system identifies a genuine user as an imposter, the person is expected to repeat the procedure. Now repeating the process again in all the stages is not an easy task to be carried out. The multi-modal biometric systems further are very expensive to build, operate and maintain.

The higher modality systems may in turn give a reasonably small or no improvement in the results. The performance in drastically increased as we increase the modalities from one to two, but one may not expect such a high transition in accuracies in the higher modal systems. Such a fact again comes from the natural world. We may recognize people better by looking at their face, seeing their face and noting their gestures simultaneously. However if we are still unable to identify them, it is likely that we may still not identify them if given more modalities.

18.2 Fusion of Speech, Lip and Face

The first model that we present in this chapter on tri-modal fusion techniques is the fusion of speech, lip and face. Speech forms a good biometric modality that has the potential of effective recognition. Here the voice of a person is recorded and the same after analysis is used for the recognition purposes. We saw the manner in which speech is analyzed in chapter 5. We further discussed about the use of speech as a biometric in uni-modal systems in chapter 6 and in bi-modal systems in chapter 17. One of the major barriers in the use of speech as a biometric is the noise that is difficult to separate from the useful signal. This makes the feature extraction erroneous. Face is another interesting biometric modality that is capable of good recognition. The face has various interesting features that enable good recognition. We saw the manner in which face can be effectively used for the recognition in chapter 8. We further discussed about the use of face in bi-modal systems in chapter 17.

Lip is another biometric modality that is capable of carrying recognition. Each person has unique features that may be used for identification of the person solely on the basis of his lips. Lip is not a conventional biometric modality that we may find a natural inspiration from. If we are shown the photograph of only lips of a person, we may not be able to identify him. The lips still hold features in themselves that has enabled the intelligent systems to differentiate people based on these features of lips.

In this section we discuss the model proposed by Fox *et al* (2003) for the biometric identification using a combination or fusion of these biometric identities. Here the three biometric modalities of speech, lips and face were three different modules of the recognition system. These may also be referred to as three experts, each of which carries the task of recognition using the expertise of its own modality. The expert decision is then used for integration. One of the key aspects of the multi-modal systems is the fusion or integration technique where we combine the outputs of the various modalities to get the final output of the system. The integration technique used in this chapter is called as the Late Integration which is primarily a post-classification integration technique that we discussed in chapter 17.

18.2.1 Face

The first modality of study is the face. The model uses the commercial face recognition system FaceIt for the task of facial recognition. This used Local Feature Analysis (LFA) for the facial recognition (Penev and Atick 1996; Penev 1998, 1999; Shukla et al 2009a, 2009b). We know that the face is an interconnection of various interesting parts like eyes, lips, nose, etc. All of these have interesting features that aid in the recognition process.

We made use of Principle Component Analysis (PCA) in chapter 8. This technique of analysis worked on the principle of extraction of good features that were automatically computed by statistical analysis from the entire face. PCA visualizes the face on the whole for extraction of the components. It tries to locate global features. The consecutive features extracted from this technique might not represent close features in reference to the physical face. The system is hence too globalized that misses out the localized features that may be source of good recognition. The LFA uses the same principles of PCA but applies it in a more localized manner where the face is seen as an image of interconnected topographical components. This makes it a much better and robust tool for recognition of face. LFA uses localized image features from multi-scale filters. These are analyzed separately by the application of PCA that does the task of dimensionality reduction.

After dimensionality reduction, the set of features is obtained that may be used for the classification. The classifier works over the feature space to make decision boundaries separating the various classes. It may be used to give the class to which any input in this feature space belongs to or output the matching score that denotes the ease by which the input belongs to the class. The probabilities of output that we used in chapter 17 are an example of the matching score.

18.2.2 Speech

The other modality used in the system is speech. In the speech module the utterances of the speaker were collected. These were pre-emphasized to increase the power at the higher frequencies by using filters. The resultant signal was passed through a Hamming window of 20ms length and a 0 ms overlap. This gave audio frames of 100 Hz that were to be used for the recognition.

The model made use of the Mel Frequency Cepstral Coefficients as features (Rabiner 1993; Reynolds 2002). 16 coefficients were extracted from each of the audio frame. The energy of each of the frames was also extracted and used as feature. The other 17 features came from the first order differences in between the adjacent frames. These are also called as the delta features. This made a total of 34 features. Cepstral mean Normalization was used to overcome the long term spectral effects of the audio channel.

The other part in the development of the complete model is the classifier. The model makes use of Hidden Markov Model for carrying out the task of classification. This is a statistical model. It gives the class to which the input belongs along with the probability of the input belonging to that class given the input feature vector.

18.2.3 Lip

The last biometric modality used in this system is lips (Kim *et al* 2004; Iwano *et al* 2001). The lips were first pre-processed and segmented to be used for the recognition purpose. The original color image was converted into a grayscale image. The segmented lips image was then histogram equalized and the mean value was subtracted. The task of dimensionality reduction was done by the implementation of Discrete Cosine Transformation (DCT). A total of 15 coefficients were selected in this model in a zig-zag manner. Here the first frame denotes the man which is zero in this case, since the mean had been subtracted out.

The modeling of the motion of the lip was done with the help of Gaussian model with a single probability density function. The training data was used for the computation of the mean feature vector and diagonal covariance matrix. The log likelihood probability was calculated for each frame of the test sentence. The scores so obtained were summed for the entire sentence.

18.2.4 Fusion

The fusion of the three modalities has been done by a technique known as the late integration. This is a post-classification technique where the individual modality decisions are fist computed and then the decision by the fused system is made. In the post-classification information fusion studied in chapter 17, we had simply taken the sum of the probabilities of the various outputs. Here we assign a weight to each of the experts or modules which denote its dominance in making the final decision. Each of the three experts gives as its output the probability of the input belonging to some class. This probability lies within the range of 0 to 1. Higher probabilities for some class denotes a more likely chance if the input being classified to that class. Using this system, the probability for some class X for some input may be given by equation (18.1)

$$P_X = a\,P^1_X + b\,P^2_X + c\,P^3_X \tag{18.1}$$

Here P^1_X is the probability of the input belonging to class X as per first expert, P^2_X is the probability of the input belonging to class X as per second expert and P^3_X is the probability of the input belonging to class X as per third expert. a, b and c are the associated weights with the three experts.

The probabilities are computed by the classifier itself. It denotes how easily the input I within the class region in the feature space. The probability would be very high if the input is reasonably far from the input space and inside the class region. The probability may be close to half if the input is close to the decision boundary.

In the specific model used by Fox *et al* (2003) the three experts correspond to the speech, face and lips biometric identification systems respectively. The equation is modified to the form given in equation (18.2).

$$P_X = \alpha P^1_X + \frac{\beta}{2} P^2_X + \frac{\beta}{2} P^3_X \tag{18.2}$$

Here P^1_X is the probability of the input belonging to class X as per speech recognition system, P^2_X is the probability of the input belonging to class X as per face recognition system and P^3_X is the probability of the input belonging to class X as per lips recognition system. α and β are the weights.

Here both face and speech have been given the same weights. This model assumes that the deviations in the recorded values of the biometric features from their ideal values are always caused by noise. The selected biometric identities are hence classified into audio and visual. Speech belongs to the first class and the face and lips belong to the second class. α denotes the total weight of the audio biometric modalities and β denotes the total contribution of the visual biometric modalities. Here the Signal to Noise Ratio (SNR) may be easily use as an indicator of the parameters α and β. In case the SNR is high, the contribution of the audio expert would be more and vice versa.

18.2.5 Experimentation

This section presents the findings of the experimentation done by Fox *et al* (2003) in their work using this model. The three biometric recognition models were built and tested independently. These were then combined in all possible manners to form bimodal system. The last experimentation combined all the three modalities to form the tri-modal system. the structure of the tri-modal system is given in figure 18.1.

As per the experimentation speech recognition system had an accuracy of 98%. The accuracy of face came out to be 93.22%. The accuracy in the use of lip was 86.37%. The bi-modal systems built by the fusion of two biometric modalities gave an accuracy of 96.8%. However the accuracy came out to be 100% when all the three biometric modalities of speech, face and lips were fused to make a tri-modal biometric identification system.

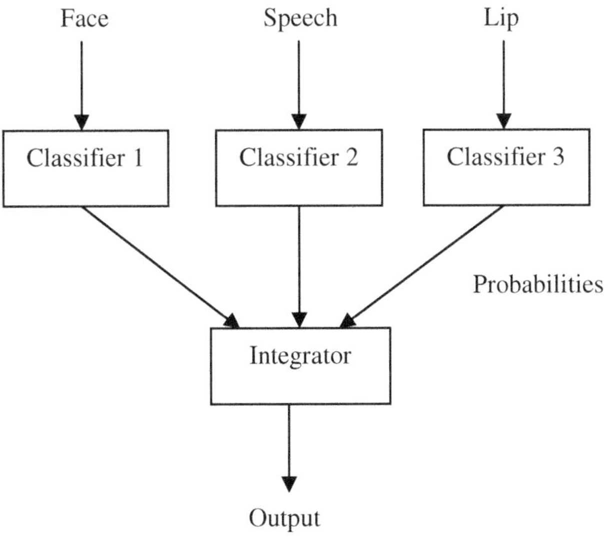

Fig. 18.1 Fusion of face, speech and lips

18.3 Fusion of Face, Fingerprint and Speech

The next fusion we carry out is between the face, fingerprint and speech. Here we would make use of a modular neural network for each of these modalities. This forms the three experts that carry out the biometric recognition. These three experts need some integration mechanism for the decision making regarding the final output of the system. This integration is done with the help of Fuzzy Inference Systems and is known as Fuzzy integration.

We have seen the recognition by face and speech numerous times throughout the book. The fingerprint is another widely used biometric modality. It has been known since quite long that each person possesses fingerprint that is unique to him. Forensics has been using fingerprints for very long for their investigation. Here manual matching of the features present in a fingerprint is done for recognition purposes. This is largely a motivation for the use of automated systems for biometric recognition and verification using fingerprint as the biometric modality. The fingerprint may be easily recorded as a digital image by the use of fingerprint readers or scanners. The fingerprint image has numerous such features that enable good recognition by the intelligent automated systems.

In this section we briefly describe the work of Melin and Castilo (2005) who developed an entire system for biometric recognition using face, fingerprint and speech as the biometric modalities. The model developed by the authors marks quite a complex model that first uses modular neural network architecture for each of the biometric modalities. This includes the independent modules and an integration unit. Then these experts, one corresponding to each module, are fused using another integrator. This integrator uses Fuzzy Rules for information fusion and decision making.

18.3.1 Face

The first biometric modality whose recognition expert is developed is face. The face is first segmented into three independent regions. These are eyes, nose and mouth. Each of these is analyzed independently by a model of the modular neural network. The system uses a Fuzzy integration technique for fusion of the decision of the three modules.

The first task of the system is hence to partition the facial images into three parts. The partition cannot be predefined as the different faces have different attributes and extents of the eyes, nose and mouth. Some may possess big nose, small eyes, etc. The task of partitioning is carried out using clustering technique. The facial image is first represented in the form of a matrix. This shows the distribution of the various points. The clustering is applied in this matrix representation of the image. We make a total of three clusters so that each of these corresponds to the eyes, nose and mouth. This may be done by any clustering technique. He authors used Competitive Neural Networks for this clustering.

The other mechanism adopted was Kohonen's Self Organizing Maps. Here a Self Organizing Feature Map of size 3x3 is taken and trained with the facial image. After training each of the cells of the feature map gets aligned. This is an indicator of the manner in which the partitions be applied on the original facial image to segment it into three regions.

Figure 18.2 shows the partitioning of the facial image into three partitions of eyes, nose and mouth for a random face.

The other task that needs to be carried out is the development of the modules of the modular neural network (MNN). Each module uses its own segment of the image and carries out the task of classification of the segment as per the available information. We know that each of the modules i.e. eyes, nose and mouth has good features that are themselves capable of good recognition by themselves. This enables us to use a large number of features by the independent modules and then integrate these by the use of an integrator. Any mechanism may be used for the dimensionality reduction. Any of the neural network models may be used for the machine learning. This would complete the architecture of all the three modules of the MNN.

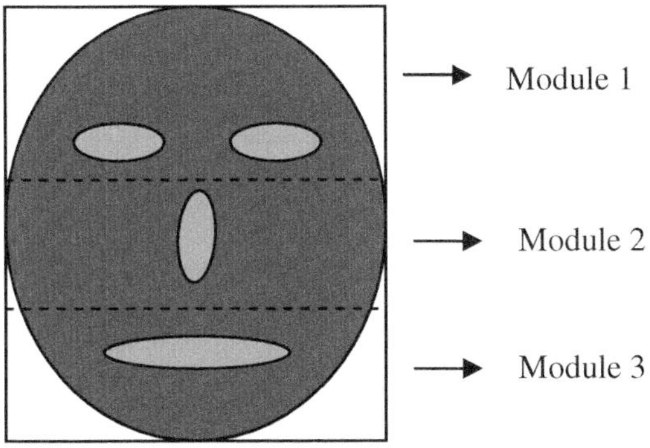

Fig. 18.2 Division of Facial Image

Each module of the MNN is itself a module. We learnt that the ensemble mechanism of problem solving in a redundant manner largely results in a performance boost because of the multiplicity of ways in which the same problem can be solved. In the recognition of each segment of the face we try to use the same principles for carrying out recognition. Each segment of speech is analyzed and classified. This forms three experts for the recognition of the face. These experts are fused by an application of Fuzzy integrator based on Type 2 Fuzzy Logic. This forms the final output of the system. The integrator uses Fuzzy measures for computing the final output once a set of decisions is available. The complete formulation and working of the Fuzzy Integration is not being taken as a part of this book.

The complete methodology of this system is shown in figure 18.3.

18.3.2 Fingerprint

The other biometric system or identification expert that we build is for the fingerprint recognition. The fingerprint contains many ridges that are in form of

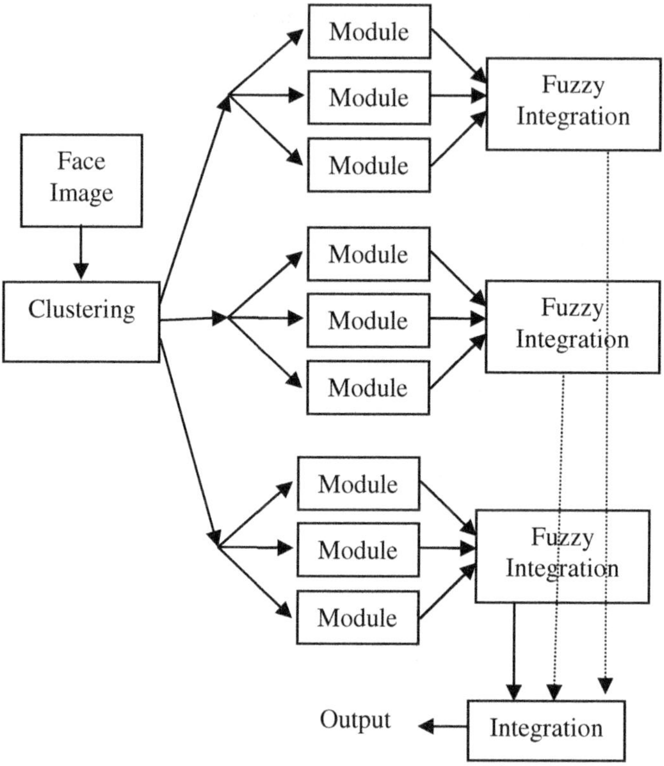

Fig. 18.3 Face Recognition System

concentric circles in the fingers. The pattern and manner in which these are arranged is unique to every person and the fingerprint recognition try to use the same properties for carrying out the classification. The fingerprint feature set that aid in recognition consists of global features and local features. The global features consist of the basic ridge patterns, pattern area, core area, delta, type lines and ridge count. These can be seen by a naked human eye. These features characterize the entire fingerprint as a whole. The local features consist of the minutia points. These points are the type of ridges and their specific location in the fingerprint. Here we are interested in how the various ridges bifurcate, combine, or end. These features clearly differentiate the people from one another as these features are unique to the person. It is possible for two people to have similar global features, but they would possess different local features.

The input to the fingerprint recognition algorithm is in form of a scanned image which may be easily obtained by the fingerprint readers. This image is preprocessed and the good features are identified and highlighted.

For the fingerprint recognition also the modular neural network (MNN) has been used. Here the entire image is divided into three equal sections. These are top sections, middle section and bottom section.

Each of these sections is given to a different module of the MNN. Each of the modules gets as input a part of the entire image that may be subjected to dimensionality reduction and machine learning techniques for the classification. The classification is done by the use of neural networks. Unlike the face, the neural networks used in this approach are optimized with the help of Genetic Algorithms.

The integrator used for the integration of the results of the three modules is same as used in case of the face. This is a Fuzzy integrator that uses fuzzy measures for computing the final output of the fingerprint recognition module.

18.3.3 Speech

The last biometric modality that we take is the speech. As per the basis of the recognition system with speech, the speech signal is first acquired, digitalized and preprocessed. This is followed by segmentation. The resulting signal is used for recognition by the recognition system.

In this modality we follow the same method of using modular neural network that was used with fingerprint. The entire signal is first broken down or partitioned in the time axis into three parts. Each of these three parts is given to a module of the modular neural network (MNN) for the task of recognition. The outputs generated by the various modules are fused together using an integrator. The integrator uses a Fuzzy integration technique for computing the final decision of the system.

Fourier coefficients are used for the purpose of dimensionality reduction and neural networks are used for the machine learning. Each of these neural networks is optimized by the application of genetic algorithms.

18.3.4 Fusion

The last job to be carried out is fusion of each of the three biometric modalities to make the final decision of the system. Each of the biometric systems does some classification based on which it is bale to decide which of the class the input belongs to. In other words, each system does some amount of person recognition. The recognition scores and decisions of each of the independent biometric systems need to be integrated to make the final system output. This is done with the help of Fuzzy rules.

The integration system is hence a Fuzzy Inference System. It takes as inputs the outputs of the three modules i.e. face recognition, fingerprint recognition and speech recognition. It then applies all the fuzzy rules to compute the individual rule outputs which are then aggregated and defuzzified to give the final output of the system.

The final architecture of the overall tri-modal biometric identification system that carries out the task of recognition is given in figure 18.4.

Each of the three inputs of face, speech and fingerprint recognition is divided into a set of membership functions. The rules are framed based on these membership functions of each of the input. A sample rule of this system is "*If Face is X*

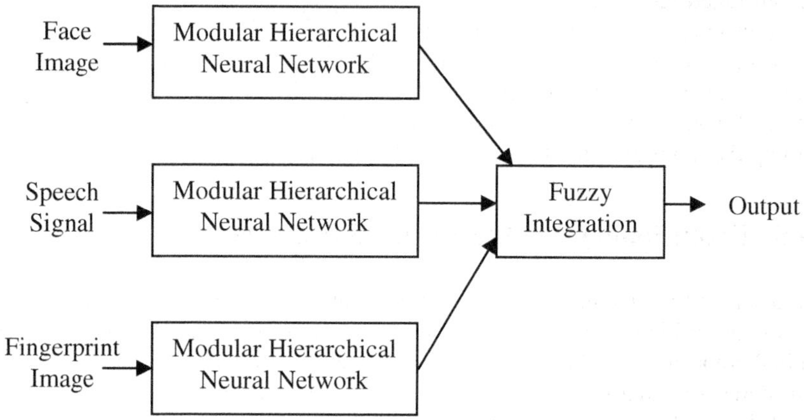

Fig. 18.4 Complete system for recognition

and Fingerprint is X and speech is Y then Person is X'. In linguistic terms this rule emphasizes on the fact that if two biometric systems support the decision in favor of some person and the third biometric system opposes it, the output is the same person. This is again what we did in the polling or voting approach where the class getting the largest votes was declared as the winning class.

18.4 Fusion of Face, Fingerprint and Hand Geometry

The last system we discuss in multi-modal biometrics is the fusion of face, fingerprint and hand geometry. We have seen the manner in which the face and fingerprint is used for recognition and the ease with which they help in recognition purposes. Hand geometry is another biometric modality that has been used in the recognition of people. The hand also carries various features that may aid in the recognition purposes. The hands of various people do not vary remarkably. Still the features present in hand are clear enough to carry out the classification or recognition.

In this section we summarize the work of Ross and Jain (2003). Here the authors used a variety of fusion methods between face, fingerprint and hand geometry. We first discuss the individual biometric methods and then present the fusion techniques.

18.4.1 Face

The first biometric modality is the face. Here the facial photograph of a person is given for recognition. The model uses Principle Component Analysis (PCA) for the purpose of dimensionality reduction. This technique extracts the best components from the given high dimensional input of face. These components are used directly for carrying out the classification.

The classifier used in this model uses a simple template based matching. In this technique the feature vector of all the recordings of a person is aggregated. This becomes the mean feature vector of the person. This feature vector is used as a template of the person. For the identification purposes we find the template closest to the input in the feature space. This person corresponding to this template is returned as the final output of the system.

18.4.2 Fingerprint

The fingerprint is recognized by means of the various ridges. The bifurcations and endings of the ridges are unique to every person and these are hence means of good recognition. The extraction of the rides along with their orientation and localization in the entire fingerprint needs to be carried. All this information serves are features which are coded in a convenient format. These encodings are then used for the task of classification (Hrechak and McHugh 1990; van der Putte and Keuning 2001; Willis and Myers 2001).

This module of biometric recognition again uses a template based matching. A template vector is made for each and every user of the system. The matching is done against the closeness of the inputs to these templates. This also gives an idea regarding the recognition score of the system.

18.4.3 Hand Geometry

The last biometric identity of use in this fused system is the hand geometry. In this module the picture of the hand is taken by a convenient scanning or pressure devise. The image is preprocessed, segmented and then used for the recognition purpose (Wong and Shi 2002;). The image processing techniques of edge detection are used for working with the hand images. A set of 14 features are used for the recognition purposes. These features include the length of the various fingers, width of the palm and width at various regions of the fingers. The various features are shown in figure 8.5.

The various computed features are then used for the task of classification. As with the case with face and fingerprint, the template based matching technique may conveniently be used for the classification purpose.

18.4.4 Fusion

The last step in the system is the fusion of the three biometric recognition systems to make a multi-modal biometric recognition system. This forms the most important part of the entire system. The aim is to use the different results and recognition scores of the different modules in the best possible manner to make a complete system. This task of information fusion or integration may be done in a variety of means depending upon the design and the problem requirements. In this

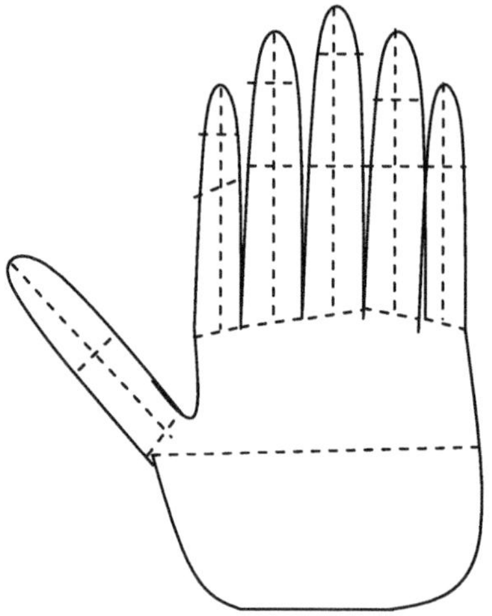

Fig. 18.5 The various features in hand

section we discuss some of these techniques of fusion. We discuss three techniques used by Ross and Jain (2003) in their work. These are sum rule, decision trees and linear discriminant function. Each of these is discussed in the next subsections.

18.4.4.1 Sum Rule

The sum rule is a simple information fusion system that computes the decision based on the probabilities returned by the various experts. The model is used for the task of biometric verification. In verification each expert gives as its output the probability that the person is genuine as per the claimed identity. Here we take a weighted addition of the various experts. This decides the final decision of the integrator. If the weighted addition is more than some threshold, the person is believed to be genuine, else the person is said to be an imposter.

18.4.4.2 Decision Trees

The other technique of integration is by using the decision trees. In this method we construct a decision tree. The nodes of the tree denote some conditions. Based on the decision of these conditions we iterate the tree from top to bottom. At the end we get the decision whether the identity is acceptable or not. One such tree is given in figure 8.6.

The attribute that best classifies the data at any node is specified and accordingly the condition is framed. In this manner the entire tree is constructed. The various parameters of the decision tree are learned by training of the tree. For this we first build a training database that contains sufficient number of instances of genuine and fake user recordings. The training tries to modify the parameters in

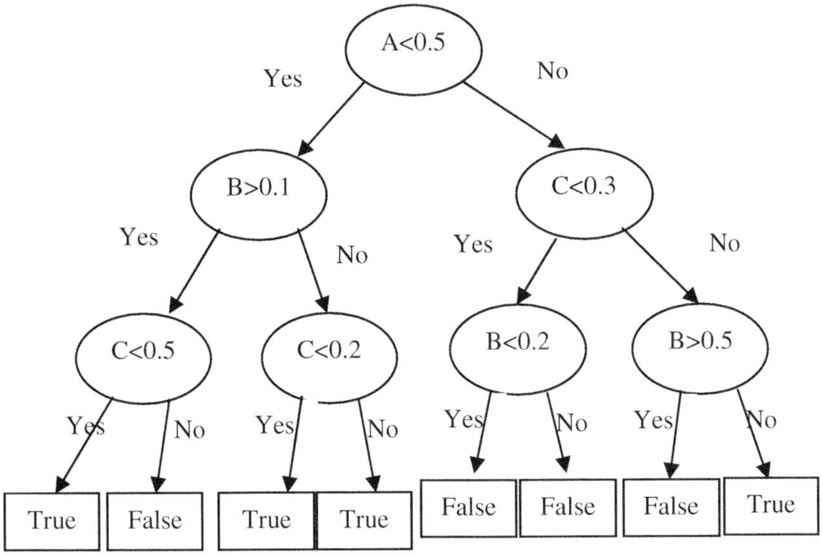

Fig. 18.6 Decision Tree

such a manner that the training data is well classified into genuine and fake identities. The training database is passed through the tree and the necessary alterations are made. In this manner the decision tree gives an optimal performance. The model used for the fusion was built using the C5.0 program Quinlan (1992) in the experiments reported by Ross and Jain (2003).

18.4.4.3 Linear Discriminant Analysis

The last method to be used is Linear Discriminant Analysis (Huang and Hwang 2005; Lu and Venetsanopoulos 2003, 2005; Shukla *et al* 2009a, 2009b). The fusion module is given three scores and based on these it is expected to give as its output the final decision of the system regarding the acceptance or rejection of the user. This is naturally a classification problem where a vector of three inputs needs to be classified into either of the two classes of genuine and fake. The input space for this problem is a three dimensional space. The input space may have the two classes mixing with each other and maintaining high intra class separation or low inter-class separation. This is naturally non-idealistic in nature. We use the Linear Discriminant Analysis to convert this space into a much better space 2-dimensional space. This has a much more idealistic nature with both the classes easily separable. This results in a better classification by the integrator based on the inputs of the recognition scores by the three recognition systems.

The integrator uses a template based matching technique for the classification. The templates are generated by using the training data which is especially used for this method based on the entire training data of the system. In this approach Mahalanobis distance is used for the computation.

18.4.5 *Experimentation*

The system consisting of the three biometric verification modules and a central integrator is used for training and testing. In this section we present the results of Fox *et al* (2003). The performance of the verification system is measured in terms of False Acceptance Ratio (FAR) and False Rejection Ratio (FRR). Based on experimentation the FAR of the tree classifier was found out to be 0.036% and FRR was found to be 9.63%. Similarly the FAR of Linear Discriminant Analysis was found to be 0.47% and FRR was found to be 0.00%. The last experiment was using the Sum rule. Using this approach the FAR was found to be 0.03% and FRR was found to be 1.78%.

18.5 Chapter Summary

This chapter focused upon the use of more than two biometric modalities for the task of recognition or verification. The use of two biometric modalities gives a high boost to the recognition scores with some complexity of the system. The urge for still larger recognition scores necessitates the fusion of more biometric modalities. The chapter presented numerous models by the fusion of various types of biometric modalities. The first model carried out the fusion of face, speech and lips. The face recognition was built over FaceIt that uses Local Feature Analysis for dimensionality reduction. The recognition of speech was performed by the Mel Frequency Cpstral Coefficients. The recognition of lips was done by the application of Discrete Cosine Transformation for dimensionality reduction. The classifiers were used to output the probabilities of the occurrence of classes. The integrator used the weighted sum for decision making. The other model used was the fusion of face, fingerprint and speech. Recordings of each of these modalities was divided into three parts. Modular neural network served as a mechanism of recognition, with fuzzy integration technique. Each module was in turn an ensemble comprising of neural network classifiers. These classifiers were optimized by Genetic Algorithm. The complete fusion of the three modalities was done by Fuzzy integrator. The last model of discussion was the fusion of face, fingerprint and hand geometry. Principle component analysis was applied for dimensionality reduction of face. Template based matching classifier was used. For fingerprint verification ridge information was used as features and template based matching as classifier. The lengths and widths of fingers were the features for hand geometry. Fusion was done using a variety of means including sum rule, decision trees and classification after linear discriminant analysis.

References

[1] Bunke, H., Kandel, A.: Hybrid Methods in Pattern Recognition Research. World Scientific, Singapore (2002)

[2] Campbell, J.P.: Speaker Recognition: A Tutorial. Proc. IEEE 85(9), 1437–1462 (1997)

[3] Er, M.J., Wu, S., Lu, J., Toh, H.L.: Face Recognition with Radial Basis Function(RBF) Neural Networks. IEEE Trans. on Neural Nets 13(3), 697–710 (1999)
[4] Fierrez-Aguilar, J., Ortega-Garcia, J., Garcia-Romero, D., Gonzalez-Rodriguez, J.: A Comparative Evaluation of Fusion Strategies for Multimodal Biometric Verification. In: Kittler, J., Nixon, M.S. (eds.) AVBPA 2003. LNCS, vol. 2688. Springer, Heidelberg (2003)
[5] Fox, N., Cohn, J.F., Gross, R., Reilly, R.B.: Person Identification Using Automatic Integration of Speech, Lip, and Face Experts. In: Proc. ACM, Multimedia Biometrics Methods and Applications Workshop, WBMA, pp. 25–32 (2003)
[6] Hrechak, A.K., McHugh, J.A.: Automated fingerprint recognition using structural matching. Pattern Recognit. 23, 893–904 (1990)
[7] Huang, S.Y., Hwang, C.R. (2005) Kernel fisher's discriminant analysis in gaussian reproducing kernel hilbert space. Technical report, Institute of Statistical Science, Academia Sinica, Taiwan (2005)
[8] Iwano, K., Tamura, S., Furui, S.: Bimodal speech recognition using lip movement measured by optical-flow analysis. In: Proc. HSC 2001, pp. 187–190 (2001)
[9] Jain, A.K.: Biometric Recognition:How do I know who you are? In: Kalviainen, H., Parkkinen, J., Kaarna, A. (eds.) SCIA 2005. LNCS, vol. 3540, pp. 1–5. Springer, Heidelberg (2005)
[10] Jain, A.K., Hong, L., Kulkarni, Y.: A Multimodal Biometric System using Fingerprint, Face and Speech. In: Proc. of Second Int. Conf. on AVBPA, March 1999, pp. 182–187 (1999)
[11] Kala, R., Shukla, A., Tiwari, R.: Fuzzy Neuro Systems for Machine Learning for Large Data Sets. In: Proc. of the IEEE Intl. Advance Comput. Conf., Patiala, India, pp. 541–545 (2009)
[12] Kala, R., Vazirani, H., Shukla, A., Tiwari, R.: Fusion of Speech and Face by Enhanced Modular Neural Network. In: Proc. of the Intl. Conf. on Info. Syst., Technol. and Manag., ICISTM 2010, Bankok, Thailand, pp. 363–372 (2010)
[13] Kim, J.O., Lee, W., Hwang, J., Baik, K.S., Chung, C.H.: Lip print recognition for security systems by multi-resolution architecture. Future Generation Computer Systems 20(2), 295–301 (2004)
[14] Konar, A.: Artificial Intelligence and Soft Computing: Behavioral and Cognitive Modeling of the Human Brain. CRC Press, Boca Raton (1999)
[15] Kumar, A., Zhang, D.: Personal recognition using hand shape and texture. IEEE Trans. Image Processing 15(8), 2454–2461 (2006)
[16] Lu, J., Plataniotis, K.N., Venetsanopoulos, A.N.: Face Recognition Using LDA-Based Algorithms. IEEE Trans. Neural Networks 14(1), 195–200 (2003)
[17] Lu, J., Plataniotis, K.N., Venetsanopoulos, A.N.: Regularization studies of linear discriminant analysis in small sample size scenarios with application to face recognition. Pattern Recognition Lett. 26(2), 181–191 (2005)
[18] Martin, A., Przybocki, M.: The NIST, Speaker Recognition Evaluation - An Overview. Digital Signal Process 10(1), 1–18 (2000)
[19] Melin, P., Acosta, M.L., Felix, C.: Pattern Recognition Using Fuzzy Logic and Neural Networks. In: Proc. Int. Conf. Artif. Intel., Las Vegas, USA, pp. 221–227 (2003)
[20] Melin, P., Castilo, O.: Hybrid Intelligent Systems for Pattern Recognition Using Soft Computing. Springer, Heidelberg (2005)
[21] Naik, J.M.: Speaker Verification: A Review. IEEE Comm. Mag. 28(1), 42–48 (1990)
[22] Penev, P.S.: Local feature analysis: a statistical theory for information representation and transmission. Ph.D. Thesis, The Rockefeller University (1998)
[23] Penev, P.S.: Dimensionality reduction by sparsification in a local-features representation of human faces. Technical Report, The Rockefeller University (1999)

[24] Penev, P.S., Atick, J.J.: Local feature analysis: a general statistical theory for object representation. Netw. Comput. Neural Syst. 7(3), 477–500 (1996)
[25] Quinlan, R.J.: C4.5: Programs for Machine Learning. In: The Morgan Kaufmann Series in Machine Learning. Morgan Kaufmann Publishers, San Francisco (1992)
[26] Rabiner, L., Juang, B.H.: Fundamentals of Speech Recognition. Prentice Hall International, Englewood Cliffs (1993)
[27] Reynolds, D.A.: An Overview of Automatic Speaker Recognition Technology. In: Proc. of the IEEE Intl. Conf. Acoustics, Speech, and Signal Process. (2002)
[28] Ross, A., Jain, A.: Information fusion in biometrics. Pattern Recognition Lett. archive, 2115–2125 (2003)
[29] Shukla, A., Tiwari, R.: Fusion of face and speech features with ANN for speaker authentication. IETE Tech. Rev. 4(5), 359–368 (2007)
[30] Shukla, A., Tiwari, R.: A Novel Approach of Speaker authentication By Fusion of Speech and Image Feature using ANN. Intl. J. Info. Comm. Technol. 1(2), 159–170 (2008)
[31] Shukla, A., et al.: Intelligent Biometric System using PCA and R-LDA. In: Proc. IEEE Intl. Global Cong. on Intel. Syst., Xiamen, China (2009a)
[32] Shukla, et al.: Face Recognition based on PCA, R-LDA and Supervised Neural networks. Intl. J. of Engg. Res. Ind. Appl. 2(3), 243–255 (2009b)
[33] Snelick, R.: Large-Scale Evaluation of Multimodal Biometric Authentication Using State-of-the-Art Systems. IEEE Trans. Pattern Analysis and Machine Intell. 27(3), 450–455 (2005)
[34] van der Putte, T., Keuning, J.: Biometrical Fingerprint Recognition: Don't Get Your Fingers Burned. In: IFIP TC8/WG8.8 Fourth Working Conf. on Smart Card Res. and Ad. Appl., pp. 289–303 (2001)
[35] Willis, A.J., Myers, L.: A cost-effective fingerprint recognition systems for use with low-quality prints and damaged fingerprints. Pattern Recognit. 34(2), 255–270 (2001)
[36] Wong, L., Shi, P.: Peg-free hand geometry recognition using hierarchical geometry and shape matching. In: IAPR Workshop on Machine Vision Appl., Nara, Japan, pp. 281–284 (2002)

List of Figures

Fig. 18.1. Fusion of face, speech and lips
Fig. 18.2. Division of Facial Image
Fig. 18.3. Face Recognition System
Fig. 18.4.Complete system for recognition
Fig. 18.5. The various features in hand
Fig. 18.6. Decision Tree

Part VI
Other Supplementary Topics

Chapter 19
Adaptive Systems

Abstract. The task of adjusting the values of the various parameters in the systems, especially the evolutionary systems, leads to a lot of sub-optimality. Besides the ideal parameter values may be different in different contexts and different stages of the algorithm. This necessitates the construction of adaptive systems that can tune their parameters themselves, leading to close to ideal performance. This chapter explores the various types of adaptive systems. The chapter first presents the types of adaptive systems based on their dynamism. Here we discuss the static, deterministic, adaptive and self-adaptive systems. We then discuss the level of adaptation in these systems. Here we present the environment, individual, population and component level adaptation. We present numerous examples of adaptive systems of various kinds.

So far we have studied abundant of intelligent systems in this book. At one end we presented the general class of neural and fuzzy systems that were effective problem solving agents. Later in the book we added an evolutionary nature to these systems. This further motivated the study of a large number of algorithms under the class of evolutionary algorithms. We further discussed various speech based simple and hybrid systems which gave a practical aspect to the entire discussion. Here the main focus was to build speech based biometric system as well as other intelligent systems in music and face.

All the intelligent systems that we studied needed to be manually designed. Even after the evolutionary nature, where these systems could evolve to the most optimal architecture, the design of the overall system consisting of the evolutionary algorithms needed to be done. These systems operate on a set of parameters. The task of working with such an intelligent framework largely lies in specifying or finding out the correct parameters. The adaptive systems are an approach where the system itself must be able to find out the parameter values based upon the present state.

Adaptation is a consequence that we get from the natural world. All the living beings we see are adaptive in nature. They are all able to adjust themselves to the changing environment. The continuous adaptation of this form led to the rise of complex living beings from the simple uni-cellular organisms. The adaptation continues to grow that is evident from the behavior of various animals who all adapt themselves to the changing living conditions very well. The adaptation in the different living animals is the reason behind their survival. We know from the

A. Shukla et al.: Towards Hybrid and Adaptive Computing, SCI 307, pp. 421–431.
springerlink.com

Darwin's principles that the natural follows the survival of the fittest mechanism where only the fit individuals survive. The others die out with time. Hence an organism needs to constantly adapt as per the requirements to enjoy a high fitness that results in its survival.

19.1 Parameters

One of the major problems with the use of neural, fuzzy and even evolutionary algorithms was identified as the specification of the design and other parameters. This needed to be done by a human expert having sound knowledge of the problem as well as the algorithm being used. This was seen as quite a human intensive and monotonous job where different parameters need to be manually fed and performance of the system needs to be analyzed. This manual task largely led to poor performance or sub-optimality in the final result. This is because of the inability of the human expert to perfectly fix the system design and parameters.

In order to improve and automate this process, these neural, fuzzy and evolutionary systems were wrapped around another layer of evolutionary algorithm (Angeline *et al* 1994; Cho and Shimohara 1998; Funabiki *et al* 1998). This algorithm was supposed to optimize the performance and find out the optimal parameter set. This largely led to an improved performance of these systems. The performance boost was especially substantial if the problem was reasonably complex that made it difficult to figure out the most optimal parameters due to complex structures. The evolutionary paradigm added to these systems relieved the humans from the task of specifying the long list of parameters as well as made the entire process automated which had greater chances of showing optimal performance subjected to complexity and computational constraints. The evolutionary nature of the systems makes us do away with the problem of setting the system parameters, but in return we are now expected to set the parameters of the build evolutionary algorithm (Lobo *et al* 2007).

19.1.1 Role

The parameters play a major role in every system we design and build. These parameters are responsible for the good or poor performance of the system. A very cleverly designed system may also give a poor performance if the parameters are not properly set. Consider the Genetic Algorithm as an example. The algorithm would showcase a reasonably poor optimization if the crossover and mutation rate are not properly specified. The manner of working with these systems is hence to engineer good systems as well as to execute them with the correct set of parameters.

Keeping constant values of the parameters may also be a reason for the sub-optimal performance by any system. In many systems it may not be wise to keep constant parameter values. The parameter values may depend upon the context in which the problem is presently at, time and other factors. Based on these the parameter values may be altered accordingly. This variable nature of the parameter values many times gives good performance in comparison to the fixed parameter values. The variable nature of the parameters where the parameter values are changed during the algorithm execution naturally marks a more flexible design of controlling a system.

19.1.2 Parameter Setting by Evolutionary Algorithms

We saw that to control a neural as well as fuzzy system, we used an evolutionary algorithm. The evolutionary algorithm may themselves be wrapped around another layer of evolutionary algorithm. This makes the system highly flexible and adaptive where the parameters of the system are set to most optimal value by an evolutionary technique. The big problem is that the parameters of the added evolutionary algorithm now need to be set by a human. These parameters may behave reasonably passive in nature. A little change of values from the ideal values may not add much of the difference to the final output of the system. This system is shown in figure 19.1.

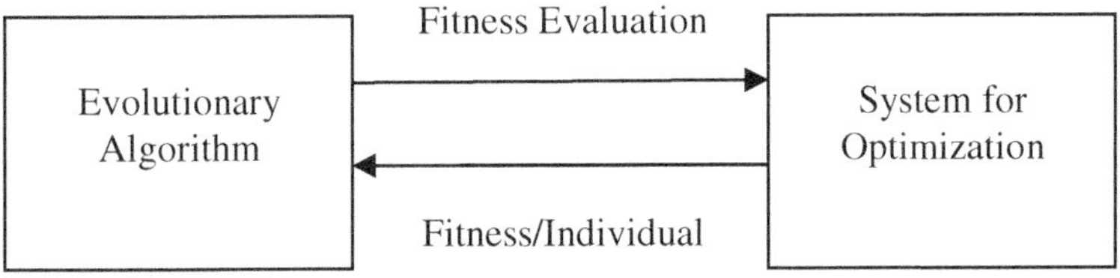

Fig. 19.1 Use of Evolutionary Algorithms for System Optimization

Additionally the parameters of this evolutionary algorithm could be set by adding another layer of evolutionary algorithm. Ideally we may keep proceeding in this manner adding layers of evolutionary algorithm to control or fix the parameters of the preceding algorithm. This is shown in figure 19.2. But the major issue is a finite computation that can handle a limited amount of complexity. Adding

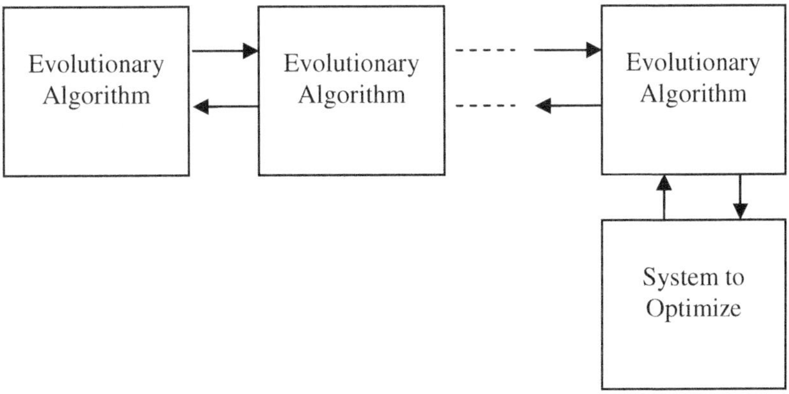

Fig. 19.2 Nested Evolutionary Algorithms for System Optimization

any layer of optimization algorithm around a system means adding of a large amount of complexity. This further requires intensive computation which might not be feasible. Hence we can only engineer systems with reasonably small complexity so as to give computation within limited amount of time.

19.1.3 Towards Parameter-Less Systems

We talk a lot about the intelligent systems where the system itself does most of the task that conventionally needed humans for the same. The task of fixing the system parameters is another intelligent task. Ideally that must be done by the system itself without the need of humans figuring out the implications of the parameters to the system and accordingly deciding the final parameters by which the system runs. This means that the users must only give the inputs and expect the output with the system entirely as black box. The system must be able to analyze the data and figure out the most optimal set of parameters on its own. This leads us towards parameter-less systems that are able to compute the optimal parameter values on their own without human intervention (Banker 1985; Chakraborty U K, Dastidar 1993; Goldberg 1989).

This however is not feasible. Every attempt to remove some parameters from the systems leads us to addition of new parameters. This now leaves with the question of how to compute the newly added parameters. In the above discussion we saw that the attempt to remove some parameters from the system made us add an evolutionary algorithm. This step resulted in the addition of another new set of parameters which could not be optimally fixed. This inability to engineer parameter-less systems is an important characteristic in the discussion of the adaptive systems, since the adaptive systems are a step towards making the systems free from parameter specifications. The addition of parameters in quest of removal of parameters and inability to automate the setting of these parameters can be attributed to the Non-Free Lunch Theorems (Wolpert and Macready 1995, 1997; Ho and Pepyne 2002).

19.2 Classification of Adaptation

During this course study in this book we have studied numerous systems that carried out some form of adaptation of various kinds and at various levels. Here we formalize our understanding of the concept of adaptation. The readers may refer to the works Bäck (1996), Bäck and Hoffmeister (1991), Deb and Goldberg (1989), Fogel (1995), Koza (1992), Michalewicz (1992), Mitchell (1999) and Schwefel (1995) for a discussion on the background systems. We first classify the systems into various kinds based on their adaptation levels.

The classification is done by two means. The first classification is done on the basis of the type of adaptation. Here we look at the manner in which the adaptation has been performed. Depending upon the manner the classification may be of either of four types i.e. static, deterministic, adaptive and self-adaptive. All these types exhibit different mechanisms of implementation of adaptation.

The other classification may be done on the basis of the level of adaptation. Here we look at the level in which the adaptation has been applied. Based on the level the adaptation may be classified in either of the four types. These are environment, population, individual, and component (Hinterding *et al* 1997; Angeline 1995, 1996).

The various types of classifications are summarized in figure 19.3.

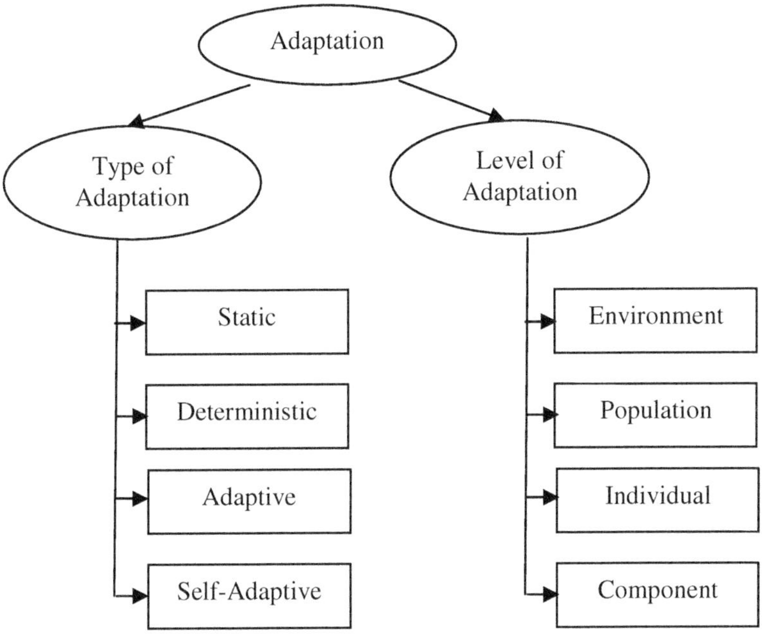

Fig. 19.3 Classification of Adaptation

The adaptive system may choose any type of adaptation along with any level of adaptation. This however is not the case of static type of adaptation which is non-adaptive system. Both the type of adaptation as well as the level of adaptation is discussed in two separate heads in the next sections.

19.3 Types of Adaptation

The first classification is on the type of the adaptation. The systems may either be static or dynamic. The static systems do not show any form of adaptation. The dynamic systems may be broadly classified into three types based on the level of adaptation. These are deterministic systems, adaptive systems and self adaptive systems. We discuss all of these in the next sub-sections.

These various systems try to adapt the strategy parameters. They try to compute the optimal values of the strategy parameters at any instance of time. The strategy parameters were discussed in chapter 11 under the Evolutionary Strategies. The strategy parameters denote the strategy by which the system carries out the optimization task. Mutation rate is one of the major strategy parameters of the evolutionary algorithms. It denotes the level of modification in the selected individuals. In this manner it contributes towards the entire strategy that carries out the evolution or the optimization task.

19.3.1 Static Systems

The static systems are completely non-adaptive systems. These systems have a fixed value of the strategy parameters that remain constant in the entire evolutionary optimization process. This means that there is no adjustment of these parameters as the algorithm executes. The entire task of fixing the values of these parameters must be performed by an expert. A considerable effort is needed for computing these parameter values. Further as per the above discussions the finally evolved system may be sub-optimal in performance.

We may be able to try out a large number of values of the strategy parameters, but the major problem in parameter setting comes from the fact that there are a number of parameters that are all independent of each other. Hence these cannot be optimized separately. Consider the mutation and crossover rate as an example. We may fix the crossover and play with the mutation rates to get the value of the mutation rate where the performance is most optimal. But the same reading may not be observed if we change the crossover rate. This causes a deep problem in the parameter setting of simultaneously fixing the values of the various parameters to achieve the optimal performance.

Using some rules of thumb to set these strategy parameter values or setting them to some default values may again not solve the purpose. The parameter values have a very high dependence on the problem. Different problems have different parameter values that ideally suit them. This again results in problems in terms of the parameter setting. We have seen that different problems have different shapes and complexities of the fitness landscapes. This largely determines the parameter setting of the algorithms.

19.3.2 Deterministic Systems

The deterministic systems are the systems that use a predefined rule that modifies the values of the various strategy parameters. This rule depends upon the time or the number of generation that the algorithm is presently at. No feedback is taken from the various individuals in the system those carry out the optimization task. The modification is only on the basis of the predefined rule. These systems are helpful in adaptation, since for many of the strategy parameters it may be guessed how their values should vary with time for an ideal performance. This may be by an understanding of the fitness landscape and its complexity and the expected convergence of the algorithm. Accordingly we may frame some rule that modifies the strategy parameters. This should lead to a better performance depending upon the design of the rules.

The rules may be formed from simple understanding of the behavior of the system. Say for the optimization we claim that the mutation rate must decrease by some definite amount in every generation. This may be given by equation (19.1).

$$m(t+1) = m(t) - \delta \tag{19.1}$$

Here *m(t)* is the mutation at generation number *t*. δ is the change in value of mutation between successive generations. Now we know that this algorithm would exhibit a high exploratory nature at start because of high mutation rate. As the algorithm continues with the optimization process, it would be more active to converge at some point by maintaining a reasonably low mutation rate.

Similar may be the case with the number of individuals. In the problem of incremental evolution of the neural network in chapter 13 we stated that as the generations increase, the complexity increases as more complex networks are allowed to be evolved. This high complexity may require more number of individuals. Hence the number of individuals N(t) are changed as per the equations (19.2) and (19.3)

$$N(t+1) = N(t) + \Delta N \tag{19.2}$$

$$\Delta N = \frac{N(G) - N(0)}{G} t \tag{19.3}$$

Here *N(t)* represents the number of individuals at generation *t*. *G* is the total number of generations.

19.3.3 Adaptive Systems

The next level of systems in this class is the adaptive systems. Here also the strategy parameters are not kept constant. They are rather changed in the evolutionary process. The change of the strategy parameters however depends upon the feedback that is obtained by the system itself. Hence the present context of the system is analyzed and based on the same analysis the optimal value of the strategy parameters is set.

An example of this adaption is the Rachenberg's 1/5 principle (Rechenberg 1973). This principle is used for the adjustment of the step size in Evolutionary Strategies. We know that the step size in Evolutionary Strategies denotes its exploration. A larger step size may result in too many changes which would make the algorithm miss out the optima. Too less of step size might result in the algorithm moving towards the optima very slowly. As per the Rechenberg's principle the step size must be fixed such that the ratio between the successful mutations to all the mutations is approximately 1/5. The application of mutation is regarded as successful in an individual if it leads to a better fitness value of the individual after the mutation has been applied. Hence at any stage we measure this ratio. If this comes out to be less than 1/5, the step size is reduced. On the other hand if it comes out to be more than 1/5, the step size is increased. In this manner we may be able to adjust or adapt the step size of the Evolutionary Strategies based on some feedback. The feedback in this approach is the ratio between the successful to all mutations. This gives some idea regarding the state of the system at present that in turn leads to the correct changes to be applied in the step size. Hence the feedback provided by the system is the driver of the alterations in the values of the strategy parameter.

19.3.4 Self-adaptive Systems

The last category in this classification of type of adaptive systems is the self-adaptive systems. In these systems the strategy parameters are embedded inside the individual representation as well. These parameters get optimized as per the evolutionary process by the conventional evolutionary operators. These systems may hence be visualized as specialized systems where the strategy of the evolution process is itself optimized in the evolutionary process. It is natural that some values of strategy parameters may be desirable and the others may be undesirable. Further these values may be different in different contexts. As the optimization proceeds, the strategy parameters also start attaining the optimal values that help in the idealistic evolution of the individual. The self adaptive systems hence benefit from the evolutionary capabilities of the evolutionary algorithm being used by making the strategy parameters embedded in the individual.

We discussed such a strategy in detail in chapter 11 under Evolutionary Strategies. The mutation is an important parameter that decides the functioning of the algorithm. The Evolutionary Strategies are a novel approach where the recombination and mutation operators were used for the optimization task. The mechanism by which the algorithm could adjust the mutation by using the information from the individual itself makes these systems self adaptive in nature.

19.4 Level of Adaptation

The other type of classification of the individual is made on the basis of the level in which the adaptation occurs. The level of adaptation emphasizes on where the adaptation is being intended to be carried out. Accordingly, we classify the adaptation into one of the four categories these are environment, individual, population, and component. Any one of these levels of adaptation may be used with any of the types of adaptation. This results in numerous possibilities in the use of adaptation for the optimization tasks or problem solving.

19.4.1 Environment

The environment level adaptation is when the environment is changed as per any of the types of adaptation. The environment has an effect on the fitness of the individual. Hence this level of adaptation changes the fitness value of the individuals with same properties as the algorithm proceeds. This means that this adaptation scheme has a variable fitness landscape. The fitness surface has regular changes as the optimization proceeds. This causes a change in the fitness and accordingly the ranks of the individuals constituting a population.

Consider the example of the evolutionary neural network. Here we evolved the neural network using the evolutionary approaches. In this mechanism we had given a penalty to the fitness based on the number of connections and neurons. Suppose that this penalty becomes a function of time and change with time. This would make the fitness landscape change with time. This is an example of the environment level adaptation.

19.4.2 Population

The population level adaptation happens when we try to adapt the parameters that apply to the entire population. The changes in these parameters hence affect all the individuals that constitute the population. In this manner the changes are usually based upon the recordings of a number of parameters which are then used for analyzing the change that needs to be carried out. In other cases the changes might be simply deterministic in nature where the various parameters may have a predefined manner in which these need to be changed in the execution process of the algorithm.

We considered the Rechenberg's principle in section 19.3.3. Here we computed the ratio between the successful mutations and all the mutations and tried to keep this value as close as possible to 1/5. Now the change in step size affects the entire population and this is computed based on the entire population. Hence this may be taken as an example of the population level of adaptation.

19.4.3 Individual

The individual level of adaptation applies to the adaptation where the parameters that affect only the specific individual are adapted. Hence using this level of adaptation every individual tries to adapt itself using the defined method and type of adaptation. In this scheme the changes made only affect the individual and not the entire population pool.

An example of this type of adaptation is when we adapt the population size of any generation t based on the equation (19.4). (Arabas *et al* 1994).

$$P(t+1) = P(t) + N(t) - D(t) \tag{19.4}$$

Here $P(t)$ is the population size at any generation t. $N(t)$ is the number of individuals produced at generation t. $D(t)$ is the number of individuals that die at generation t.

This model assumes that every individual has an age. The age of the individual increases as it continues to survive in the evolutionary process. After some age the individual dies. Usually a maximum age may be defined after which the individual has to die. This contributes to the factor $D(t)$. The total number of individuals produced $N(t)$ is directly proportional to the population size at that generation.

19.4.4 Component

The last level of adaptation we discuss is at the component level. Here the various strategy parameters are adapted as per the some gene or component in an individual that is local to some individual. This type of adaptation is evident into the manner of functioning of the Evolutionary Strategies. Here the mutation step sizes and rotational angles are adapted as per every gene in the individual.

19.5 Chapter Summary

This chapter formally introduced the paradigm of adaptation. The chapter first presented the essence of parameters and their role in effective system optimization. The use of evolutionary algorithms was seen as a means of system optimization. The chapter focused upon the classification of adaptation. The classification took place based upon type of adaptation and level of adaptation. On the basis of type of adaptation, the system could be classified into static, deterministic, adaptive and self-adaptive systems. The static systems display constant behavior and do not showcase any adaptation. The deterministic adaptation is displayed by a change of the parameter value in a pre-specified manner. The adaptive systems are able to themselves adapt to the situation and attain optimal parameter values. Self-adaptive systems have the strategy parameter embedded into the individual itself. As the evolution continues, these parameters also attain their optimal values. On the basis of level of adaptation the systems are classified into environment, population, individual and component. An environment level adaptation means a change in the fitness function that changes the complete scenario. Population level adaptation is an adaptation that effects the entire population. The individual level adaptation affects the individual. Component level adaptation is found when the strategy parameters are adapted as per the component in the individual.

References

[1] Angeline, P.J.: Adaptive and self-adaptive evolutionary computation. In: Palaniswami, M., Attikiouzel, Y., Marks, R.J., Fogel, D., Fukuda, T. (eds.) Computational Intelligence: A Dynamic System Perspective, pp. 152–161. IEEE Press, New York (1995)
[2] Angeline, P.J.: Two self-adaptive crossover operators for genetic programming. In: Angeline, P.J., Kinnear, K.E. (eds.) Advances in Genetic Programming, vol. 2, pp. 89–109. MIT Press, Cambridge (1996)
[3] Angeline, P.J., Saunders, G.M., Pollack, J.B.: An Evolutionary Algorithm that Constructs Recurrent Neural Networks. IEEE Trans. Neural Netw. 5(1) (1994)
[4] Arabas, J., Michalewicz, Z., Mulawka, J.: GAVaPS—A genetic algorithm with varying population size. In: Proc. 2nd IEEE Conf. Evol. Comput., pp. 73–78 (1994)

[5] Bäck, T.: Evolutionary Algorithms in Theory and Practice: Evolution Strategies, Evolutionary Programming, Genetic Algorithms, Oxford (1996)
[6] Bäck, T., Hoffmeister, F.: Extended selection mechanisms in genetic algorithms. In: Belew, R.K., Booker, L.B. (eds.) Proc. of the Fourth Intl' Conf. Genetic Algorithms. Morgan Kaufmann, San Francisco (1991)
[7] Baker, J.E.: Adaptive selection methods for genetic algorithms. In: Grefenstette, J.J. (ed.) Proc. of the First Intl. Conf. on Genetic Algorithms and their Appl., Erlbaum, Mahwah (1985)
[8] Chakraborty, U.K., Dastidar, D.G.: Using reliability analysis to estimate the number of generations to convergence in genetic algorithm. Inf. Process. Lett. 46, 199–209 (1993)
[9] Cho, S.B., Shimohara, K.: Evolutionary Learning of Modular Neural Networks with Genetic Programming. Applied Intelligence 9, 191–200 (1998)
[10] Deb, K., Goldberg, D.E.: An investigation of niche and species formation in genetic function optimization. In: Schaffer, J.D. (ed.) Proc. of the Third Intl. Conf. on Genetic Algorithms. Morgan Kaufmann, San Francisco (1989)
[11] Fogel, D.B.: Evolutionary Computation: Toward a New Philosophy of Machine Intelligence. IEEE Press, Los Alamitos (1995)
[12] Funabiki, N., Kitamichi, J., Nishikawa, S.: An Evolutionary Neural Network Approach for Module Orientation Problems. IEEE Trans. Syst. Man Cybern. – Part B Cybern. 28(6), 849–855 (1998)
[13] Goldberg, D.E.: Sizing populations for serial and parallel genetic algorithms. In: Schaffer, J.D. (ed.) Proc. of the Third Intl. Conf. on Genetic Algorithms. Morgan Kaufmann, San Francisco (1989)
[14] Hinterding, R., Michalewicz, Z., Eiben, A.E.: Adaptation in Evolutionary Computation: A Survey. In: Proc. of the Fourth IEEE Conf. on Evol. Comput., pp. 65–69 (1997)
[15] Ho, Y., Pepyne, D.: Simple explanation of the no-free-lunch theorem and its implications. J. Opt. Theory Appl. 155, 549–570 (2002)
[16] Koza, J.R.: Genetic Programming: On the Programming of Computers by Means of Natural Selection. MIT Press, Cambridge (1992)
[17] Lobo, F.G., Lima, C.F., Michalewicz, Z.: Parameter Setting in Evolutionary Algorithms. Springer, Heidelberg (2007)
[18] Michalewicz, Z.: Genetic Algorithms + Data Structures = Evolution Programs. Springer, Berlin (1992)
[19] Mitchell, M.: An Introduction to Genetic Algorithms. MIT Press, Cambridge (1999)
[20] Rechenberg, I.: Evolutionsstrategie: Optimierung technischer Systeme nach Prinzipien der biologischen Evolution. Frommann–Holzboog Verlag, Stuttgart (1973)
[21] Schwefel, H.P.: Evolution and Optimum Seeking. Wiley, Chichester (1995)
[22] Wolpert, D.H., Macready, W.G.: No free lunch theorems for search. Santa Fe Institute, Tech. Rep. SFI-TR-95-02-010 (1995), `http://citeseer.nj.nec.com/wolpert95no.html`
[23] Wolpert, D.H., Macready, W.G.: No Free Lunch Theorems for Optimization. IEEE Trans. Evol. Comput. 1(1), 67–82 (1997)

List of Figures

Fig. 19.1. Use of Evolutionary Algorithms for System Optimization
Fig. 19.2. Nested Evolutionary Algorithms for System Optimization
Fig. 19.3. Classification of Adaptation

Chapter 20
A Taxonomy of Models

Abstract. The paradigm of hybrid computing deals with the fusion of numerous computing methods, where each method contributes towards the removal of the limitations of the other methods. In this manner we try to magnify the advantages of all methods and remove the limitations of each of the methods. The hybridization deals with the fusion of Artificial Neural Networks, Evolutionary Algorithms, Fuzzy Logic along with heuristics. The large number of models of all of these present a large set of options using which the various methods may be fused. The application areas present further options where each of the models may be customized as per the problem requirements. The same is true with adaptive computing as well, where different adaptation techniques or adaptive algorithms may be used for a better system performance. Complexity largely limits the millions of possibilities of creating different types of hybrid and adaptive systems, which is further emphasized by a limited computation and data availability. This chapter gives a broad overview of the systems presenting the immense varieties in which the systems can be engineered. The basic question that hence arises is how to make a judicious choice amongst these immense options.

Many of the most essential problems in computing may not be solvable in any conventional manner, where the algorithms are framed to process the inputs and give precise or hard outputs. Each of the problems is highly complex to model and work over. The inability of these conventional methods in problem solving opens door for Soft Computing approaches. Here the answers may not be perfect, but these may be best as per problem, computation, data or other relevant constraints. We saw throughout the book that the systems were never perfect. The accuracies were close to cent percent, but mostly never cent percent. This happens when the systems were not tested against large volumes of synthetic data that can be much critical than the live data. Consider the problem of person identification with the use of speech. A genuine person may be correctly identified. But what would happen if a good mimic speaks in front of the system. Now suppose that we start making digital alterations to the speech signal of a person who was identified clearly. The system might still recognize the speech signal to be belonging to the correct person. After some fine amount of changes, the system may no longer identify the person. The amount of changes that may be digitally made in order to make a person lose his individuality in front of the intelligent system is an important and unanswerable question. The fact remains that the systems are approximate rather than perfect. This approximation or imprecise character goes in favor of the

A. Shukla et al.: Towards Hybrid and Adaptive Computing, SCI 307, pp. 433–439.
springerlink.com

intelligent systems using soft computing approaches, since we know that the input data is always imprecise and cannot be precisely determined. Hence any computing technique must account for this impreciseness.

The paradigm of Soft Computing calls for a complete change in the methodology of problem solving. This largely differs from the conventional manner of using a series of predetermined steps for input processing in a sequential manner, framed as a hard bound algorithm. Many times in the book we solved problems assuming these systems as black boxes that could themselves extract interesting features and associated patterns out of the training data and generalize the knowledge in the testing data. This made us solve problems that were unimaginable. At many other instances we had to specifically highlight the features that could assist the system in the extraction. In many other systems we have to give as inputs the specific features that completely aid in the problem solving.

The knowledge of the basic working of the system is another main aspect. If some idea is available regarding the manner or the logic, it may be supplied to the system. The system in this case tries to make the best use of the available concept and further tries to build over or further optimize the same. We saw how the fuzzy systems carry out such a task. Many times we may not have any idea regarding the manner in which the system works. Here we expect the systems themselves to figure out all relations, simple or complex, and solve the problem. This was usually the approach adopted in the neural systems. Optimization is another of the discussed paradigm. We saw the manner in which any system could be optimized for the best performance, no manner how hard or complex the optimization be. This was carried out by the use of the Evolutionary Algorithms.

Each of the systems that we discussed had an associated biological analogy. The ease with which the natural systems solve the most complex problems is simply fascinating. The various phenomenons that we observe all around, especially in the biological domain, seem to be perfectly carved. It is natural to believe that these systems were developed over a large period of time and attained a highly optimal nature.

Heuristics may play a vital role in driving the systems. It tells us about the goodness of some state that helps in the correct state selection and accordingly driving the system from one good state to an even better state. This technique may be applied to many problems. The heuristics hence plays a major role in problem solving. It may be used as a local search strategy in evolutionary algorithms, as a time effective search strategy in search based problems, or for finding the optimal state of a system, out of some possible discrete states.

Hybrid computing is the art of using of a combination of two or more soft computing approaches for achieving better performances. Here we try to mix the neural networks, fuzzy logic, evolutionary algorithms and heuristic approaches. The various systems intelligently combine with one of the system removing the weakness of the other system. As a result the weaknesses of the various systems get reduced and the advantages get magnified.

Whether some concept is known regarding the problem or not, the intelligent system framework can themselves carry out the task of building the system and setting its various parameters. This was carried out with the adoption of the

Evolutionary Algorithms. This marks a fabulous system, where we engineer a system to develop a system. Naturally there is an additional complexity that would normally demand an additional computing. Many times it may not be possible to meet the large computing demands to finish the evolution with an optimal structure in finite time. This calls for a good use of the optimization of the evolutionary algorithms along with a good modeling of the problem.

The real world problems may not be adequately solved by the use of a single soft computing approach. The system for its practical usage may require an extremely high accuracy. This would be on highly scalable inputs and in a real time environment. This requires certain measures to be taken for better performance. Many of the problems may be naturally complex by their nature. For this we need multiple approaches that together solve the bigger and more complex problem.

The fusion of one system with other creates possibilities of a large variety of systems. The neural and fuzzy systems themselves integrated in a variety of ways to solve many problems that existed with the conventional neural and fuzzy systems. This included the neural representation and training of the fuzzy systems, fuzzy modeling of the neural network and fuzzy rule extraction of neural networks.

The use of hybrid architecture results in effective problem solving and the resultant systems give optimal results. This however adds a lot of complexity in the entire system. Solving complex problem always has limitations in terms of computation, data requirements, or like. It may hence not be too simple to engineer and work with hybrid systems. Many times the evolution of intelligent systems may need to be carried guided. Many times some other restrictions may need to be applied, in order to enable the evolutionary approach have finite and computable evolution in the evolutionary search space.

The entire paradigm of hybrid computing holds numerous possible techniques of fusion or integration of the neural networks, fuzzy logic, evolutionary algorithms and heuristics. We saw some of these methods in our discussion in various chapters. This creates a large amount of options for the designer to engineer systems as per the problem behavior, complexity and requirements. The selection of the correct option is always a big decision.

Each of the neural networks, evolutionary algorithms and fuzzy systems has more varieties to offer. This means that the selection of hybrid computing is not only on deciding the hybridizing technique, rather it also lies in selection of the best model of the system being hybridized. Take the example of any model of evolutionary neural network. The evolution may be of any of the models of neural network, and it may be carried out using any of the evolutionary approaches. This itself creates a huge possibilities where any of the neural models may be evolved using any of the evolutionary approaches.

One would normally expect some model of all of these to be known to give a high performance. We may further expect high performing hybrid systems mean fusion of the best performing individual systems. This means that an ideal evolutionary neural network would make use some known to be best neural network model and some known to be best evolutionary model. This however is not possible.

From literature one may find that different models are being used for different problems. The results of all of these models for different problems may be reasonably distinct which would point to the fact that no generalization may be made. A model may suite some specific problem and some other may suite some other problem. Even when the problem is same, the different models may behave differently. Hence consider the problem where some speech signal features are to be used for speaker recognition. Here we cannot say with surety whether learning vector quantization, or radial basis function networks, or even fuzzy systems might give good results. The hybridization further makes it even more difficult to select the models out of all the possible models.

The application areas further add a new dimension to this paradigm of hybrid computing. Many times the algorithm may be customized as per the problem requirements. This customization changes the system operations and behavior so as to better solve the problem. A clear understanding of the problem along with the behavior of the system largely results in modeling the problem and the applied algorithm so as to maximize its performance over the time or other feasibility constraints. In case any evolutionary, neural or other approach is being modified, the underlying manner of use and role of the various parameters may change. This further creates a pool of opportunities in which the different algorithms and systems may be used in a more customized format.

Adaptation is another important point of concern in the system design. The adaptation may be implemented by various means and various ways. Adaptation in most of the evolutionary algorithms is inbuilt in the algorithm itself. Hence the selection of the evolutionary technique would decide for the adaptation as well. In the customized evolutionary algorithms the decision may have to be made which would require appropriate problem modeling as per the adaptation requirements. Not always it may be possible to implement all types of adaptations at all levels. This may depend upon the understanding of the system and the behavior of the various parameters. Many other algorithms may also require similar decisions to adaptation. The adaptation again depends upon the manner in which the algorithm behaves and the mechanism by which the task is carried out.

It is naturally not an easy task to model the problem and use hybrid problem solving approach to optimally solve the problem, given the large amount of options. If it had been such a simple task, we could have engineered intelligent systems to carry out this task. The design of an effective hybrid and adaptive computing system requires a deep understanding of the various models and systems. It further requires an understanding of the advantages and disadvantages of the systems that we wish to hybridize. For these reasons the book emphasized upon the conceptualizations in the initial chapters where we also discussed the role of various parameters. In the later chapters the book presented the motivation and advantages of the fusion.

One of the key aspects of hybrid computing is that the problem must be complex or difficult enough for the application of hybrid computing. Application of hybrid concepts into trivial problems may even result in more poor performance by the additional overheads. Further it may require more time. These problems could have simply been solved by any simple soft computing technique. Hence even before designing or using any hybrid computing technique, one must gather enough motivation to ensure that hybrid systems would get a performance boost and would make effective use of computation. It is hence better to use simple approaches and note their limitations. One may then think of designing systems that remove these limitations. This way we can ensure that the different systems fuse in an ideal manner so as to maximize the effects of both of them. Working with simple or hybrid systems would only return training graphs, performance or other metrics. A prior conceptualization would enable to understand the actual scenario and what precisely is limiting the performance. Similar would be the case with the change in parameters. We know that however hard we try, the system always has some parameters that need to be optimally set to get the best performance. An understanding of the underlying system enables understanding of the changes that must be brought about by noting the performance and other measures and analyzing the background working of the algorithm.

The phenomenon of getting the motivated behind the use of system and then designing it is true for the entire soft computing paradigm as a whole. Many problems we believe are to be solved by Soft Computing techniques may be rather trivially solved by simple techniques. There might be simple logic behind the inputs and outputs that may be easily framed out by the problem analysis. Using any neural or fuzzy methods may not be of use. On the contrary the impreciseness associated with these approaches would make the system have reduced performance. Similar is the case with the Evolutionary Algorithms. The search space might be exhaustively searchable or might be solvable by simple statistical analysis. In such a case using Evolutionary Algorithms would make no use. Further the results of the Evolutionary Algorithm would not be as precise as an exhaustive search or statistical analysis.

Each of the neural network models that we studied had a different architecture and manner of problem solving. Further there were different parameters for each of the different models. Mostly the problem being solved by the neural networks would be a classification problem or the functional prediction problem. Any of the models may be used for any of the problems after understanding of the manner of working of the model and the associated parameters. This may be fixed as a base system or a model. In worst case when the performance is below expectations, another model may be tried. The other problem solving methodology states that one may try out a variety of models and compare their performances. Based on these performances, the selection of the proper network may be done. It is less likely that a particular model behaves very poor and another behaves very well for the same training and testing data. Large differences may sometimes be involved if a network primary used for functional prediction problem is use for classificatory problems. The use of Multi-Layer Perceptron with Back Propagation Algorithm requires special problem modeling to be used for the classificatory problems. It may further face the problem of high dimensionality of the outputs, where the system has too many outputs. Besides most of the models must have comparable performances for the same data.

Once the base method is identified, we may note the limitations, the associated reasons for the limitations and then we may frame suitable technique for removing these weaknesses. The different weaknesses may be removed by different hybrid models that we discussed. Additionally custom problems may require a custom hybridization. The scope for hybridization is never closed with the standard models only. Every problem opens new scope for different models of hybrid computing. This largely depends upon the problem complexity, computational requirements, data availability, etc.

The case with evolutionary algorithms is as well quite similar in nature. The different approaches use different analogies and modeling scenarios. Many of the problems may be easily solvable by a variety of techniques. The algorithm may many times be selected based on the ease of modeling the problem. With evolutionary computation, much like the neural and genetic approaches, the proper modeling of the problem may be more important than the selection of the evolutionary model. We have discussed numerous examples of good problem modeling with the use of evolutionary algorithms. From the point of view of model selection, any of the approaches may be used. The performance of the different techniques largely depends upon the fitness landscape which may not be easy to visualize. Hence the evolutionary approaches may vary to some degree in their performances. It would still be good to stick with some evolutionary model and to try to optimize it to the largest extent. The approach of using multiple models and compare the performance is always possible. This however may require different modeling techniques which might make the work very cumbersome.

In literature numerous comparisons of the different neural models for different problems and data sets has been conducted. However the comparison in context of different evolutionary algorithms is not very vast. Most of the comparisons are restricted to the synthetic problems of some optimization which may not necessarily follow the patterns of real life applications. The task of comparing different evolutionary models for different systems in different applications is naturally cumbersome. The best way to work out may hence be taken to selection of a few models and their application in solving the concerned problem. It may be better to try various modeling scenarios to best represent the problem as per the evolutionary model demands and to use the selected model in the best possible manner. Once the simple approaches do not work, the hybrid ones may be tried. This would enable understanding of the limitations of the system.

Throughout the text we have given a lot of emphasis on the understanding of the system. Soft Computing is a highly inter-disciplinary domain which requires a prominent collaboration between people from various disciples to contribute towards the engineering of the intelligent systems. The selection of parameters, attributes, inputs, outputs may largely require a problem domain expertise with an understanding of the mechanism of the soft computing approaches. The task of making out the best from this modeling scenario may be done by the soft computing experts with relevant problem domain knowledge. The interdisciplinary nature of the problem sometimes makes it a difficult field to work in.

Due to these limitations, many times a complete black box approach may be used. Here we not know much about the problem and system behavior, we rather leave it entirely over the system to try to make the best use of available data or other inputs and form an intelligent system. The intelligent system may or may not work. If the system does not work, we may then try to understand the problem and frame it in a better manner. In many cases the system would work well. This makes the task of system design very simple, where the entire soft computing approach has been taken as a black box that does what it claims to do.

Even though the complete picture of hybrid and adaptive computing systems may sound very complex and difficult to engineer, especially because of the taxonomy of models, a conceptual understanding of the technique and problem is usually enough to get high performing intelligent systems.

Chapter 21
A Programmer's Approach

Abstract. The entire text in this book talks a lot about various kinds of intelligent systems. One of the associated problems with hybrid and adaptive intelligent systems is that these may be complex to code and implement. The lack of implementation may many times lead to a majority of people unable to execute and appreciate much of whatever we present in various chapters. It would hence be wise to invest a few pages speaking about the systems from a programmer's point of view. This chapter gives some of the implementation details with code snippets that may assist practitioners in engineering these systems. We first have a look at problem solving with neural systems. We would then touch upon the modular neural networks. This would be followed by a brief discussion over the evolutionary neural networks.

Soft Computing is being used for problem solving for a reasonably long time. The great ease with which this approach problems makes it play a prominent role in a wide variety of problems. The paradigm of hybrid and adaptive computing adds an even more glitter to the applicability. The hybrid systems are able to solve many of the problems that could not have been solved by the traditional systems. This further makes it possible to achieve a higher degree of performance which is needed in every system. The manner in which the various soft computing systems and heuristics fuse with each other creates possibilities for a lot of development in various application domains for making better systems.

It might be easy to design systems on paper and to analyze their working once these are available. But the major aspect is that the system ultimately needs to be typed as a set of programming instructions using a programming language along with the available tools and libraries. Only then would the computer be able to execute it and process the inputs to give outputs so that the designer may analyze the working of the resultant system. The ultimate reality that executes is always the code.

From our experience we understand that many times development and coding happen to restrict people in implementation of any of the systems that they may easily study and understand from a theoretical point of view. Giving a series of online advises and having numerous mail exchanges with enthusiasts trying to

A. Shukla et al.: Towards Hybrid and Adaptive Computing, SCI 307, pp. 441–460.
springerlink.com

depict imitate piece of work, we find it obligatory to write a few pages from a development point of view. This is done to exhibit information along with code snippets that we were unable to do in the main text. This would allow the readers to themselves develop and test many of the systems discussed throughout the book in various chapters.

One of the main problems from a development aspect is usually the availability of tools. Most of the primitive systems discussed in the book have good tools where the problem may be directly keyed in to make the complete system. This would then have mechanism to give the input and collect the outputs. Numerous tools exist for the simulation of the neural networks, fuzzy logic and most of the evolutionary algorithms. These may be available in form of Graphical User Interface (GUI) rich simulation tools or as programs of C, C++ or JAVA that may be easily extended as per the specific problem. While the GUIs are very useful to carry out most of the trivial computations, the codes might be helpful in extending the algorithms and changing the basic algorithm, or many times to customize it as per the problem needs which may not be possible without the code control.

MATLAB is one of the most commonly used software for most of the scientific computations. It contains inbuilt simulators for neural networks, fuzzy logic and genetic algorithms. Most of the tools that the application domain would use image processing, signal processing, etc. are also embedded into MATLAB which may hence be used as an integrated development tool. The manner in which any system may be used is clearly given in the individual MATLAB toolbox guide which comes as a part of its help.

From the domain of Free/Open Source Software SkiLab and Octave may be used as a powerful tool for most of the computing operations. These provide an integrated platform for numerous scientific operations. An easy implementation of numerous custom codes as libraries or software in multiple platforms can also be searched at sourceforge.net. The implementation of Genetic Algorithm as a simple C code is available in the work of Deb (Available for download at http://www.iitk.ac.in/kangal/codes.shtml). The Co-variance Matrix Adaptation Evolutionary Strategy implementation along with description is available from the work of Hansen (Available for download at http://www.lri.fr/~hansen/cmaes_inmatlab.html). NeuroSolutions is a good simulator of neural networks that provides all the needed functionalities in a GUI rich application.

Much of the help is available from the documentation of the individual tools themselves that helps in solving any problem by these tools. We take some of the systems discussed in this text and discuss them from an implementation view in the coming sections. We specifically make use of the MATLAB platform for these discussions. We specifically do not take neural networks, fuzzy systems, genetic algorithms and adaptive neuro fuzzy inference systems (ANFIS) in this text, since they are well explained in the MATLAB guides. This includes the use with both command lines as well as GUI tools.

Apart from the discussions in this text, we also advise the users to visit the website http://softcomputing.iiitm.ac.in/. The various code snippets discussed in this text may be available in a direct downloadable format at the site. These may be easily modified as per the problem demands and used for the problem solving from any application domain. All the codes that we give here are written for MATLAB 7.0. These may require alterations before being used for other versions of MATLAB or other MATLAB compatible simulators.

21.1 Neural Systems

In this section we present a prototype for problem solving with neural networks using MATLAB. Here we would assume that the complete database is available in the form of a table which has been imported into the MATLAB workspace with appropriate variable name. We further assume that the table contains all inputs at various columns with a single output that is the last column of the table. The first task here is the normalization of all the inputs and outputs.

21.1.1 Training and Testing Data Sets

In the first method we randomly split the entire database randomly into training and testing data sets. Here we assume that 70% of the database would be used for training and the rest 30% data would be used for testing. This is then followed by the splitting of the inputs and outputs in the training and testing data sets respectively. Finally the network is constructed, initialized and finally trained. This completes the training phase. The second stage is the testing phase. Here the training and testing data is given to the system to compute the output. The output is matched against the known targets to compute the accuracy of the system. This entire procedure may be applied a number of times and the final aggregate of the results may be taken.

The complete code spited in stages as per the discussion is given in Box 21.1.

```
co=1; % Testing takes place for co runs and average is returned
meanTrain=0; %mean performance of training data
meanTest=0; %mean performance of testing data

% variable 'dataset' contains complete data which may
% be easily imported we assume the first column contains the
% outputs with 0 and 1 denoting the two classes
% all other columns denote inputs
in=dataset(:,2:31); %in contains all inputs
ou=dataset(:,1:1); %ou contains all outputs

% normalization of all inputs by dividing by maximum value
for i=1:size(in,2)
   in(:,i)=in(:,i)/max(in(:,i));
end
```

```
for i=1:co

  % division of inputs and outputs in training and testing data
  trainInput=[];
  trainOutput=[];

  testInput=[];
  testOutput=[];

  for j=1:size(in,1)
    if rand<.7 % Let 70% data be for training and 30% for testing
      trainInput=[trainInput;in(j,:)];
      trainOutput=[trainOutput;ou(j)];
    else
      testInput=[testInput;in(j,:)];
      testOutput=[testOutput;ou(j)];
    end
  end

  % network initialization
  net=newff(minmax(in'),[18,1],{'tansig','purelin'},'traingd');
  net.trainParam.show = 50;
  net.trainParam.lr = 0.05;
  net.trainParam.mc = 0.7;
  net.trainParam.epochs = 3500;
  net.trainParam.goal = 1e-2;

  % training
  [net,tr]=train(net,trainInput',trainOutput');

  % testing on training data
  a = sim(net,trainInput');
  a=a';
  corr=0;
  for j=1:size(trainInput,1)
    if a(j)<0.5 ans=0; else ans=1; end
    if trainOutput(j)==ans corr=corr+1; end
  end

  corr
  corr/size(trainInput,1)*100 % percent train accuracy
  meanTrain=meanTrain+corr;

  % testing on testing data
  a = sim(net,testInput');
  a=a';
```

```
    corr=0;
    for j=1:size(testInput,1)
        if a(j)<0.5 ans=0; else ans=1; end
        if testOutput(j)==ans corr=corr+1; end
    end

    corr
    meanTest=meanTest+corr;
    corr/size(testInput,1)*100 % percent test accuracy
end

s=size(trainInput,1);
meanTrain/co
(meanTrain/co)/s*100 % mean percent train accuracy

s=size(testInput,1);
meanTest/co
(meanTest/co)/s*100 % mean percent test accuracy
```

Box. 21.1 Multi-Layer Perceptron with Back Propagation Algorithm

21.1.2 k-Fold Testing

The other manner of working with neural networks that we discussed in the book was k-fold testing. In this mechanism we divided the entire dataset into k divisions. Now k-1 divisions were used for training and the last one was used for testing. The system was simulated for the testing data set. This entire step was repeated k number of times with a different testing data set at each time. The training data set may also be simulated and compared with target for errors in case the training accuracy needs to be measured as well.

The complete code for this system is shown in Box 21.2.

```
meanTrain=0; %mean performance of training data
meanTest=0; %mean performance of testing data

% variable 'dataset' contains the complete data which may
% be easily imported  we assume the first column contains the
% outputs with 0 and 1 denoting the two classes
% all other columns denote inputs
in=dataset(:,2:31); %in contains all inputs
ou=dataset(:,1:1); %ou contains all outputs

% normalization of all inputs by simply dividing by maximum value
for i=1:size(in,2)
```

```
    in(:,i)=in(:,i)/max(in(:,i));
end

% division of inputs and outputs in k-folds
k=10; % no of folds
inputs=cell(k,1);
outputs=cell(k,1);

for i=1:size(in,1)
    f=floor(rand*k+1);
    inputs{f}=[inputs{f};in(i,:)];
    outputs{f}=[outputs{f};ou(i,:)];
end

for i=1:k
    % network initialization
    net=newff(minmax(in'),[18,1],{'tansig','purelin'},'traingd');
    net.trainParam.show = 50;
    net.trainParam.lr = 0.05;
    net.trainParam.mc = 0.7;
    net.trainParam.epochs = 3500;
    net.trainParam.goal = 1e-2;

    % training with all-'i'th data set
    inps=[];outs=[];
    for j=1:k
        if j==i continue; end
        inps=[inps;inputs{j}];
        outs=[outs;outputs{j}];
    end

    [net,tr]=train(net,inps',outs');

    % testing on training data
    a = sim(net,inps');
    a=a';
    corr=0;
    for j=1:size(inps,1)
        if a(j)<0.5 ans=0; else ans=1; end
        if outs(j)==ans corr=corr+1; end
    end

    corr
    meanTrain=meanTrain+corr;
    corr/size(inps,1)*100 % percent train accuracy
```

```
    % testing on testing data
    a = sim(net,inputs{i}');
    a=a';
    corr=0;
    for l=1:size(inputs{i},1)
        if a(l)<0.5 ans=0; else ans=1; end
        if outputs{i}(l)==ans
            corr=corr+1;
        end
    end

    corr
    corr/size(inputs{i},1)*100 % percent test accuracy on 'j'th data set
    meanTest=meanTest+corr;
end

meanTrain
meanTrain/(size(in,1)*(k-1))*100
% mean percent train accuracy.
% Each item in dataset is tested k-1 times

meanTest
meanTest/size(in,1) *100
% mean percent test accuracy.
% Each item in dataset is tested once
```

Box. 21.2 Multi-Layer Perceptron with Back Propagation Algorithm with k-Fold Testing

21.2 Modular Neural Networks

Into the domain of Modular Neural Networks, we covered two approaches. The first approach made use of redundant solution to the problem by a number of modules or neural networks. The second approach made use of modularity at the computational level, so that only one of the modules could be activated. In both these approaches we made use of independent neural networks as modules. This means that these were trained and tested separately.

21.2.1 Ensembles

In ensembles all the modules work over the same data consisting of the training and testing inputs and outputs. We hence follow the same methodology that was used while working with conventional neural networks. The data set is fetched, normalized, broken into inputs and outputs and further into training data set and testing data set. Then this training data set is used for the training of all the networks. The different networks are either on different models or have some difference in training parameters.

The other major task is testing. The problem being considered is classificatory in nature. Each class outputs the probability which is then integrated by an integrator. In this approach the integrator simply computes the average. Other models may be adequately implemented in the same manner. The initialization code is given in Box 21.3. The code for training of a single neural network is given in Box 21.4. This must be repeated for all neural networks in design. Box 21.5 gives the testing code.

```
in=dataset(:,2:31); %in contains all inputs
ou=dataset(:,1:1); %ou contains all outputs

% normalization of all inputs by simply dividing by maximum value
for i=1:size(in,2)
   in(:,i)=in(:,i)/max(in(:,i));
end

% division of inputs and outputs in training and testing data
trainInput=[];
trainOutput=[];

testInput=[];
testOutput=[];

for j=1:size(in,1)
   if rand<.7 % Let 70% data be for training and 30% for testing
      trainInput=[trainInput;in(j,:)];
      trainOutput=[trainOutput;ou(j)];
   else
      testInput=[testInput;in(j,:)];
      testOutput=[testOutput;ou(j)];
   end
end

nmods=3; %number of modules to be formed
nets=cell(nmods,1);
```

Box 21.3 Initialization code for ensemble

```
% this process may be repeated for each network i.
% presently we only use MLP with BPA as the architecture with
% different parameters. different modules may have different models
i=3;

% network initialization
nets{i}=newff(minmax(in'),[20,1],{'tansig','purelin'},'traingd');
nets{i}.trainParam.show = 50;
nets{i}.trainParam.lr = 0.05;
nets{i}.trainParam.mc = 0.7;
nets{i}.trainParam.epochs = 3500;
nets{i}.trainParam.goal = 1e-2;

% training
[nets{i},tr]=train(nets{i},trainInput',trainOutput');

% testing on training data with a single module.
% for setting of parameters such that individual performance
% is maximized
a = sim(nets{i},trainInput');
a=a';
corr=0;
for j=1:size(trainInput,1)
   if a(j)<0.5 ans=0; else ans=1; end
   if trainOutput(j)==ans corr=corr+1; end
end

corr
corr/size(trainInput,1)*100 % percent train accuracy

% testing on testing data
a = sim(nets{i},testInput');
a=a';
corr=0;
for j=1:size(testInput,1)
   if a(j)<0.5 ans=0; else ans=1; end
   if testOutput(j)==ans corr=corr+1; end
end

corr
corr/size(testInput,1)*100 % percent test accuracy
```

Box. 21.4 Training of each neural network in ensemble

```
% find accuracy for training data
corr=0;
for i=1:size(trainInput,1)
   %solve by each module and take mean
   ans=0;
   for j=1:nmods
      a = sim(nets{j},trainInput(i,:)');
      a=a';
      ans=ans+a(1);
   end
   ans=ans/nmods;
   if ans<0.5 ans=0; else ans=1; end
   if trainOutput(i)==ans corr=corr+1; end
end
corr
corr/size(trainInput,1)*100 % percent train accuracy

% find accuracy for testing data
corr=0;
for i=1:size(testInput,1)
   %solve by each module and take mean
   ans=0;
   for j=1:nmods
      a = sim(nets{j},testInput(i,:)');
      a=a';
      ans=ans+a(1);
   end
   ans=ans/nmods;
   if ans<0.5 ans=0; else ans=1; end
   if testOutput(i)==ans corr=corr+1; end
end
corr
corr/size(testInput,1)*100 % percent test accuracy
```

Box. 21.5 Testing of the ensemble

21.2.2 Modularity in Computation

The other model that we studied was to introduce modularity at the computational level. Here we cluster the data set using Fuzzy C Means clustering and then use these clusters for problem space division. The entire training data set consisting of all training inputs is passed through Fuzzy C Means clustering.

After clustering, the next step to be performed is to divide the entire training dataset into these modules. Each module or ANN is trained with its own training data set. Similarly the testing data set is divided and each testing data set goes to the designated module for the computation of the output. The division is done on the basis of the least distance of the input to the data centers.

The initialization code is given in Box 21.6. The training code for each of the neural networks is similar to the one discussed in Box 21.4. The testing code is given in Box 21.7.

```
in=dataset(:,2:31); %in contains all inputs
ou=dataset(:,1:1); %ou contains all outputs

% normalization of all inputs by simply dividing by maximum value
for i=1:size(in,2)
   in(:,i)=in(:,i)/max(in(:,i));
end

% division of inputs and outputs in training and testing data
trainInput=[];
trainOutput=[];

testInput=[];
testOutput=[];

for j=1:size(in,1)
   if rand<.7 % Let 70% data be for training and 30% for testing
      trainInput=[trainInput;in(j,:)];
      trainOutput=[trainOutput;ou(j)];
   else
      testInput=[testInput;in(j,:)];
      testOutput=[testOutput;ou(j)];
   end
end

% cluster using the training data using Fuzzy C Means
% Clustering and get cluster centers
nclus=3; %number of clusters to be formed
centers = fcm(trainInput,nclus);

trainInput1=cell(nclus,1);
trainOutput1=cell(nclus,1);
testInput1=cell(nclus,1);
testOutput1=cell(nclus,1);

for j=1:size(trainInput,1)
   cn=1;
   for l=2:nclus
```

```
        if norm(trainInput(j)-centers(l)) <
                    norm(trainInput(j)-centers(cn))
            cn=l;
        end
    end
    trainInput1{cn}=[trainInput1{cn};trainInput(j,:)];
    trainOutput1{cn}=[trainOutput1{cn};trainOutput(j,:)];
end

for j=1:size(testInput,1)
    cn=1;
    for l=2:nclus
        if norm(testInput(j)-centers(l)) < norm(testInput(j)-centers(cn))
            cn=l;
        end
    end
    testInput1{cn}=[testInput1{cn};testInput(j,:)];
    testOutput1{cn}=[testOutput1{cn};testOutput(j,:)];
end

nets=cell(nclus,1);
```

Box. 21.6 Initialization Code for Modular Neural Network

```
% find accuracy for training data
corr=0;
for i=1:size(trainInput,1)
    %find cluster for the ith training input
    cn=1;
    for l=2:nclus
        if norm(trainInput(j)-centers(l)) <
                            norm(trainInput(j)-centers(cn))
            cn=l;
        end
    end

    % find output by respective module
    a = sim(nets{cn},trainInput(i,:)');
    a=a';
    if a(1)<0.5 ans=0; else ans=1; end
    if trainOutput(i)==ans corr=corr+1; end
end
corr
corr/size(trainInput,1)*100 % percent train accuracy
```

```
% find accuracy for testing data
corr=0;
for i=1:size(testInput,1)
   %find cluster for the ith test input
   cn=1;
   for l=2:nclus
      if norm(testInput(j)-centers(l)) < norm(testInput(j)-centers(cn))
         cn=l;
      end
   end

   % find output by respective module
   a = sim(nets{cn},testInput(i,:)');
   a=a';
   if a(1)<0.5 ans=0; else ans=1; end
   if testOutput(i)==ans corr=corr+1; end
end
corr
corr/size(testInput,1)*100 % percent test accuracy
```

Box. 21.7 Testing Code for Modular Neural network

21.3 Evolutionary Neural Networks with Fixed Architecture

The evolutionary neural networks are neural networks that are trained and evolved by the evolutionary examples. In this section we only discuss the mechanism of training a neural network using evolutionary algorithm. The integration of the genetic algorithm (GA) and artificial neural network (ANN) is done at the level of the fitness function of the GA. Here the individual is decoded from the genotype to the phenotype representation. This forms the ANN that may be used. The training of this network may be done for a few generations if some local search strategy is to be used. The performance is measured on the training data set. The training data set needs to be loaded from the workspace before it can be used in the program. The divided training data set may be saved as a variable in the hard disk which may be loaded at the time of evaluation.

The other part is the fixing of the GA parameters. This is done using the GA toolbox of MATLAB. Let *hn* be the number of neurons in hidden layer and *in* be the number of neurons in input layer. Let there be a single neuron in the output layer. Box 21.8 gives the initialization code. Table 21.1 gives the calculation for the total number of individuals in the GA. Box 21.9 shows the fitness function. Figure 21.1 shows the genetic optimization code. The testing code with the most optimal genetic individual is given in Box 21.10. Here it is assumed that the results of GA have been exported to workspace with name *garesults*.

Table 21.1 Number of genes for genetic individual

S. No.	Item	Number of genes
1.	Weights: Input Layer to Hidden Layer	hn x in
2.	Weights: Hidden Layer to Output Layer	hn x 1
3.	Hidden Layer Biases	hn
4.	Output Layer Bias	1
	Total	hn x in +hn x 1 +hn + 1

```
global trainInput trainOutput

in=dataset(:,2:31); %in contains all inputs
ou=dataset(:,1:1); %ou contains all outputs

% normalization of all inputs by simply dividing by maximum value
for i=1:size(in,2)
   in(:,i)=in(:,i)/max(in(:,i));
end

% division of inputs and outputs in training and testing data
trainInput=[];
trainOutput=[];

testInput=[];
testOutput=[];

for j=1:size(in,1)
   if rand<.7 % Let 70% data be for training and 30% for testing
      trainInput=[trainInput;in(j,:)];
      trainOutput=[trainOutput;ou(j)];
   else
      testInput=[testInput;in(j,:)];
      testOutput=[testOutput;ou(j)];
   end
end
```

Box. 21.8 Initialization Code for Evolutionary Neural Network

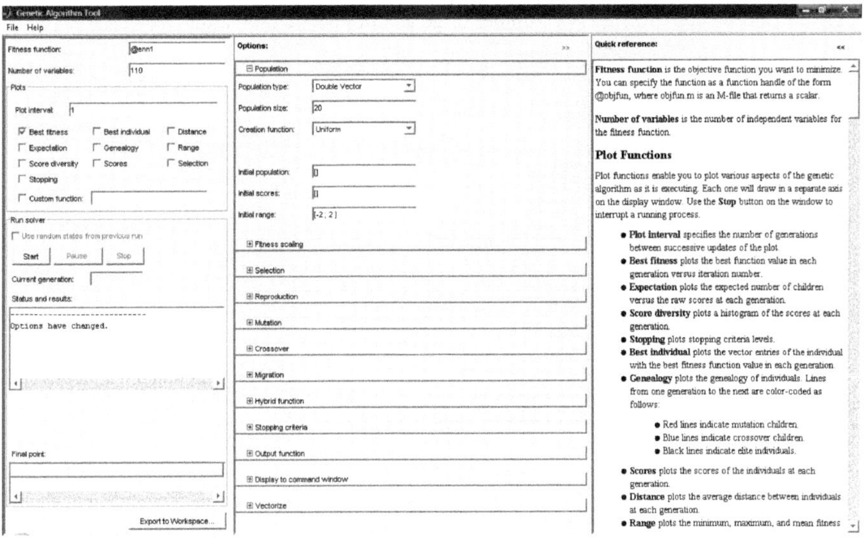

Fig. 21.1 MATLAB Genetic Algorithm Toolbox

```
function fit=enn(X)

global trainInput trainOutput;

in=30; %number of input layer neurons
hn=18; %number of hidden layer neurons

% network initialization
net=newff(minmax(trainInput'),[hn,1],{'tansig','purelin'},'traingd');

c=1; %counter for the GA individual genes

%set input to hidden layer weights
for i=1:hn
   for j=1:in
      net.IW{1}(i,j)=X(c);c=c+1;
   end
end

%set hidden layer to output weights
for i=1:hn
   net.LW{2}(i)=X(c);c=c+1;
end

%set hidden layer neuron biases
```

```
for i=1:hn
   net.b{1}(i)=X(c);c=c+1;
end

%set output layer neuron bias
net.b{2}(1)=X(c);c=c+1;

% training as local search strategy
net.trainParam.show = NaN;
net.trainParam.lr = 0.05;
net.trainParam.mc = 0.1;
net.trainParam.epochs = 30;
net.trainParam.goal = 1e-2;
[net,tr]=train(net,trainInput',trainOutput');

% testing on training data as a measure of fitness
a = sim(net,trainInput');
a=a';
corr=0;
for j=1:size(trainInput,1)
   if a(j)<0.5 ans=0; else ans=1; end
   if trainOutput(j)==ans corr=corr+1; end
end
corr
size(trainInput,1)
fit=-corr/size(trainInput,1)*100; % percent train accuracy as fitness

%note that the number of variables in GA are:
%Weights(input to hidden): hn*in
%Weights(hidden to output): hn*1
%Biases(hidden): hn
%Biases(output): 1

% Total: hn*in+hn*1+hn+1
% all usually lie within -2 to 2
```

Box. 21.9 Fitness Function of the genetic Algorithm

```
X=garesults.x;

in=30; %number of input layer neurons
hn=18; %number of hidden layer neurons

% network initialization
net=newff(minmax(trainInput'),[hn,1],{'tansig','purelin'},'traingd');
```

```
c=1; %counter for the GA individual genes

%set input to hidden layer weights
for i=1:hn
   for j=1:in
      net.IW{1}(i,j)=X(c);c=c+1;
   end
end

%set hidden layer to output weights
for i=1:hn
   net.LW{2}(i)=X(c);c=c+1;
end
%set hidden layer neuron biases
for i=1:hn
   net.b{1}(i)=X(c);c=c+1;
end

%set output layer neuron bias
net.b{2}(1)=X(c);c=c+1;

% training as local search strategy
net.trainParam.show = NaN;
net.trainParam.lr = 0.05;
net.trainParam.mc = 0.1;
net.trainParam.epochs = 30;
net.trainParam.goal = 1e-2;
[net,tr]=train(net,trainInput',trainOutput');

% testing on training data
a = sim(net,trainInput');
a=a';
corr=0;
for j=1:size(trainInput,1)
   if a(j)<0.5 ans=0; else ans=1; end
   if trainOutput(j)==ans corr=corr+1; end
end
corr
size(trainInput,1)
corr/size(trainInput,1)*100 % percent train accuracy

% testing on testing data
a = sim(net,testInput');
a=a';
corr=0;
```

```
for j=1:size(testInput,1)
   if a(j)<0.5 ans=0; else ans=1; end
   if testOutput(j)==ans corr=corr+1; end
end
corr/size(testInput,1)*100 % percent test accuracy
```

Box 21.10 Code for Testing Evolutionary Neural Network

21.4 Evolutionary Neural Networks with Connectionist Architecture

In this approach we not only try to optimize the weights and biases of the neural network, we rather also try to optimize the connections between the neurons. The GA is responsible for finding which neuron is connected to which other neurons as well. Here also we assume a single hidden layer. The maximum number of neurons is fixed. The individual of this GA is supposed to store the weights, biases as well as the connections between the input to hidden layer and hidden layer to output layer. The connection information precedes the weights. If the connection is found, the corresponding gene stores a value greater than 0.5. The fitness function puts an additional penalty proportional to the number of connections. This stops the GA from construction of highly connected networks.

Let *hn* be the number of neurons in hidden layer and *in* be the number of neurons in input layer. Let there be a single neuron in the output layer. The initialization code of these networks is similar to the code given in Box 21.8. Table 21.2 gives the calculation for the total number of individuals in the GA. Box 21.11 shows the fitness function. The testing code with the most optimal genetic individual is similar to the code given in Box 21.10 and Box 21.11.

Table 21.2 Number of genes for genetic individual

S. No.	Item	Number of genes
1.	Connections: Input Layer to Hidden Layer	hn x in
2.	Weights: Input Layer to Hidden Layer	hn x in
3.	Connections: Hidden Layer to Output Layer	hn x 1
4.	Weights: Hidden Layer to Output Layer	hn x 1
5.	Hidden Layer Biases	hn
6.	Output Layer Bias	1
	Total	2 x hn x in + 2 x hn x 1 + hn + 1

```
function fit=enn(X)

global trainInput trainOutput;
ALPHA = 0.01; %connection penalty
in=30; %number of input layer neurons
hn=30; %number of hidden layer neurons

% network initialization
net=newff(minmax(trainInput'),[hn,1],{'tansig','purelin'},'traingd');

c=1; %counter for the GA individual genes
nc=0;%number of connections counter
%set input to hidden layer weights
for i=1:hn
   for j=1:in
      con=X(c);c=c+1;
      if(con<0)  net.IW{1}(i,j)=0;  else  net.IW{1}(i,j)=X(c);nc=nc+1;
end
      c=c+1;
   end
end

%set hidden layer to output weights
for i=1:hn
   con=X(c);c=c+1;
   if(con<0) net.LW{2}(i)=0; else net.LW{2}(i)=X(c);nc=nc+1; end
   c=c+1;
end

%set hidden layer neuron biases
for i=1:hn
   net.b{1}(i)=X(c);c=c+1;
end

%set output layer neuron bias
net.b{2}(1)=X(c);c=c+1;

% training as local search strategy
net.trainParam.show = NaN;
net.trainParam.lr = 0.05;
net.trainParam.mc = 0.1;
net.trainParam.epochs = 30;
net.trainParam.goal = 1e-2;
[net,tr]=train(net,trainInput',trainOutput');

% testing on training data as a measure of fitness
```

```
a = sim(net,trainInput');
a=a';
corr=0;
for j=1:size(trainInput,1)
   if a(j)<0.5 ans=0; else ans=1; end
   if trainOutput(j)==ans corr=corr+1; end
end

fit=-corr/size(trainInput,1)*100+ALPHA*nc; %fitness

%note that the number of variables in GA are:
%Connections(input to hidden): hn*in
%Weights(input to hidden): hn*in
%Connections(hidden to output): hn*1
%Weights(hidden to output): hn*1
%Biases(hidden): hn
%Biases(output): 1

% Total: 2*hn*in+2*hn*1+hn+1
% all usually lie within -2 to 2
```

Box. 21.11 Fitness function for Genetic Algorithm

List of Boxes

Box. 21.1. Multi-Layer Perceptron with Back Propagation Algorithm
Box. 21.2. Multi-Layer Perceptron with Back Propagation Algorithm with k-Fold Testing
Box 21.3. Initialization code for ensemble
Box. 21.4. Training of each neural network in ensemble
Box. 21.5. Testing of the ensemble
Fig. 21.6. Initialization Code for Modular Neural Network
Box. 21.7. Testing Code for Modular Neural network
Box. 21.8. Initialization Code for Evolutionary Neural Network
Box. 21.9. Fitness Function of the genetic Algorithm
Box 21.10. Code for Testing Evolutionary Neural Network
Box. 21.11. Fitness function for Genetic Algorithm

List of Figures

Fig. 21.1. MATLAB Genetic Algorithm Toolbox

List of Tables

Table 21.1. Number of genes for genetic individual
Table 21.2. Number of genes for genetic individual

GPSR Compliance

The European Union's (EU) General Product Safety Regulation (GPSR) is a set of rules that requires consumer products to be safe and our obligations to ensure this.

If you have any concerns about our products, you can contact us on ProductSafety@springernature.com

In case Publisher is established outside the EU, the EU authorized representative is:

Springer Nature Customer Service Center GmbH
Europaplatz 3
69115 Heidelberg, Germany

Batch number: 10614938

Printed by Printforce, the Netherlands